YinYang Bipolar Relativity:

A Unifying Theory of Nature, Agents and Causality with Applications in Quantum Computing, Cognitive Informatics and Life Sciences

Wen-Ran Zhang
Georgia Southern University, USA

INFORMATION SCIENCE REFERENCE
Hershey · New York

Senior Editorial Director: Kristin Klinger
Director of Book Publications: Julia Mosemann
Editorial Director: Lindsay Johnston
Acquisitions Editor: Erika Carter
Development Editor: Joel Gamon
Production Coordinator: Jamie Snavely
Typesetters: Keith Glazewski & Natalie Pronio
Cover Design: Nick Newcomer

Published in the United States of America by
Information Science Reference (an imprint of IGI Global)
701 E. Chocolate Avenue
Hershey PA 17033
Tel: 717-533-8845
Fax: 717-533-8661
E-mail: cust@igi-global.com
Web site: http://www.igi-global.com

Library of Congress Cataloging-in-Publication Data

Zhang, Wen-Ran, 1950-
Yinyang bipolar relativity : a unifying theory of nature, agents and causality with applications in quantum computing, cognitive informatics and life sciences / by Wen-Ran Zhang.
p. cm.
Summary: "This book presents real-world applications of YinYang bipolar relativity that focus on quantum computing and agent interaction. This unique work makes complex theoretical topics, such as the ubiquitous effects of quantum entanglement, logically comprehendible to a vast audience"-- Provided by publisher.
Includes bibliographical references and index.
ISBN 978-1-60960-525-4 (hardcover) -- ISBN 978-1-60960-526-1 (ebook) 1. Unified field theories. I. Title.
QC794.6.G7Z43 2011
530.14'2--dc22

2010054424

British Cataloguing in Publication Data
A Cataloguing in Publication record for this book is available from the British Library.

Dedication

To
My Parents,
My Former Professor Michael N. Huhns,
My Dear Friend Faye Flanagan Heath and
My Dear Friend Beatrice Rosenthal Buller

(Note: My mother passed away in 1982 and my father passed away in 1998; Bea was a Holocaust survivor, and she passed away in 2008.)

Table of Contents

Part 2
Set Theoretic Foundation

Part 3
YinYang Bipolar Relativity and Bipolar Quantum Computing

Foreword

I returned to my undergraduate alma mater, Georgia Southern University, in 2000 to develop programs in biostatistics and to establish a school of public health. Subsequently, I met Wen-Ran Zhang—the author of this book—and soon learned that he was highly regarded both as a teacher and a researcher. Some colleagues remarked: "Wen-Ran is a very hard worker and a prolific contributor to the literature, but few, if any, understand his research."

Over the next few years Wen-Ran and I often met to discuss each other's research interests and some common research interests. I must say that I was taken aback when I once asked him to tell me what his research was about in a few words. He said: "Multiagent Brain Modeling and YinYang Bipolar Logic." Somewhere in my undergraduate studies I learned logic and having had many Chinese friends over the years, I had a layman's understanding of Yin and Yang. Further, having designed and analyzed clinical trials in the development of drugs to treat bipolar disorder, I had some understanding of the word bipolar. But how these concepts could be brought together in a unifying theory was indeed puzzling.

But that is what the author has done in this book. YinYang Bipolar Relativity presents a logical unification of the two complementary opposites – the negative and positive energies of nature. By bringing the two sides together the book claims to provide an equilibrium-based computing paradigm for applications in physical, social, and life sciences especially in quantum computing.

It is well said by someone that *"Real innovation has no peers."* Therefore, I will not try to judge the book as a peer. On the other hand, as a colleague and friend, I am determined to stay impartial. So I asked the author to pass me the anonymous review comments on an earlier draft of his book from a double-blind review process sponsored by the publisher together with his response. In the following I provide a summary of both the negative and the positive review comments using direct quotes. Interestingly, the negative and positive sides themselves together present, to a certain extent, a vindication of YinYang bipolar relativity in a balanced and impartial manner.

THE POSITIVE SIDE WROTE

"The strengths of this book perhaps are best described in its contribution to understanding of YinYang bipolar relativity as 'an equilibrium based unifying computing paradigm that (1) logically defines causality, (2) logically unifies gravity with quantum theory, (3) brings relativity and quantum theory to the real-world for scientific computation and exploratory knowledge discovery in microscopic and macroscopic agent interaction, coordination, decision, and global regulation in physical, social, and life sciences especially in quantum computing and communication.' The objective defined by the author(s) was clearly met in the book."

"The author(s) have perhaps under-emphasized the importance of conceptualizing the concept of the YinYang and the power of the symbol as continual movement of two energies, etc. to better explain and lead into YinYang bipolar relativity. Such emphasis is given in the manuscript for Aristotelian science and logic and also to Einstein's theories of relativity. Yet, as the author(s) state, 'the word 'YinYang' indicates that the main idea is philosophically rooted in the ancient Chinese YinYang cosmology.' Since the manuscript establishes the bipolar theory and the importance of the opposite poles, the theory is different from Einstein's theory of relativity."

"Yes, the information in this manuscript does illustrate the issues, problems, and trends related to the theme or argument according to the author(s) is 'that equilibrium or non-equilibrium, as a physical state of any dynamic agent or universe at the system, molecular, genomic, particle, or sub-atomic level, forms a philosophical chicken and egg paradox with the universe because no one knows which one created the other in the very beginning.' This analogy serves as a guide to understanding the major point or argument in the manuscript. This analogy is further expanded in the YinYang concept when the author(s) argue that 'it is undoubtedly necessary to bridge the gap between the Western positivist thinking and the Eastern balanced thinking for solving unsolved scientific problems.' It is strongly supported in the literature (Ebrey, 1993, et al) in that the symbol YinYang represents an understanding of how things work in the universe. In fact, Ebrey noted in Chinese Civilization: A Sourcebook, 2nd ed. (New York: Free Press, 1993, pp. 77-79) that the concepts of Yin and Yang and the Five Agents 'provided intellectual framework of much of Chinese scientific thinking especially in the fields of biology and medicine.' This adds further credence to the argument made in the manuscript. In addition, it supports the contention that YinYang bipolar relativity has, in fact, opened an eastern road toward quantum gravity which, as noted, is Einstein's unfinished scientific unification of general relativity and quantum mechanics."

"As I have noted earlier, the target audience was defined in the Preface of the text. ... Given the nature of the subject, it might be a valuable research in other areas particularly in medicine and biology."

"The organization and/or flow of the book is a strength of this text. The chapters are well-illustrated for better understanding of major concepts. The chapters contain both a summary of important ideas and separate references for each chapter. Finally and perhaps most importantly, the chapters follow a logical sequence of ideas in the manuscript."

THE NEGATIVE SIDE WROTE

"The book is easy to read, except for extensive logic derivations and proofs which are too great for the reviewer to validate during a short review process and are considered by the reviewer to be secondary for this review purpose."

"The strength of this book is the mathematic work for bipolar logic. The reviewer believes that some models (not all though) might be useful for cases where bipolar agents do exist."

"As a researcher in the areas of information sciences, management science, and computer science, the reviewer would like to comment on modeling which is actually the theme of the book."

"The fatal mistake of this book is over-claiming. In reviewer's view, the bipolar theory proposed by the author(s) is nothing more than a modeling technique. The reviewer does not want to argue whether bipolar agents exist in the real world which is not important at all, but want to see the evidence to support your claims."

"This book over-emphasizes the following issues which do not add values to the research community.

1 Debate on philosophies. YinYang or not does not really matter - bipolar matters.
2 Unifying theory. Do not over-claim your theory for the fields you do not really know much.
3 Parallel with Einstein. 'Relativity' for your book is unnecessary."

"This book under-emphasizes the following problems which are important to the research community.

1 Practical problem solving.
2 Objective evidences beyond ideas and mathematics derivations.
3 Comparison of your models and other existing commonly used models in the fields."

"A sustainable theory must be supported by evidences of problem solving. Subjective modeling and interpretation of phenomena or events are not good enough. People do not really mind whether YinYang is in their body, but they do care whether the theory helps medical doctors cure the patients."
"'The author(s)' term 'application' means 'to apply the bipolar theory to explain the world.' To the reviewer, 'application' means 'to apply the model to solve a real problem.'"
"The book over-claims the target audience in the Preface."
"The reviewer does (not) find any problem with the organization of the book. The presentation flow is quite smooth, and easy to follow, except for lengthy mathematical derivations and proofs. Literally, the manuscript is well written."

After I finished reading the last draft of the book and the anonymous reviews with the author's response, it is quite clear to me which side of the above review comments is more objective. But, as I promised earlier, I shall not divulge in order to stay impartial. Instead, I leave it to readers of the book to make their own judgment. Of course, reviewer concerns have helped the author to improve his book.

In any case, I can positively say that the author is an outstanding research scientist. The breadth and depth of his knowledge in many areas, particularly in the computer, mathematical and physical sciences, and his capacity for original thinking and advancing knowledge are awe-inspiring. He is to be commended for his efforts in writing this book and his efforts to venture into uncharted, if not controversial scientific territories.

I am honored and flattered to have the opportunity to write the foreword of this remarkable book.

Karl E. Peace

Karl E. Peace *is the Georgia Cancer Coalition (GCC) Distinguished Cancer Scholar, Senior Research Scientist and Professor of Biostatistics in the Jiann-Ping Hsu College of Public Health (JPHCOPH) at Georgia Southern University. He is the architect of the MPH in Biostatistics and Founding Director of the Karl E. Peace Center for Biostatistics in the JPHCOPH. Dr. Peace holds the Ph.D. in Biostatistics from the Medical College of Virginia, the M.S. in Mathematics from Clemson University, the B.S. in Chemistry from Georgia Southern College, and a Health Science Certificate from Vanderbilt University. Dr. Peace's first career was academic teaching and research. He previously taught Mathematics at Georgia Southern College, Clemson University, Virginia Commonwealth University, and Randolph-Macon College, where he was a tenured professor. He also holds or has held numerous adjunct professorships at the Medical College of Virginia, the University of Michigan, Temple University, the University of North Carolina, and Duke University. Dr. Peace's second career was in research, technical support and management in the pharmaceutical industry. He held the positions of Senior Statistician at Burroughs-Wellcome, Manager of Clinical Statistics at A.H. Robins, Director of Research Statistics at SmithKline and French Labs, Senior Director of GI Clinical Studies, Data Management and Analysis at G.D. Searle, and Vice President of World-Wide Technical Operations at Warner Lambert/Parke-Davis. He then founded Biopharmaceutical Research Consultants, Inc. (BRCI), where he held the positions of President, Chief Executive Officer, and Chief Scientific Officer. Dr. Peace has made pivotal contributions in the development and approval of drugs to treat Alzheimer's disease, to prevent and treat gastrointestinal ulcers, to reduce the risk of myocardial infarction, to treat anxiety, depression and panic attacks, to treat hypertension and arthritis, and several antibiotics. Dr. Peace is or has been a member of several professional and honorary societies, including the Drug Information Association (DIA), the Biometric Society, Technometrics, the American Society for Quality Control (ASQC), the American Statistical Association (ASA), and Kappa Phi Kappa (KPK). He is a past member of the Committee on Applied and Theoretical Statistics, National Research Council, National Academy of Science. He is the recipient of numerous citations and awards: (1) Georgia Cancer Coalition Distinguished Cancer Scholar, (2) Fellow of the ASA, (3) the Distinguished Service Award of the DIA, (4) Star and Featured Alumnus, School of Basic Sciences, and Founder's Society Medal from the Medical College of Virginia, (5) College of Science and Technology Alumnus of the year, Alumnus of the year in private enterprise, Presidential Fellowship Award, Researcher of the year awards, and the First Recipient of the prestigious President's Medal for outstanding service and extraordinary contributions, all from Georgia Southern University, (6) 2007 APHA Statistics Section Award, (7) 2008 Shining Star and HR #2118 recognition by GA House of Representatives, (8) US Congress Citation for contributions to drug research and development and to Public Health, (9) the Tito Mijares Lifetime Achievement Award, (10) the Deen Day Smith Humanitarian Award and (11) several meritorious service awards from the ASQC, BASS and the Georgia Cancer Coalition. He is or has been Chair of: the Biostatistics Subsection of the Pharmaceutical Manufacturers Association (BSPMA), the Training Committee of the BSPMA, the Biopharmaceutical Section of the ASA, the Statistics Section of the APHA, and is Founder of the Biopharmaceutical Applied Statistics Symposium (BASS) in 1994. Dr. Peace is the author or editor of nine books and the author or co-author of 200 articles. His primary research interests are in drug research and development, clinical trial methodology, time-to-event methodology, and public health applications of biostatistics. Dr. Peace is a renowned philanthropist. He has created 21 scholarship endowments across 5 educational and public health institutions. Further details of Dr. Peace's accomplishments and contributions may be found in his autobiography entitled Paid In Full available at www.plowboy-press.com.*

Preface

A few years away from the centennial celebration of general relativity, the author of this monograph feels blessed for having the opportunity to present *YinYang Bipolar Relativity* to readers of the world. It seems surreal but, hopefully, the book can serve three purposes: (1) to add a piece of firework to the centennial celebration; (2) to introduce a deeper theory that transcends spacetime; (3) to reveal the ubiquitous effects of quantum entanglement in simple, logically comprehendible terms. Certainly, whether it is indeed an applicable deeper theory or just a piece of firework is ultimately up to the readers to make a judgment. As pointed out by Einstein: *"Experience remains, of course, the sole criterion of the physical utility of a mathematical construction. But the creative principle resides in mathematics."* (Einstein, 1934)

In this book we refer to relativity theories defined in spacetime as *spacetime relativity*. Thus, all previous relativity theories by Galileo, Newton, Lorenz, and Einstein are classified as spacetime relativity. This terminological treatment is for distinguishing *YinYang bipolar geometry* from *spacetime geometry*.

Regarding judgment, believe it or not, in the world-wide scientific community there may be more Chinese who emotionally resent the word "YinYang" due to misinformation or misunderstanding than Westerners who scientifically oppose the YinYang cosmology. This may sound ironic but is actually a historical phenomenon with socioeconomic reasons. First, most modern day Chinese want China to be integrated into the modern world and don't care much about YinYang, deemed an unscientific concept of the old school. Secondly, some overseas Chinese are concerned that the word "YinYang" might offend Western colleagues.

Subsequently, while the word "YinYang" has appeared in numerous Western publications spanning almost the whole spectrum of arts and sciences including but not limited to the prestigious journals *Science, Nature,* and *Cell*, some Chinese scholars including some researchers in traditional Chinese medicine (TCM) tend to shun YinYang. For instance, a few years ago a well-established Chinese American friend strongly advised (or demanded) the author to drop the word "YinYang" from titles of future submissions to avoid *"hurting the others' feelings."*

Western scholars, on the other hand, are free from carrying the above historical or socioeconomical baggage and are curious about YinYang. While many Westerners regard "YinYang" objectively as a philosophical word related to nature, society, and TCM, some Western scientists expect YinYang to play a critical or even unifying role in modern science. Here are a few examples:

1. Regarding the *"hurting the others' feelings"* matter, the author consulted a few "Westerner" colleagues and was given exactly the opposite advice: YinYang symbolizes the two energies of dynamic equilibrium, harmony, and complementarity; bipolarity without YinYang is often used in the West to indicate disorder, chaos, and dichotomy. (Note: Bipolarity is used in this book as YinYang bipolarity.)

2. Legendary German mathematician Leibniz – co-founder of calculus – invented the modern binary numeral system in the 17th century and attributed his invention to YinYang hexagrams recorded in the oldest *Chinese Book of Change – I Ching* (Leibniz 1703) (Karcher 2002).
3. Legendary Danish physicist Niels Bohr, father figure of quantum mechanics, brought YinYang into quantum theory for his particle-wave complementarity principle. When he was awarded the Order of the Elephant by the Danish government in 1947, he designed his own coat of arms which featured in the center a YinYang logo (or Taiji symbol) with the Latin motto *"contraria sunt complementa"* or *"opposites are complementary."*
4. Following Einstein's lead that history and philosophy provide the context for science and should be a significant part of physics education (Smolin 2006 p310-311), a group of renowned scientists and linguists noticed that different philosophies and cosmologies could result in different cultures and linguistic terms which in turn could make a major difference in the interpretation and understanding of space, time, and the quantum world (Alford 1993). Specifically, the word *"YinYang"* is deemed a most suitable noun for characterizing quantum interaction. As stated by linguist Alford (Alford 1993), YinYang *"represents a higher level of formal operations" "which lies beyond normal Western Indo-European development."*
5. A widely referenced genetic agent (protein) discovered at Harvard Medical School is named Yin-Yang 1 (YY1) (Shi *et al.* 1991) due to its ubiquitous repressor-activator (YinYang) functionalities in gene expression regulation in all cell types of living species (Jacobsen & Skalnik 1999).
6. A YinYang Pavilion created by American artist Dan Graham is dedicated to MIT and housed in Simmons Hall on the MIT campus (MIT News 2004).
7. A New York Times science report (Overbye 2006) described a subatomic particle discovered at the Fermi National Accelerator Laboratory as a *"YinYang dance"* that can change polarity three trillion times per second (Fermilab 2006).

While Western science and media don't seem to have problem with the word "YinYang", the word is, nevertheless, largely mysterious, albeit extremely pervasive. Its pervasive and mysterious nature can be characterized with a famous quote from Einstein: *"After a certain high level of technical skill is achieved, science and art tend to coalesce in esthetics, plasticity, and form. The greatest scientists are always artists as well."*

Evidently, a resolution to the *"science and art"* YinYang mystery bears great significance and has become imperative for the advancement of science and humanity. Unfortunately, such a resolution has been deemed scientifically impossible by many. This monograph is intended to accomplish the mission impossible based on the following observations and assertions:

1. The *"science and art"* YinYang paradox is similar to particle-wave quantum duality in Niels Bohr's complementarity principle. However, quantum mechanics has so far only recognized YinYang complementarity but failed to identify the essence of YinYang bipolarity. Without bipolarity, any complementarity is less fundamental due to the missing "opposites." In one word, the negative and positive poles such as action-reaction forces and particle-antiparticle pairs are the most fundamental opposites of Mother Nature but science-art, particle-wave, and truth-falsity are not exactly YinYang bipolar opposites.
2. Resolving the YinYang mystery is essentially the same as logically defining Aristotle's causality principle, axiomatizing all of physics (Hilbert 1901), resolving the EPR paradox (Einstein, Podolsky

& Rosen 1935), or providing a logical foundation for the grand unification of general relativity and quantum mechanics.

3. The *"higher level" "post-formal"* YinYang operation entails a philosophically different logical foundation that does indeed lie *"beyond normal Western Indo-European development"* and such a logical foundation is attainable in formal mathematical terms.

OBJECTIVE

The objective of this monograph is to present YinYang bipolar relativity as an equilibrium-based unifying computing paradigm with a minimal but most general axiomatization of physics that (1) logically defines bipolar quantum causality; (2) logically unifies gravity with quantum theory; (3) brings relativity and quantum entanglement to the real-world of microscopic and macroscopic agent interaction, coordination, decision, and global regulation in physical, social, and life sciences especially in quantum computing and exploratory knowledge discovery.

INTENDED AUDIENCE

The intended audience of the book includes, but is not limited to,

1. Students, professors, and researchers in mathematics, computer science, artificial intelligence, information science, information technology, data mining and knowledge discovery. These readers may find bipolar mathematical abstraction, bipolar sets, bipolar dynamic logic, bipolar quantum linear algebra, bipolar quantum cellular automata and their applications useful in their fields of teaching, learning, and research.
2. Students, professors, and researchers in quantum computing, physical sciences, nanotechnology, and engineering. These readers may find both of the theoretical and application aspects useful in their field of teaching, learning, and research. It is expected that quantum computing will be a major interest to these readers.
3. Students, professors, and researchers in bioinformatics, computational biology, genomics, bioeconomics, psychiatry, neuroscience, traditional Chinese medicine, and biomedical engineering. These readers may use the book material as an alternative holistic approach to problem solving in their fields of teaching, learning, research, and development.
4. Students, professors, and researchers in socioeconomics, bioeconomics, cognitive science, and decision science. These readers may find the mathematical tools and the quantum computing view useful in their fields of teaching, learning, research, and development.
5. Industrial researcher/developers in all fields who are interested in equilibrium-based modeling, analysis, and exploratory knowledge discovery in quantum computing, cognitive informatics, and life sciences. These readers may actually apply the theory of bipolar relativity for dealing with uncertainties and resolving unsolved problems in uncharted territories.

Limited logical and mathematical proofs of related theorems are included in Chapters 3-8. The proofs are for the convenience of logicians and mathematicians. They can be skipped by non-mathematical readers who are only interested in using the mathematical results for practical applications.

ORIGIN

While YinYang bipolar relativity can trace its philosophical origin back to ancient Chinese YinYang cosmology which claimed that everything has two sides or two opposite but reciprocal poles or energies, the formal theory presented in this monograph, however, is not the result of experimentation or elaboration of ancient Chinese YinYang but the result of free invention in the following spirit:

1. According to Einstein logical axiomatization of physics is possible: *"Physics constitutes a logical system of thought which is in a state of evolution, whose basis (principles) cannot be distilled, as it were, from experience by an inductive method, but can only be arrived at by free invention."* (Einstein 1916).
2. According to Einstein: *"Evolution is proceeding in the direction of increasing simplicity of the logical basis (principles)." "We must always be ready to change these notions – that is to say, the axiomatic basis of physics – in order to do justice to perceived facts in the most perfect way logically."* (Einstein 1916)
3. According to Einstein: *"... pure thought can grasp reality, as the ancients dreamed"* and *"nature is the realization of the simplest conceivable mathematical ideas."* (Einstein 1934)
4. According to Einstein the grand unification of general relativity and quantum mechanics needs a new logical foundation: *"For the time being we have to admit that we do not possess any general theoretical basis for physics which can be regarded as its logical foundation."* (Einstein 1940)
5. According to Einstein: *"Put your hand on a hot stove for a minute, and it seems like an hour. Sit with a pretty girl for an hour, and it seems like a minute. That's relativity."*

In the last quote, Einstein used sorrow and joy to hint the two sides of YinYang in general. Symbolically, the two sides can be paired up as a bipolar variable and generalized to action-reaction forces denoted $(-f, +f)$, negative-positive electromagnetic charges denoted $(-q, +q)$, matter-antimatter particles $(-p, +p)$ or the equilibrium-based bipolar variable (e^-, e^+) in a YinYang bipolar dynamic logic (BDL) for the theory of YinYang bipolar relativity (Zhang 2009a,b,c,d).

While space and time are not symmetrical to each other, not quantum entangled with each other, and not bipolar interactive, the concept of YinYang bipolarity is symmetrical and applicable in both microscopic and macroscopic worlds of physical and social sciences for characterizing agent interaction and bipolar quantum entanglement. Arguably, if space is expanding, spacetime has to be caused by something more fundamental; if YinYang bipolarity can survive a black hole due to particle-antiparticle emission or Hawking radiation, the logical foundation of physics has to be bipolar in nature; if particle-antiparticle pairs and nature's basic action-reaction forces are the most fundamental components of our universe, YinYang bipolar relativity has to be more fundamental than spacetime relativity. These arguments provide a basis for the transcending and unifying property of YinYang bipolar relativity beyond spacetime geometry.

Historically, even though YinYang has been the philosophical basis in the actual practice of TCM for thousands of years in China, it has failed to enter the arena of modern science until recent decades. It is a living proof to Einstein's assertion that *"Physics constitutes a logical system of thought which is in a state of evolution, whose basis (principles) cannot be distilled, as it were, from experience by an inductive method, but can only be arrived at by free invention."* (Einstein 1916).

Here are a few major modern developments in YinYang research:

1. **Biological YinYang.** The most noticeable result in this category is the discovery of the genetic regulator protein Yin Yang 1 (YY1) in 1991 at Harvard Medical School (Shi *et al.* 1991). YY1 exhibits ubiquitous repressor-activator functionalities in gene expression regulation in all types of cells of living species. The discovery of YY1 marks the formal entry of the ancient YinYang into genomics – a core area of bioinformatics. Since then, YY1 has been widely referenced by the top research institutions in the US and the world.
2. **Bayesian YinYang (BYY)**. BYY harmony learning (Xu 2007) has been widely cited and has become a well-established area in neural networks.
3. **Binary or Boolean YinYang**. Boolean YinYang (Zhang 1992; Kandel & Zhang 1998) follows Leibniz binary interpretation of YinYang. The binary interpretation provides a basis for all digital technologies.
4. **Bipolar YinYang.** Bipolar YinYang consists of YinYang bipolar sets, bipolar dynamic logic, bipolar quantum linear algebra, YinYang-N-Element quantum cellular automata, bipolar quantum entanglement, and the theory of YinYang bipolar relativity for applications in quantum computing, socioeconomics, and brain and life sciences (Zhang and coauthors, 1989-2009) (Zhang 1996-2010). Bipolar YinYang follows the YinYang cosmology that claims everything in the universe including the universe itself has two opposite reciprocal poles or energies.

This book follows the direction of bipolar YinYang. However, it should be remarked that the above different approaches to YinYang are interrelated or overlapped with each other. The repression and activation regulatory properties of Yin Yang 1 are bipolar in nature; YinYang equilibrium is essential in YinYang harmony; the two poles of YinYang are truth objects plus reciprocal bipolarity. From a physical science perspective, (-,+) bipolarity and symmetry in particle physics can also be considered evidence that supports the YinYang bipolar cosmology. From a decision science perspective, YinYang has been an influential philosophy in business management, socioeconomics, and international relations especially in Eastern countries. Noticeably, the national flag of South Korea is featured with a YinYang logo.

Indeed, YinYang has entered every aspect of the Western as well as the Eastern societies. Due to its lack of a unique formal logical basis, however, YinYang theory has remained largely mysterious. This book is to fill this gap. Although the technical ideas have been partially reported in refereed journal and conference articles, they have never been systematically presented as a coherent relativity theory in a monograph.

CENTRAL THEME

It is well-known that microscopic and macroscopic agents and agent interactions are essential in physics, socioeconomics, and life sciences. Unifying logical and mathematical axiomatization of agent interaction in microscopic and macroscopic worlds including but not limited to quantum, molecular, genetic, and neurobiological worlds is needed for scientific discoveries and for the coordination and global regulation of both non-autonomous and autonomous agents. Since agent interactions are governed by physical and social dynamics, the difficulty of axiomatizing agent interactions can be traced back to Hilbert's effort in axiomatizing physics, Aristotle's causality principle, the concept of singularity, and bipolar equilibrium.

English mathematical physicist Roger Penrose described two mysteries of quantum entanglement (Penrose 2005, p591). The first mystery is identified as the phenomenon itself. The second one, according to Penrose, is *"How are we to come to terms with quantum entanglement and to make sense of it in terms of ideas that we can comprehend, so that we can manage to accept it as something that forms an important part of the workings of our actual universe? .. The second mystery is somewhat complementary to the first. Since according to quantum mechanics, entanglement is such a ubiquitous phenomenon – and we recall that the stupendous majority of quantum states are actually entangled ones – why is it something that we barely notice in our direct experience of the world? Why do these ubiquitous effects of entanglement not confront us at every turn? I do not believe that this second mystery has received nearly the attention that it deserves, people's puzzlement having been almost entirely concentrated on the first."*

A major argument of this monograph is that equilibrium or non-equilibrium, as a physical state of any dynamic agent or universe at the system, molecular, genomic, particle, or subatomic level, forms a philosophical chicken and egg paradox with the universe because no one knows exactly which one created the other in the very beginning. Since bipolar equilibrium (or non-equilibrium) is a generic form of equilibrium (or non-equilibrium), any multidimensional model in spacetime geometry is not fundamental. *It is postulated that the most fundamental property of the universe is YinYang bipolarity.* Based on this postulate, bipolar relativity is presented that extends YinYang cosmology from *"Everything has two reciprocal poles"* to a formal logical foundation for physical and social sciences which claims that *"Everything has two reciprocal poles and nature is the realization of YinYang bipolar relativity or bipolar quantum entanglement."*

The main idea of the book starts with the paradox *"logical axiomatization for illogical physics" (LAFIP)* (Zhang 2009a) on Hilbert's Problem 6. It is observed that without bipolarity the bivalent truth values 0 for false and 1 for true are incapable of carrying any shred of direct physical syntax and semantics, let alone illogical physical phenomena such as chaos, particle-wave duality, bipolar disorder, equilibrium, non-equilibrium, and quantum entanglement. Therefore, truth-based (unipolar) mathematical abstraction as a basis for positivist thinking cannot avoid the LAFIP paradox. It is suggested that this is the fundamental reason why there is so far no truth-based logically definable causality, no truth-based axiomatization of all physics, no decisive battleground in the quest for quantum gravity, and no logic for particle-wave duality, bipolar disorder, economic depression, big bang, black hole, and quantum entanglement.

Furthermore, it is pointed out that, while no physicist would say *"electron is isomorphic to positron"*, it is widely considered in logic and mathematics that "-1 is isomorphic to +1" and (-,+) bipolar symmetry, equilibrium, or non-equilibrium is not observable. If we check the history of *negative numbers*, we would find that the ancient Chinese and Indians started to use negative numbers thousands of years ago but European mathematicians resisted the concept of negative numbers until the 18th centuries (Temple 1986, pp.141) (Bourbaki 1998) (Martinez 2006).

Regardless of the great achievement of Western science and technology, it is undoubtedly necessary to bridge the gap between the Western positivist thinking and the Eastern balanced thinking for solving unsolved scientific problems. As *"passion for symmetry"* can *"permeate the Standard Model of elementary particle physics"* and can unify *"the smallest building blocks of all matter and three of nature's four forces in one single theory"* (The Royal Swedish Academy of Sciences 2008), it is not only reasonable but also inevitable to explore the bipolar equilibrium-based computing paradigm (Note: Equilibrium-based is to equilibrium and non-equilibrium as truth-based is to truth and falsity with fundamentally different syntax and semantics).

SYNOPSIS

YinYang bipolar relativity is intended to be a logical unification of general relativity and quantum mechanics. The monograph can be considered the first step to address the gigantic topic with real-world applications in both natural and social sciences focused on quantum computing and agent interaction in socioeconomics, cognitive informatics and life sciences. Subjects opened in the book can be further addressed in succeeding volumes in depth.

The main body of the book starts with a new set-theoretic logical foundation. To avoid LAFIP, bipolar set theory is introduced with a holistic equilibrium-based approach to mathematical abstraction. Bipolar sets lead to YinYang bipolar dynamic logic (BDL). A key element of BDL is bipolar universal modus ponens (BUMP) that provides, for the first time, logically definable causality. It is shown that BDL is a non-linear bipolar dynamic fusion of Boolean logic and quantum entanglement. The non-linearity, however, does not compromise the basic law of excluded middle (LEM) and bipolar computability. Soundness and completeness of a bipolar axiomatization are asserted. Bipolar sets and BDL are extended to bipolar fuzzy sets, bipolar dynamic fuzzy logic (BDFL) and equilibrium relations.

With the emergence of space, time, and bipolar agents, a completely background independent theory of bipolar relativity, the central theme of the book, is formally introduced based on bipolar sets and BDL. It is shown that, with bipolar agents and bipolar relativity, causality is logically definable; a real-world bipolar string theory is scalable, and an equilibrium-based minimal but most general axiomatization of physics, socioeconomics, and life sciences, as a partial solution to Hilbert's Problem 6, is logically provable.

It is shown that YinYang bipolar relativity is rich in predictions. Predictions are presented, some of which are expected to be falsifiable in the foreseeable future. In particular, it is shown that bipolar relativity provides the unified logical form for both gravity and quantum entanglement. It is conjectured that all forces in the universe are bipolar quantum entanglement in nature in large or small scales and in symmetrical or asymmetrical forms; the speed of gravity is not necessarily limited by the speed of light as it could well be limited by the speed of quantum entanglement.

Due to bipolar quantum entanglement, YinYang bipolar relativity leads to a logically complete theory for quantum computing with digital compatibility. The bipolar quantum computing paradigm is ideal for modeling non-linear bipolar dynamic oscillation and interaction such as non-local connection and particle-wave duality in quantum mechanics as well as self-negation/self-assertion abilities in cognitive informatics and competition-cooperation in socioeconomics. In particular, it is shown that bipolar quantum entanglement makes quantum teleportation theoretically possible without conventional communication between Bob and Alice. Furthermore, it is shown that bipolar quantum-digital compatibility and bitwise cryptography have the potential to make obsolete both prime number based encryption and quantum factorization.

Based on the logical foundation, limited mathematical construction is presented. Specifically, bipolar quantum linear algebra (BQLA) and YinYang-N-Element bipolar quantum cellular automata (BQCA) are introduced with illustrations in biosystem simulation and equilibrium-based global regulation. It is shown that the dimensional view, bipolar logical view, and YinYang-N-Element BQCA view are logically consistent. Therefore, bipolar set theory, bipolar dynamic logic, BQLA, bipolar agents, bipolar causality, and BQCA are all unified under YinYang bipolar relativity.

It is contended that YinYang bipolar relativity is an Eastern road toward quantum gravity. It is argued that it would be hard to imagine that quantum gravity as the grand unification of gravity and quantum mechanics would not be the governing theory for all sciences. This argument leads to five sub-theories of quantum gravity: *physical quantum gravity, logical quantum gravity, social quantum gravity, biological quantum gravity, and mental quantum gravity that form a Q5 quantum computing paradigm.* The Q5 paradigm is then used as a vehicle to illustrate the ubiquitous effects of bipolar quantum entanglement that confronts us at every turn of our lives in comprehendible logical terms.

LIMITATIONS

Mathematically, the theory of YinYang bipolar relativity as a pure invention is not derived from general relativity or quantum theory. Instead, it presents a fundamentally different approach to quantum gravity. As a first step, the monograph is focused on the logical level of the theory and its applications in physical, social, brain, biological, and computing sciences with limited mathematical or algebraic extensions. Thus, equilibrium-based bipolar logical unification of gravity and quantum mechanics is within the scope of the book; the quantization of YinYang bipolar relativity and the mathematical unification of Einstein's equations of general relativity and that of quantum mechanics have to be left for future research efforts because *"For the time being we have to admit that we do not possess any general theoretical basis for physics which can be regarded as its logical foundation"* (Einstein 1940).

Theoretically, YinYang bipolar relativity presents an open-world and open-ended approach to science that is not *"a theory of everything."* In this approach, the author doesn't attempt to define the smallest fundamental element such as strings in string theory. Instead, it is postulated that YinYang bipolarity is the most fundamental property of the universe based on well-established observations in physical and social sciences. With the basic hypothesis, equilibrium-based logical constructions are developed with a number of predictions for experimental verification or falsification. This approach actually follows the principle of exploratory scientific knowledge discovery.

Practically, YinYang bipolar relativity is expected to be applicable wherever bipolar equilibrium or non-equilibrium is central (e.g. Zhang 2003a,b; Zhang 2006; Zhang, Pandurangi & Peace 2007; Zhang *et al.* 2010). As a quantum logic theory it is recoverable to Boolean logic and, therefore, is computational. As a relativity theory, its major role is to provide predictions and interpretations about nature, agents, and causality. Since it is not *"a theory of everything"*, it does not claim universal applicability. Simulated application examples are presented in quantum computing, cognitive informatics, and life sciences to illustrate the utility of the theory. The examples, however, are not intended to be systematic and comprehensive applications but only sufficient illustrations. While the theory is logically proven sound, predictions or interpretations made in the book can be either verified or falsified in the future, as usual.

CITING

References to others in this monograph are focused on important relevant works related to the logical foundation of this work. Since the formal system presented in the book is a free invention, not a philosophical elaboration of YinYang or an extension of other quantum gravity theories, references to YinYang literature are limited to the well-known basic concepts related to the logical foundation and references to relativity and quantum theory are limited to the basic concepts of spacetime geometry, particle physics, quantum entanglement, and teleportation. Selected references are mostly published scientific works in peer reviewed books, journals, or conference proceedings. Non-peer reviewed Web articles cited are strictly limited to well-known historical facts or philosophical non-technical viewpoints. This treatment ensures that all technical references are from peer-reviewed scientific sources but undisputed well-known historical facts available online, and freely expressed, non-peer reviewed philosophical viewpoints published in the Web by related experts could be taken into account for readers' convenience.

SIGNIFICANCE

To the author's knowledge, this is the first monograph of its kind to introduce logically definable causality into physical and social sciences and to make the ubiquitous effects of quantum entanglement logically comprehendible. While Leibniz binary YinYang provided a technological basis for digital technologies, YinYang bipolar relativity is expected to bring quantum gravity into logical, physical, social, biological, and mental worlds for quantum computing.

The significance of YinYang bipolar relativity lies in its four equilibrium-based logical unifications: (1) the unification of unipolar positivist truth with bipolar holistic truth, (2) the unification of classical logic with quantum logic, (3) the unification of quantum entanglement with microscopic and macroscopic agent interaction in simple logical terms, and (4) the unification of general relativity with quantum mechanics under bipolar equilibrium and symmetry. Despite its limited mathematical depth, it is shown that YinYang bipolar relativity constitutes a deeper theory beyond spacetime geometry tailored for open-world open-ended exploratory knowledge discovery in all scientific fields where equilibrium and symmetry are central.

ORGANIZATION

The book consists of twelve chapters which can be roughly divided into the following five sections:

Part 1. Introduction and Background. This part consists of Chapter 1 and Chapter 2. Chapter 1 is an introduction; Chapter 2 is a background review.

Part 2. Set Theoretic Logical Foundation. This part consists of Chapters 3-5. This part lays out the set-theoretic logical foundation for bipolar relativity including YinYang bipolar sets, bipolar dynamic logic, bipolar quantum lattices, bipolar dynamic fuzzy logic, bipolar fuzzy sets and equilibrium relations.

Part 3. YinYang Bipolar Relativity and Quantum Computing. This part consists of Chapters 6-8 which are focused on the central theme of the book. Chapter 6 presents the theory of agents, causality, and YinYang bipolar relativity with a number of predictions. Chapter 7 presents bipolar quantum entanglement for quantum computing. Chapter 8 presents YinYang bipolar quantum linear algebra (BQLA), bipolar quantum cellular automata (BQCA), and a unifying view of YinYang bipolar relativity in logical, geometrical, algebraic, and physical terms.

Part 4. Applications. This part consists of Chapters 9-11. Chapter 9 is focused on biosystem simulation with BQLA and BQCA. Chapter 10 is focused on bipolar computational neuroscience and psychiatry. Chapter 11 is focused on bipolar cognitive mapping and decision analysis.

Part 5. Discussions and Conclusions. This part consists of the last chapter (Chapter 12) in which discussions and conclusions are presented.

CHAPTER OUTLINE

Chapter 1. Introduction: Beyond Spacetime

This chapter serves as an introduction to bring readers from spacetime relativity to YinYang bipolar relativity. Einstein's assertions regarding physics, logic, and theoretical invention are reviewed and his hint of YinYang bipolar relativity is identified. The limitations of general relativity and quantum mechanics are briefly discussed. It is concluded that logically definable causality, axiomatization of physics, axiomatization of agent interaction, and the grand unification of general relativity and quantum theory are essentially the same problem at the fundamental level. A paradox on Hilbert's Problem 6 – Logical Axiomatization for Illogical Physics (LAFIP) – is introduced. Bipolarity is postulated as the most fundamental property of nature transcending spacetime. The theoretical basis of agents, causality and YinYang bipolar relativity is highlighted and distinguished from established theories. The main ideas of the book are outlined.

Chapter 2. Background Review: Quest for Definable Causality

This chapter presents a review on the quest for logically definable causality. The limitation of observability and truth-based cognition is discussed. The student-teacher philosophical dispute between Aristotle and Plato is revisited. Aristotle's causality principle, David Hume's challenge, Lotfi Zadeh's *"Causality Is Undefinable"* conclusion, and Judea Pearl's probabilistic definability are reviewed. Niels Bohr's particle-wave complementarity principle, David Bohm's causal interpretation of quantum mechanics, and Sorkin's causal set program are discussed. Cognitive-map-based causal reasoning is briefly visited. YinYang bipolar logic and bipolar causality are previewed. Social construction and destruction in science are examined. It is asserted that, in order to continue its role as the doctrine of science, the logical definability of Aristotle's causality principle has become an ultimate dilemma of science. It is concluded that, in order to resolve the dilemma, a formal system with logically definable causality has to be developed, which has to be logical, physical, relativistic, and quantum in nature. The formal system has to be applicable in the microscopic world as well as in the macroscopic world, in the physical world as well as in the social world, in cognitive informatics as well as in life sciences, and, above all, it has to reveal the ubiquitous effects of quantum entanglement in simple comprehendible terms.

Chapter 3. Bipolar Sets and YinYang Bipolar Dynamic Logic (BDL)

In this chapter an equilibrium-based set-theoretic approach to mathematical abstraction and axiomatization is presented for resolving the LAFIP paradox (Ch. 1) and for enabling logically definable causality (Ch. 2). Bipolar set theory is formally presented, which leads to YinYang bipolar dynamic logic (BDL). BDL in zeroth-order, 1st-order, and modal forms are presented with four pairs of dynamic DeMorgan's laws and a bipolar universal modus ponens (BUMP). BUMP as a key element of BDL enables logically definable causality and quantum computing. Soundness and completeness of a bipolar axiomatization are asserted; computability is proved; computational complexity is analyzed. BDL can be considered a non-linear bipolar dynamic generalization of Boolean logic plus quantum entanglement. Despite its non-linear bipolar dynamic quantum property, it does not compromise the basic law of excluded middle. The recovery of BDL to Boolean logic is axiomatically proved through depolarization and the computability of BDL is asserted. A redress on the ancient paradox of the liar is presented with a few observations on Gödel's incompleteness theorem. Based on BDL, bipolar relations, bipolar transitivity, and equilibrium relations are introduced. It is shown that a bipolar equilibrium relation can be a non-linear bipolar fusion of many equivalence relations. Thus, BDL provides a logical basis for YinYang bipolar relativity – an equilibrium-based axiomatization of social and physical sciences.

This chapter is based on ideas presented in (Zhang & Zhang 2003, 2004) (Zhang 2003a,b; 2005a,b; 2007; 2009a,b,c,d). Early works of this line of research can be found in (Zhang, Chen & Bezdek 1989) (Zhang *et al.* 1992) (Zhang, Wang & King 1994)

Chapter 4. Bipolar Quantum Lattices and Dynamic Triangular Norms

Bipolar quantum lattice (BQL) and dynamic triangular norms (t-norms) are presented in this chapter. BQLs are defined as special types of bipolar partially ordered sets or posets. It is shown that bipolar quantum entanglement is definable on BQLs. With the addition of fuzziness, BDL is extended to a bipolar dynamic fuzzy logic (BDFL). The essential part of BDFL consists of bipolar dynamic triangular norms (t-norms) and their co-norms which extend their truth-based counterparts from a static unipolar fuzzy lattice to a bipolar dynamic quantum lattice. BDFL has the advantage in dealing with uncertainties in bipolar dynamic environments. With bipolar quantum lattices (crisp or fuzzy), the concepts of bipolar symmetry and quasi-symmetry are defined which form a basis toward a logically complete quantum theory. The concepts of strict bipolarity, linearity, and integrity of BQLs are introduced. A recovery theorem is presented for the depolarization of any strict BQL to Boolean logic. The recovery theorem reinforces the computability of BDL or BDFL.

This chapter is based on the ideas presented in (Zhang & Zhang 2004) (Zhang 1996, 1998, 2003, 2005a,b, 2006a,b, 2007, 2009b). Early works of this line of research can be found in (Zhang, Chen & Bezdek 1989) (Zhang *et al.* 1992) (Zhang, Wang & King 1994)

Chapter 5. Bipolar Fuzzy Sets and Equilibrium Relations

Based on bipolar sets and quantum lattices, the concepts of bipolar fuzzy sets and equilibrium relations are presented in this chapter for bipolar fuzzy clustering, coordination, and global regulation. Related theorems are proved. Simulated application examples in multiagent macroeconomics are illustrated. Bipolar fuzzy sets and equilibrium relations provide a theoretical basis for cognitive-map-based bipolar decision, coordination, and global regulation.

This chapter is based on the ideas presented in (Zhang 2003a,b, 2005a,b, 2006a). Early works of this line of research can be found in (Zhang, Chen & Bezdek 1989) (Zhang *et al.* 1992) (Zhang, Wang & King 1994)

Chapter 6. Agents, Causality, and YinYang Bipolar Relativity

This chapter presents the theory of bipolar relativity – a central theme of this book. The concepts of YinYang bipolar agents, bipolar adaptivity, bipolar causality, bipolar strings, bipolar geometry, and bipolar relativity are logically defined. The unifying property of bipolar relativity is examined. Space and time emergence from YinYang bipolar geometry is proposed. Bipolar relativity provides a number of predictions. Some of them are domain dependent and some are domain independent. In particular, it is conjectured that spacetime relativity, singularity, gravitation, electromagnetism, quantum mechanics, bioinformatics, neurodynamics, and socioeconomics are different phenomena of YinYang bipolar relativity; microscopic and macroscopic agent interactions in physics, socioeconomics, and life science are directly or indirectly caused by bipolar causality and regulated by bipolar relativity; all physical, social, mental, and biological action-reaction forces are fundamentally different forms of bipolar quantum entanglement in large or small scales; gravity is not necessarily limited by the speed of light; graviton does not necessarily exist.

This chapter is based on the ideas presented in (Zhang 2009a,b,c,d; Zhang 2010).

Chapter 7. YinYang Bipolar Quantum Entanglement: Toward a Logically Complete Theory for Quantum Computing and Communication

YinYang bipolar relativity leads to an equilibrium-based logically complete quantum theory which is presented and discussed in this chapter. It is shown that bipolar quantum entanglement and bipolar quantum computing bring bipolar relativity deeper into microscopic worlds. The concepts of bipolar qubit and YinYang bipolar complementarity are proposed and compared with Niels Bohr's particle-wave complementarity. Bipolar qubit box is compared with Schrödinger's cat box. Since bipolar quantum entanglement is fundamentally different from classical quantum theory (which is referred to as unipolar quantum theory in this book), the new approach provides bipolar quantum computing with the unique features: (1) it forms a key for equilibrium-based quantum controllability and quantum-digital compatibility; (2) it makes bipolar quantum teleportation theoretically possible for the first time without conventional communication between Alice and Bob; (3) it enables bitwise encryption without a large prime number that points to a different research direction of cryptography aimed at making prime-number-based cryptography and quantum factoring algorithm both obsolete; (4) it shows potential to bring quantum computing and communication closer to deterministic reality; (5) it leads to a unifying Q5 paradigm aimed at revealing the ubiquitous effects of bipolar quantum entanglement with the sub theories of logical, physical, mental, social, and biological quantum gravities and quantum computing.

This chapter is based on ideas presented in (Zhang 2003a, 2005a, Zhang 2009a,b,c,d; 2010).

Chapter 8. YinYang Bipolar Quantum Linear Algebra (BQLA) and Bipolar Quantum Cellular Automata (BQCA)

This chapter brings bipolar relativity from the logical and relational levels to the algebraic level. Following a brief review on traditional cellular automata and linear algebra, bipolar quantum linear algebra (BQLA) and bipolar quantum cellular automata (BQCA) are presented. Three families of YinYang-N-Element bipolar cellular networks (BCNs) are developed, compared, and analyzed; YinYang bipolar dynamic equations are derived for YinYang-N-Element BQCA. Global (system level) and local (element level) energy equilibrium and non-equilibrium conditions are established and axiomatically proved for all three families of cellular structures that lead to the concept of collective bipolar equilibrium-based adaptivity. The unifying nature of bipolar relativity in the context of BQCA is illustrated. The background independence nature of YinYang bipolar geometry is demonstrated with BQLA and BQCA. Under the unifying theory, it is shown that the bipolar dimensional view, cellular view, and bipolar interactive view are logically consistent. The algebraic trajectories of bipolar agents in YinYang bipolar geometry are illustrated with simulations. Bipolar cellular processes in cosmology, brain and life sciences are hypothesized and discussed.

This chapter is based on earlier chapters and the ideas presented in (Zhang 1996, 2005a, 200ba, Zhang 2009a,b,c,d, 2010; Zhang & Chen 2009; Zhang *et al.* 2009).

Chapter 9. Bipolar Quantum Bioeconomics for Biosystem Simulation and Regulation

As a continuation of Chapter 8, this chapter presents a theory of bipolar quantum bioeconomics (BQBE) with a focus on computer simulation and visualization of equilibrium, non-equilibrium, and oscillatory properties of YinYang-N-Element cellular network models for growing and degenerating biological processes. From a modern bioinformatics perspective, it provides a scientific basis for simulation and regulation in genomics, bioeconomics, metabolism, computational biology, aging, artificial intelligence, and biomedical engineering. It is also expected to serve as a mathematical basis for biosystem inspired socioeconomics, market analysis, business decision support, multiagent coordination and global regulation. From a holistic natural medicine perspective, diagnostic decision support in TCM is illustrated with the YinYang-5-Element bipolar cellular network; the potential of YinYang-N-Element BQCA in qigong, Chinese meridian system, and innate immunology is briefly discussed.

This chapter is based on earlier chapters and the ideas presented in (Zhang 1996, 2005a, 2006a, Zhang 2009a,b,c,d, 2010; Zhang & Chen 2009; Zhang *et al.* 2009).

Chapter 10. MentalSquares: An Equilibrium-Based Bipolar Support Vector Machine for Computational Psychiatry and Neurobiological Data Mining

While earlier chapters have focused on the logical, physical, and biological aspects of the Q5 paradigm, this chapter shifts focus to the mental aspect. MentalSquares (MSQs) − an equilibrium-based dimensional approach is presented for pattern classification and diagnostic analysis of bipolar disorders. While a support vector machine is defined in Hilbert space, MSQs can be considered a generic dimensional approach to support vector machinery for modeling mental balance and imbalance of two opposite but bipolar interactive poles. A MSQ is dimensional because its two opposite poles form a 2-dimensional background independent YinYang bipolar geometry from which a third dimension – equilibrium or non-equilibrium – is transcendental with mental fusion or mental separation measures. It is generic because any multidimensional mental equilibrium or non-equilibrium can be deconstructed into one or more bipolar equilibria which can then be represented as a mental square. Different MSQs are illustrated for bipolar disorder (BPD) classification and diagnostic analysis based on the concept of mental fusion and separation. It is shown that MSQs extend the traditional categorical standard classification of BPDs to a non-linear dynamic logical model while preserving all the properties of the standard; it supports both classification and visualization with qualitative and quantitative features; it serves as a scalable generic dimensional model in computational neuroscience for broader scientific discoveries; it has the cognitive simplicity for clinical and computer operability. From a broader perspective, the agent-oriented nature of MSQs provides a basis for multiagent data mining (Zhang & Zhang 2004) and cognitive informatics of brain and behaviors (Wang 2004).

This chapter is based on earlier chapters and the ideas presented in (Zhang 2007; Zhang, Pandurangi & Peace 2007; Zhang & Peace 2007)

Chapter 11. Bipolar Cognitive Mapping and Decision Analysis: A Bridge from Bioeconomics to Socioeconomics

The focus of this chapter is on cognitive mapping and cognitive-map-based (CM-based) decision analysis. This chapter builds a bridge from mental quantum gravity to social quantum gravity. It is shown that bipolar relativity, as an equilibrium-based unification of nature, agent and causality, is naturally the unification of quantum bioeconomics, brain dynamics, and socioeconomics as well. Simulated examples are used to illustrate the unification with cognitive mapping and CM-based multiagent decision, coordination, and global regulation in international relations.

This chapter is based on earlier chapters and the ideas presented in (Zhang, Chen & Bezdek 1989) (Zhang *et al.* 1992) (Zhang, Wang & King 1994) (Zhang & Zhang 2004) (Zhang 1996, 1998, 2003, 2005a,b, 2006a,b).

Chapter 12. Causality is Logically Definable: An Eastern Road toward Quantum Gravity

This is the conclusion chapter. Bertrand Russell's view on logic and mathematics is briefly reviewed. An enjoyable debate on bipolarity and isomorphism is presented. Some historical facts related to YinYang are discussed. Distinctions are drawn between BDL from established logical paradigms including Boolean logic, fuzzy logic, multiple-valued logic, truth-based dynamic logic, intuitionist logic, paraconsistent logic, and other systems. Some major comments from critics on related works are answered. A list of major research topics is enumerated. The ubiquitous effects of YinYang bipolar quantum entanglement are summarized. Limitations of this work are discussed. Some conclusions are drawn.

REFERENCES

Alford, D.M. (1993). *A report on the Fetzer Institute-sponsored dialogues between Western and indigenous scientists*. A presentation for the Annual Spring Meeting of the Society for the Anthropology of Consciousness, April 11, 1993. Retrieved from http://www.enformy.com/dma-b.htm

Bourbaki, N. (1998). *Elements of the history of Mathematics*. Berlin, Heidelberg, New York: Springer-Verlag.

Einstein, A. (1916). The foundation of the general theory of relativity. Originally published in Annalen der Physik (1916), *Collected Papers of Albert Einstein*, English Translation of Selected Texts, Translated by A. Engel, Vol. 6, (pp 146-200).

Einstein, A. (1934). On the method of theoretical Physics. *The Herbert Spencer lecture*, delivered at Oxford, June 10, 1933. Published in Mein Weltbild, Amsterdam: Querido Verlag.

Einstein, A. (1940). Considerations concerning the fundamentals of theoretical Physics. *Science, 2369*(91), 487-491.

Einstein, A., Podolsky, B. & Rosen N. (1935). *Can quantum-mechanical description of physical reality be considered complete? Physics Review, 47*, 777.

Fermi National Accelerator Laboratory. (2006). *Press Release 06-19*, September 25, 2006. Retrieved from http://www.fnal.gov/ pub/presspass/press_releases/CDF_meson.html

Hilbert, D. (1901). Mathematical problems. *Bulletin of the American Mathematics Society, 8*, 437-479.

Jacobsen, B.M. & Skalnik, D.G. (1999). YY1 binds five cis-elements and trans-activates the myeloid cell-restricted gp91phox promoter. *Journal of Biological Chemistry, 274*, 29984-29993.

Kandel, A. & Zhang, Y. (1998). Intrinsic mechanisms and application principles of general fuzzy logic through Yin-Yang analysis. *Information Science, 106*(1), 87-104.

Karcher, S. (2002). *I Ching: The classic Chinese oracle of change: The first complete translation with concordance*. London: Vega Books.

Leibniz, G. (1703). *Explication de l'Arithmétique Binaire* (*Explanation of Binary Arithmetic*); Gerhardt, *Mathematical Writings* VII.223.

Martinez, A.A. (2006). *Negative math: How mathematical rules can be positively bent*. Princeton University Press.

MIT News. (2004). Yin/Yang Pavilion dedicated. Retrieved from http://web.mit.edu/newsoffice/2004/yinyang-0505.html

Overbye, D. (2006). A real flip-flopper, at 3 trillion times a second. *The New York Times – Science Report*. Retrieved from http://www.nytimes.com/2006/04/18/science/18find.html

Penrose, R. (2005). *The road to reality: A complete guide to the laws of the universe*. New York: Alfred A. Knopf.

Shi, Y., Seto, E., Chang, L.-S. & Shenk, T. (1991). Transcriptional repression by YY1, a human GLI-Kruppel-related protein, and relief of repression by adenovirus E1A protein. *Cell, 67*(2), 377-388.

Smolin, L. (2000). *Three road to quantum gravity.* Basic Books.

Smolin, L. (2006). *The trouble with physics: The rise of string theory, the fall of a science, and what comes next?* New York: Houghton Mifflin Harcourt.

Temple, R. (1986). *The genius of China: 3,000 years of science, discovery, and invention*. New York: Simon and Schuster.

The Royal Swedish Academy of Sciences. (2008). The Nobel Prize in Physics 2008. *Press Release*, 7 October. Retrieved from http://nobelprize.org/nobel_prizes/physics/laureates/2008/press.html

Wang, Y. (2004). On cognitive informatics. *Brain and Mind, 4*(2), 151-167.

Woit, P. (2006). *Not even wrong: The failure of string theory and the search for unity in physical law.* New York: Basic Book.

Xu, L. (2007). Bayesian Ying Yang learning. *Scholarpedia, 2*(3), 1809. Retrieved from http://www.scholarpedia.org/article/Bayesian_Ying_Yang_learning

Zadeh, L.A. (2001). Causality is undefinable–toward a theory of hierarchical definability. *Proceedings of FUZZ-IEEE,* (pp. 67-68).

Zhang, W.-R., Chen, S. & Bezdek, J.C. (1989). POOL2: A generic system for cognitive map development and decision analysis. *IEEE Transactions on SMC*, *19*(1), 31-39.

Zhang, W.-R., Chen, S., Wang, W. & King, R. (1992). A cognitive map based approach to the coordination of distributed cooperative agents. *IEEE Transactions on SMC*, *22*(1), 103-114.

Zhang, W.-R., Wang, W. & King, R. (1994). An agent-oriented open system shell for distributed decision process modeling. *Journal of Organizational Computing*, *4*(2), 127-154.

Zhang, W.-R. (1996). NPN fuzzy sets and NPN qualitative algebra: A computational framework for bipolar cognitive modeling and multiagent decision analysis. *IEEE Transactions on SMC*, *16*, 561-574.

Zhang, W.-R. (1998). YinYang bipolar fuzzy sets. *Proceedings of IEEE World Congress on Computational Intelligence, Fuzz-IEEE*, (pp. 835-840). Anchorage, AK, May 1998.

Zhang W.-R. & Zhang, L. (2003). Soundness and completeness of a 4-valued bipolar logic. *International Journal on Multiple-Valued Logic*, *9*, 241-256.

Zhang, W.-R. (2003a). Equilibrium relations and bipolar cognitive mapping for online analytical processing with applications in international relations and strategic decision support. *IEEE Transactions on SMC, Part B, 33*(2), 295-307.

Zhang, W.-R. (2003b). Equilibrium energy and stability measures for bipolar decision and global regulation. *International Journal of fuzzy systems, 5*(2), 114-122.

Zhang, W.-R. & Zhang, L. (2004a). YinYang bipolar logic and bipolar fuzzy logic. *Information Sciences*, *165*(3-4), 265-287.

Zhang, W.-R. & Zhang, L. (2004b). A Multiagent Data Warehousing (MADWH) and Multiagent Data Mining (MADM) approach to brain modeling and NeuroFuzzy control. *Information Sciences, 167(1-4),* 109-127.

Zhang, W.-R. (2005a). YinYang bipolar lattices and L-sets for bipolar knowledge fusion, visualization, and decision making. *International Journal of Information Technology and Decision Making*, *4*(4), 621-645.

Zhang, W.-R. (2005b). YinYang bipolar cognition and bipolar cognitive mapping. *International Journal of Computational Cognition, 3*(3), 53-65.

Zhang, W.-R. (2006a). YinYang bipolar fuzzy sets and fuzzy equilibrium relations for bipolar clustering, optimization, and global regulation. *International Journal of Information Technology and Decision Making*, *5*(1), 19-46.

Zhang, W.-R. (2006b). YinYang bipolar T-norms and T-conorms as granular neurological operators. *Proceedings of IEEE International Conference on Granular Computing*, (pp. 91-96). Atlanta, GA.

Zhang, W.-R (2007). YinYang bipolar universal modus ponens (bump)–a fundamental law of non-linear brain dynamics for emotional intelligence and mental health. *Walter J. Freeman Workshop on Nonlinear Brain Dynamics, Proceedings of the 10th Joint Conference of Information Sciences*, (pp. 89-95). Salt Lake City, Utah, July 2007.

Zhang, W.-R., Pandurangi, A. & Peace, K. (2007). YinYang dynamic neurobiological modeling and diagnostic analysis of major depressive and bipolar disorders. *IEEE Transactions on Biomedical Engineering, 54*(10), 1729-39.

Zhang, W.-R. & Peace, K.E. (2007). YinYang MentalSquares–an equilibrium-based system for bipolar neurobiological pattern classification and analysis. *Proceedings of IEEE BIBE*, (pp. 1240-1244). Boston, Oct. 2007.

Zhang, W.-R., Wang, P., Peace, K., Zhan, J. & Zhang, Y. (2008). On truth, uncertainty, equilibrium, and harmony–a taxonomy for YinYang scientific computing. *International Journal of New Mathematics and Natural Computing, 4*(2), 207 – 229.

Zhang, W.-R, Zhang, H.J., Shi, Y. & Chen, S.S. (2009). Bipolar linear algebra and YinYang-N-Element cellular networks for equilibrium-based biosystem simulation and regulation. *Journal of Biological Systems, 17*(4), 547-576.

Zhang, W.-R & Chen, S.S. (2009). Equilibrium and non-equilibrium modeling of YinYang WuXing for diagnostic decision analysis in traditional Chinese medicine. *International Journal of Information Technology and Decision Making, 8*(3), 529-548.

Zhang, W.-R. (2009a). Six conjectures in quantum physics and computational neuroscience. *Proceedings of 3rd International Conference on Quantum, Nano and Micro Technologies (ICQNM 2009)*, (pp. 67-72). Cancun, Mexico, February 2009.

Zhang, W.-R. (2009b). YinYang Bipolar Dynamic Logic (BDL) and equilibrium-based computational neuroscience. *Proceedings of International Joint Conference on Neural Networks (IJCNN 2009)*, (pp. 3534-3541). Atlanta, GA, June 2009.

Zhang, W.-R. (2009c). YinYang bipolar relativity–a unifying theory of nature, agents, and life science. *Proceedings of International Joint Conference on Bioinformatics, Systems Biology and Intelligent Computing (IJCBS)*, (pp. 377-383). Shanghai, China, Aug. 2009.

Zhang, W.-R. (2009d). The logic of YinYang and the science of TCM–an Eastern road to the unification of nature, agents, and medicine. *International Journal of Functional Informatics and Personal Medicine (IJFIPM), 2*(3), 261–291.

Zhang, W.-R. (2010). YinYang bipolar quantum entanglement–toward a logically complete quantum theory. *Proceedings of the 4th International Conference on Quantum, Nano and Micro Technologies (ICQNM 2010)*, (pp. 77-82). February 2010, St. Maarten, Netherlands Antilles.

Zhang, W.-R., Pandurangi, K.A., Peace, K.E., Zhang, Y.-Q. & Zhao, Z. (2010). MentalSquares–a generic bipolar support vector machine for psychiatric disorder classification, diagnostic analysis and neurobiological data mining. *International Journal on Data Mining and Bioinformatics.*

Zhang, Y.-Q. (1992). Universal fundamental field theory and Golden Taichi. *Chinese Qigong, Special Issue 3*, 242-248. Retrieved from http://www.cs.gsu.edu/~cscyqz/ (in Chinese)

Acknowledgment

Special thanks go to the anonymous reviewers who provided critical and/or constructive review comments on the first draft of this monograph. In particular, I acknowledge an expert reviewer in the field of quantum gravity who made authoritative and decisive judgment on the central theme of this book.

Special thanks go to Development Editor Joel Gamon at IGI Global for his effective and timely double-blind review and decision process.

I acknowledge all my co-authors of published or accepted article(s) in more than 20 years for their contributions in different aspects related to the chapters of this monograph. In particular, I acknowledge Professor Anand K. Pandurangi (MD), Professor Karl E. Peace, Professor Yan-Qing Zhang, Professor Zhongming Zhao, Professor Su-Shing Chen, Professor Yong Shi, Dr. Stefan Jaeger, Ms. Jane H. Zhang, Mr. Hongzhao Zang, Professor Paul P. Wang, and Professor Justin Zhan, who have been my co-authors of recent journal or conference article(s).

I acknowledge all anonymous reviewers who have been critical and/or constructive on accepted or unaccepted submissions related to this monograph in more than 20 years. Special thanks go to one authoritative anonymous reviewer who was antagonistic toward this line of research but unwilling to reveal his name for a public debate. The antagonistic attitude provided a surviving environment for this work to grow stronger and hopefully healthier.

I acknowledge all journal Editors who handled my submissions and provided review comments in the last 20+ years. In particular, I acknowledge Professor Roger Jean, Professor Lawrence O. Hall, Professor Toshiyo Tamura, Professor Yong Shi, Professor Henri Prade, Professor Hao Ying, Professor Anew P. Sage, Professor Andrew B. Whinston, Professor Petre Dini, Professor Ivan Stojmenovic, Professor Tao Yang, Professor Robert Kozma, Professor Guo-Zheng Li, Professor Mingyu You, Professor Dan Xi, Professor Xiaohua (Tony) Hu, Professor Paul P. Wang, and Professor Feiyue Wang for their positive review decisions on my related submissions. The accepted publications laid the groundwork for this monograph.

I acknowledge my former students in the College of Information Technology, Georgia Southern University, who took the computer science course CSCI 5436/5436G – Web Programming and Design in the summer of 2007. They helped me in the design and development of a software prototype for the illustrations in Chapter 10. Special thanks go to Mr. Wisley Howard, *2007 Student of the Year in Computer Science*, for his remarkable contribution to the coding and installation of the prototype.

I acknowledge my employer Georgia Southern University for my tenure and professorship that enabled me to conduct related research work and to write this monograph in addition to regular teaching. If one day YinYang bipolar relativity is accepted by the world scientific community, it should be alternatively called *A Georgia Southern University Interpretation of Nature, Agents, and Causality.*

I acknowledge all authors who have cited my related publications in their articles for their inspiration. In particular, I acknowledge Professor Lotfi A. Zadeh, founder of fuzzy set theory, for being the first academic authority to recognize (YinYang) bipolar fuzzy sets in *Scholarpedia.*

I acknowledge Professor Walter J. Freeman III for his legendary inspiration. I got to know Walter at the *Panel Discussion on Brain-Like Computer Architecture* at *JCIS-2003*, Research Triangle Park, NC, where he served as a distinguished panelist and I was in the audience. On the argument between a hierarchical brain model presented by a keynote speaker and a semiautonomous multiagent society brain model with coordinated autonomy suggested by someone unknown in the audience, Walter gave his authoritative word: "*I agree with the audience.*" Walter's comment inspired many to join the fruitful debate in the packed meeting room. Four years later, the unknown person in the audience resurfaced and dedicated a paper to Walter at *Walter J. Freeman (80 Year Birthday) Workshop on Nonlinear Brain Dynamics*, Salt Lake City, Utah (July 2007). The paper was entitled "YinYang bipolar universal modus ponens (BUMP) – a fundamental law of non-linear brain dynamics for emotional intelligence and mental health." The author happened to be myself. After my talk on the workshop, Walter gave me encouraging comment and suggested the possibility of developing mathematical models for the classical concepts of YinYang-5-agents and Chinese Meridian System in Traditional Chinese Medicine (TCM). Now, three years later, YinYang-N-element cellular networks have been developed that form a basis for Chapters 8 and 9 of this book.

I acknowledge Professor Yang Shi for his discovery of the Yin Yang 1 (YY1) ubiquitous genetic agent. YY1 has been a major inspiration to my work.

I acknowledge Professor Yan-Qing Zhang for his leadership role in YinYang scientific research and long time support to my work. Yan-Qing pioneered original works in YinYang scientific research and successfully organized the first Workshop on YinYang Computation for Brain, Behavior and Machine Learning, Atlanta, GA, Aug. 2005. It was a well-organized delightful event. Its fruitful discussions will continue to show long lasting scientific impact on a wide spectrum of research topics for many years to come.

I acknowledge Professor Lei Xu, founder of Bayesian YinYang machinery for his inspiration. In particular, Dr. Xu made encouraging comment on my YinYang Bipolar Universal Modus Ponens (BUMP) at the aforementioned workshop back in 2005. Specifically, Dr. Xu asserted that BUMP is a generalization of MP to a different domain.

I acknowledge Professor Tingsen Xu, Taiji Grand Master, who was a speaker at the aforementioned workshop. In addition to his inspiring talk and illustration, our conversation has become my long lasting memory. Five years later, I still remember vividly an elegant point made by Master Xu on YinYang theory at the dinner table: Yin leads Yang because Yin can reproduce Yang. This key point is reflected in this monograph.

I acknowledge my colleague Professor Robert P. Cook for his advice on YinYang and bipolarity. A few years ago, a well established Chinese American friend strongly urged me to drop the word "YinYang" from "YinYang Bipolarity" "to avoid hurting the others' feelings". On this matter, I consulted with Robert for his opinion from a Westerner's perspective. Robert unequivocally stated: "YinYang is to your advantage; bipolarity is not."

I acknowledge Professor Xing Fang, Professor Don Fausett, Professor Hassan Kazemian, and Professor Huei Lee for their long time support to my related research works.

Special thanks go to Professor David (Yang) Gao for his encouraging comments on my recent works.

Deep appreciation is due to my friend and colleague Professor Karl E. Peace. As a distinguished scientist and educator, accomplished entrepreneur, and renowned philanthropist, Karl has been a major inspiration to my research. In particular, Karl generously proofread the last draft of my manuscript, corrected typographical errors, and suggested enhancements. I am honored and grateful for his innovative, impartial, and insightful foreword.

And last but not least, I acknowledge my wife and children for their love and unfailing support during the years and months it took to bring this monograph to the world.

Wen-Ran Zhang

Part 1
Introduction and Background

Chapter 1
Introduction:
Beyond Spacetime

ABSTRACT

This chapter serves as an introduction to bring readers from spacetime relativity to YinYang bipolar relativity. Einstein's assertions regarding physics, logic, and theoretical invention are reviewed and his hint of YinYang bipolar relativity is identified. The limitations of general relativity and quantum mechanics are briefly discussed. It is concluded that logically definable causality, axiomatization of physics, axiomatization of agent interaction, and the grand unification of general relativity and quantum theory are essentially the same problem at the fundamental level. A paradox on Hilbert's Problem 6—Logical Axiomatization for Illogical Physics (LAFIP)—is introduced. Bipolarity is postulated as the most fundamental property of nature transcending spacetime. The theoretical basis of agents, causality and YinYang bipolar relativity is highlighted and distinguished from established theories. The main ideas of the book are outlined.

(Note: In this book we refer to relativity theories defined in spacetime geometry as spacetime relativity. Thus, all previous relativity theories by Galileo, Newton, Lorenz, and Einstein belong to spacetime relativity. This terminological treatment is for distinguishing YinYang bipolar geometry from spacetime.)

INTRODUCTION

Ever since Aristotelian science was established together with Aristotelian bivalent truth-based syllogistic (or classical) logic 2300 years ago, scientists have been devoting their lifetime efforts to the noble cause of seeking truths from the universe. Boolean logic (Boole, 1854) reinforced the truth-based tradition and

DOI: 10.4018/978-1-60960-525-4.ch001

eventually led to modern digital computer technologies that, in turn, significantly extended the reach of scientific explorations by mankind into both macroscopic and microscopic *agent* worlds.

Scientific explorations, unfortunately, have not been able to escape the delicate balance of Mother Nature. In the microscopic agent world, the painstaking quest for *quantum gravity*—Einstein's unfinished unification of general relativity and quantum mechanics—has so far failed to find a decisive battleground (Smolin, 2006; Woit, 2006); quantum entanglement remains a mystery (Penrose, 2005, p. 591) that is hindering the development of quantum computers; mental equilibrium and disorders are unexplained at the neurobiological and neurophysiologic levels; and, despite one insightful surprise after another that the genome has yielded to biologists, the primary goal of the Human Genome Project—to ferret out the genetic roots of common diseases like cancer and Alzheimer's and then generate treatments—has been largely elusive (Wade, 2010). In the macroscopic agent world, "big bang" so far came from nowhere and was caused by nothing to our knowledge; economic recession has been a recurring problem; and global warming is threatening the very existence of human civilization including the scientific establishment itself.

Conceivably, truth is subjected to observability and limited to certain spacetime but equilibrium or non-equilibrium, as a central concept of thermodynamics—the ultimate physical source of existence, energy, life, and information, is ubiquitous and ruthless. Despite the proven incompleteness of truth-based reasoning (Gödel, 1930) and the mounting evidence for action-reaction forces, negative-positive electromagnetic charges, matter-antimatter particles, mental depression and mania, economic recession and expansion, genomic repression and activation, social competition and cooperation, global cooling and global warming, big bang and black holes, or Yin and Yang of nature in general that overwhelmingly suggest a *bipolar equilibrium-based universe* (including equilibrium and non-equilibrium states), few scientists have asked the difficult question: *Whether the universe is actually truthful and whether the truth-based tradition is adequate for furthering scientific explorations?*

A central theme of this book is that the universe is not truthful but bipolar. This theme leads to a paradox on Hilbert's Problem 6—*"Axiomatize all of Physics"* (Hilbert, 1901). The paradox states: *"Logical Axiomatization for Illogical Physics"* (*LAFIB* or *LAFIP*) (Zhang, 2009a, 2009b, 2009c, 2009d). LAFIP manifests the inconvenient truth that truth-based logical reasoning is inadequate for axiomatizing the illogical aspects of physics. This monograph is, therefore, not for seeking truth from the universe but for resolving the LAFIP paradox in modern science. The resolution to be presented is *YinYang bipolar relativity* which is shown to be a deeper unifying logical foundation transcending spacetime and *spacetime relativity* including relativity theories by Galileo, Newton, Lorenz, and Einstein.

In front of the historical giants of science and philosophy, every living scientist or philosopher is entitled to feel humble and respectfully follow the established scientific tenet. No wonder the editor of an influential logic journal once posted a slogan on his website that read *"Never Question the Logic of Aristotle."* Evidently, this editor became "too humble" to realize that at Aristotle's time air and water were deemed the most fundamental elements and the Earth was believed the center of the universe; Copernicus would have not been able to discover the solar system had he not questioned Aristotle's cosmology and Einstein would have not been able to develop his general theory of relativity had he not questioned Aristotle's ether theory.

Despite his great contribution to science and philosophy, Aristotle's logic as well as his philosophy was inevitably subjected to the scientific and technological limitations at his time. For instance, while Aristotle's causality principle has been widely considered the doctrine of all sciences for more than two thousand years, the principle, however, is irreducible to regularity as asserted by 18th century Scottish philosopher David Hume (Blackburn, 1990) or *"Causality Is Undefinable"* by classical (Aristotelian) logic as concluded by Lotfi Zadeh – founder of fuzzy logic (Zadeh, 2001).

Actually, without bipolarity, causality is undefinable by any truth-based logic due to the LAFIP paradox. Unfortunately, bipolarity such as electron-positron pairs were beyond Aristotle's imagination and history did not make negative numbers and zero available in ancient Greece at Aristotle's disposal. Consequently, following Aristotelian truth-based doctrine, all established logical systems in more than two thousand years have so far failed to provide logically definable causality. Such established systems include but are not limited to Boolean logic (Boole, 1854), fuzzy logic (Zadeh, 1965), many-valued logic (Łukasiewicz, 1920, 1970), paraconsistent logic (Belnap, 1977), intuitionistic logic (Brouwer, 1912), and quantum logic (Birkhoff & von Neumann, 1936).

In the same light, Einstein pointed out the limitation of his own glorious establishment. A quarter century after conquering his general theory of relativity, Einstein unequivocally stated that (Einstein, 1940) *"For the time being we have to admit that we do not possess any general theoretical basis for physics which can be regarded as its logical foundation."* By this statement, not only did Einstein make it abundantly clear that his general theory of relativity could not serve as the unifying logical foundation for physics, but also hinted that a fundamentally different mathematical abstraction beyond classical set theory and truth-based bivalent tradition might be necessary for the grand unification of general relativity and quantum mechanics.

As an electrical engineer turned to computer scientist, the author of this monograph has been deeply humbled by even thinking about writing a book on relativity. In the meantime, however, we have to admit that the new logical foundation as envisioned by Einstein in the last century is still missing today. Therefore, we have to recall Newton's favorite slogan *"If I have seen further, it is by standing on the shoulders of giants."* It is this slogan that encouraged the author to write this monograph regarding scientific unification from a logical and computational perspective. The author must admit though that he is not a theoretical physicist but believes that the next scientific unification or *"The Next Superstring Revolution"* (Smolin, 2006, p. 274) is not necessarily to come from theoretical physics and could well be from another field that encompasses logical and mathematical computation as well as physical, social, biological, and brain sciences. It is well-known that Berners-Lee as a physicist invented the World Wide Web and revolutionized computer applications. Hopefully, computer scientists can do something in return to physics especially for the logical foundation of *quantum computing, cognitive informatics*, and lifesciences.

Nevertheless, readers may have deep doubts regarding the intended unifying role of this work. Such doubts are natural and understandable because the book title *YinYang Bipolar Relativity* is extremely unusual by any means from a classical scientific perspective. But the purpose of this monograph is, indeed, to present a *unifying theory* of nature, agents, and causality with applications illustrated in both charted and uncharted territories of quantum computing, cognitive informatics, and life sciences.

This chapter serves as an introduction to bring readers from spacetime relativity to YinYang bipolar relativity. The introduction is organized in the following sections:

- **Einstein and YinYang Bipolar Relativity.** This section presents a review on a few of Einstein's assertions regarding physics, logic, and theoretical invention. Surprisingly, it is shown that, not only is YinYang bipolarity supported by action-reaction forces in classical Newtonian mechanics but also hinted by Einstein in one of his famous quotes. By this review, the discussion of YinYang bipolar relativity is positioned on the shoulders of giants.

- **General Relativity and Quantum Theory.** This section discusses the incomplete nature of general relativity and quantum theory.
- **Logical Axiomatization for Illogical Physics (LAFIP).** This section identifies the limitation of truth-based cognition and characterizes it with the LAFIP paradox.
- **Observation and Postulation.** This section introduces equilibrium-based cognition and bipolar agents, identifies *nature's most fundamental property*, and introduces the basic concept of YinYang bipolar geometry *beyond spacetime*.
- **Book Overview.** This section presents an outline of the monograph.
- **Summary.** This section summarizes the major points in the chapter and draws a few conclusions.

EINSTEIN AND YINYANG BIPOLAR RELATIVITY

Regardless of the great achievement of Western science, without logically definable causality, it can be argued that Western science is an incomplete and evolving science. Actually, the incomplete and evolving nature was already elaborated by Einstein. According to him:

Physics constitutes a logical system of thought which is in a state of evolution, whose basis (principles) cannot be distilled, as it were, from experience by an inductive method, but can only be arrived at by free invention. The justification (truth content) of the system rests in the verification of the derived propositions (a priori/logical truths) by sense experiences (a posteriori/empirical truths). ... Evolution is proceeding in the direction of increasing simplicity of the logical basis (principles). .. We must always be ready to change these notions – that is to say, the axiomatic basis of physics – in order to do justice to perceived facts in the most perfect way logically. (Einstein, 1916)

Regarding reality, experience, and human thought, Einstein pointed out that

If, then, it is true that the axiomatic basis of theoretical physics cannot be extracted from experience but must be freely invented, can we ever hope to find the right way? Nay, more, has this right way any existence outside our illusions? Can we hope to be guided safely by experience at all when there exist theories (such as classical mechanics) which to a large extent do justice to experience, without getting to the root of the matter? I answer without hesitation that there is, in my opinion, a right way, and that we are capable of finding it. Our experience hitherto justifies us in believing that nature is the realization of the simplest conceivable mathematical ideas. I am convinced that we can discover by means of pure mathematical constructions the concepts and the laws connecting them with each other, which furnish the key to the understanding of natural phenomena. Experience may suggest the appropriate mathematical concepts, but they most certainly cannot be deduced from it. Experience remains, of course, the sole criterion of the physical utility of a mathematical construction. But the creative principle resides in mathematics. In a certain sense, therefore I hold it true that pure thought can grasp reality, as the ancients dreamed. (Einstein, 1934)

Regarding the next unification Einstein stated:

The development during the present century is characterized by two theoretical systems essentially independent of each other: the theory of relativity and the quantum theory. The two systems do not directly contradict each other; but they seem little adapted to fusion into one unified theory (Einstein, 1940).

Einstein unequivocally stated: *"For the time being we have to admit that we do not possess any general theoretical basis for physics which can be regarded as its logical foundation"* (Einstein, 1940).

Clearly, after inventing his most celebrated general theory of relativity, Einstein believed that:

1. Logical axiomatization of physics is possible.
2. The grand unification of general relativity and quantum mechanics needs a new logical foundation transcending spacetime relativity and quantum theory.
3. The axiomatic basis of theoretical physics cannot be extracted from experience but must be freely invented.
4. Pure thought can grasp reality as the ancients dreamed and nature is the realization of the simplest conceivable mathematical ideas.

YinYang bipolar cosmology was an ancient dream. It claims that everything has two sides or two opposite but reciprocal poles or interactive energies. The bipolar reciprocal and interactive nature makes YinYang attractive to scientific unification. That is why some Western scientists expect YinYang to play a unifying role to make modern science more complete.

While YinYang bipolar relativity can trace its philosophical origin back to the ancient Chinese dream, the formal logical theory presented in this monograph is neither resulted from elaboration or experimentation of the ancient YinYang nor *"extracted from experience"*. It is *"freely invented"* that is a *"must"* for the axiomatic basis of theoretical physics according to Einstein.

YinYang bipolar relativity is intended to provide an equilibrium-based unifying logical foundation for the grand unification. The unifying properties discussed are focused primarily on the logical level with limited mathematical extensions including a bipolar quantum linear algebra and a bipolar quantum cellular automation theory. Further investigations into the possibility of an equilibrium-based unification of Einstein's classical equations of general relativity and that of classical quantum mechanics are left for future research efforts because *"For the time being we have to admit that we do not possess any general theoretical basis for physics which can be regarded as its logical foundation."* Hopefully, YinYang bipolar relativity can serve as the first step forward.

Historically, YinYang as a relativistic theory has survived more than 5000 years of recorded human history without a formal logical foundation (Moran & Yu, 2001; Karcher, 2002). It is a living proof to Einstein's assertion that *"the axiomatic basis of theoretical physics cannot be extracted from experience but must be freely invented"* and such invention requires *"pure thought"* to *"grasp reality"*.

The author has to say that his grasp of reality has been largely by accident but not by intention. Twenty years ago he entered bipolar cognitive mapping research (Zhang, Chen & Bezdek, 1989; Zhang et al., 1992) and later developed a YinYang bipolar dynamic logical system (Zhang, 2003a, 2005; Zhang & Zhang, 2004b). YinYang eventually brought the author into the quantum world (Zhang, 2009a, 2009b, 2009c, 2009d, 2010). For most of these years, however, the author was unaware of the fact that he was on the path to a formal logical theory beyond spacetime relativity until recent years.

There is an old saying: *"Difference in profession makes one feel world apart."* That is still true in modern days. As a computer scientist, the author didn't pay much attention to the advances in physics until late 2008 and early 2009.

There were two remarkable events in 2008. One was the Nobel Prize in Physics. Japanese American physicist Yoichiro Nambu shared the Prize with Japanese physicists Makoto Kobayashi and Toshihide Maskawa *"for the discovery of the mechanism of spontaneous broken symmetry in subatomic physics"* (The Royal Swedish Academy of Sciences, 2008). According to the press release, it was their *"passion for symmetry"* that permeated *"the Standard Model of elementary particle physics"* and unified *"the smallest building blocks of all matter and three of natures four forces in one single theory"* (The Royal Swedish Academy of Sciences, 2008).

The impact of the 2008 Nobel Prize is far reaching because bipolar dynamic equilibrium and non-equilibrium or symmetry and broken-symmetry form the theoretical basis of the oldest Eastern YinYang philosophy that is well-known to the world and has been extremely pervasive in all fields of physical and social science in the West as well as in the East especially in Japan. In this regard, 1949 Nobel Laureate Japanese Physicist Hideki Yukawa once said: *"You see, we in Japan have not been corrupted by Aristotle"* (Rosenfeld, 1963).

Another remarkable event in 2008 was a report on the speed of quantum entanglement. A quantum physics experiment performed in Geneva, Switzerland, determined that the *"speed" of the quantum non-local connection* or *quantum entanglement* what Einstein called *"spooky action at a distance"* has a minimum lower bound of 10,000 times the speed of light (Salart et al., 2008). Thus, the geometry of a logical foundation for physics has to transcend spacetime because spacetime has been based on the hypothesis that no speed could go beyond the speed of light.

With well-observed space expansion and quantum entanglement, it is easy to assert that space and time cannot serve as the most fundamental properties of a quantum universe. Instead, the most fundamental property has yet to be determined that would cause space expansion and the minimum lower bound of 10,000 times the speed of light. Consequently, in order to achieve the grand unification of general relativity and quantum mechanics, spacetime relativity has to be superseded and science has to advance beyond spacetime geometry.

Since strong and weak forces have been unified with electromagnetic force per 2008 Nobel Prize in Physics, gravitational and electromagnetic forces can be considered nature's two basic forces that can be denoted as action-reaction forces $(-f, +f)$ and negative-positive electromagnetic forces $(-q, +q)$, respectively. On the other hand, particles and antiparticles are the only known tangible stuff in the universe that form a broken symmetry or quasi-equilibrium denoted $(-p, +p)$. Evidently, all nature's basic forces and subatomic particles are bipolar in nature. Consequently, it is natural to ask the question: *If nature cannot be the realization of space and time could the universe including spacetime be the realization of YinYang bipolar relativity?*

Understandably, the road to YinYang bipolar relativity has been and has to be "an odd path." The author did not select the path intentionally but was destined to "hit" the road accidentally and to stumble upon the uncharted territory. He has no choice but to face the challenges in writing this monograph with a trembling heart yearning for falsifiability and fear of humiliation.

The title *"YinYang Bipolar Relativity"* tells readers that the book is about the relativity of the two poles or the interactive energies of nature that are conceptualized as the Yin and Yang in general. The word *"YinYang"* indicates that the main idea is philosophically rooted in the ancient Chinese cosmology. On the one hand, it symbolizes the holistic reciprocal interactive nature of the two poles of an agent to distinguish them from bipolar dichotomy. On the other hand, the word "YinYang" distinguishes this work from bipolar (transistor) logic that actually stands for "bivalent digital logic implemented by bipolar transistors." (Note: *For simplicity, we use the short term "bipolar" alternatively for "YinYang bipolar" in this work without further specification.*)

Since the new theory is YinYang bipolar in nature, it is fundamentally different from Einstein's *general theory of relativity* or *spacetime relativity* in general. While general relativity is about spacetime and gravity, bipolar relativity is about the opposite poles of agents and the manifestations of bipolar fusion, interaction, oscillation, equilibrium, non-equilibrium, and quantum entanglement. An *agent* in this book can be any microscopic or macroscopic physical or social entity such as quantum agent, molecular agent, genomic agent, human agent, organization, society, celestial object, galaxy, or the universe itself.

Einstein's general relativity and Bohr's quantum mechanics are undoubtedly the deepest theories in the history of science. Now, we have been living in Einstein-Bohr's universe for almost a century, which, on the macro scale, is defined by general relativity and, on the micro scale, is defined by quantum mechanics. The two theories, however, have found no unification, and both are so far limited to spacetime geometry. Few have tried to bring relativity and quantum theory to the mental, social, and biological worlds of the agent societies on the earth. It is suggested by many that causal-effect relation is a key for a complete quantum theory and for its grand unification with general relativity. Unfortunately, causality is logically undefinable with classical logic (Zadeh, 2001).

Can YinYang bipolar relativity serve as a holistic unifying logical theory that directly or indirectly governs all microscopic and macroscopic agents and agent interactions in physical, social, and life sciences beyond spacetime geometry? Although the question sounds unrealistic, it is actually a legitimate question because *"YinYang bipolar relativity"* was already hinted by the scientific giant – Albert Einstein. The hint is in his famous quote about his own relativity theory: *"Put your hand on a hot stove for a minute, and it seems like an hour. Sit with a pretty girl for an hour, and it seems like a minute. That's relativity."*

If readers failed to see the hint of YinYang bipolar relativity from the quote at first glance, it would be understandable. Our modern science education has been based on truth and *singularity* (Hawking & Penrose, 1970), which are both unipolar in nature. And for thousand years bipolar equilibria have been excluded from observables in science textbooks even though equilibrium is a central concept in thermodynamics – the ultimate source of energy, existence, life, and information.

To be more specific, the Yin and the Yang hinted by Einstein in his quote are *sorrow* and *joy* that can be paired up as a bipolar equilibrium or non-equilibrium variable (s, j) and generalized to action-reaction forces denoted $(-f, +f)$, electromagnetic forces denoted $(-q, +q)$, matter-antimatter particles denoted $(-p, +p)$, repression-activation abilities in gene expression regulation denoted *(repression, activation)*, competition-cooperation relations in socioeconomics denoted *(competition, cooperation)*, or, symbolically, the formal logical function or predicate $\phi = (Yin, Yang) = (\phi^-, \phi^+)$ in a YinYang bipolar dynamic logic for the theory of equilibrium-based YinYang bipolar relativity (Zhang, 2009a, 2009b, 2009c, 2009d). Hopefully, with Einstein's hint and Newton's laws of classical mechanics regarding action-reaction forces, YinYang bipolarity is positioned "*on the shoulders of giants.*"

It should be remarked that, legendary Danish physicist Niels Bohr, father figure of quantum mechanics besides Einstein, was the first one to bring YinYang into quantum theory for his particle-wave duality. When he was awarded the Order of the Elephant by the Danish government, he designed his own coat of arms which featured in the center a YinYang logo and the Latin motto: *opposites are complementary*. Based on YinYang, Niels Bohr developed his complementarity principle in the Copenhagen Interpretation of quantum mechanics (Hilgevoord & Uffink, 2006; Faye 2008).

Bohr's principle recognized particle-wave complementarity but stopped short of recognizing YinYang bipolarity – a central theme of YinYang cosmology and a fundamental property of the quantum world. Without YinYang bipolarity, the two complementary sides are not exactly opposites and could not form a bipolar equilibrium or non-equilibrium. Thus, Bohr's complementarity principle didn't go beyond Aristotle's truth-based logical reasoning and his particle-wave duality has often been considered a paradox.

Through this book, the word *"equilibrium-based"* is to equilibrium and non-equilibrium as the word "truth-based" is to truth and falsity but with fundamentally different syntax and semantics. One is rooted in physical YinYang bipolarity; the other is rooted in bivalent logical singularity. Thus, the word "equilibrium-based" has full equilibrium, quasi-equilibrium, and non-equilibrium concepts self-contained regarding the two poles of an agent as the word "truth-based" has truth, partial truth, falsity, and contradiction self-contained regarding a logical concept.

To be honest, Einstein himself perhaps did not realize his hint of YinYang bipolar relativity in his famous quote either. Although he didn't promote YinYang bipolarity explicitly, his special relativity predicted electron-positron pair production that can be denoted as a YinYang bipolar element $e = (e^-,e^+)$. As the most fundamental example of the materialization of energy, electron-positron pair production has been accurately described by quantum electrodynamics (QED) – the jewel of physics (Dirac, 1927,1928) (Feynman, 1962,1985).

On the other hand, although Einstein never believed in the theory of singularity and even regarded the theory as *"bizarre"*, resisting the logic of his own theory right up to his death in 1955, his equations of general relativity did eventually lead to the flourish of singularity after his death following the discovery of black holes (Hawking & Penrose, 1970). Based on singularity and partially observable truth, it is now a commonly accepted theory that spacetime as well as the universe was created by a big bang and will end in one or more black holes. So far as we know, the big bang came from nowhere and caused by nothing.

To reconcile the inconsistency between singularity and the second law of thermodynamics, Stephen Hawking (Hawking, 1974) proposed the remedy that a black hole should have particle and/or antiparticle emission or Hawking radiation. While Hawking radiation has been a hot topic of discussion in quantum theory, its far-reaching consequence has so far been overlooked by the scientific community. The consequence is that when the universe ends matter-antimatter bipolarity will miraculously survive. Therefore, singularity is not a contradiction but a vindication of YinYang bipolarity. Unfortunately, the vindication has so far been largely ignored.

Even though bipolarity is so essential that can survive a black hole, scientists equipped with truth-based logical systems tend to deny the unavoidability of bipolarity. For instance, digital logic devices implemented with bipolar transistors led to the commercialization of modern computers, but the truth values 0 and 1 don't directly support bipolar equilibrium or symmetry. Instead, digital logic is widely referred to as *"bipolar logic"* while it really meant to be *"bipolar transistor logic."*

In clinical psychiatry, bipolarity has been deemed a notorious concept due to its link to bipolar disorder which is not treated as the loss of bipolar equilibrium (the order) but a *"mental big bang"* from nowhere. Consequently, a medical intervention to bipolar disorder may not be focused on recovering the disorder to a calm healthy energetic reciprocal bipolar equilibrium but to treat bipolar symptoms to zero. Although, it is logically correct to eliminate the bipolar symptoms through a medical intervention, but zero symptom does not mean a healthy bipolar equilibrium as it carries no life signal and can also be used to characterize eternal equilibrium or brain death. Why not bipolar mental equilibrium? Based on unipolar set theory and truth-based logic, *bipolar equilibrium* is non-existence and outside the realm of science except for a few rare "out of the box" models (Zhang & Zhang, 2004b; Zhang, 2005, 2007; Zhang, Pandurangi & Peace, 2007; Zhang *et al.*, 2010).

To be honest, the above problem did not originate from Western science though. Truth-based unipolar cognition as a Western tradition has triumphed and made glorious achievements in all fields of science and technology. A key for the success is its formal logical and mathematical foundation. Even though YinYang has survived more than 5000 years of recorded human history (Moran & Yu, 2001) and equilibrium-based traditional Chinese medicine (TCM) has been practiced in China for at least 2000 years, the Eastern philosophy has failed to provide a systematic formal logical and mathematical foundation for its physical, social, and biological claims until recent years. Subsequently, while truth-based cognition stopped short of offering logically definable causality, YinYang could not step up with a complementary solution.

Conceptually, the two sides of YinYang are inseparable. Without the concept Yin there would be no such a concept as Yang. Mixed feelings such as intermingled sorrow and joy of a person do exist in a single physical entity. That is why we have the paradoxical phrase *"bittersweet"*. To maintain mental equilibrium a person should exercise positive thinking when depressed and exercise negative thinking when excited. That is, actually, common public knowledge for emotional intelligence followed by most people in the world. Otherwise, more people would suffer bipolar disorder.

Consequently, we may add a second part to the famous quote from Einstein:

Part 1. *"Put your hand on a hot stove for a minute, and it seems like an hour. Sit with a pretty girl for an hour, and it seems like a minute. That's relativity."*

Part 2. *"Let Yin and Yang be self-negation and self-assertion abilities of an agent: Yin without Yang brings depression; Yang without Yin brings mania; balanced YinYang brings healthy bipolar equilibrium. That is YinYang bipolar relativity."*

While in Part 1 Einstein focused on space, time, and mental gravity that led to the great theory of general relativity, in Part 2 our focus is on the relativity of the two poles of agents and their equilibrium. The second focus leads to the theory of YinYang bipolar relativity to be discussed in this book.

Intuitively, the two poles of nature such as (-q,+q), (-f,+f), (-p,+p), and (self-negation, self-assertion) could be more fundamental physical concepts than space and time. It is natural to ask: *"Can YinYang bipolar relativity be a deeper theory applicable in physical, social and life sciences?" "Can YinYang bipolar relativity lead to a complete quantum theory for quantum computing and communication?" "Can YinYang bipolar relativity be a logical foundation for scientific unification?"*

GENERAL RELATIVITY AND QUANTUM THEORY

Until Einstein a single idea had dominated physics: *the world is made of nothing but ether or matter.* Even electricity and magnetism were considered aspects of matter or ether. This beautiful picture was smashed by Einstein's special relativity, because if the whole thing of being at rest or in motion is completely relative, Aristotle's ether theory has to be a fiction. If fields are not made of matter, matter might be made of fields. Then, fields could be the fundamental makings of the universe (Smolin, 2006, p. 38). Following this line of thinking, Einstein unified electromagnetism with his special theory of relativity in which electromagnetic wave and light became the same thing and both travel at the same speed in spacetime.

With his general theory of relativity, Einstein succeeded in unifying all kinds of motion with gravitational field. First, *the principle of equivalence* states that *the effects of acceleration are indistinguishable from the effects of gravity*. Secondly, gravitational field is unified with spacetime geometry.

Einstein's general theory of relativity has led to unprecedented predictions and experimental verifications. These predictions and verifications have resulted in revolutionary fundamental understandings about our universe. They include but are not limited to:

1. Atomic clocks would slow down in a gravitational field;
2. Light would bend toward a gravitational source;
3. Two parallel light rays could meet in spacetime geometry.

In Euclidian geometry, two parallel straight lines can never meet; according to Maxwell's theory of electromagnetism, light rays move in straight lines; how could two parallel light rays meet in spacetime geometry? Einstein's answer is that, if they pass on each side of a gravitational source such as a star, the two light rays will bend toward each other in the real world. Euclidian geometry is, therefore, not true in the real world. Since matter is constantly moving, spacetime geometry is constantly changing. It is incredibly dynamic with great waves of gravitational field.

Ever since 1915 when Einstein's general relativity theory was born, the quest for further unification has not been so lucky. Although the broken symmetry model unified strong and weak forces with electromagnetic force, the search for quantum gravity – the quest for the unification of general relativity and quantum mechanics – has so far found no decisive battleground. The unification of quantum gravity with agents, socioeconomics, cognitive informatics, and life sciences is rarely mentioned by anyone. Quantum non-local connection – the *"spooky action at a distance"* so-called by Einstein remains unexplained. Despite the tremendous academic and industrial research and development efforts, quantum computer is still not ready for practical application.

The EPR paradox (Einstein, Podolsky & Rosen, 1935) challenged long-held ideas about the relation between the observed values of physical quantities and the values that can be accounted for by a physical theory. It was their theoretical possibility or impossibility that led Einstein to reject the idea that quantum mechanics might be a fundamental physical law. Instead, Einstein deemed quantum mechanics as an incomplete theory. The missing parts are often referred to as *"hidden variables"* in the literature. Bell's theorem (Bell 1964) extended the argument of the EPR paradox and proved the validity of quantum entanglement in statistical terms. The theorem, however, so far hasn't led to a unification of quantum mechanics and general relativity. Until this day, the so-called *"hidden variables"* have not really surfaced in simple logically definable terms.

English mathematical physicist Roger Penrose described the above unsolved problem as two mysteries of quantum entanglement (Penrose, 2005, p. 591). The first mystery is characterized as the phenomenon itself. The second one, according to Penrose, is *"How are we to come to terms with quantum entanglement and to make sense of it in terms of ideas that we can comprehend, so that we can manage to accept it as something that forms an important part of the workings of our actual universe? .. The second mystery is somewhat complementary to the first. Since according to quantum mechanics, entanglement is such a ubiquitous phenomenon – and we recall that the stupendous majority of quantum states are actually entangled ones – why is it something that we barely notice in our direct experience of the world? Why do these ubiquitous effects of entanglement not confront us at every turn? I do not believe that this sec-*

ond mystery has received nearly the attention that it deserves, people's puzzlement having been almost entirely concentrated on the first."

Nevertheless, since 1980s, many successful experiment results have shown the physical existence of quantum entanglement and the "spooky" quantum phenomenon has become a fundamental concept in quantum computing even though, until recently, physicists have only been able to demonstrate quantum entanglement in laboratory conditions. Without a decisive victory in the quest for quantum gravity, it is fair to say that something fundamental must still be missing from the big picture.

Many influential authors have pointed out that the missing fundamental is the causal-effect relationship in quantum mechanics that is part of Aristotle's causality principle – the doctrine of all sciences. Unfortunately, the doctrine was stated by Aristotle only in words. After more than 2300 years, the principle still has not been adapted to logical formula. Without making clear the cause-effect relation research on quantum mechanics and quantum computing can only move forward slowly in theoretical darkness.

As Einstein asserted that *"Physics constitutes a logical system of thought.."*, *"the axiomatic basis of theoretical physics cannot be extracted from experience but must be freely invented"* and *"pure thought can grasp reality"*, the light at the end of the quantum tunnel is not likely to come nearer until someone grasps the reality with pure thought to invent a new logic beyond truth-based spacetime that leads to logically definable causality. Is that possible at all? Einstein thought so. He asserted that *"nature is the realization of the simplest conceivable mathematical ideas."* Unfortunately, scientists who dare to try to use their pure thought to grasp reality at such a theoretical magnitude would be most likely faced with failure and unemployment or being squarely rejected with the deepest humiliation.

LOGICAL AXIOMATIZATION FOR ILLOGICAL PHYSICS – THE LAFIP PARADOX

Since agent interactions in both macroscopic and macroscopic worlds are governed by physical and social dynamics that provides the basis for causality, Einstein's unfinished unification is essentially the axiomatization of agent interactions, the axiomatization of physics, or the logical definition of causality. Thus, the grand unification, Hilbert's Problem 6, axiomatization of agent interaction, and logical definition of Aristotle's causality principle – can all be deemed the same problem at the fundamental level.

As one of Hilbert's 23 mathematical questions, Problem 6 (Hilbert, 1901) – *"axiomatize all of physics"* – the most general mathematical physics problem is widely believed "unsolvable" now after a century but Einstein's grand unification is believed achievable by many due to its focused terms of description. Based on discussions in earlier sections, we can conclude that without a different mathematical abstraction that is relativistic and quantum in nature applicable in microscopic agent world as well as in macroscopic agent world, in physical as well as in social world, in brain science as well as in life sciences, logically definable causality, axiomatization of physics, and the grand unification would all be mission impossible.

The principle of classical mathematical abstraction itself is, therefore, a major barrier preventing a possible solution to the above three problems. Consequently, the grand unification depends on a thorough reexamination of the fundamental limitation of classical truth-based mathematical abstraction.

From a logical perspective, it can be observed that a basic limitation of a truth-based logic lies in bivalency. Regardless of the vital importance of bivalency in modern science, without directly carrying a shred of basic physical semantics such as bipolarity or equilibrium, the logical values {false, true} or {0,1} in classical logic or any of its truth-based extensions including quantum models are unipolar in nature and inadequate for axiomatizing physics. The bivalent limitation could be the fundamental reason why there is so far no truth-based logical axiomatization for physics, no logically definable causality, and no grand unification.

From a set-theoretic perspective, the truth-based limitation can be attributed to the principle of mathematical abstraction in classical set theory. The principle states that the concept of element in a set is self-evident without the need for proof. In addition, it is commonly interpreted that the properties of a set are independent of the nature of its elements. This principle is so fundamental in information and computation that it is rarely challenged and commonly considered *"unquestionable."* It can be argued, however, that:

1. If an element is a bipolar equilibrium-based agent, it can be particle or wave, static or dynamic, linear or non-linear, orderly or disorderly, generous or greedy, and logical or illogical that can't be simply characterized as true or false.
2. The property of a set of bipolar equilibria could be physically dependent on the property of its elements.

For instances, with quantum entanglement a split photon can be in two places at the same time; a subatomic particle can change polarity trillion times per second (Fermilab, 2006); some genetic agent exhibits YinYang bipolar repression-activation abilities in gene expression regulation (Shi *et al.,* 1991); the bipolar interactive cellular nature of Yin and Yang are essential in synthetic biology (Gore & van Oudenaarden, 2009); a bipolar disorder can be oscillatory or mixed (American Psychiatric Association, 2000); two competitor agents can also be cooperative. We need syntactic and semantic representations for these seemingly "illogical" but nevertheless physical or natural phenomena for scientific logical solutions such as medicine for mental disorders, particle-wave duality for quantum mechanics, gene expression regulation for genomics, agent emotion modeling for mental health, and axiomatization of agent interactions or axiomatization of physics.

With classical mathematical abstraction and truth-based logic, however, we have the 3-fold dilemma:

1. If we treat each pole of a bipolar equilibrium or non-equilibrium as a self-evident element we will lose holistic bipolar fusion, binding, coupling, or quantum entanglement.
2. If we treat a bipolar equilibrium or non-equilibrium as a self-evident element its membership in a set can only be true or false where polarity cannot be represented.
3. If we preserve the independence rule between a set of equilibria and its elements we will not be able to link the global equilibrium/non-equilibrium to local ones.

Due to the 3-fold dilemma, classical set theory and truth-based logic provide no operation for equilibrium-based bipolar fusion, interaction, oscillation, symmetry, and quantum entanglement. This can be further illustrated with some intuitive examples.

Example 1. (a) How can depression, mania, mental equilibrium, and eternal equilibrium (or brain death) be directly characterized with logical values? How can the negative effect, positive effect, balancing effect, and deadly side effect of a bipolar disorder medicine be characterized with logical expressions? *(b)* A depressed patient took a positive antidepressant drug and regained mental equilibrium; a second patient took the same drug but became manic; a third patient took the drug and died of side effect; patients in deep depression tend to become suicidal. How to characterize the neurobiological reactions of the patients for personalized biomedicine? *(c)* A combo treatment with a negative medicine (for un-exciting manic nervous system) and a positive medicine (for un-depressing depressed nervous system) may have the outcomes: balancing effect, negative effect, positive effect, or deadly side effect. How to characterize the combo and the combo effects?

Example 2. (a) How can economic recession, overheat, normal equilibrium, and collapse be directly characterized with logical values? How can the negative effect, positive effect, balancing effect, and disastrous effect of an economic intervention be characterized with logical expressions? *(b)* A government intervention to an economic recession can regain economic equilibrium, cause economic overheat, lead to deeper recession, or lead to economic collapse. How to logically characterize the reactions of the economy to the intervention? *(c)* A combo economic intervention with government regulation for economic overheating and a stimulating package for consumer spending may have the outcomes: balancing effect, cooling down effect, warming up effect, or disastrous side effect. How to characterize the combo and the combo effects?

Example 3. How can a unifying logical model of the universe be developed that enables the fusion of black hole and big bang, electron and positron, matter and antimatter, action and reaction forces, or Yin and Yang in general for quantum computing and communication.

Evidently, without bipolarity, there is no way to define the above seemingly *"illogical"* but nevertheless natural non-linear dynamic logical values and operators in the unipolar bivalent lattice {0,1} or any of its truth-based extensions. This dilemma is characterized as the truth-based paradox: *logical axiomatization for illogical physics* (LAFIP or LAFIB) (Zhang, 2009a).

Now it is imperative to answer the question: *How can physics be formally proven illogical?* This seems to be an extremely difficult question at first glance. The answer, however, is rather simple: Aristotle's causality principle has been widely followed as the doctrine of all sciences but truth-based logic has failed to provide logically definable causality for more than 2300 years. Since axiomatization of physics or agent interaction, logically definable causality, and the grand unification are equivalent, history has made the LAFIP paradox self-evident.

LAFIP manifests the inconvenient truth:

1. Without YinYang bipolarity a truth value is incapable of carrying any physical semantics;
2. Without YinYang bipolarity any truth-based logic is "too logical" to be natural for axiomatizing the "illogical" aspects together with the logical aspects of non-linear bipolar dynamics;
3. YinYang bipolarity is indispensable in an equilibrium-based (or symmetry-based) axiomatization of physics because particle-antiparticle pairs form the basis of energy equilibrium in the universe;
4. Truth-based mathematical abstraction is itself a major barrier to the effort of axiomatizing physics;
5. A different mathematical abstraction is necessary for logically definable causality and the grand unification of general relativity and quantum theory.

OBSERVATION AND POSTULATION

Equilibrium-Based Cognition

The concepts of equilibrium and non-equilibrium are central in both physical and social sciences as manifested by the laws of thermodynamics, Nash equilibrium in macroeconomics (Nash, 1950), and the spontaneous broken symmetry model in physics (The Royal Swedish Academy of Sciences, 2008). On the one hand it is well-known that electromagnetic and gravitational forces both support the theory of an equilibrium-based dynamic open-world as evidenced by the "perfect" equilibria of negative-positive electromagnetic particles (-q,+q) and action-reaction forces (-f,+f) in modern physics textbooks and. On the other hand, the broken symmetry model suggests an "imperfect" quasi-equilibrium of matter-antimatter particles (-p,+p). In any case, the universe is either an equilibrium, quasi- or non-equilibrium. Here we use equilibrium as a synonym to symmetry and quasi-equilibrium as a synonym to broken symmetry. A broken symmetry can also be considered as non-equilibrium depending on the context.

As natural reality, different equilibria can be considered as forms of dynamic holistic truth where two local non-equilibria can form a global equilibrium. Since a multidimensional equilibrium or non-equilibrium can be deconstructed into one or more bipolar equilibria and/or non-equilibria (Figure 1), bipolar equilibrium can serve as a generic form of equilibria and bipolarity as an integral and inherent part of bipolar truth is inseparable from equilibrium-based holistic truth. In addition to the two basic action-reaction forces (-q,+q) and (-f,+f) in modern physics, the bipolar fusion of self-negation and self-assertion abilities denoted (self-negation, self-assertion) in neuroscience and mental health (Zhang, 2007; Zhang, Pandurangi & Peace, 2007), the (oxidation, antioxidation) pairs in biochemistry and traditional Chinese medicine (TCM) (Ou *et al.*, 2003), repression-activation abilities denoted (repression, activation) (Vasudevan, Tong & Steitz, 2007) of genetic regulator protein such as YinYang1 (YY1) (Shi *et al.*, 1991; Park & Atchison, 1991) that can act as a transcriptional activator or repressor (Ai, Narahari & Roman, 2000; Kim, Faulk & Kim, 2007; Palko *et al.*, 2004; Zhou & Yik, 2006; Santiago *et al.*, 2007; Liu *et al.*, 2007; Wilkinson *et al.*, 2006; Gore & van Oudenaarden, 2009), and the YinYang bipolar subatomic particle (Overbye, 2006) discovered at the Fermi National Accelerator Laboratory (Fermilab, 2006) that can change polarity trillion times a second show typical bipolar equilibrium/non-equilibrium properties . Furthermore, it is becoming scientifically evident that brain bioelectromagnetic or quantum field is crucial for neurodynamics and different mental states (Carey, 2007) where bipolarity is unavoidable. Indeed, without equilibrium such as mental equilibrium any disorder such as mental disorder would be "big bang" from nowhere (Zhang, 2009b) and the universe would be caused by nothing, from nowhere, and to nowhere.

Intuitively, equilibrium-based bipolar sets and logic seem to be inevitable for a probable axiomatization of physics. Such an axiomatization could lead to a new knowledge engineering paradigm for exploratory knowledge discovery especially in uncharted territories of global regulation and coordination, bioinformatics and biomedicine, information and decision, quantum mechanics and other domains.

Figure 1. Multidimensional equilibrium or non-equilibrium deconstructed to bipolar equilibria/non-equilibria

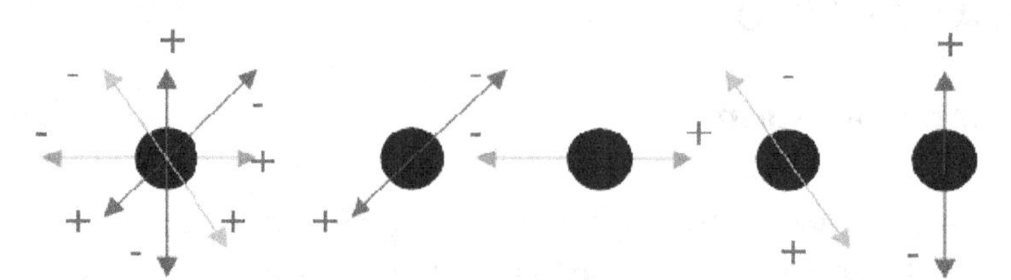

Practically, while the word "YinYang" has appeared in many articles published in prestigious journals including but not limited to *Nature*, *Science*, and *Cell*, some scientists including some Chinese scientists still tend to shun "YinYang" due to misunderstanding. Actually YinYang is unavoidable because:

1. The YinYang1 (YY1) genetic regulator protein discovered at Harvard Medical School (Shi *et al.*, 1991) is ubiquitous in all cell types of biological species (Jacobsen & Skalnik, 1999) that no one can avoid.
2. YinYang is about equilibrium and non-equilibrium or symmetry and asymmetry without which there would be no universe and no mental health.
3. Leibniz, co-founder of calculus, invented the first binary numeral system (Leibniz, 1703) (Figure. 2) and he attributed his invention to YinYang hexagrams recorded in the oldest Chinese *Book of Change* or *I Ching* (Moran & Yu, 2001; Karcher, 2002), and now binary numeral system is a technological basis for all digital technologies which no one can avoid.

Based on Leibniz interpretation we have Yin = 0; Yang = 1; and the well-known YinYang logo (or Taiji symbol) is one bit information. Using a broken line for Yin and an unbroken line for Yang, two binary bits would be corresponding to YinYang-4-images which clearly form a 4-valued diamond lattice; three binary bits would be corresponding to YinYang-8-trigrams; and six binary bits would be corresponding to the 64 YinYang hexagrams where six broken lines would be equivalent to the six bit binary number $000000_2 = 0_{10}$ and six unbroken lines would be equivalent to the binary number $111111_2 = 63_{10}$. By his invention and attribution, not only did Leibniz exhibit unprecedented creativity of a great mathematician but also exemplary academic integrity.

To most Chinese and Western YinYang scholars, however, it is well-known that the binary interpretation by Leibniz is only one of two or more major interpretations (Zhang, 1992; Zhang *et al.*, 2008). According to the Daoist cosmology, YinYang stands for everything has two sides or two opposite poles. We have to ask the honest question: *What would be the consequence if YinYang is interpreted as the two opposite energies of nature in continual movement such as (-f,+f), (-q,+q), and particle-antiparticles (-p,+p)?"*

Bipolar relativity follows the bipolar interpretation of YinYang as symbolically sketched in Figure 3. From the figure it is clear that the bipolar interpretation leads to a multidimensional universe in bipolar equilibrium similar to an electromagnetic or quantum field. Interestingly, bipolar universal modus ponens (BUMP) (Zhang, 2005a, 2007) provides the first logical formula for the reciprocal and interactive nature of the two opposite energies. It is shown that BUMP leads to logically definable causality from a mathematical physics or biophysics perspective and brings the ubiquitous effects of quantum entanglement to the real world (Chs. 3-11).

A comparison of Figures 2, 3, and 4 reveals that bipolar equilibria can form a quantum representation for a binary numeral system but binary numbers carry neither bipolar physical semantics nor geometric properties (Zhang, 2009a). That is why we call the Leibnizian interpretation *binary YinYang* – a prelude of Boolean algebra. The two sides in binary YinYang are not physical and not interactive. *Bipolar YinYang*, on the other hand, is a modern quantum physics interpretation that is central in the equilibrium-based approach to mathematical abstraction and axiomatization of physics, cognitive informatics, and life sciences presented in the book.

"YinYang is unscientific" has been a widely held view in the scientific community. Now it can be argued that, if Leibniz derivation of binary numeral system from YinYang hexagrams is scientific, if hundreds of papers appeared in *Science*, *Nature*, *Cell*, and other top scientific journals are also scien-

Figure 2. Binary YinYang – Leibniz Interpretation: (a) YinYang; (b) YinYang-4-images; (c) YinYang-8-trigrams; (d) 64 hexagrams. Adapted from (Zhang, 2009d)

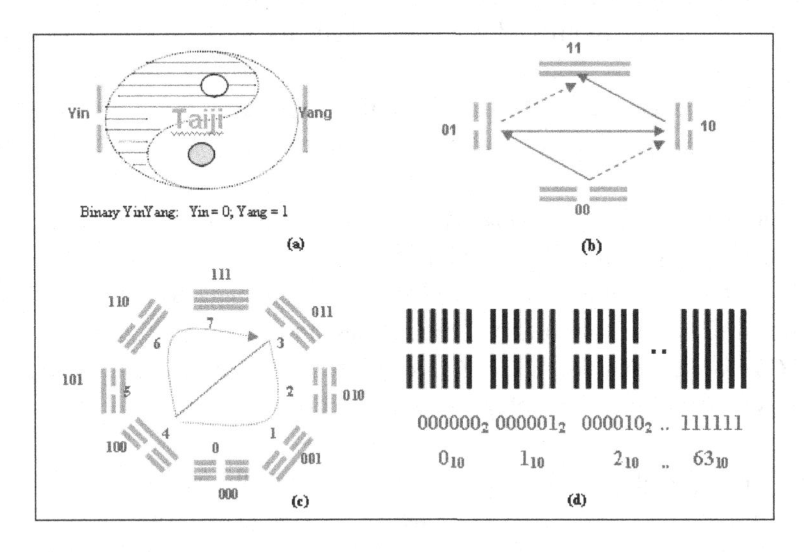

tific, how could the original YinYang be unscientific? Of course, the fortune telling interpretations of YinYang in ancient Chinese literature are indeed unscientific. But that should not be a reason to ban YinYang from being an evolving scientific theory just as Aristotelian science should not have been forbidden to evolve to Western science because Aristotle once claimed the Earth was the centre of the universe and his cosmology later was borrowed for the brutal fire execution of Bruno and the jailing of Galileo.

"YinYang is dichotomy" is perhaps a major misunderstanding of YinYang. Bipolar YinYang is a bipolar reciprocal, interactive, oscillatory, entangled, and unifying equilibrium or non-equilibrium. The non-exclusive nature of the Yin and Yang is clearly shown in the Taiji symbol or YinYang logo. Since the two poles are reciprocal and interdependent, YinYang is not a dichotomy. It is shown in later chapters that such interdependent bipolar reciprocal nature is mathematically definable with bipolar dynamic logic (BDL) (Ch. 3), bipolar dynamic fuzzy logic (BDFL) (Ch. 4, Ch. 5), bipolar relativity (Ch. 6), bipolar quantum entanglement (Ch. 7) and bipolar quantum liner algebra (BQLA) (Ch. 8) for charactering bipolar agents in microscopic and macroscopic worlds in quantum mechanics (Ch. 7), cognitive informatics (Ch. 10, Ch. 11), and life sciences (Ch. 8, Ch. 9).

Agents and Agent Interaction

It is a long held idea that microscopic and macroscopic agents and agent interactions are essential in physics, socioeconomics, and life sciences. While autonomous agents, agent interaction, and agent organization have been intensively studied by the research communities of distributed artificial intelligence (DAI) and multiagent systems (MAS) (Huhns & Shehory, 2007; Singh & Huhns, 2005; Huhns & Singh, 1997; Huhns, 1987, 2000; Bond & Gasser, 1988; Zhang, 1998b; Zhang & Zhang, 2004a) for applications in macroscopic worlds or societies, agent interactions in microscopic worlds including but not limited to quantum, neural, molecular, and genomic worlds are studied in physical, biological, neural, and life sciences.

Figure 3. Bipolar YinYang: (a) Everything has two poles (Taiji (one) generates YinYang (two)); (b) The two poles can form four patterns (YinYang generates four images); (c) YinYang four images generate 8 trigrams (Wilhelm & Baynes, 1967, pp. 318-319)

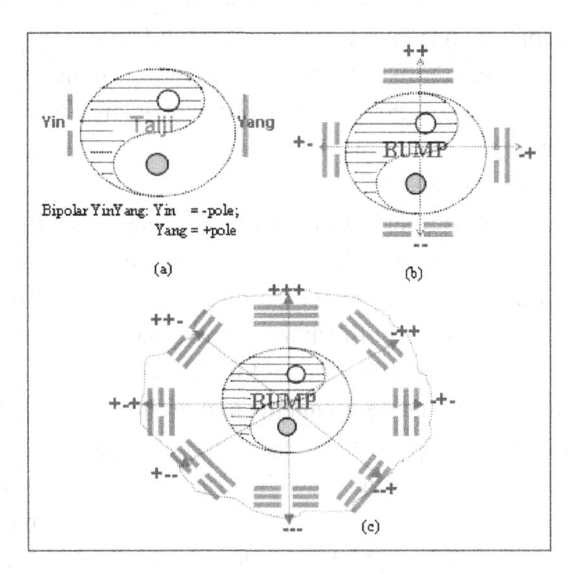

Unifying mathematical axiomatization of agent interaction in both microscopic and macroscopic worlds is needed for the coordination and global regulation of non-autonomous and autonomous agents as well as for scientific discoveries in all scientific fields. Such applications range from biorobotics and nanobiomedicine to quantum brain dynamics, astrophysics, strategic decision and coordination, axiomatization of physics and the grand unification. Since agent interactions are governed by physical or social dynamics, the difficulty of axiomatizing agent interactions can be traced back to Hilbert's effort in axiomatizing physics (Hilbert, 1901), Aristotle's causality principle, the concept of *singularity*, and Einstein's unfinished grand unification.

Arguably, if bipolar equilibrium or non-equilibrium is a generic form of any multidimensional equilibrium or non-equilibrium (Figure 1), *bipolarity* has to be a fundamental property of any physical and biological agent. Thus, without bipolarity, a multidimensional model of the universe such as the 11-dimensional superstring theory is not fundamental.

Bipolarity is sometimes deemed an antonym to *singularity*. That is actually incorrect. YinYang bipolarity is actually a fundamental property of nature for the understanding of our body, brain, and the

Figure 4. Bipolar quantum representation for binary numbers

Decimal	Bipolar	Sign	Binary	Decimal	Bipolar	Sign	binary
0	(-1,0)(-1,0)(-1,0)	---	000	4	(0,1)(-1,0)(-1,0)	+--	100
1	(-1,0)(-1,0)(0,1)	--+	001	5	(0,1)(-1,0)(0,1)	+-+	101
2	(-1,0)(0,1)(-1,0)	-+-	010	6	(0,1)(0,1)(-1,0)	++-	110
3	(-1,0)(0,1)(0,1)	-++	011	7	(0,1)(0,1)(0,1)	+++	111

universe that enables mathematically definable causality (Zhang, 2009a, 2009b, 2009c). Singularity, on the other hand, is not a property but a point in spacetime characterized with an infinite measure of gravity such as big bang and black hole. Singularity does not lead to logically definable causality.

Even though bipolar equilibrium and non-equilibrium are fundamental physical concepts, they are not strictly logical in unipolar truth-based terms. For instance, quantum entanglement, particle-wave duality, bipolar disorder, economic recession and expansion, greed and generosity can be described with bipolar equilibria or non-equilibria, but they are illogical observables. It is reasonable to ask the question: *Can there be equilibrium-based bipolar agents?*

Every scientist would agree that, without a certain form of dynamic equilibrium or quasi-equilibrium, any agent or universe would be in total disorder or chaos and we would have no observables, no observers, and no universe. Similarly, without mental equilibrium, mental disorder would be "big bang" from nowhere. Therefore, the concept of equilibrium is unavoidable in agent interaction at the system, molecular, genomic, neurobiological, particle, or quantum level.

Since without equilibrium there would be no agents and no universe, we humans are practically all bipolar, either in bipolar equilibrium, quasi-equilibrium, or non-equilibrium physically, mentally, or biologically because there are no other options. Such YinYang bipolarity as fundamental natural reality is evidently essential in science regardless of the truthfulness of a person in terms of trustworthiness. A typical example is the YinYang 1 (YY1) regulatory genomic agent (Shi *et al.,* 1991; Park & Atchison, 1991) which is ubiquitous in all cell types of biological species.

Different from established theories on agents and agent interactions, this book assumes that agents are all bipolar either in equilibrium or non-equilibrium. It is shown in later chapters that bipolar agents, bipolar adaptivity, and bipolar causality are logically definable with bipolar relativity (Ch. 6).

Nature's Most Fundamental Property

While the expansion of the universe is partially observable and, therefore, *singularity* and *big bang* are imaginable and understandable as well as derivable from Einstein's equations of general relativity, it is equally imaginable and understandable, however, that singularity and the big bang could not have been from nowhere, to nowhere, and caused by nothing. If space is expanding, the cause behind the expansion has to be more fundamental than the effect. This leads us to Aristotle's causality principle. Sadly, *causality is undefinable* in classical truth-based logic (Zadeh, 2001).

Subsequently, Hilbert's Problem 6—*"Axiomatize All of Physics"* (Hilbert, 1901)—has remained unsolved for more than a century; after Einstein the world has witnessed no widely accepted theory deeper than general relativity; the quest for quantum gravity has so far stalled in spacetime geometry and found no decisive battleground. Even though the broken symmetry theory (The Royal Swedish Academy of Sciences, 2008) is influential, it so far has not led to quantum gravity. String theory is *"theory of everything"* but it is not experimentally testable, far from reality, and its many dimensions are questionable. It is not surprising to notice the courageous book titles *The Trouble with Physics: The Rise of String Theory, The Fall of a Science, and What Comes Next* (Smolin, 2006) and *Not Even Wrong: The Failure of String Theory and the Search for Unity in Physical Law* (Woit, 2006).

Practically speaking, modern science has been entirely trapped within spacetime. For furthering scientific exploration, now it is time for science to advance beyond spacetime. But no one knows how. Consequently, the following fundamental questions are unavoidable:

1. What caused spacetime and spacetime relativity?
2. Could we have spacetime and singularity without bipolarity?
3. What is the ultimate cause of economic depression, mental disorder, big bang, and black holes?
4. Is equilibrium-based causality logically and physically definable?
5. Is there an equilibrium-based unifying logical axiomatization of physics, socioeconomics, and life sciences?
6. Is there an equilibrium-based logical unification of gravity and quantum mechanics for quantum computing and exploratory knowledge discovery?
7. Is there an equilibrium-based logical unification of agent interaction for decision, coordination, and global regulation?
8. What are the real-world binding strings of nature, agents, and society?
9. Is quantum neuroscience and quantum bioinformatics a reality?
10. Is socioeconomics quantum in nature?
11. What comes next?

Evidently, the above questions are all gigantic ones and any concrete scientific answer will have far reaching impact on science especially information and life sciences.

Arguably, equilibrium or non-equilibrium, as a physical state of any dynamic agent or universe at the system, molecular, genomic, particle, or subatomic level, forms a philosophical chicken and egg paradox with the universe because no one knows exactly which one created the other in the very beginning. If bipolar equilibrium (or non-equilibrium) is a generic form of equilibrium (or non-equilibrium) (Figure 1), spacetime is not a fundamental model in equilibrium-based terms because space and time do not form a bipolar equilibrium.

According to a 2008 Nobel Prize in Physics press release:

Nature's laws of symmetry are at the heart of this subject: or rather, broken symmetries, both those that seem to have existed in our universe from the very beginning and those that have spontaneously lost their original symmetry somewhere along the road. In fact, we are all the children of broken symmetry. It must have occurred immediately after the Big Bang some 14 billion years ago when as much antimatter as matter was created. The meeting between the two is fatal for both; they annihilate each other and all that is left is radiation. Evidently, however, matter won against antimatter, otherwise we would not be here. But we are here, and just a tiny deviation from perfect symmetry seems to have been enough – one extra particle of matter for every ten billion particles of antimatter was enough to make our world survive. This excess of matter was the seed of our whole universe, which filled with galaxies, stars and planets – and eventually life. But what lies behind this symmetry violation in the cosmos is still a major mystery and an active field of research. (The Royal Swedish Academy of Sciences, 2008)

Conceivably, the universe is either in equilibrium, quasi-equilibrium or non-equilibrium from a thermodynamics perspective. It would convey no physical information if we say that the universe is true, false, fuzzy, intuitionistic, or paraconsistent. Furthermore, it is alright if we say *"the universe is an equilibrium or non-equilibrium"* but it would be odd if we say *"equilibrium is a universe"*. Intuitively, equilibrium or non-equilibrium is a governing or at least a regulating property of the universe but not vice versa.

Based on the above, we have the following postulate:

Postulate 1.1. *The most fundamental property of the universe is not truth, not fuzziness, not space, not time, not spacetime relativity, not matter, not ether, not strings, and not singularity; the most fundamental property of the universe should, instead, be YinYang bipolarity.* (Zhang, 2009d, 2010)

Despite the supporting evidence, the postulate may still sound like an unbelievable hype full of absurdity but it is not. On the one hand, the universe is physical but truth and fuzziness are not physical properties but logical properties subjected to the LAFIP paradox; spacetime and spacetime relativity are not quantum in nature, not symmetrical, not bipolar interactive, and not quantum entangled; matter, ether, and strings are things but not physical properties; singularity is an infinite point or location in spacetime geometry and not a property either. On the other hand, as discussed in earlier sections, we have the following undisputable scientific observations:

1. Electromagnetic force between two particles denoted (-q,+q) and gravitational force between action and reaction objects denoted (-f,+f) are nature's two basic forces and both happen to be bipolar (Note: Strong and weak forces have been unified with electromagnetic force per 2008 Nobel Prize Award in physics).
2. Particle-antiparticle pairs denoted (-p,+p) are the only known tangible stuff in the universe that happen to form a near perfect bipolar quasi-equilibrium (Sather, 1999).
3. Electron-positron pair (e^-, e^+) production as the most fundamental example of the materialization of energy predicted in special relativity has been accurately described by quantum electrodynamics (QED) – *"the jewel of physics"* (Dirac, 1927, 1928; Feynman, 1962, 1985).

Based on the above facts, it can be concluded that, if space is expanding, spacetime has to be caused by something more fundamental; if YinYang bipolarity can survive a black hole due to particle-antiparticle emission or Hawking radiation (Hawking, 1974), the logical foundation of physics has to be bipolar in nature; if particle-antiparticle pairs and nature's basic action-reaction forces are the most fundamental components of our universe, YinYang bipolar relativity has to be more fundamental than spacetime relativity. It would be hard or even absurd to deny the scientific validity of Postulate 1.1.

Postulate 1.2. *Nature is not the realization of spacetime relativity; instead, matter, energy, spacetime, or the universe itself is the realization of YinYang bipolar relativity.*

It is evident that Postulate 1.2 is a direct consequence of Postulate 1.1. The two postulates provide a theoretical basis for the transcending and unifying property of YinYang bipolar relativity beyond spacetime geometry to be presented in this monograph.

It should be remarked that

1. Einstein's general theory of relativity has been validated numerous times on the scale of the solar system and, recently, on a galactic or cosmic scale (Reyes *et al.*, 2010). It has been proven the deepest theory regarding spacetime and gravity in macro scales. Someone may ask: "*Then, how can it be wrong?*" The answer is simple: *"It has nothing wrong to our best knowledge but we still need a deeper theory for its unification with quantum mechanics in micro scales as envisioned by Einstein."*

2. Einstein posited that a person inside a falling elevator would feel weightless because gravity's tug is canceled by the downward acceleration. This insight, called the equivalence principle of gravity and acceleration, led Einstein to develop the theory of general relativity. But just how general relativity applies to objects on the quantum scale remains a mystery of quantum gravity (van Zoest *et al.*, 2010).

YinYang Bipolar Geometry

Agents, causality, and bipolar relativity form a central theme of this book. While Einstein's general relativity theory is a unification of spacetime and gravity, bipolar relativity is intended to be a logical unification of the four aspects:

1. the negative and positive poles of agents (or the Yin and the Yang energies of nature),
2. the energy equilibrium or non-equilibrium of the two poles (or energies),
3. the agents and agent interactions involved in the equilibrium or non-equilibrium, and
4. the fusion, separation, and quantum entanglement of the two poles or energies.

YinYang geometry is to support the unification in simple mathematical terms. The new geometry has three dimensions: the Yin and the Yang are the two reciprocal and interdependent opposite dimensions from which a third dimension – equilibrium – is transcendental. The agents are the actors in the geometry; space and/or time are emerging factors following the arrivals of the agents; bipolar agent is to Yin and Yang in bipolar relativity as gravity is to space and time in general relativity with fundamentally different syntax and semantics as shown in Figures 5a and 5b. While Figure 5a shows the magnitudes of the Yin and the Yang of a bipolar agent explicitly, YinYang equilibrium of a bipolar agent can be shown in Figure 5b. Following graphical user interface (GUI) convention, Figure 5b can be converted to Figure 5c.

Since the Yin and the Yang are two opposite reciprocal and interdependent poles or energies that are completely *background-independent* (Smolin, 2005, 2006), YinYang bipolar geometry is fundamentally different from Euclidian, Hilbert, and spacetime geometries. With the *background-independent* property, YinYang geometry makes quadrants irrelevant because bipolar identity, interaction, fusion, separation, and equilibrium can be accounted for in YinYang geometry even without quadrants. In later discussions of this book, however, quadrants are used in case where observer is involved and background dependent information is needed.

Figure 5. YinYang bipolar geometry: (a) Magnitudes of Yin and Yang; (b) Growing curve in YinYang bipolar geometry; (c) YinYang bipolar geometry following GUI convention

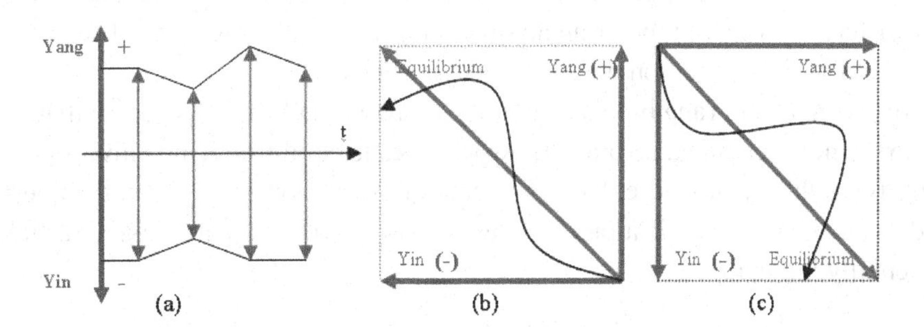

Logically speaking, without YinYang bipolarity, quantum gravity defined in spacetime geometry does not transcend spacetime because space and time are not bipolar interactive energies and can't provide bipolar fusion, interaction, quantum entanglement, and definable causality. Therefore, relativity and quantum mechanics defined in spacetime geometry can't be the unifying logical foundation for gravity, quantum mechanics, socioeconomics, and life sciences. This could well be the fundamental reason that propelled Einstein to conclude (Einstein, 1940) that: *"For the time being we have to admit that we do not possess any general theoretical basis for physics which can be regarded as its logical foundation."* Unfortunately, in his late years some of Einstein's works were deemed junk and even some of his colleagues tried to avoid him (Smolin, 2006, pp. 49-50). Now in the eve of the centennial celebration of general relativity, we are still struggling with Einstein's vision.

Different from space and time, the Yin and Yang in YinYang bipolar geometry can be any reciprocal opposite poles of nature including but not limited to physical, biological, molecular, and quantum agents such as electron-positron pairs in the microscopic agent world and competition-cooperation relations in the macroscopic agent world. The bipolar agents can form equilibrium-based bipolar quantum entanglement (Zhang, 2010) in YinYang geometry. The unification can be considered manifold at the logical level (not at the mathematical level yet):

1. Since YinYang bipolarity is a common property of nature's basic forces of action and reaction, bipolar relativity can be considered a logical unification of nature's basic forces including gravity, electromagnetism, and quantum mechanics.
2. Since space and time are associated with agents and they can emerge in YinYang bipolar geometry following the arrivals of agents, YinYang bipolar geometry can be considered a logical unification of spacetime geometry and microscopic/macroscopic agents.
3. Since YinYang bipolar equilibrium or non-equilibrium is a generic property of nature with logically definable causality, bipolar relativity can be deemed a logical unification of nature, agents, and causality.
4. Since all agent interaction in physics, socioeconomics and life sciences are governed or regulated by social or physical dynamics under equilibria or non-equilibria, bipolar relativity can be deemed a logical unification of physics, socioeconomics, and life sciences.

With logically definable causality, bipolar relativity leads to numerous predictions (Ch. 6). Some of them are domain dependent and some are domain independent. In particular, it is conjectured that spacetime relativity, singularity, gravity, electromagnetism, quantum mechanics, bioinformatics, and socioeconomics are different *phenomena of YinYang bipolar relativity*; microscopic and macroscopic agent interactions in physics, socioeconomics, and life sciences are directly or indirectly caused and regulated by bipolar relativity; all physical, social, mental, and biological action-reaction forces are fundamentally different forms of bipolar quantum entanglement in large or small scales; gravity is not limited by the speed of light; graviton does not necessarily exist.

Consequently, without YinYang bipolar relativity, there would be no bipolar oscillation, no particle-wave duality, no memory scanning, no brain, no spacetime, no equilibrium, no universe, and no quantum gravity. Any unifying theory of nature, therefore, cannot avoid bipolarity. Then, the quest for quantum gravity should, indeed, be the unification of nature's basic forces and particle-antiparticle pairs which are all fundamentally bipolar.

Someone may argue that quantum mechanics uses vector operations; bipolar geometry is simply a vector space V that has already been accounted for in quantum computing. The argument is false. Vector operations are based on truth-based reasoning where for every vector $u \in V$ there exists an additive inverse of u such that u+(-u) = (-u)+u =0. The problem is that the coexistence of (-u,+u) in bipolar equilibrium, quasi-equilibrium, or non-equilibrium is replaced with 0 in classical quantum mechanics as well as in psychiatry. Using 0 to characterize a reciprocal energetic interactive or oscillatory bipolar dynamic equilibrium of the Yin and Yang of nature is evidently a denial of the coexistence, fusion, and interaction of the two sides.

Furthermore, since acceleration is equivalent to gravitation under general relativity, it can be argued that any natural, social, economical, and mental acceleration or growth are qualified to be a kind of quantum gravity. Thus, as a most fundamental scientific unification, not only can quantum gravity be applied in physical science, but also in computing science, social science, brain science, and life sciences as well. It would be hard to imagine that quantum gravity as the grand unification would not be the unifying theory for all sciences.

While general relativity and quantum theory has so far failed to go deeper beyond spacetime geometry, YinYang bipolar relativity leads to five equilibrium-based sub-theories which are: *physical quantum gravity, logical quantum gravity, social quantum gravity, biological quantum gravity, and mental quantum gravity*. The five sub-theories form a Q5 paradigm (see Ch. 7). In this paradigm, the theory of physical quantum gravity is concerned with bipolar quantum entanglement (Chs. 3-7); the theory of logical quantum gravity is focused on bipolar quantum computing (Ch. 7); the theory of social quantum gravity spans social sciences and economics for bipolar equilibrium and harmony (Ch. 11); the theory of mental quantum gravity is focused on the interplay of quantum mechanics and brain dynamics for computational neuroscience and mental equilibrium (Ch. 10); the theory of biological quantum gravity is focused on life sciences for biological equilibrium (Ch. 8 & Ch. 9).

The Q5 paradigm may sound like a mission impossible. Surprisingly, it is shown that the different quantum gravities and their interplay are all unified under YinYang bipolar relativity. It actually follows a single undisputable observation and a single condition: (1) bipolar equilibrium or non-equilibrium is a ubiquitous or pervasive concept from which nothing can escape not even spacetime; (2) equilibrium-based bipolar causality and bipolar quantum entanglement can be logically defined with bipolar universal modus ponens in simple comprehendible terms (Zhang 2005a, 2007, 2009 a, 2009b, 2009c, 2009d, 2010) (see Chs. 3 & 6).

BOOK OVERVIEW

Theoretically, this book can be deemed an equilibrium-based holistic approach to physical and social sciences; technically, it can be considered a resolution to the LAFIP paradox. First, it is observed that without bipolarity the bivalent truth values 0 (false) and 1 (true) are incapable of carrying any shred of direct physical syntax and semantics, let alone illogical physical phenomena such as quantum chaos, bipolar disorder, equilibrium, non-equilibrium, black hole, big bang, and quantum entanglement. Therefore, truth-based (unipolar) mathematical abstraction as a basis for positivist thinking cannot avoid the LAFIP paradox. A new logical foundation is a necessity.

Historically, while no physicist would say *"electron is isomorphic to positron"* or *"matter is isomorphic to antimatter"*, it is widely considered in logic and mathematics that *"-1 is isomorphic to +1"* and

(-,+) bipolar symmetry, equilibrium, and non-equilibrium are not observable. If we check the history of *negative numbers*, we would find that the ancient Chinese and Indians started to use negative numbers in the ancient time (Temple, 1986, p. 141; Bourbaki, 1998, p. 49; Martinez, 2006), but European mathematicians resisted the concept of negative numbers until the 18th centuries due to their so-called "absurdity". Here is an excerpt from BBC:

In 1759 the British mathematician Francis Maseres wrote that negative numbers 'darken the very whole doctrines of the equations and make dark of the things which are in their nature excessively obvious and simple.' Because of their dark and mysterious nature, Maseres concluded that negative numbers did not exist, as did his contemporary, William Friend. However, other mathematicians were braver. They took a leap into the unknown and decided that negative numbers could be used during calculations, as long as they had disappeared upon reaching the solution.

The history of negative numbers is one of stops and starts. The trailblazers were the Chinese who by 100 BC were able to solve simultaneous equations involving negative numbers. The Ancient Greeks rejected negative numbers as absurd, by 600 AD, the Indians had written the rules for the multiplication of negative numbers and 400 years later, Arabic mathematicians realized the importance of negative debt. "But it wasn't until the Renaissance that European mathematicians finally began to accept and use these perplexing numbers.

Why were negative numbers considered with such suspicion? Why were they such an abstract concept? And how did they finally get accepted? (BBC, 2006)

Regardless of the great achievement of Western science and technology, it is undoubtedly necessary to bridge the gap between the Western positivist thinking and the Eastern balanced thinking for solving unsolved scientific problems such as resolving the quantum entanglement mystery with logically definable causality. Because *"passion for symmetry"* can *"permeate the Standard Model of elementary particle physics"* and can unify *"the smallest building blocks of all matter and three of nature's four forces in one single theory"* (The Royal Swedish Academy of Sciences, 2008), it is not only reasonable but also inevitable to explore the equilibrium-based computing paradigm.

To avoid LAFIP, *bipolar set* theory (including bipolar crisp and fuzzy set theories) is introduced in Chapters 3-5 based on a holistic equilibrium-based approach to mathematical abstraction. It is shown that,

1. bipolar sets lead to YinYang *bipolar dynamic logic* (BDL) and *bipolar dynamic fuzzy logic (BDFL)* that exhibit the unifying property of *truth-based reasoning* and *bipolar quantum entanglement*;
2. following the arrivals of bipolar agents, *the emergence of space and time* are practically attainable (Ch. 6);
3. based on bipolar agents and agent interaction, *causality* becomes logically and physically definable (Chs. 6-8).

Consequently, a completely *background independent theory of bipolar relativity*, central theme of the book, is formally defined with BDL. Due to bipolar quantum entanglement, bipolar relativity presents a logically complete equilibrium-based holistic paradigm for *quantum mechanics, quantum comput-*

ing, quantum socioeconomics, quantum brain dynamics, and quantum bioinformatics (Q5 paradigm). The new paradigm is ideal for modeling non-linear bipolar dynamic agent oscillation and interaction. Applications are illustrated in quantum computing and communication, bioeconomics, system biology and TCM, computational psychiatry, bipolar cognitive mapping, decision analysis, coordination, and global regulation. A number of axioms and conjectures are posted as predictions for dynamic exploratory knowledge discovery and decision analysis in microscopic and macroscopic agent worlds.

In particular, it is shown that YinYang bipolar relativity leads to equilibrium-based logically definable causality, quantum entanglement, teleportation, and cryptography with quantum-digital compatibility and an interpretation of the hidden variables in the EPR paradox. It is shown with simulation that bipolar teleportation is theoretically possible for the first time without conventional communication between Alice and Bob. It is shown that bipolar cryptography enables the unique feature of bitwise encryption. Thus, bipolar quantum entanglement has the potential to bring quantum computer closer to reality.

The book consists of twelve chapters which can be divided into five sections:

Section 1. Introduction and Background. This section consists of Chapter 1 and Chapter 2. Chapter 1 is an introduction; Chapter 2 is a background review.

Section 2. Set Theoretic Logical Foundation. This section consists of three chapters, Chapters 3-5. This section layouts the set-theoretic and logical foundation for this book including YinYang bipolar sets, bipolar dynamic logic, bipolar quantum lattices and bipolar dynamic fuzzy logic, bipolar fuzzy sets and fuzzy equilibrium relations.

Section 3. YinYang Bipolar Relativity and Quantum Computing. This section consists three chapters, Chapters 6-8, which are focused on the central theme of the book. Chapter 6 presents the theory of agents, causality, and YinYang bipolar relativity. Chapter 7 is focused on the logical and physical aspects of quantum computing. Chapter 8 presents YinYang bipolar quantum linear algebra (BQLA), bipolar quantum cellular automata (BQCA), and a unifying view of YinYang bipolar relativity in logical, geometrical, algebraic, and physical terms. YinYang bipolar relativity leads to the notion of *bipolar quantum gravity*.

Section 4. Applications. This section consists of three chapters, Chapters 9-11. Chapter 9 is focused on biosystem simulation with bipolar quantum linear algebra and bipolar quantum cellular automata. Chapter 10 is focused on bipolar computational neuroscience and psychiatry. Chapter 11 is focused on bipolar cognitive mapping and decision analysis.

Section 5. Discussions and conclusions. This section consists of the last chapter—Chapter 12. This chapter presents a debate related to the central theme of this book, answers to critics regarding bipolarity, and a summary of the key points of the book with a few conclusions.

SUMMARY

This chapter is an introduction to the monograph. Some of Einstein's assertions regarding physics, logic, and theoretical invention have been reviewed. Einstein's hint on YinYang bipolar relativity has been identified. Limitations of general relativity and quantum mechanics have been briefly discussed. It is concluded that logically definable causality, axiomatization of physics or agent interaction, and the logical foundation for the grand unification are essentially the same problem. A paradox *"Logical*

Axiomatization for Illogical Physics" (LAFIP) – has been introduced. Bipolarity has been postulated as the most fundamental property of Mother Nature. The theoretical basis of agents, causality and YinYang bipolar relativity has been highlighted and distinguished from general relativity. The main ideas of the book have been outlined. While Leibniz binary YinYang provided a technological basis for all digital technologies, YinYang bipolar relativity is expected to make equilibrium-based quantum computing a logical, physical, social, biological, and mental reality.

Since the universe is either an equilibrium or non-equilibrium, it is argued that spacetime as part of the universe is governed or regulate by the laws of equilibrium but not vice versa. Practically, it is observed that equilibrium or non-equilibrium is a ubiquitous concept at the system, molecular, genetic, and quantum levels of all physical systems. Spacetime geometry, on the other hand, is essential in space exploration but not necessarily the most fundamental concept of Mother Nature. Thus, the equilibrium-based approach effectively brings scientific explorations beyond spacetime into YinYang bipolar geometry for applications in quantum computing, cognitive informatics and life sciences.

The significance of YinYang bipolar relativity lies in its four equilibrium-based logical unifications (not at mathematical yet): (1) the unification of unipolar positivist truth with bipolar holistic truth, (2) the unification of classical logic with quantum logic, (3) the unification of quantum entanglement with microscopic and macroscopic agent interaction in simple logical terms, and (4) the unification of general relativity with quantum mechanics under equilibrium and symmetry. On the one hand, these unifications present a minimal equilibrium-based axiomatization of physics, socioeconomics, and life sciences. On the other hand, they lead to a real-world scalable bipolar binding string theory for nature, agents, and society. Despite its limited mathematical depth, it is shown that bipolar relativity qualifies as a deeper logical theory beyond spacetime geometry tailored for open-world open-ended exploratory knowledge discovery in all scientific fields where equilibrium and symmetry are central.

REFERENCES

Ai, W., Narahari, J., & Roman, A. (2000). Yin yang 1 negatively regulates the differentiation-specific E1 promoter of Human Papillomavirus type 6. *Journal of Virology, 74*(11), 5198–5205. doi:10.1128/JVI.74.11.5198-5205.2000

Aiton, E. J. (1985). Leibniz [Hilger, UK.]. *Biography*, 245–248.

(2000). *American Psychiatric Association* (4th ed.). Diagnostic and Statistical Manual of Mental Disorders.

Bell, J. S. (1964). On the Einstein Podolsky Rosen paradox. *Physics, 1*(3), 195–200.

Belnap, N. (1977). A useful 4-valued logic . In Epstein, G., & Dunn, J. M. (Eds.), *Modern uses of multiple-valued logic* (pp. 8–37).

Birkhoff, G., & von Neumann, J. (1936). The logic of quantum mechanics. *The Annals of Mathematics, 37*(4), 823–843. doi:10.2307/1968621

Blackburn, S. (1990). Hume and thick connexions. *Philosophy and Phenomenological Research, 50*, 237–250. doi:10.2307/2108041

Bond, A. H., & Gasser, L. (Eds.). (1988). *Readings in distributed artificial intelligence* Morgan Kaufmann, 3-35.

Boole, G. (1854). *An investigation of the laws of thoughts.* London: MacMillan. Reprinted by New York: Dover Books.

Bourbaki, N. (1998). *Elements of the history of mathematics.* Berlin: Springer-Verlag.

Carey, B. (2007, Aug. 1). Man regains speech after brain stimulation. *The New York Times – Health.* Retrieved from http://www.nytimes.com/2007/08/01/health/01cnd-brain.html

Dirac, P. A. M. (1927). The quantum theory of the emission and absorption of radiation. *Proceedings of the Royal Society of London, 114*(767), 243–265. doi:10.1098/rspa.1927.0039

Dirac, P. A. M. (1928). The quantum theory of the electron. *Proceedings of the Royal Society of London, 117*(778), 610–624. doi:10.1098/rspa.1928.0023

Einstein, A. (1916). The foundation of the general theory of relativity. Originally published in *Annalen der Physik* (1916). *Collected papers of Albert Einstein, English translation of selected texts,* Translated by A. *Engel, 6,* 146–200.

Einstein, A. (1934). *On the method of theoretical physics. The Herbert Spencer lecture, delivered at Oxford, June 10, 1933. Published in Mein Weltbild.* Amsterdam: Querido Verlag.

Einstein, A. (1940). Considerations concerning the fundamentals of theoretical physics. *Science, 91*(2369), 487–491. doi:10.1126/science.91.2369.487

Einstein, A., Podolsky, B., & Rosen, N. (1935). Can quantum-mechanical description of physical reality be considered complete? *Physical Review, 47*(10), 777–780. doi:10.1103/PhysRev.47.777

Faye, J. (2008). Copenhagen interpretation of quantum mechanics. *Stanford Encyclopedia of Philosophy.* Retrieved from http://plato.stanford.edu/entries/qm-copenhagen/

Fermi National Accelerator Laboratory. (2006).). *Press Release 06-19,* Sept. 25, 2006. Retrieved from http://www.fnal.gov/pub/ presspass/press_releases/CDF_meson.html

Feynman, R. P. (1962). *Quantum electrodynamics.* Addison Wesley.

Feynman, R. P. (1985). *QED: The strange theory of light and matter.* Princeton University Press.

Gödel, K. (1930). Uber formal unentscheidbare Sätze der Principia Mathematica und verwandter Systeme. *Monats hefte fur Math. und Phys., 37, 349-360,* 1930. Repr. in M. Davis, *The Undecidable.* 4-38.

Gore, J., & van Oudenaarden, A. (2009). Synthetic biology: The yin and yang of nature. *Nature, 457*(7227), 271–272. doi:10.1038/457271a

Hawking, S. (1974). Black-hole evaporation. *Nature, 248,* 30–31. doi:10.1038/248030a0

Hawking, S., & Penrose, R. (1970). The singularities of gravitational collapse and cosmology. *Proceedings of the Royal Society of London, 314*(1519), 529–548. doi:10.1098/rspa.1970.0021

Hilbert, D. (1901). Mathematical problems. *Bulletin of the American Mathematical Society, 8*, 437–479. doi:10.1090/S0002-9904-1902-00923-3

Hilgevoord, J., & Uffink, J. (2006). *The uncertainty principle*. Stanford Encyclopedia of Philosophy.

Huhns, M. N. (Ed.). (1987). *Distributed artificial intelligence*. London: Pitman.

Huhns, M. N. (2000). An agent-based global economy. *IEEE Internet Computing, 4*(6), 83–84.

Huhns, M. N., & Shehory, O. (Eds.). (2007). *Proceedings of the Sixth International Joint Conference on Autonomous Agents and Multi-Agent Systems (AAMAS 07)*, International Foundation for Autonomous Agents and Multiagent Systems, USA, 2007.

Huhns, M. N., & Singh, M. P. (1997). *Readings in agentsagents*. Morgan Kaufmann.

Jacobsen, B. M., & Skalnik, D. G. (1999). YY1 binds five cis-elements and trans-activates the myeloid cell-restricted gp91phox promoter. *The Journal of Biological Chemistry, 274*, 29984–29993. doi:10.1074/jbc.274.42.29984

Karcher, S. (2002). *I Ching: The classic Chinese oracle of change: The first complete translation with concordance*. London: Vega Books.

Kim, J. D., Faulk, C., & Kim, J. (2007). Retroposition and evolution of the DNA-binding motifs of YY1, YY2, and REX1. *Nucleic Acids Research, 35*(10), 3442–3452. doi:10.1093/nar/gkm235

Leibniz, G. (1703). *Explication de l'Arithmétique Binaire* (*Explanation of Binary Arithmetic*); Gerhardt . *Mathematical Writings, VII*, 223.

Liu, H., Schmidt-Supprian, M., Shi, Y., Hobeika, E., Barteneva, N., & Jumaa, H. (2007). Yin Yang 1 is a critical regulator of B-cell development. *Genes & Development, 21*, 1179–1189. doi:10.1101/gad.1529307

Łukasiewicz, J. (1920). O logice trojwartosciowej. [English translation in: Łukasiewicz]. *Ruch Filozoficny, 5*, 170–171.

Łukasiewicz, J. (1970). *Selected Works. North-Holland*. Amsterdam, Warsaw: PWN.

Martinez, A. A. (2006). *Negative math: How mathematical rules can be positively bent*. Princeton University Press.

Moran, E., & Yu, J. (2001). *The complete idiot's guide to the I Ching*. New York: Alpha Books.

Nash, J. (1950). Equilibrium points in n-person games. *Proc. of the Nat'l Academy of the USA, 36*(1), 48–49. doi:10.1073/pnas.36.1.48

Ou, B., Huang, D., Hampsch-Woodill, M., & Flanagan, J. A. (2003). When East meets West: The relationship between yin-yang and antioxidation-oxidation. *The FASEB Journal, 17*, 127–129. doi:10.1096/fj.02-0527hyp

Overbye, D. (2006). A real flip-flopper, at 3 trillion times a second. *The New York Times – Science Report*. Retrieved from http://www.nytimes.com/2006/04/18/science/18find.html

Palko, L., Bass, H. W., Beyrouthy, M. J., & Hurt, M. M. (2004). The Yin Yang-1 (YY1) protein undergoes a DNA-replication-associated switch in localization from the cytoplasm to the nucleus at the onset of S phase. *Journal of Cell Science, 117*, 465–476. doi:10.1242/jcs.00870

Penrose, R. (2005). *The road to reality: A complete guide to the laws of the universe*. New York: Alfred A. Knopf.

Reyes, R., Mandelbaum, R., Seljak, U., Baldauf, T., Gunn, J. E., Lombriser, L., & Smith, R. E. (2010). Confirmation of general relativity on large scales from weak lensing and galaxy velocities. *Nature, 464*, 256–258. doi:10.1038/nature08857

Rosenfeld, L. (1963). Niels Bohr's contribution to epistemology. *Physics Today, 16*(10), 47. doi:10.1063/1.3050562

Santiago, F. S., Ishii, H., Shafi, S., Khurana, R., Kanellakis, P., & Bhindi, R. (2007). Yin Yang-1 inhibits vascular smooth muscle cell growth and intimal thickening by repressing p21WAF1/Cip1 transcription and p21WAF1/Cip1-Cdk4-Cyclin D1 assembly. *Circulation Research, 101*, 146–155. doi:10.1161/CIRCRESAHA.106.145235

Shi, Y., Seto, E., Chang, L.-S., & Shenk, T. (1991). Transcriptional repression by YY1, a human GLI-Kruppel-related protein, and relief of repression by adenovirus E1A protein. *Cell, 67*(2), 377–388. doi:10.1016/0092-8674(91)90189-6

Singh, M. P., & Huhns, M. N. (2005). *Service-Oriented Computing: Semantics, processes, agents*. West Sussex, UK: John Wiley & Sons, Ltd.

Smolin, L. (2005). *The case for background independence*.

Smolin, L. (2006). *The trouble with physics: The rise of string theory, the fall of a science, and what comes next?* New York: Houghton Mifflin Harcourt.

Temple, R. (1986). *The genius of China: 3,000 years of science, discovery, and invention*. New York: Simon and Schuster.

The Royal Swedish Academy of Sciences. (2008, October 7). The Nobel Prize in Physics, 2008 [Press Release]. Retrieved from http://nobelprize.org/nobel_prizes/physics/laureates/2008/press.html

van Zoest, T., Gaaloul, N., Singh, Y., Ahlers, H., Herr, W., & Seidel, S. T. (2010). Bose-Einstein condensation in microgravity. *Science, 328*(5985), 1540–1543. doi:10.1126/science.1189164

Vasudevan, S., Tong, Y., & Steitz, J. A. (2007). Switching from repression to activation: MicroRNAs can up-regulate translation. *Science, 318*(5858), 1931–1934. doi:10.1126/science.1149460

Wade, N. (2010, June 12). A decade later, human gene map yields few new cures. *New York Times*, Retrieved from http://www.nytimes.com/2010/06/13/health/research/13genome.html?partner=rss&emc=rss

Wilhelm, R., & Baynes, C. F. (1967). *The I Ching or book of changes. Bollingen Series XIX*. Princeton University Press.

Wilkinson, F. H. Park, K. & Atchison, M.L. (2006). Polycomb recruitment to DNA in vivo by the YY1 REPO domain. *Proceedings of the National Academy of Sciences, USA, 103*, 19296-19301.

Woit, P. (2006). *Not even wrong: The failure of string theory and the search for unity in physical law.* New York: Basic Book.

Zadeh, L. A. (2001). Causality is undefinable–toward a theory of hierarchical definability. *Proc. of FUZZ-IEEE,* 67-68

Zhang, W.-R. (1996). NPN fuzzy sets and NPN qualitative algebra: A computational framework for bipolar cognitive modeling and multiagent decision analysis. *IEEE Trans. on SMC*, *16*, 561–574.

Zhang, W.-R. (1998a). Yinyang bipolar fuzzy setssets. *Proceedings of IEEE World Congress on Computational Intelligence – Fuzz-IEEE,* (pp. 835-840). Anchorage, AK, May 1998.

Zhang, W.-R. (1998b). Nesting, safety, layering, and autonomy: A reorganizable multiagent Cerebellar architecture for intelligent control–with application in legged locomotion and gymnastics. *IEEE Transactions on Systems, Man, and Cybernetics . Part B, 28*(3), 357–375.

Zhang, W.-R. (2003a). Equilibrium relations and bipolar cognitive mapping for online analytical processing with applications in international relations and strategic decision support. *IEEE Trans. on SMC . Part B, 33*(2), 295–307.

Zhang, W.-R. (2003b). Equilibrium energy and stability measures for bipolar decision and global regulation. *International Journal of Fuzzy Systems*, *5*(2), 114–122.

Zhang, W.-R. (2005a). YinYang bipolar lattices and l-sets for bipolar knowledge fusion, visualization, and decision. *International Journal of Information Technology and Decision Making*, *4*(4), 621–645. doi:10.1142/S0219622005001763

Zhang, W.-R. (2005b). YinYang bipolar cognition and bipolar cognitive mapping. *International Journal of Computational Cognition*, *3*(3), 53–65.

Zhang, W.-R. (2006a). YinYang bipolar fuzzy sets and fuzzy equilibrium relations for bipolar clustering, optimization, and global regulation. *International Journal of Information Technology and Decision Making, 5*(1),), 19-46.

Zhang, W.-R. (2006b). YinYang bipolar tt-norms and tt-conorms as granular neurological operators. *ProceedingsProceedings of IEEE International Conference on Granular Computing* (pp. 91-96). Atlanta, GA.

Zhang, W.-R. (2007). YinYang bipolar universal modus ponens (BUMP) – a fundamental law of nonlinear brain dynamics for emotional intelligence and mental health. *Walter J. Freeman Workshop on Nonlinear Brain Dynamics, Proceedings of the 10th Joint Conference of Information Sciences* (pp. 89-95). Salt Lake City, Utah, USA.

Zhang, W.-R. (2009a). Six conjectures in quantum physics and computational neuroscience. *Proceedings of 3rd International Conference on Quantum, Nano and Micro Technologies (ICQNM 2009)*, (pp. 67-72). Cancun, Mexico.

Zhang, W.-R. (2009b). YinYang Bipolar Dynamic Logic (BDL) and equilibrium-based computational neuroscience. *Proceedings of International Joint Conference on Neural Networks (IJCNN 2009)*, (pp. 3534-3541). Atlanta, GA.

Zhang, W.-R. (2009c). YinYang bipolar relativity–a unifying theory of nature, agents, and life science. In *Proceedings of International Joint Conference on Bioinformatics, Systems Biology and Intelligent Computing (IJCBS),* Shanghai, China (pp. 377-383).

Zhang, W.-R. (2009d). The logic of YinYang and the science of TCM–an Eastern road to the unification of nature, agents, and medicine. [IJFIPM]. *International Journal Functional Informatics and Personal Medicine*, *2*(3), 261–291. doi:10.1504/IJFIPM.2009.030827

Zhang, W.-R. (2010). YinYang bipolar quantum entanglement–toward a complete quantum theory. *Proceedings of the 4th Int'l Conference on Quantum, Nano and Micro Technologies (ICQNM 2010)*, (pp. 77-82). St. Maarten, Netherlands Antilles.

Zhang, W.-R., Chen, S., & Bezdek, J. C. (1989). POOL2: A generic system for cognitive map development and decision analysis. *IEEE Trans. on SMC*, *19*(1), 31–39.

Zhang, W.-R., Chen, S., Wang, W., & King, R. (1992). A cognitive map based approach to the coordination of distributed cooperative agents. *IEEE Trans. on SMC*, *22*(1), 103–114.

Zhang, W.-R., & Chen, S. S. (2009). Equilibrium and non-equilibrium modeling of YinYang WuXing for diagnostic decision analysis in traditional Chinese medicine. *International Journal of Information Technology and Decision Making*, *8*(3), 529–548. doi:10.1142/S0219622009003521

Zhang, W.-R., Pandurangi, A., & Peace, K. (2007). Yinyang dynamic neurobiological modeling and diagnostic analysis of major depressive and bipolar disorders. *IEEE Transactions on Bio-Medical Engineering*, *54*(10), 1729–1739. doi:10.1109/TBME.2007.894832

Zhang, W.-R., Pandurangi, K.A., Peace, K.E., Zhang, Y. & Zhao, Z. (In press). MentalSquares–a generic bipolar support vector machine for psychiatric disorder classification, diagnostic analysis and neurobiological data mining. *International Journal on Data Mining and Bioinformatics.* In press, 2010.

Zhang, W.-R., & Peace, K. E. (2007). YinYang mentalsquares–an equilibrium-based system for bipolar neurobiological pattern classification and analysis. *Proceedings of IEEE BIBE* (pp. 1240-1244). Boston.

Zhang, W.-R., Wang, P., Peace, K., Zhan, J., & Zhang, Y. (2008). On truth, uncertainty, equilibrium, and harmony–a taxonomy for YinYang scientific computing. *International Journal of New Mathematics and Natural Computing*, *4*(2), 207–229. doi:10.1142/S1793005708001033

Zhang, W.-R., Wang, W., & King, R. (1994). An agent-oriented open system shell for distributed decision process modeling. *Journal of Organizational Computing*, *4*(2), 127–154. doi:10.1080/10919399409540220

Zhang, W.-R., Zhang, H. J., Shi, Y., & Chen, S. S. (2009). Bipolar linear algebra and YinYang-N-element cellular networks for equilibrium-based biosystem simulation and regulation. *Journal of Biological System*, *17*(4), 547–576. doi:10.1142/S0218339009002958

Zhang, W.-R., & Zhang, L. (2003). Soundness and completeness of a 4-valued bipolar logic. *International Journal on Multiple-Valued Logic*, *9*, 241–256.

Zhang, W.-R., & Zhang, L. (2004a). Yin yang bipolar logic and bipolar fuzzy logic. *Information Sciences*, *165*(3-4), 265–287. doi:10.1016/j.ins.2003.05.010

Zhang, W.-R., & Zhang, L. (2004b). A Multiagent Data Warehousing (MADWH) and Multiagent Data Mining (MADM) approach to brain modeling and neuro-fuzzy control. *Information Sciences, 167*, 109–127. doi:10.1016/j.ins.2003.05.011

Zhang, Y.-Q. (1992). Universal fundamental field theory and Golden Taichi. *Chinese Qigong, 3*, 242-248. Retrieved from http://www.cs.gsu.edu/~cscyqz/

Zhou, Q., & Yik, J. H. N. (2006). The Yin and Yang of P-TEFb regulationregulation: Implications for human immunodeficiency virus gene expression and global control of cell growth and differentiation. *Microbiology and Molecular Biology Reviews, 70*(3), 646–659. doi:10.1128/MMBR.00011-06

ADDITIONAL READING

Anthony, C. K., & Moog, H. (2002). *I Ching: The Oracle of the Cosmic Way*. Stow, MA: Anthony Publishing Company, Inc.

Balkin, J. M. (2002). *The Laws of Change: I Ching and the Philosophy of Life*. New York: Schocken Books.

Lloyd, S. (2006). *Programming the Universe*. New York: Alfred A. Knopf, Inc.

Petoukhov, S., & He, M. (2009). *Symmetrical Analysis Techniques for Genetic Systems and Bioinformatics: Advanced Patterns and Applications*. Hershey, PA: IGI Global.

Wang, Y. (2004). On Cognitive Informatics. *Brain and Mind, 4*(2), 151–167. doi:10.1023/A:1025401527570

KEY TERMS AND DEFINITIONS

YinYang: In ancient Chinese philosophy, the concept of Yin Yang is used to describe how seemingly opposing forces are reciprocal and interdependent existing in equilibrium or harmony in the natural world, giving rise to each other in turn. The concept lies at the heart of many branches of classical Chinese science and philosophy as well as being a primary guideline of traditional Chinese medicine (TCM), and a central principle of different forms of Chinese martial arts and exercise, such as Taiji box and qigong. According to the philosophy, yin and yang are complementary opposite energies within a greater whole. Everything has both yin and yang aspects, which constantly interact, never existing in absolute stasis. This book extends the YinYang cosmology into modern scientific worlds of action-reaction, matter-antimatter particles, genomic repression-activation, mental self-negation and self-assertion, social competition-cooperation, and all other YinYang pairs in physical and social sciences.

Binary YinYang: Leibniz interpretation of YinYang where Yin = 0 and Yang = 1 where physical semantics are not assumed. The modern binary numeral system was fully documented by Leibniz in the 17th century in his article *Explication de l'Arithmétique Binaire* (Leibniz, 1703). Leibniz's system used 0 and 1 like the modern binary numeral system. As a Sinophile, Leibniz was aware of the *I Ching* and noted with fascination how its 64 hexagrams correspond to the 64 6-bit binary numbers from 000000 to 111111. (Aiton, 1985, pp. 245-248).

Bipolar YinYang: An equilibrium-based interpretation of YinYang where Yin is the negative pole or energy and Yang is the positive pole or energy of nature or agent where physical semantics are assumed. In this interpretation the Yin and Yang are reciprocal but not dichotomy. The short term "bipolar" is used alternatively for "YinYang bipolar" in this book for simplicity without further specification.

Quantum Gravity: Or quantum theory of gravity, an evolving field of theoretical physics attempting to unify quantum mechanics with general relativity in a self-consistent manner.

Unifying Theory: This book follows Lee Smolin's approach to scientific unification: *"The main unifying idea is simple to state: Don't start with space, or anything moving in space. Start with something that is purely quantum-mechanical and has, instead of space, some kind of purely quantum structure. If the theory is right, then space must emerge, representing some average properties of the structure – in the same sense that temperature as a representation of the average motion of atoms."* (Smolin, 2006, p. 240)

YinYang Bipolar Relativity: The intra-relativity of the two reciprocal poles or energies Yin and Yang of a bipolar agent and the inter-relativity of different bipolar agents (Zhang, 2009a, 2009b, 2009c, 2009d). The theory is based the basic postulate that YinYang bipolarity is the most fundamental property of the universe which leads to logically definable causality and bipolar quantum entanglement,

Bipolar Quantum Gravity: An equilibrium-based approach to quantum gravity based on bipolar relativity with a Q5 paradigm of logical, physical, social, biological, and mental quantum gravities.

Agent: Any microscopic or macroscopic physical entity which is capable of action or reaction such as quantum agent, molecular agent, genomic agent, human agent, society, celestial object, galaxy, or the universe itself. (Zhang, 2009c, 2009d)

LAFIP: Logical axiomatization for illogical physics (Zhang, 2009a). It can be alternatively denoted as LAFIB (Zhang, 2009a).

Equilibrium-Based: Equilibrium-based is to equilibrium and non-equilibrium as truth-based is to truth and falsity but with fundamentally different syntax and semantics. One is free from LAFIP and the other is subjected to LAFIP.

Nature's Most Fundamental Property: The physical property that best characterizes the universe at the most general and fundamental level.

Spacetime Relativity: Relativity theories defined in spacetime geometry.

Beyond Spacetime: YinYang bipolar geometry is beyond spacetime in the sense that space and time are parameters in the geometry. In the bipolar geometry, big bang and spacetime can be caused by equilibrium-based bipolar causality and relativity; on the other hand, bipolarity is able to survive a black hole. While the former is a postulate, the latter is corroborated by particle-antiparticle emission or Hawking radiation.

Quantum Computing: Any mathematical, physical, mental, social, or biological computing in the broadest terms where quantum entanglement is involved in one way or another. Note that, this is a move away from the classical definition of quantum computing.

Cognitive Informatics: A interdisciplinary research area that tackles the common root problems of modern informatics, computation, software engineering, artificial intelligence (AI), neural psychology, and cognitive science. Cognitive informatics studies the internal information processing mechanisms and natural intelligence of the brain. (Wang, 2004)

Life Sciences: Comprising all fields of science that involve the scientific study of living organisms, like plants, animals, and human beings.

Chapter 2
Background Review:
Quest for Definable Causality

ABSTRACT

This chapter presents a review on the quest for logically definable causality. The limitation of observability and truth-based cognition is discussed. The student-teacher philosophical dispute between Aristotle and Plato is revisited. Aristotle's causality principle, David Hume's challenge, Lotfi Zadeh's "Causality Is Undefinable" conclusion, and Judea Pearl's probabilistic definability are reviewed. Niels Bohr's particle-wave complementarity principle, David Bohm's causal interpretation of quantum mechanics, and Sorkin's causal set program are discussed. Cognitive-map-based causal reasoning is briefly visited. YinYang bipolar logic and bipolar causality are previewed. Social construction and destruction in science are examined. It is asserted that, in order to continue its role as the doctrine of science, the logical definability of Aristotle's causality principle has become an ultimate dilemma of science. It is concluded that, in order to resolve the dilemma, a formal system with logically definable causality has to be developed, which has to be logical, physical, relativistic, and quantum in nature. The formal system has to be applicable in the microscopic world as well as in the macroscopic world, in the physical world as well as in the social world, in cognitive informatics as well as in life sciences, and, above all, it has to reveal the ubiquitous effects of quantum entanglement in simple, comprehensible terms.

INTRODUCTION

While the EPR paradox (Einstein, Podolsky & Rosen, 1935) challenged long-held ideas about the relation between the observed values of physical quantities and the values that can be accounted for by a physical theory, Bell's theorem (Bell, 1964) extended the argument of the EPR paradox and proved

DOI: 10.4018/978-1-60960-525-4.ch002

the validity of quantum entanglement in statistical terms but hasn't led to unification beyond quantum theory. On the one hand, the search for quantum gravity has so far failed to find a decisive battleground; on the other hand, practical research and development of quantum computers have been moving forward slowly with countless difficulties and setbacks. Evidently, a logically complete theory of quantum mechanics is inevitable for its ultimate unification with general relativity and for its practical application in quantum computing.

Per the EPR paradox, scientists are faced with the dilemma that if observability is questionable truth-based reasoning could be at fault especially in the quest for the grand unification. This prompted us to ask the legitimate questions: *Could equilibrium be the unifying force for the difficult and sometimes painful quest? Is bipolar equilibrium or non-equilibrium observable? Could YinYang bipolarity be the hidden fundamental?*

From a truth-based unipolar perspective, equilibria or non-equilibria are not observables; from a holistic equilibrium-based YinYang bipolar perspective, they are observables because equilibrium is a fundamental scientific concept in thermodynamics – the ultimate source of existence, energy, life, and information. Actually, observable or not may no longer be the key when we are faced with the faultline of observability in truth-based reasoning. Some additional scientific principle seems to be missing from the big picture of science.

What could be the missing scientific principle? In a court of law, a conviction of a crime needs two elements: evidence and motive. Evidence is based on scientific observation in criminal investigation. Motive, on the other hand, doesn't seem to find a match in science. Is there such a matching principle in science? If so, what is it?

Although agent and agent interaction is the source of any motivation, the legal term "motive" in criminal science is not a mathematical concept. To find a match to the concept in mathematical terms, therefore, requires a unifying mathematical axiomatization of agents and agent interaction in both microscopic and macroscopic worlds. As discussed in the last chapter, since agent interactions are governed or regulated by physical and social dynamics, the difficulty of axiomatizing agent interactions can be traced back to Hilbert's effort in axiomatizing physics, Aristotle's causality principle, and the concept of singularity and equilibrium. Among these concepts, causality is the only one related to motive. *Could causality be the missing principle from the big picture of science?*

In Lee Smolin's influential book *"The Trouble with Physics: The Rise of String Theory, The Fall of a Science, and What Comes Next"*, the author outlined five scientific unifications (Smolin, 2006):

1. Combine general relativity and quantum theory into a single theory that can claim to be the complete theory of nature. This is called the problem of quantum gravity.
2. Resolve the problems in the foundations of quantum mechanics, either by making sense of the theory as it stands or by inventing a new theory that does make sense.
3. Determine whether or not the various particles and forces can be unified in a theory that explains them all as manifestations of a single fundamental entity.
4. Explain how the values of the free constants in the standard model of particle physics are chosen in nature.
5. Explain dark matter and dark energy. Or, if they don't exist, determine how and why gravity is modified on large scale. More generally, explain why the constants of the standard model of cosmology, including dark energy, have the values they do.

After discussing the difficult and sometimes painful journey in the quest for quantum gravity (Problem 1), the author wrote: *"These days, many of us working on quantum gravity believe that causality itself is fundamental – and is thus meaningful even at a level where the notion of space has disappeared."* (Smolin, 2006, p. 241241)

Causality denotes a necessary relationship between one event called *cause* and another event called *effect* which is the direct consequence of the cause. Without making clear the causal-effect relation any scientific theory is incomplete. For instance, a letter published in Nature reported that a 2008 quantum physics experiment performed in Geneva, Switzerland determined that the "speed" of the quantum non-local connection or quantum entanglement (what Einstein called "spooky action at a distance") has a minimum lower bound of 10,000 times the speed of light (Salart et al., 2008). The minimum lower bound makes the concept of time questionable as time is defined based on the hypothesis that no speed can go beyond the speed of light. However, modern quantum physics cannot expect to determine the maximum or to interpret the cause of the lower bound given that we do not know the sufficient causal-effect relation of quantum entanglement.

Aristotle's principle of causality is commonly called the doctrine of the four causes. For Aristotle, a firm grasp of what a cause is, and how many kinds of causes there are, is essential for a successful investigation of the world around us. Although the principle of causality is widely considered the cornerstone of science, it was only stated in words by Aristotle 2300 years ago. Due to its problem of formal definability, the principle became historically controversial in the 18th century after Scottish philosopher David Hume challenged it from an empiricist perspective. Hume believed that causation is empirical in nature and irreducible to pure regularity (Blackburn, 1990). Extending Hume's argument, Lotfi Zadeh, founder of fuzzy logic, concluded that *"Causality Is Undefinable"* in classical logic (Zadeh, 2001) and, therefore, uncertainty is unavoidable.

Extending Hume's argument in a different direction, Judea Pearl, founder of probabilistic artificial intelligence (AI), presented influential models for causal reasoning through Bayesian networks (Pearl, 1988, 2000). When being combined with subjective knowledge, Pearl's models can be very useful in decision analysis by philosophers, economists, psychologists, epidemiologists, computer scientists, lawyers, and other decision makers alike. Pearl's probabilistic approach to AI, however, doesn't attempt to answer the fundamental question on logical definability of causality from a mathematical physics perspective.

In quantum physics the definability of causality is at the center of debate. Niels Bohr, father figure of quantum mechanics besides Einstein, believed that a causal description of a quantum process cannot be attained and we have to content ourselves with particle-wave complementary descriptions (Bohr, 1948). Einstein, on the other hand, refused to recognize quantum non-local connection or quantum entanglement as a basic law of physics due to its unclear cause-effect relation. He once called it *"spooky action at a distance"* and concluded that *"God does not play dice."*

Following Einstein's lead, David Bohm proposed the theory of implicate and explicate cosmological orders (Bohm, 1980) with a causal interpretation of quantum mechanics (Bohm, 1957). It is observed, however, that Bohm's causal interpretation is based on a wave function that so far hasn't led to a set-theoretic logical foundation for physics in general as envisioned by Einstein. Without such a logical foundation the ubiquitous effects of quantum entanglement cannot be revealed in simple comprehendible logical terms.

Among the various approaches in the quest for quantum gravity, the causal set hypothesis is distinguished by its logical simplicity and by the fact that it incorporates the assumption of underlying spacetime discreteness from the very beginning (Bombelli *et al.*, 1987). As a set-theoretic model, however, causal sets did not go beyond classical truth-based set theory and so far didn't lead to a formal logical system beyond spacetime geometry.

Consequently, in order to continue its role as the cornerstone of science, logical definability of Aristotle's causality principle has become an *ultimate dilemma of science*. In order to apply the principle in macroscopic and microscopic agent interactions such as quantum and molecular interactions as well as competition-cooperation relations in socioeconomics, it has to be formulated into logical or mathematical equations. Simply speaking, causality in words is insufficient for furthering modern science. A deeper fundamental concept of science requires a deeper philosophy. To resolve the problem, a formal system with logically definable causality has to be developed with a different mathematical abstraction. The formal system has to be logical, physical, relativistic, and quantum in nature. It has to host the "illogical" aspects of physics for logical reasoning to avoid the LAFIP paradox (Ch. 1). Furthermore, it has to be applicable in microscopic world as well as in macroscopic world, in physical world as well as in social world, in cognitive informatics as well as in life sciences, and, above all, it has to reveal the ubiquitous effects of quantum entanglement in a meaningful comprehendible way. Is such a new logic possible at all? According to Hume the answer is No; but according to Einstein the answer is yes.

Einstein clearly stated:

1. *"Physics constitutes a logical system of thought which is in a state of evolution, whose basis (principles) cannot be distilled, as it were, from experience by an inductive method, but can only be arrived at by free invention."* (Einstein, 1916)
2. *"Our experience hitherto justifies us in believing that nature is the realization of the simplest conceivable mathematical ideas. I am convinced that we can discover by means of pure mathematical constructions the concepts and the laws connecting them with each other, which furnish the key to the understanding of natural phenomena... In a certain sense, therefore I hold it true that pure thought can grasp reality, as the ancients dreamed."* (Einstein, 1934)

In the case of Einstein vs. Hume, we take Einstein's side. As postulated in the last chapter, YinYang bipolarity can be regarded as the most fundamental property of the universe. Since it is common knowledge in computer and cognitive sciences that without semantics there would be no need for syntax and without syntactic representation there would be no semantic processing, without (-,+) bipolar syntax and semantics it would be virtually impossible to reason on (-,+) bipolarity. Consequently, the limitation of classical truth-based mathematical abstraction in set theory has to be re-examined and the inevitability of an equilibrium-based bipolar axiomatization for logically definable causality has to be explored.

In this chapter we review the literature related to causality and layout the background for the later chapters. First, we discuss on the faultline of observability and truth-based thinking followed by a discussion on YinYang bipolarity vs. singularity. After that, we present a literature review on Aristotle's causality principle followed by David Hume's critique, Lotfi Zadeh's conclusion, Judea Pearls probabilistic definability, Niels Bohr and Einstein on quantum causality, David Bohm's causal interpretation of quantum theory, causal set theory, cognitive-map-based causal reasoning, and equilibrium-based logically definable *YinYang bipolar causality*. We then introduce the concepts of social construction and destruction in science followed by a chapter summary. The topics are organized in the following sections:

- Faultline of Observability and Truth-Based Thinking
- Bipolarity vs. Singularity

FAULTLINE OF OBSERVABILITY AND TRUTH-BASED THINKING

Since the beginning of human history, ancestors of mankind began to record their observations and hypotheses regarding the most fundamental elements of nature. Without modern communication technologies such as telephones, television, and the WWW (World Wide Web), ancestors who inhabited in different continents could not share ideas with worldwide collaboration. Nevertheless, the Eastern and Western philosophies on the fundamental elements of nature exhibit striking similarities. Both schools of philosophers used a set of archetypal *classical elements* that are assumed to be the simplest essential parts upon which the constitution and fundamental powers of anything are based (Strathern, 2000).

The most frequently occurring theory of classical elements, held by the Hindu, Japanese, India, and Greek systems of thought, was that there were five elements, namely Earth, Water, Air, Fire, and a fifth element known variously as space, *Idea*, *Void* "quintessence" (the term "quintessence" derives from "quint" meaning "fifth"). In Greek thought Aristotle added aether (or ether) as the quintessence as an unchangeable heavenly substance (Lloyd, 1968). The Greek Classical Elements (Earth, Water, Air, Fire, and Aether) deeply influenced European thought and culture (Strathern, 2000).

Rather than using the Western notion of different kinds of materials, the basic elements or agents of nature in ancient China were named *YinYang WuXing* or *YinYang-5-Elements* consisting of the set {Metal, Wood, Water, Fire, Earth}, which were understood as different phases of nature in a state of constant interaction with one another to cause the state to change (Eberhard, 1986). Yin and Yang were considered the two opposite reciprocal and interactive poles of the system and each element or agent. Although it is usually translated to "YinYang-5-Elements" or "YinYang-5-Agents", the Chinese word *Wu* stands for "five" and *Xing* for "walks". Together, *YinYang WuXing* literally means something like "YinYang five walks" or "five phases of the universe." Thus, *YinYang WuXing* provided the earliest process-based theory of the universe.

In Traditional Chinese Medicine (TCM) the five elements are matched to the set of five subsystems represented by the main organs {liver, lung, kidney, heart, spleen} of the human body. Thus, *YinYang WuXing* was intended to represent the unity of the universe and the human body. The vital energy circulating the five elements is called *qi* in TCM. *Qigong* (or *qi gong*) refers to a wide variety of traditional cultivation practices that involve methods of developing, circulating, regulating, and working with *qi* for healing and health. The vital energy *qi* also has the meaning of "air" in Chinese. It can be imagined that the ancient Chinese believed that qi (vital energy) is to the human body as air is to nature. Historian Professor Ebrey noted in *Chinese Civilization: A Sourcebook* (Ebrey, 1993, pp. 77-79) that the concepts of YinYang and the Five Agents *"provided intellectual framework of much of Chinese scientific thinking especially in the fields of biology and medicine."*

Except for ether, all other classical elements of the East and the West were proven composite by modern science before Newton. After Newton's time, however, all scientific ideas on the fundamental makings of nature were focused on nothing but matter or ether until Einstein's special and later general relativity theories crushed the ether. *Ether theory* is superseded by relativity theory because, if fields are not made of matter, fields are perhaps the more fundamental makings of nature (Smolin, 2006).

Among modern scientific states of matter, we have the periodic table of elements, the concept of combustion (fire), and, more recently, subatomic particles. While the periodic table of elements and the concept of combustion can be considered a secular successor of earlier models, elementary subatomic particles have shown no substructures like composite particles which are particles with substructure or particles that are made of other particles. Elementary subatomic particles are divided into three classes: quarks and leptons (particles of matter), and gauge bosons (force carriers, such as the photon). Elementary particles of the Standard Model include six different types of quark ("up", "down", "bottom", "top", "strange", and "charm"), as well as six different leptons ("electron", "electron neutrino", "muon", "muon neutrino", " tauon", " tauon neutrino"), force carriers ("photon", the W^+, W^-, and Z bosons, gluons), as well as the Higgs boson. It was predicted by the SU(5) symmetry theory that there had to be processes by which quarks can decay into electrons and neutrinos (Smolin, 2006, p. 63). The simultaneous broken symmetry model unified *"the smallest building blocks of all matter and three of nature's four forces in one single theory"* (The Royal Swedish Academy of Sciences, 2008).

Notably, string theory had been the dominating theory in theoretical physics for nearly three decades in the difficult and sometimes painful quest for quantum gravity. In this theory, strings are proposed as very tiny vibrating lines, points, or surfaces which are hypothesized as the smallest and most fundamental elements of the universe. The theory was once regarded as the only candidate for "*theory of everything*." In recent years, however, string theory has become controversial due to its failure to live up to its promised unification. Strong criticisms from insiders and outsiders of theoretical physics have put string theory on the defensive (Smolin, 2006; Woit, 2006). Three major criticisms of string theory include:

1. it is not observable or experimentally testable;
2. its extra dimensions (up to eleven) lead to a landscape of theories of many universes; and
3. it has so far failed to make falsifiable predictions.

Evidently, the striking similarities of different failed ideas regarding the basic elements of nature between the East and West and between ancient history and modern times resulted from either intuitive observations or incorrect assumptions about the existence of a smallest fundamental element. A wrong assumption will definitely lead to a wrong conclusion. A wrong observation can also lead to a wrong assumption. Moreover, observations and assumptions are often influenced by socioeconomics.

As a principle, observability or experimental testability is a cornerstone of science. However, there is no doubt that human understanding of nature in ancient time was based on superficial observations and shallow understanding. Nevertheless, such observations and understanding could translate into strong beliefs or even doctrines of powerful institutions such as government, religion or other socioeconomical, political, monetary, and cultural systems. No wonder some prominent scientists in history were persecuted or even brutally executed due to their scientific beliefs about the universe.

It is simply a fact that some tradition tends to kill new ideas if not the lives of their holders. Or should we say superficial intuitive observations and beliefs can also indirectly kill? Of course, a killed idea would sooner or later be rediscovered if it is really scientific in nature even though the original holders of such ideas were persecuted. A well-known example is the discovery of the solar system. For thousands of years, humanity in general did not recognize the existence of the Solar System. Following Aristotle's cosmology people believed the Earth to be stationary at the center of the universe. Now everyone knows the truth. Unfortunately, Bruno could not escape fire execution and Galileo could not avoid prison for believing in Copernicanism and telling the truth.

From the above discussion we can conclude that observability is a foundation for science but it is not absolute but a relative term. This is not a discovery of this book but part of Einstein's theory of relativity. According to Einstein, observation is a relative term between an observer and the observed. Thus, two different observers at different space and time may have different observations of the same thing. This is particularly true in quantum mechanics where the relation between the observed values of physical quantities and the values that can be accounted for has been a debatable problem.

The quantum observability controversy remains unresolved today. While industrial researchers and developers in quantum computing have been trying hard to avoid quantum disturbance involved in a measurement or observation, Lee Smolin, author of *The Trouble with Physics* asserted that *"If several of the best living theoretical physicists feel compelled to question the basic assumptions of relativity and quantum theory, there must be others who come to this position from the beginning. There are indeed people who, early in their studies, began to think quantum theory must be wrong. They learn it, and they can carry out its arguments and calculations as well as anyone. But they don't believe it. What happens to them?"* Smolin's view is significantly strengthened by his blunt confession: *"There are roughly two kinds of such people: the sincere ones and the insincere ones. I am one of those who never found a way to believe in quantum mechanics, but I am one of the insincere ones..."* (Smolin, 2006, p. 319).

While Einstein's special theory of relativity tells us that all light travels at the same speed, no matter in what frequency, his general theory of relativity (or general relativity) tells us space and time are relative and gravitation forms a field in spacetime. With general relativity, a quantum clock slows down in outerspace and light can bend in a gravitational field. These predictions have been proven observable in astrophysics.

General relativity is not free from problems though. One problem is that it cannot avoid infinities. That is, inside a black hole, the density of matter and the strength of the gravitational field quickly become infinite and the equations of general relativity breakdown. This same problem is also associated with the birth of the universe or the big bang theory. Lee Smolin states: *"Some people interpret this as time stopping, but a more sober view is that the theory is just inadequate. For a long time, wise people have speculated that it is inadequate because the effects of quantum physics have been neglected"* (Smolin, 2006, p. 5).

Quantum theory also has a problem with infinities (Smolin, 2006). That is, a quantum field has values at every point in space. This leads to the so-called infinite number of variables problem. Although quantum mechanics has been extremely successful at explaining a vast realm of phenomena, there are many experts who are convinced that quantum theory hides something essential about nature we need to know (Smolin, 2006, p. 6).

From a cause-effect perspective, a major problem of general relativity and quantum theory is that they are both positioned in spacetime geometry. Unfortunately, space is expanding, space and time are not symmetrical, not quantum entangled and, therefore, not most fundamental. If space is expanding as observed, the cause of such expansion has to be more fundamental than the effect. If the speed of quantum entanglement is far more than the speed of light, time is no longer a mature concept.

From a logical perspective, a basic limitation of general relativity and quantum theory lies in truth-based reasoning. Regardless of the vital importance of truth-based bivalency in information and computation, without directly carrying a shred of basic physical semantics such as equilibrium and bipolarity, the logical values {false, true} or {0,1} in a truth-based logic or any of its extensions including many-valued logic (Łukasiewicz, 1920, 1970), intuitionistic logic (Brouwer, 1912), fuzzy logic (Zadeh, 1965), paraconsistent logic (Belnap, 1977), and quantum logic (Birkhoff & von Neumann, 1936), are inadequate for axiomatizing physics. It is suggested in Chapter 1 that the bivalent limitation is the fundamental reason why there is so far no truth-based logical axiomatization for physics, no logically definable causality, and no decisive battle ground in the quest for quantum gravity (Zhang, 2009a, 2009b, 2009c, 2009d).

The inadequacy of truth-based thinking can be characterized by a combination of the EPR paradox (Einstein, Podolsky & Rosen, 1935) and Bell's theorem (Bell, 1964) in quantum mechanics. While the EPR paradox challenged the relation between observed values of physical quantities and the values that can be accounted for by a physical theory, Bell's theorem proved the validity of quantum mechanics from a statistical perspective. However, the validity so far hasn't led to further unification beyond quantum mechanics.

While the EPR paradox tells us that there is a faultline of observability, the LAFIP paradox (Ch. 1) tells us that there is a faultline for truth-based reasoning. Actually, both observability and truth-based reasoning belong to the realm of positivism. Positivism holds that the only authentic knowledge is based on actual observable truth and metaphysical speculation is avoided. The positivist approach has been a recurrent theme in the history of Western thought from the Ancient Greeks to the present day. However, it became more controversial in the last quarter of the 20th century. Some published work even claimed that the era of positivism ended in 1970s (Henrickson & McKelvey, 2002).

Regardless of the controversy on positivism, both observability and truth-based reasoning do indeed have their limitations. Besides the observability problem of quantum states, it is a common belief that equilibrium (including quasi- or non-equilibrium) and the universe form a philosophical chicken-egg paradox because no one had ever answered the fundamental question: *"Equilibrium and universe, which one created the other in the very beginning?"* (Zhang & Zhang, 2004).

BIPOLARITY VS. SINGULARITY

Truth and Singularity

Singularity could be gravitational or technological. The first is defined in the context of cosmology and the second is defined in technology. They are similar in philosophical thinking but different in context.

A *gravitational singularity* or *spacetime singularity* is a location where the quantities that are used to measure the gravitational field become infinite in a way that does not depend on the coordinate system (Hawking & Penrose, 1970). These quantities are the scalar invariant curvatures of spacetime, some of which are a measure of the density of matter. The two major types of spacetime singularities are *curvature singularities* and *conical singularities*. Singularities can also be categorized according to whether they are covered by an *event horizon* or not (*naked singularities*). According to general relativity, the initial state of the universe, at the beginning of the big bang, was a singularity. Another type of singularity predicted by general relativity is inside a black hole. Any star collapsing beyond a certain point would form a black hole, inside which a singularity (covered by an event horizon) would be formed, as all matter would flow into a certain point (or a circular line, if the black hole is rotating).

These singularities are also known as curvature singularities. The exact details are not currently well understood in all circumstances.

Technological singularity, on the other hand, is the theoretical future point which is hypothesized to take place during a period of *accelerating change* sometime after the creation of *superintelligence*. Superintelligence is a speculative artificially enhanced human brain, a computer program or a device that is much smarter, more creative and wiser than any current or past existing human brain. Thus, technological singularity is an analogy of gravitational singularity in technology.

Even though Einstein never believed in the theory of gravitational singularity and considered the theory as *"bizarre"*, resisting the logic of his own theory right up to his death in 1955, his equations of general relativity did eventually lead to the flourish of the gravitational singularity theory following the discovery of black holes after his death (Hawking & Penrose, 1970). Based on singularity and partially observable truth, it is now a commonly accepted theory that the universe was created by the big bang that is caused by nothing that we know.

It is interesting to examine the nature of Einstein's dilemma and its relationship with the dilemma of science in general on logically undefinable causality. First, Einstein's dilemma is a contradiction between his belief and his logic. Based on his belief, he rejected the idea of singularity. Based on his equations of general relativity, singularity should be a logical consequence.

Secondly, our classical logic is defined in the bivalent unipolar lattice {0,1} which forms the logical basis for classical set theory, computing, decision, cognition, and modern science. The truth-based unipolar approach to science has been proven the most effective way of thinking with countless glorious achievements in all fields of science, technology, and sociology. Einstein himself was certainly associated with the truth-based glories and greatness, and, unfortunately, also the truth-based dilemma in science.

It has been observed that nature's basic forces form bipolar action and reaction pairs; all matter and antimatter particles form a bipolar quasi-symmetry. Although *bipolarity* is essential in our understanding of nature as well as in implementing digital and quantum logic devices, all truth-based logical systems implemented with such bipolar devices tend to deny the unavoidability of bipolarity. If a mentally healthy person later got bipolar disorder, the disorder is evidently the loss of mental equilibrium, but modern science refuses to recognize the observable fact that we are all bipolar, either in calm and energetic healthy bipolar equilibrium or in unhealthy bipolar disorder (or in between).

With bipolarity, a black hole can be explained as physical depression; big bang can be explained as physical mania; a normal universe can be described as an equilibrium or quasi-equilibrium of matter and antimatter or action and reaction forces. Consequently, these bipolar explanations lead to logically definable causality and a process-based universe governed or regulated by YinYang bipolar relativity (Zhang, 2009a, 2009b, 2009c, 2009d2009d). The problem is that, on the one hand, truth-based cognition denies the existence of bipolar equilibrium and its observability; on the other hand, equilibrium is nevertheless a fundamental concept in science as manifested by the 2nd law of thermodynamics.

YinYang vs. Greek Philosophy

Greek philosophy in science and nature has profoundly impacted scientific advances of the West for more than 2300 years. The scientific foundation of Greek philosophy, however, is arguably imperfect (Zhang *et al.,* 2008). For instance, the long-standing dispute (*Aristotle-Plato Dispute*) between the two leading Greek philosophers Aristotle (384 BC – 322 BC) and his teacher Plato (428/427 BC – 348/347 BC) remains unresolved until today (Theosophy, 1939). It is argued that, since a solid philosophical foundation is crucial to scientific computation, we must not and should not shun the unresolved historical dispute between Aristotle and Plato (Zhang *et al.*, 2008).

It is asked (Zhang *et al.*, 2008) that: *Can some other philosophy on nature, such as equilibrium- or harmony-based YinYang theory of the ancient Chinese Daoist philosophy, be adopted as a valuable inspiration for science? If yes, in what capacity? Can it be used as a unifying force for resolving the two thousand year old Greek dispute?*

It is well-known that Aristotle refused to recognize supersensible cognition as the source of knowledge and he refuted his teacher Plato's doctrines, maintaining that the Platonic method is fatal to science. "But, is there, in the universe or outside of it, an underlying reality which is eternal, immovable, unchanging as claimed by *Platonic realism*?" (Theosophy, 1939)

If our answer is "No" to the above question, we will be faced with the follow-up questions *"What is equilibrium (including dynamic quasi- or non-equilibrium through this book)?" "Is equilibrium an underlying reality in the universe or outside of it which is eternal, immovable, unchanging?" "Is equilibrium qualified to be a Platonic universal?"*

Furthermore, we have the provocative question: *"Did the universe create equilibrium or did equilibrium create the universe?"* (Zhang & Zhang, 2004)

According to Aristotle the answer would be *"The universe created equilibrium."* Aristotle wrote extensively in criticism of Plato's doctrine of Ideas (Theosophy, 1939) – the Forms of things, affirming that *"no universals exist over and above the individual objects and separate from them."* He refused any substantial reality to *"the unity which is predicated of many individual things."* Universal principles, he held, are real, and are the objects of our reason, as distinguished from the physical objects of sense-perception. Yet universals are real only as they exist in individuals. *"It is,"* he said, *"apparently impossible that any of the so-called universals should exist as substance"* (Theosophy, 1939). Evidently, Aristotle's philosophy of science is a bottom-up positivist cognition where the universe is perceived consisting of a set of truth objects that can be mapped to his bivalent space {false, true} or {0, 1} based on observed evidence.

According to Plato the answer would be *"Equilibrium created the universe."* Plato held that the Ideas, the Forms of things, are self-existent, and not dependent upon the ever-changing objects of the senses. The noumenon, according to Plato, is the real, the phenomenon only appearance (Theosophy, 1939). Presumably, Plato could have used *YinYang bipolar equilibrium* as a universal to support his idealistic reality. Without modern communication technology such as the web, it was highly unlikely that he had been aware of the development of YinYang theory or even had contact with the ancient Chinese YinYang philosophers. Nevertheless, Platonic realism could have qualified as a top-down cognition had he identified an ultimate universal called *"equilibrium."* In that case he could have claimed that the universe is a form of equilibrium and everything else is simply the appearance of this form.

Most scientists would probably take side and agree with Aristotle – the widely recognized father of science. However, this conflict and dispute between Plato and Aristotle on the subject of reality has led to almost infinite controversy and confusion. As a major historical source of uncertainty it continues to feed the temptation to debate among later generations of philosophers, scientists, and mathematicians alike (Theosophy, 1939).

Although Aristotle is widely recognized the father of modern science, many renowned scientists and mathematicians also have claimed themselves as modern realists whose life-long efforts have been devoted to the noble cause of discovering universals or mathematical truths. For instances, Kurt Gödel was a well-known modern Platonic realist who formulated and defended mathematical Platonism (Kennedy,

2007); despite his argument on local reality in the EPR paradox, even Albert Einstein once sided with Plato and bluntly claimed that (Einstein, 1934) *"Our experience hitherto justifies us in believing that nature is the realization of the simplest conceivable mathematical ideas. ... In a certain sense, therefore I hold it true that pure thought can grasp reality, as the ancients dreamed."* It has become quite clear that the controversy results in a major philosophical root of uncertainty that is still holding the key for many unresolved scientific problems.

Can the dispute between Aristotle and Plato ever be resolved? It seems to be extremely difficult if not impossible because the dispute on equilibrium and universe is essentially a philosophical *"chicken and egg"* paradox. No one knows exactly which one created the other in the very beginning, and there is no definite answer even from a religious perspective. For instance, if we say *"God created the universe"*, we still have the follow-up questions *"Does equilibrium belong to God?" "Did God create equilibrium?" "Did God use his magic power equilibrium to create the universe?"* (Zhang *et al.*, 2008)

Although the dispute seems to be irresolvable, the main characteristics on both sides are clear. On one side is the intuitive or perceived supersensible realist cognition as founded by Plato and his teacher Pythagoras; on the other side is the observed, sensed, or evidenced bottom-up positivist scientific cognition as founded by Aristotle. While the former doesn't seem to have a chance to compete with the latter, the YinYang philosophy seems to play a unifying role.

According to YinYang theory every matter consists of two sides: Yin is the feminine or negative side and Yang is the masculine or positive side; the fusion of the two sides in equilibrium and/or harmony is a key for the mental or physical health of a person, a society, or any dynamic system. Since the concept of *equilibrium* (including quasi- or non-equilibrium) is central in thermodynamics – the ultimate physical source for all existence, energy, life, and information, it should become a key part of the philosophy of science.

An analog can be made in mathematics as well. Although Greeks recognized the importance of *"symbols"* a long time ago, the concept of *negative numbers* and *zero* clearly was not recognized then. While the Chinese and Indians started to use negative numbers in ancient time, influenced by Greek philosophy European mathematicians resisted negative numbers until the 17th – 18th centuries (Temple, 1986, p.141; Bourbaki, 1998, p.49; Martinez, 2006). With the introduction of negative numbers and zero, symmetry is preserved and the picture of mathematics is completed. Unfortunately, the symmetrical, interactive, and dynamic nature of the universe has not been taken into consideration in logical representation until recently. It is pointed out that the unipolar bivalent cognition is inadequate and equilibrium-based bipolar cognition is inevitable (Zhang *et al.*, 2008).

It is shown in later chapters that equilibrium-based bipolar dynamic logic, definable bipolar causality, and bipolar relativity play a unifying role between top-down and bottom-up cognitions. Indeed, nothing can practically exist without equilibrium. Therefore, the two coexisting sides, top-down cognition and bottom-up cognition, do not only oppose each other but also depend on each other in a mutually beneficial equilibrium, quasi-equilibrium, or non-equilibrium. There would be no "bottom-up cognition" if there were no "top-down cognition" and vice versa. Thus, YinYang bipolar cognition builds a bridge between the two sides for bipolar interaction. Consequently, we never need to resolve the dispute between Aristotle and Plato. Instead, we can promote bipolar interaction, equilibrium, balance, and harmony in mathematical, philosophical, and scientific terms for handling many different kinds of uncertainties such as logically definable causality with complementary and joint forces.

QUEST FOR DEFINABLE CAUSALITY

Aristotle's Principle of Causality

Aristotle's principle of Causality is deemed essential in all sciences. It denotes a necessary relationship between one event called *cause* and another event called *effect* where the effect is the direct consequence of the cause. Each *Aristotelian science* consists in the causal investigation of a specific world of reality. If such an investigation is successful, it results in causal knowledge; that is, knowledge of the relevant or appropriate causes for certain type of effects. The emphasis on the concept of cause explains why Aristotle developed a theory of causality which is commonly known as the doctrine of the four causes (Falcon, 2008):

- The material cause: "that out of which", e.g., the bronze of a statue.
- The formal cause: "the form", e.g., the shape of a statue.
- The efficient cause: "the primary source of the change or rest", e.g., the man who gives advice or the father of the child.
- The final cause: "the end, that for the sake of which a thing is done", e.g., health is the end of walking, losing weight, purging, drugs, and surgical tools.

According to Aristotle, all the four types of causes may enter in the explanation of something. However, Aristotle made it very clear that all his predecessors merely touched upon these causes. That is to say, they did not engage in their causal investigation with a firm grasp of these four causes. They lacked a complete understanding of the range of possible causes and their systematic interrelations. Put differently, and more boldly, their *use* of causality was not supported by an adequate *theory* of causality. According to Aristotle, this explains why their investigation, even when it resulted in important insights, was not entirely successful (Falcon, 2008).

Despite the great significance of Aristotle's causality principle, it is only stated in words. Unlike his predecessor, Aristotle made causality a principle for all sciences. Like his predecessors, Aristotle failed to provide formal logical or mathematical formulation for his principle. Without a formal definition, causal reasoning would be largely dependent on human understanding, subjective judgment, cognitive and intellectual abilities. This leads to a number of historical critiques.

David Hume's Critique on Aristotle's Causality Doctrine

Scottish philosopher David Hume (Morris, 2009) was a strong critic of Aristotle's philosophy. Among David Hume's criticisms of Aristotle's philosophy of science the historically most famous is the critique of the principle of causality.

According to Aristotle, the principle of causality consists in a relationship of *necessary connection* between cause and effect, in virtue of which the effect cannot exist without the presence of the cause. Hence the formulation of the principle of causality: *"Everything that begins to exist must have a cause for its existence."* Followed as a principle by many generations in all sciences, this absolute necessity of connection between effect and its cause is anything but exempt from doubt. Analytical a priori reasoning is such that it implies a proposition whose predicate is derivable from the idea of the subject as in the example, "Three times five is equal to fifteen."

According to Hume, however, the mind can never find the effect by examination of the supposed cause; for the effect is totally different from the cause, and consequently can never be discovered in it. Hume shows that our causal inferences are not due to reasoning or any operation of the understanding (Morris, 2009). Hume establishes that, whatever assures us that a causal relation obtains, it is not reasoning concerning relations between ideas. Effects are distinct events from their causes: we can always conceive of one such event occurring and the other not. So causal reasoning can't be *a priori* reasoning.

Hume claimed that causes and effects are discovered through experience, when we find that particular objects are constantly conjoined with one another (Morris, 2009). Thus, according to Hume, our causal expectations aren't formed on the basis of reason. But we do form and improve them through experience.

The necessary connection upon which the principle of causality is based is not demonstrable, according to Hume, even by experience. Hume states: *"The repetition of perfectly similar instances can never alone give rise to the original idea, different from what is to be found in any particular instance"* (Norton & Norton, 2000). Not even our activity and the effect of the will upon the movements of our body and our spirit can give us the impression of causality: *"No relationship is more inexplicable,"* Hume adds, *"than that which exists between the faculties of thought and the essence of matter."*

Evidently, according Hume, it is necessary to give up attributing any objective value to the idea of cause. But from where comes the idea of causality which undeniably exists? According to Hume it arises from a psychological fact formed in the following manner (Morris, 2009):

- Experience has shown that fact B has constantly followed fact A.
- This stability, never contradicted by experience, shows indeed that the two facts, A and B, are associated with one another, so that the one evokes the other.
- Through force of association there arises in me the trusting expectation, and hence the habit of expecting, that also in the future, and necessarily, granted fact A, fact B must follow.
- Thus the necessary connection is not a bond which regulates reality, but is a manner of feeling on the part of the subject, a new law which the subject places in regard to his impressions.

Hume's critique on Aristotle's principle of causality is the first major challenge to the doctrine from an empiricist perspective. His reasoning concludes with the collapse of all rational understanding and leads inevitably to uncertainty and skepticism.

Lotfi Zadeh on Causality

Extending Hume's argument on Aristotle's principle of causality, Lotfi Zadeh, founder of fuzzy logic, has become the strongest critic of the principle in modern days. Zadeh plainly concluded in the title of his paper that *"Causality Is Undefinable"* (Zadeh, 2001).

Zadeh wrote:

Attempts to formulate mathematically precise definitions of basic concepts such as causality, randomness, and probability, have a long history. The concept of hierarchical definability that is outlined in this project suggests that such definitions may not exist." Zadeh observed that "In essence, definability is concerned with whether and how a concept, X, can be defined in a way that lends itself to mathematical analysis and computation. In mathematics, definability of mathematical concepts is taken for granted.

But as we move farther into the age of machine intelligence and automated reasoning, the issue of definability is certain to grow in importance and visibility, raising basic questions that are not easy to resolve.

To be more specific, let X be the concept of, say, a summary, and assume that I am instructing a machine to generate a summary of a given article or a book. To execute my instruction, the machine must be provided with a definition of what is meant by a summary. It is somewhat paradoxical that we have summarization programs that can summarize, albeit in a narrowly prescribed sense, without being able to formulate a general definition of summarization. The same applies to the concepts of causality, randomness, and probability. Indeed, it may be argued that these and many other basic concepts cannot be defined within the conceptual framework of classical logic and set theory. (Zadeh, 2001)

Zadeh presented a hierarchical approach to the definability of causality. The root of his hierarchy is undefinability which is decomposed into different types of definability at lower levels. He wrote:

In this perspective, the highest level of definability hierarchy,... is that of undefinability or amorphicity. A canonical example of an amorphic concept is that of causality. More specifically, is it not possible to construct a general definition of causality such that given any two events A and B and the question, 'Did A cause B?', the question could be answered based on the definition. Equivalently, given any definition of causality, it will always be possible to construct examples to which the definition would not apply or yield counterintuitive results. In general, definitions of causality are non-operational because of infeasibility of conducting controlled experiments.

The theory of hierarchical definability is not a theory in the traditional spirit. The definitions are informal and conclusions are not theorems. Nonetheless, it serves a significant purpose by raising significant questions about a basic issue – the issue of definability of concepts that lie at the center of scientific theories. (Zadeh, 2001)

Judea Pearl on Causality

Extending David Hume's empiricist perspective on causality in a different direction, Judea Pearl has significantly advanced the understanding of causality in statistics, psychology, medicine, and social sciences. While his book *Probabilistic Reasoning in Intelligent Systems* (Pearl, 1988) is widely considered among the single most influential works in shaping the theory and practice of knowledge-based systems in artificial intelligence, his book *Causality: Models, Reasoning, and Inference* (Pearl, 2000) went beyond computer science and engineering and has made a major impact on the definability of causality.

Contrary to Zadeh's assertion that *"Causality is undefinable"*, Judea Pearl's book *Causality: Models, Reasoning, and Inference* offers a complete axiomatic characterization – a structural causal model (Pearl, 2000, Ch. 7). As Pearl described:

Full Pearl calls for framing problems the way an investigator perceives nature to work, regardless of whether one can estimate the parameters involved, then, once you set your model, you mark down what you know and what you do not know, what you are sure about and what you are not sure about, and do the analysis in this space of 'Nature models'. If it so happens that you are only sure about the structure of the graph and nothing else, then and only then, you do 'graphs, colliders, and do operators.' (Pearl, 2009)

It is clear from the above description that, Pearl's structural causal model is a probabilistic approach to reasoning and inference. It doesn't entail a fundamentally different mathematical abstraction with a new set theory and a new logic. In order to effectively use the model, a user needs to mark down what he/she knows and what he/she does not know, what he/she is sure about and what he/she is not sure about. These treatments clearly follow the empirical spirit in line with David Hume using probability and conventional statistics, where causes and effects are formed and improved through knowledge and experience but not on the basis of mathematical physics and logical deduction. Therefore, Pearl's definability of causality, albeit very useful in decision analysis, doesn't answer the basic question: *Is causality logically definable from a mathematical physics or biophysics perspective?*

Niels Bohr's Particle-Wave YinYang Complementarity

Niels Bohr, father figure of quantum mechanics besides Einstein, was the first to bring YinYang into quantum theory for his complementarity principle regarding particle-wave duality. When Bohr was awarded the Order of the Elephant by the Danish government, he designed his own coat of arms which featured in the center a YinYang logo (or Taijit symbol) and the Latin motto *contraria sunt complementa* or opposites are complementary.

Bohr believed that a causal description of a quantum process cannot be attained and we have to content ourselves with complementary descriptions. Bohr wrote (Bohr, 1948): "The viewpoint of complementarity presents itself as a rational generalization of the very ideal of causality." As a result, Bohr stopped short of offering logically definable causality for quantum mechanics because dualism without bipolarity doesn't account for the fundamental property of nature. Inevitably, his Copenhagen interpretation of quantum mechanics has to be subjected to Heisenberg uncertainty principle (Hilgevoord & Uffink, 2006; Faye 2008).

David Bohm's Causal Interpretation of Quantum Theory

Following Einstein's lead, his onetime associate David Bohm continued the quest for the grand unification of general relativity and quantum theory. In his book *Wholeness and the Implicate Order*, Bohm proposed a cosmological order radically different from generally accepted conventions, which he expressed as a distinction between *implicate and explicate order*. Bohm wrote:

In the enfolded (or implicate) order, space and time are no longer the dominant factors determining the relationships of dependence or independence of different elements. Rather, an entirely different sort of basic connection of elements is possible, from which our ordinary notions of space and time, along with those of separately existent material particles, are abstracted as forms derived from the deeper order. These ordinary notions in fact appear in what is called the "explicate" or "unfolded" order, which is a special and distinguished form contained within the general totality of all the implicate orders. (Bohm, 1980, p. xv)

A major motivation of Bohm's cosmological order was to resolve the incompatibility between quantum theory and relativity theory. He observed:

...in relativity, movement is continuous, causally determinate and well defined, while in quantum mechanics it is discontinuous, not causally determinate and not well-defined. Each theory is committed to its own notions of essentially static and fragmentary modes of existence (relativity to that of separate events connectible by signals, and quantum mechanics to a well-defined quantum state). One thus sees that a new kind of theory is needed which drops these basic commitments and at most recovers some essential features of the older theories as abstract forms derived from a deeper reality in which what prevails is unbroken wholeness. (Bohm, 1980, p. xv)

Clearly Bohm believed that the new concept of cosmological order can be derived from a deeper reality in which undivided wholeness prevails. Central to the cosmological order is a hidden variable theory of quantum physics (Bohm, 1957). Bohm's hidden variable theory forms a causal interpretation of quantum mechanics that contains a wave function – a function on the space of all possible configurations. The theory is explicitly non-local and deterministic. The velocity of any one particle depends on the value of the wave function, which depends on the whole configuration of the universe. Relativistic variants require a preferred frame. Variants which handle spin and curved spaces are known. It can be modified to handle quantum field theory. It is reported that Bell's theorem (Bell, 1964) was inspired by Bell's discovery of the early work of David Bohm (Bohm, 1957) and his own subsequent wondering if the obvious non-locality of the theory could be removed.

It is very interesting to notice that, Bohm's implicate and explicate cosmological order forms a pair denoted *(im,ex)* and *(enfolding, unfolding)*. If the pairs can be defined as a YinYang bipolar variable or predicate, Bohm's interpretation would be essentially a YinYang bipolar approach to quantum gravity. A closer examination reveals that the first pair can be defined as a harmonic YinYang pair but it is not exactly bipolar in nature. The second pair, however, is exactly a YinYang bipolar pair like input-output, sorrow-joy, action-reaction, negative-positive electromagnetic charges, and matter-antimatter particles. Evidently, YinYang are generally interpreted as the two sides of one thing but the two sides have to be bipolar interactive opposites for bipolar YinYang variables.

With the *(enfolding, unfolding)* pair, Bohm's approach to cosmological order is essentially a YinYang bipolar approach. Bohm's *implicate and explicate* cosmological order is an important theory and his interpretation of quantum mechanics is the first causal interpretation of its kind. It can be observed, however, Bohm's causal interpretation and Bell's theorem stopped short of providing a new set-theoretic mathematical abstraction and a different logical foundation for physics as envisioned by Einstein. Since a new mathematical abstraction is imperative for a new logical foundation such that the ubiquitous effects of quantum entanglement would confront us at every turn in both physical and social sciences, many influential researchers believe that some fundamental concept is still missing from the big picture. Without the missing fundamental, the quest for quantum gravity has so far failed to find a definitive battleground (Smolin, 2006; Woit, 2006; Guizzo, 2010).

Causal Set Theory

Causal set theory is a set theoretic approach to quantum gravity. The causal set research program was initiated by Rafael Sorkin (Bombelli *et al.*, 1987) who continues to be the main proponent of the program. Its founding principle is that *spacetime* is fundamentally discrete and that the spacetime events are related by a *partial order*. This partial order has the physical meaning of the *causality relations* between spacetime events.

The causal set program is based on a theorem by *DavidMalament* (1977) which states that if there is a *bijective* map between two past and future distinguishing spacetimes which preserves their causal structure then the map is a conformal isomorphism. The conformal factor that is left undetermined is related to the volume of regions in the spacetime. This volume factor can be recovered by specifying a volume element for each spacetime point. The volume of a spacetime region could then be found by counting the number of points in that region.

Among different approaches to definable causality, causal set theory specifically targets physical causality and quantum gravity. In recent years, Bombelli (in Bombelli, Henson & Sorkin, 2009; Bombelli, Corichi & Winkler, 2009) and Dowker (2006) presented new results that further examined causal sets related to the deep structure of spacetime.

Although the causal set hypothesis is distinguished by its logical simplicity, as a set-theoretic model it didn't go beyond classical truth-based set theory and didn't lead to a formal logical system beyond spacetime geometry.

Cognitive-Map-Based Causal Reasoning

Cognitive-map-based causal reasoning relies on conceptual graphs called cognitive maps (CMs). In this context, a CM is defined as a representation of relations that are perceived to exist among the attributes and/or concepts of a given environment (Axelrod, 1976). Since relations in a CM can be neutral, negative and/or positive, CM is a more general representation than classical binary relation. CM-based causal reasoning has been applied by researchers in the fields of international relations (e.g. Bonham, Shapiro & Trumble, 1979), decision and coordination (e.g. Zhang et al., 1989,, 1992; Zhang, Wang & King, 1994; Zhang, 1996, 2003a, 2003b, 2006a), operational research (e.g. Klein & Cooper, 1982; Montibeller *et al.,* 2008; Montibeller & Belton, 2009), management science (Kwahk & Kim, 1999; Lee & Kwon, 2008), neural networks (e.g. Kosko, 1986), and knowledge representation (e.g. Wellman, 1994; Noh *et al.,* 2000; Chaib-draa, 2002).

CM-based causal reasoning has been application-oriented where the word "causal" has been used loosely in most of the cases and logical definability of cause-effect relation has not been a focus until recent years. Notably, a formal bipolar logic with a bipolar axiomatization is presented in (Zhang, 2003a, 2003b. 2005a2005a; Zhang & Zhang, 2004). The formal logical approach leads to equilibrium-based logically definable YinYang bipolar causality, bipolar relativity, and bipolar quantum entanglement (Zhang, 2009a, 2009b, 2009c, 2009d, 2010).

Equilibrium-Based YinYang Bipolar Causality

The discovery of the ubiquitous genetic regulator protein YinYang1 (YY1) in 1991 at Harvard Medical School (Shi *et al.,* 1991) brought the ancient Chinese YinYang into modern genomics – a core area of bioinformatics and life sciences. From that point on, YinYang has reemerged as a unifying philosophical foundation for holistic thinking and global regulation in both Traditional Chinese Medicine (TCM) and modern science in microscopic as well as in macroscopic terms. In less than two decades, hundreds of important works have been reported on YY1 by the top research institutions in the US and the world.

According to YinYang, everything consists of two sides or two poles. The coexistence of the two sides in equilibrium and/or harmony is considered a key for the mental and physical health of any dynamic system. This principle has played an essential role in TCM where symptoms are often diagnosed as the loss of balance and/or harmony of the two sides.

In genomics, the Yin of YY1 is its repression ability and the Yang is its activation ability in gene expression regulation. The two sides are hypothesized to form certain equilibrium, quasi-equilibrium, or non-equilibrium (Vasudevan, Tong & Steitz, 2007). It is, however, not clear so far how to mathematically characterize such bipolar equilibrium, quasi-equilibrium, or non-equilibrium for global regulation in epigenomics because classical mathematical abstraction does not support physical and biological bipolarity.

Since equilibrium (including quasi- or non-equilibrium through this book) is natural reality, it can be considered a form of polarized holistic truth that is essential to both physical and social sciences including thermodynamics, socioeconomics, information sciences, nanoscience, bioinformatics, optimization, decision, and coordination. Indeed, YinYang has entered every aspect of the Eastern as well as the Western societies; due to its lack of a formal mathematical basis, however, YinYang theory has remained largely mysterious.

Modern scientific research efforts have been made in recent decades to unravel the logical, statistical, and mathematical mystery underlying YinYang. Lei Xu (2007) introduced YinYang into statistics and successfully developed his harmony learning machinery with significant applications. Xu characterized his YinYang theory as *Bayesian YinYang* (BYY) which is *harmony-based.* Kandel and Y. Zhang (1998) introduced YinYang analysis into fuzzy logic and they characterized their approach as *Boolean YinYang* which is *truth-based.* Boolean YinYang follows Leibniz binary interpretation of YinYang.

While Bayesian YinYang and Boolean YinYang are largely consistent with classical bivalent cognition, bipolar YinYang (Zhang *et al.,* 1989-2009; Zhang, 1998-2010) started at a more fundamental level. Despite the great success of classical bivalent cognition in modern scientific and technological advances, it is argued that the bivalent cognition is static, unipolar, and bottom up in nature that lacks the direct representation for bipolar equilibria with a top-down holistic visualization. Simply, it is pointed out that true-false is inadequate and (-, +) bipolarity is inevitable for representing bipolar equilibrium and non-equilibrium as bipolar holistic truth (Zhang, 2005a).

To circumvent the representational limitations of unipolar cognition, YinYang bipolar dynamic logic (BDL), bipolar causality, and bipolar relativity are necessary (Zhang, 2009a, 2009b, 2009c, 2009d2009d). The bipolar approach is characterized as *bipolar YinYang* for its bipolar equilibrium-based nature (Zhang, 1996, 1998, 2003a, 2003b, 2005a, 2005b, 2006a, 2006b, 2007, 2009a, 2009b, 2009c, 2009d, 2010; Zhang & Chen, 2009; Zhang *et al.,* 2009; Zhang & Zhang, 2003, 2004).

Bipolar YinYang is fundamentally different from classical truth-based unipolar cognition in a cosmological perspective. While unipolar cognition views the universe as a set of truth objects that can be mapped to the bivalent {false, true} or {0,1} lattice where the two sides are not physical and not interactive, bipolar cognition views the universe as a set of coexisting bipolar dynamic equilibria (including full- quasi- or non-equilibria) with a focus on the fusion, interaction, oscillation, and quantum entanglement of the negative and positive poles or the two energies.

The bipolar view is based on the observation that bipolar equilibrium is natural reality that exists in microcosms (e.g. quantum world) as well as in macrocosms (e.g. the universe), in social worlds (e.g. competition and cooperation) as well as in physical worlds (e.g. centripetal and centrifugal forces), and no one knows whether the universe of unipolar sets belongs to equilibrium or, on the contrary, equilibrium of bipolar sets belongs to the universe. This is evidently a philosophical "chicken and egg" paradox because no one knows exactly which one created the other in the very beginning. Since equilibrium is central in thermodynamics – the ultimate physical source of energy, life, existence, and information, there

Figure 1. A YinYang classification of different mathematics models (Adapted from Zhang et al., 2008)

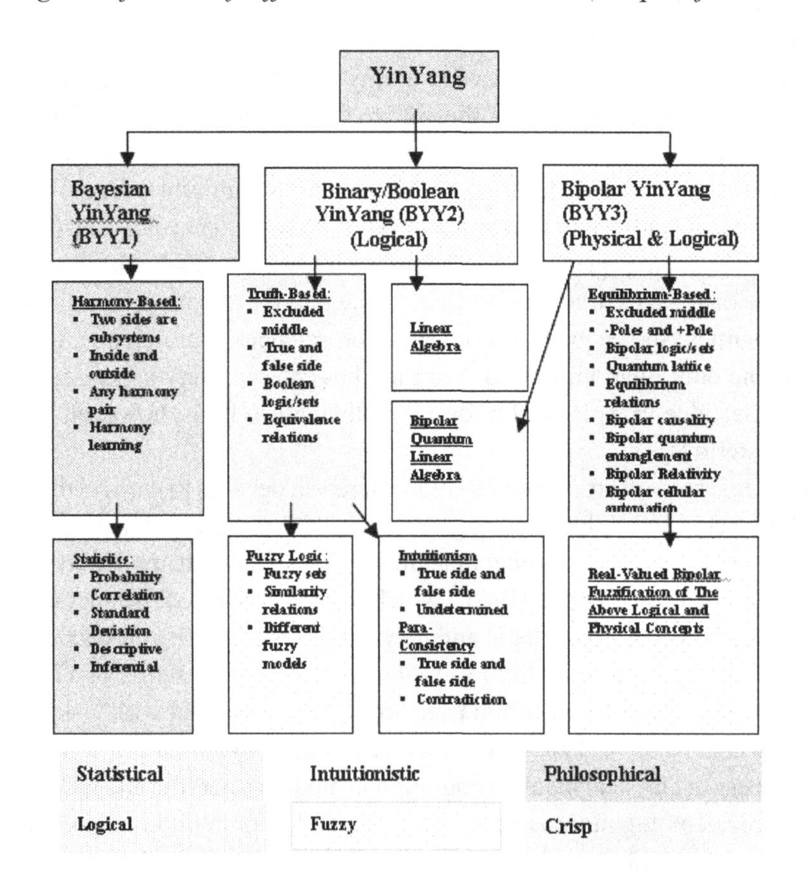

would be no universe without equilibrium. Therefore, equilibrium-based bipolar YinYang is a scientific extension or bipolar generalization of its unipolar counterpart.

The three different scientific approaches Bayesian YinYang (BYY1), Binary or Boolean YinYang (BYY2), and bipolar YinYang (BYY3) can all trace their origin back to different translations of ancient YinYang from Chinese to English. In English the word "positive" and "negative" are used for the logical values "true" and "false" or "sure" and "unsure" as well as for the parity signs "+" and "–". In Chinese the semantics of "positive" and "negative" are limited to the parity signs "+" and "–" where "true" and "false" are distinguished from "positive" and "negative" with different Chinese words. When YinYang is translated into English, it means the two opposing sides of one matter. While Boolean YinYang takes the "true side" and "false side" semantics and Bayesian YinYang takes the "inside" and "outside" semantics, bipolar YinYang takes the "positive pole" and "negative pole" semantics that leads to the polarized truth space or bipolar lattice $\{-1,0\} \times \{0,1\} = \{(0,0)(0,1)(-1,0)(-1,1)\}$ and the bipolar fuzzy lattice $[-1,0] \times [0,+1]$ (Zhang 2005a).

With the three different mathematical YinYang models, we have the YinYang classification as depicted in Figure 1. The essential features and relationships of truth, polarity, intuitionism, and fuzziness among different logical models are roughly summarized with the concise set of formulae as shown in Figure 2.

It is interesting to notice that, based on the classification in Figure 1, Niels Bohr's YinYang complementarity principle can be classified as an example of Bayesian YinYang which focuses on the har-

Figure 2. Truth, polarity, and fuzziness: a concise comparison (Adapted from Zhang et al., 2008)

<table>
<tr><td>Boolean Logic
Unipolar and Originated from Bivalent Cognition
With the Laws of Excluded Middle (LEM) and Non-Contradiction
Generic Inference Rule is Modus Ponens (MP)
Crisp and Static with the Closed-World Assumption</td></tr>
<tr><td>Intuitionistic Logic = Classical Logic - LEM
Originated from Vague Philosophy
Unipolar, Crisp or Fuzzy</td></tr>
<tr><td>Zadeh's Fuzzy Logic = Boolean Logic + Fuzziness – LEM
Originated from Vague Philosophy
Unipolar (Can not satisfy LEM: –0.5 = 1-0.5 = 0.5)</td></tr>
<tr><td>YinYang BDL = Boolean Logic + Bipolar Equilibrium
Generic Inference Rule is BUMP
Originated from Yijing or I Ching – The Book of Change
Equilibrium-Based, Dynamic, Quantum in Nature</td></tr>
<tr><td>YinYang BDFL = Fuzzy Logic + Bipolar Equilibrium
= YinYang Bipolar Crisp Logic + Fuzziness - LEM
= Boolean Logic + Fuzziness + Bipolar Equilibrium - LEM</td></tr>
</table>

monic complementary nature of particle-wave duality. Since particle and wave are not bipolar opposites, they are complementary but not bipolar complementary. Since YinYang (-,+) bipolar equilibrium or non-equilibrium was not recognized and formally defined in logical terms, Bohr's principle didn't go beyond Aristotle's truth-based logical reasoning.

This book follows the direction of bipolar YinYang. Bipolar YinYang can be said *physically logical* because certain physical aspects are illogical from a truth-based perspective but perfectly logical from an equilibrium-based physical perspective such as bipolar quantum entanglement (Zhang, 2010). The physically logical property enables bipolar variables to carry equilibrium-based physical semantics and syntax directly. Such direct semantics and syntax play an essential role in resolving the LAFIP paradox and lead to equilibrium-based logically definable causality (Zhang, 2009a, 2009b, 2009c, 2009d2009d, 2010).

A key element of BDL for logically definable causality is bipolar universal modus ponens (BUMP). It is well-known that, classical modus ponens (MP) has been the only generic inference rule for truth-based logical deduction for thousands of years. BUMP generalized MP from the bivalent lattice {0,1} to the bipolar lattice $B_1 = \{-1,0\}\times\{0,1\}$. While, without bipolarity, the truth-values 0 and 1 in the bivalent lattice are incapable of carrying any shred of direct physical semantics and syntax, the equilibrium-based bipolar truth values (0,0), (0,1), (-1,0), and (-1,1) provide direct fundamental physical syntax and semantics, which stand, respectively, for bipolar false or eternal equilibrium, negative pole false and positive pole true or non-equilibrium (e.g. mania), negative pole true and positive pole false or another non-equilibrium (e.g. depression), and bipolar true or equilibrium.

The bipolar logical syntax and equilibrium-based physical semantics of BDL provide a unifying representation for both classical truth-based logical values and certain illogical physical properties such as quantum entanglement (Zhang, 2009a, 2009b, 2009c, 2009d, 2010). From a bivalent static truth-based logical point of view, causality is indeed undefinable as concluded by Zadeh (2001). For instance, following classical modus ponens (MP) we have IF A→B and A, THEN B, where cause and effect have no position because the consequent B could be true regardless of the premise A. With BDL and BUMP, however, we have IF A⇒C, B⇒D, and A*B, THEN C*D, which stands for *"For all bipolar equilibrium variables A, B, C, and D, if A implies C, B implies D, and A is bipolar interactive with B, we must have the same bipolar interaction between C and D."*

Thus, from an equilibrium or non-equilibrium-based point of view, BUMP provides for the first time logically definable bipolar dynamic causality. Furthermore, it is shown in later chapters that bipolar causality, in turn, leads to a completely background independent theory of YinYang bipolar relativity where space and time are emerging factors following the arrivals of agents. Thus, BDL and BUMP are meaningful even without space and time. Surprisingly, despite its non-linear dynamic quantum entanglement features, BDL does not compromise the basic law of excluded middle (LEM) and possesses four pairs of dynamic De Morgan's laws.

Thus, bipolar relativity provides a unique logical foundation for the unification of general relativity and quantum mechanics. It is shown in Chapter 6 that bipolar relativity leads to a minimal but most general equilibrium-based axiomatization for physics. Consequently, as the first relativity theory supported with logically definable causality, YinYang bipolar relativity presents itself as the first contender for the missing logical foundation as envisioned by Einstein in his quote *"For the time being we have to admit that we do not possess any general theoretical basis for physics which can be regarded as its logical foundation."* (Einstein, 1940).

SCIENCE VS. SOCIAL CONSTRUCTION-DESTRUCTION

Constructionists claim that reality is socially constructed. *Constructionism* is founded by Peter Berger and Thomas Luckmann with their book entitled *The Social Construction of Reality* (Berger & Luckmann, 1967) and became prominent in the U.S. in the last quarter of the 20th century. Berger and Luckmann argue that all knowledge, including the most basic, taken-for-granted common sense knowledge of everyday reality, is derived from and maintained by social interactions. When people interact, they do so with the understanding that their respective perceptions of reality are related, and as they act upon this understanding their common knowledge of reality becomes reinforced. Since this common sense knowledge is negotiated by people, human typifications, significations and institutions come to be presented as part of an objective reality. It is in this sense that it can be said that reality is socially constructed. (NationMaster – Encyclopedia)

Following Berger & Luckmann, social constructionism established itself in sociology in 1970s-1980s and advanced to the front of science and technology studies. It is argued that, despite its common perception as objective, science and mathematics are not immune to social constructivist accounts. The "provocative" arguments "incited" heated debates between the scientific community and the social constructionism community.

It is, of course, very difficult for scientists to swallow that their "beloved" science is no more than a social construction. The objectivity of science, however, is undoubtedly influenced by social construction and/or social destruction denoted as a YinYang pair (destruction, construction). Such influences are evidenced with numerous historical and contemporary events. For instances,

1. Aristotle's cosmology was followed for about two thousand years, which described the Earth as the stationary center of the universe. Even a half century after Copernicus discovery of the solar system, Aristotle's cosmology was borrowed for the brutal fire execution of Bruno and jail sentence to Galileo for their "crimes" of believing in the solar system.
2. While Lorentz ether theory became influential in the West scientific community at the end of the 19th and the beginning of the 20th century, a YinYang bipolar ether theory proposed by a prominent

Chinese physicist Sitong Tan didn't get a chance to be fully developed before Tan was beheaded at the age of 30 by Empress Cixi of the late Qing Dynasty in 1898 for his "crime" of being one of the six famous reformers who have been referred to as "The Six Gentlemen" in China. (Note: Tan refused to follow the other five reformers to flee for his life but decided to stay and sacrifice his life for his cause to "wake up" the Chinese people.) (Wright, 1994, p. 57).

3. After string theory dominated theoretical physics for nearly three decades in terms of funding, faculty hiring, and number of academic publications, it was sharply criticized in recent years by two popular books entitled *The Trouble with Physics: The Rise of String Theory, the Fall of a Science, and what Comes Next?* (Smolin, 2006) and *Not Even Wrong: The Failure of String Theory and the Search for Unity in Physical Law.* (Woit, 2006).
4. Nowadays no parents would be willing to ask their children to learn math without negative numbers, but negative numbers were once considered by many as absurd. European mathematicians resisted them until the 17th – 18th centuries due to their so-called "absurdity" (Martinez, 2006).
5. For thousands of years, modus ponens (MP) has been the only generic inference rule in logical deduction; YinYang bipolar universal modus ponens (BUMP) (Zhang, 2005a, 2007) presents an equilibrium-based generalization of MP. But BUMP has been said "unscientific" by quite a few critics because it is not truth-based. The irony is that there would be no universe and no unipolar truth without YinYang bipolar equilibrium.
6. Bipolar disorder is clearly the loss of mental equilibrium but in clinical psychiatry bipolar equilibrium is deemed not observable and unscientific.
7. It is common knowledge in computer/cognitive science that without semantics there would be no need for syntax and without syntactic representation there would be no semantic processing. Thus, without (-,+) bipolar syntax and semantics it would be virtually impossible to reason on (-,+) bipolarity. In logic and mathematics, however, it is often said that "-1 is isomorphic to +1, the 2nd quadrant is isomorphic to the 1st, and using (-,+) bipolarity as part of a new logic is unscientific." Ironically, no physicist would say "electron is isomorphic to positron" or "black hole is isomorphic to big bang"; no physician or economist would say "depression is isomorphic to mania"; no logician and mathematician would be willing to ask their children to learn math without the negative sign; no philosopher would say "Yin is isomorphic to Yang"; no electrical engineer would use "+,+" to label the two poles of a battery. Someone has to wonder whether the above so-called "isomorphism" is *"a scientific principle"* that enhances the strictness of science or *"a kind of socially constructed entrenched noble hypocrisy"* that hinders the development of a new logical foundation for scientific computation and discovery.
8. Science is great and greatness always lies in the past but most importantly in the future. Scientists, therefore, should keep an open mind. It is amazing to observe, however, as a "foot soldier" or an academic researcher of the last century, someone might have fought courageously against academic bashing and won triumphantly through public debate. Once the soldier became a "general" or an authority of an established academic field of this century, the same scientist could raise an academic "bashing stick" – the one he/she seized from his/her last century opponent. Even worse, only this time, he/she might close the door for any public debate. This kind of phenomenon seems to be common in academic research generation after generation. It is actually a kind of "social quantum gravity." Once a field is established, some authorities tend to define the boundaries of their territory and close the doors to maintain their established position. They dislike new ideas especially when their academic authorities are challenged.

9. If space is expanding as observed, the cause for the expansion has to be more fundamental; if quantum entanglement is at least 10,000 times faster than the speed of light (Salart et al., 2008), the concept of time might have to be reexamined. Space and time do not provide logically definable causality and are clearly less fundamental than nature's basic forces and matter-antimatter particles, but science has so far failed to go beyond spacetime geometry because space and time are to science as air and water were to humans in the ancient time. They are essential for life and social construction/destruction but not necessarily the most fundamental property of the universe. For instance, human civilization including science itself has been threatened by global warming – a byproduct of social construction/destruction.
10. Professor Smolin was shocked to learn that a Nobel Prize laureate was under pressure from his department and funding agency to keep doing normal science rather than spending time on his new ideas about space, time, and gravity (Smolin, 2006, Chapter 18). Smolin wrote: *"Having done Nobel Prize-level physics – or even having won the prize itself – apparently doesn't protect you when you question universally held assumptions such as the special and general theories of relativity."* The irony is that, a quarter century after conquering his most celebrated general theory of relativity, Einstein himself unequivocally stated: *"For the time being we have to admit that we do not possess any general theoretical basis for physics which can be regarded as its logical foundation."* (Einstein, 1940). By his statement, not only did Einstein made it abundantly clear that his general theory of relativity could not serve as the unifying logical foundation for physics, but also hinted that a new mathematical abstraction beyond classical set theory and truth-based bivalent tradition might be necessary for the next scientific unification. As Smolin wrote

"... what we need is a return to a revolutionary kind of science. Once again we need a few seers. The problem is that there are now very few around, as a result of science having been done so long in a way that rarely recognize and barely tolerate them... Even if everyone can see that a revolution is necessary, the most powerful parts of our community have forgotten how to make one." (Smolin, 2006, p. 312)

Consequently, Smolin wrote *"The paradoxical situation of string theory – so much promise, so little fulfillment – is exactly you get when a lot of highly trained master craftspeople try to do the work of seers."* Smolin went further to ask his so-called cliché question (Smolin, 2006, p. 328) *"whether a young Einstein would now be hired by a university"* and his answer is *"obviously no."*

SUMMARY

Following a discussion on the faultline of observability and truth-based reasoning, an introduction to the concepts of singularity and YinYang bipolarity has been presented. Philosophical, mathematical, empirical, and set theoretic approaches to causality have been introduced including Aristotle's causality principle, David Hume's challenge, Lotfi Zadeh's conclusion, Judea Pearl's empirical definability, Niels Bohr's particle-wave complementarity principle, Bohm's causal interpretation of quantum mechanics, Sorkin's causal set program for quantum gravity, bipolar cognitive mapping, and equilibrium-based bipolar causality.

It has been asserted that, in order to continue its role as the doctrine of science, the logical definability problem of Aristotle's causality principle has become an ultimate dilemma of science. To overcome the dilemma, the LAFIP paradox has to be resolved and a formal system has to be developed with a different mathematical abstraction. The formal system is expected to be logical, physical, relativistic, and quantum in nature. Furthermore, it has to be applicable in microscopic agent world as well as in macroscopic agent world, in physical world as well as in social world, in cognitive informatics as well as in life sciences.

Social construction and destruction in science has been briefly discussed. It is concluded that just as it is hard for a scientist to swallow that his/her "beloved" science is no more than a social construction, it is also hard to deny the fact that the objectivity of science is frequently compromised by social construction and destruction – a phenomenon of bipolar relativity. In particular, it has been pointed out that space and time do not provide logically definable causality and are clearly less fundamental than nature's basic forces and matter-antimatter particles, but science has so far failed to go beyond spacetime geometry because space and time are to modern science as air and water were to the ancestors of mankind in the ancient time.

With the first two chapters we have laid out a background for the remaining presentations and discussions of this monograph. It is shown in later chapters that bipolar dynamic logic (BDL) can resolve the LAFIP paradox and lead to bipolar equilibrium-based logically definable causality and bipolar relativity for quantum computing, cognitive informatics, and life sciences.

REFERENCES

Axelrod, R. (1976). *Structure of decision*. Princeton, NJ: Princeton University Press.

Bell, J. S. (1964). On the Einstein Podolsky Rosen paradox. *Physics*, *1*(3), 195–200.

Belnap, N. (1977). A useful 4-valued logic. In Epstein, G., & Dunn, J. M. (Eds.), *Modern uses of multiplemultiple-valued logic* (pp. 8–37).

Berger, P. L., & Luckmann, T. (1967). *The social construction of Reality: A treatise in the sociology of knowledge*. Anchor.

Birkhoff, G., & von Neumann, J. (1936). The logic of quantum mechanics. *The Annals of Mathematics*, *37*(4), 823–843. doi:10.2307/1968621

Blackburn, S. (1990). Hume and thick connexions. *PhilosophyPhilosophy and Phenomenological Research*, *50*, 237–250. doi:10.2307/2108041

Bohm, D. (1957). *Causality and chance in modern physics*. Philadelphia: University of Pennsylvania Press. doi:10.4324/9780203201107

Bohm, D. (1980). *Wholeness and the implicate order*. London: Routledge.

Bohr, N. (1948). On the notions of causality and complementarity. *Dialectica*, *2*(3-4), 312–319. doi:10.1111/j.1746-8361.1948.tb00703.x

Bombelli, L., Corichi, A., & Winkler, O. (2009). Semiclassical quantum gravity: Obtaining manifolds from graphs. *Classical and Quantum Gravity*, *26*(24). doi:10.1088/0264-9381/26/24/245012

Bombelli, L., Henson, J., & Sorkin, R. D. (2009). Discreteness without symmetry breaking: A theorem. *Modern Physics Letters A*, *24*, 2579–2587. doi:10.1142/S0217732309031958

Bombelli, L., Lee, J., Meyer, D., & Sorkin, R. D. (1987). Spacetime as a causal set. *Physical Review Letters, 59*, 521–524. doi:10.1103/PhysRevLett.59.521

Bonham, G. M., Shapiro, M. J., & Trumble, T. L. (1979). The October war. *International Studies, 23*, 3–14. doi:10.2307/2600273

Bourbaki, N. (1998). *Elements of the history of mathematics*. Berlin, Heidelberg, New York: Springer-Verlag.

Brouwer, L. E. (1912). Intuitionism and formalism. *Bulletin of the American Mathematical Society, 20*, 81–96. doi:10.1090/S0002-9904-1913-02440-6

Chaib-draa, B. (2002). Causal maps: Theory, implementation, and practical applications in multiagent environments. *IEEE Transactions on KDE, 14*(6), 1201–1217.

Dowker, F. (2006). Causal sets as discrete spacetime. *Contemporary Physics, 47*(1), 1–9. doi:10.1080/17445760500356833

Ebrey, P. (1993). *Chinese civilization: A sourcebook* (2nd ed.). New York: Free Press.

Einstein, A. (1916). The foundation of the general theory of relativity. Originally published in *Annalen der Physik* (1916). (pp. 146-200).).

Einstein, A. (1934). *On the method of theoretical physics. The Herbert Spencer lecture, delivered at Oxford, June 10, 1933. Published in Mein Weltbild.* Amsterdam: Querido Verlag.

Einstein, A., Podolsky, B., & Rosen, N. (1935). Can quantum-mechanical description of physical reality be considered complete? *Physical Review, 47*(10), 777–780. doi:10.1103/PhysRev.47.777

Falcon, A. (2008). *Aristotle on causality*. Stanford Encyclopedia of Philosophy.

Faye, J. (2008). *Copenhagen interpretation of quantum mechanics*. Stanford Encyclopedia of Philosophy.

Guizzo, E. (2010). *IEEE Spectrum's special report: Winners & loserslosers VII.*

Hawking, S., & Penrose, R. (1970). The singularities of gravitational collapse and cosmology. *Proceedings of the Royal Society of London. Series A, 314*(1519), 529–548. doi:10.1098/rspa.1970.0021

Henrickson, L., & McKelvey, B. (2002). Foundations of new social science: Institutional legitimacy from philosophy, complexity science, postmodernism, and agent-based modeling. *Proceedings of the National Academy of Sciences of the United States of America, 99*(3), 7288–7295. doi:10.1073/pnas.092079799

Hilgevoord, J., & Uffink, J. (2006). *The uncertainty principle*. Stanford Encyclopedia of Philosophy.

Kandel, A., & Zhang, Y. (1998). Intrinsic mechanisms and application principles of general fuzzy logic through Yin-Yang analysis. *Information Science, 106*(1), 87–104. doi:10.1016/S0020-0255(97)10007-X

Kennedy, J. (2007). *Kurt Gödel*. Stanford Encyclopedia of Philosophy. Retrieved from http://plato.stanford.edu/entries/goedel/#GodRea

Klein, J. H., & Cooper, D. F. (1982). Cognitive maps of decision-makers in a complex game. *The Journal of the Operational Research Society, 33*(1), 63–71.

Kosko, B. (1986). Fuzzy cognitive maps. *International Journal of Man-Machine Studies, 24*(1), 65–75. doi:10.1016/S0020-7373(86)80040-2

Kwahk, K.-Y., & Kim, Y.-G. (1999). Supporting business process redesign using cognitive maps. *Decision Support Systems, 25*(2), 155–178. doi:10.1016/S0167-9236(99)00003-2

Lee, K. C., & Kwon, S. (2008). A cognitive map-driven avatar design recommendation DSS and its empirical validity. *Decision Support Systems, 45*(3), 461–472. doi:10.1016/j.dss.2007.06.008

Lloyd, G. E. R. (1968). *Aristotle: The growth and structure of his thought* (pp. 133–139). Cambridge University Press. doi:10.1017/CBO9780511552595.008

Łukasiewicz, J. (1920). O logice trojwartosciowej. *Ruch Filozoficny, 5*, 170–171.

Łukasiewicz, J. (1970). *Selected works. Amsterdam: North-Holland Publishing Co*. Warsaw: PWN.

Malament, D. (1977). The class of continuous timelike curves determines the topology of spacetime. *Journal of Mathematical Physics, 18*(7), 1399–1404. doi:10.1063/1.523436

Martinez, A. A. (2006). *Negative math: How mathematical rules can be positively bent*. Princeton University Press.

Montibeller, G., & Belton, V. (2009). Qualitative operators for reasoning maps: Evaluating multi-criteria options with networks of reasons. *European Journal of Operational Research, 195*, 829–840. doi:10.1016/j.ejor.2007.11.015

Montibeller, G., Belton, V., Ackermann, F., & Ensslin, L. (2008). Reasoning maps for decision aid: An integrated approach for problem-structuring and multi-criteria evaluation. *The Journal of the Operational Research Society, 59*(5), 575–589. doi:10.1057/palgrave.jors.2602347

Morris, W. E. (2009). *David Hume*. Stanford Encyclopedia of Philosophy. Retrieved from http://plato.stanford.edu/ entries/hume/

NationMaster Encyclopedia. (2003). *Social constructionism*. Retrieved from http://www.statemaster.com/encyclopedia/Social-constructionism

Norton, D. F., & Norton, M. J. (2000). *A treatise of human nature*. Oxford: Oxford University Press.

O'Keefe, J., & Nadal, L. (1979). *The hippocampus as a cognitive map*. Oxford: Claredon Press.

Pearl, J. (1988). *Probabilistic reasoning in intelligent systems: Networks of plausible inference*. San Mateo, CA: Morgan Kaufmann Publishers.

Pearl, J. (2000). *Causality: Models, reasoning, and inference*. Cambridge: Cambridge University Press.

Pearl, J. (2009). *More on pearls*. Retrieved from http://www.stat.columbia.edu/~cook/movabletype/archives/2009/07/ more_on_pearls.html

Salart, D., Baas, A., Branciard, C., Gisin, N., & Zbinden, H. (2008). Testing the speed of spooky action at a distance. *Nature, 454*, 861–864. doi:10.1038/nature07121

Shi, Y., Seto, E., Chang, L.-S., & Shenk, T. (1991). Transcriptional repression by YY1, a human GLI-Kruppel-related protein, and relief of repression by adenovirus E1A protein. *Cell, 67*(2), 377–388. doi:10.1016/0092-8674(91)90189-6

Smolin, L. (2006). *The trouble with physics: The rise of string theory, the fall of a science, and what comes next?*New York: Houghton Mifflin Harcourt.

Temple, R. (1986). *The genius of China: 3,000 Years of science, discovery, and invention*. New York: Simon and Schuster.

The Royal Swedish Academy of Sciences. (2008). The Nobel Prize in Physics, 2008. *Press Release*, 7 October. Retrieved from http://nobelprize.org/nobel_prizes/physics/laureates/2008/press.html

Theosophy. (1939). Ancient landmarks: Plato and Aristotle. *Theosophy,, 11*(55), 483-491.

Vasudevan, S., Tong, Y., & Steitz, J. A. (2007). Switching from repression to activation: MicroRNAs can up-regulate translation. *Science, 318*(5858), 1931–1934. doi:10.1126/science.1149460

Wellman, M. P. (1994). Inference in cognitive maps. *Mathematics and Computers in Simulation, 36*, 137–148. doi:10.1016/0378-4754(94)90028-0

Woit, P. (2006). *Not even wrong: The failure of string theory and the search for unity in physical law*. New York: Basic Books.

Wright, D. (1994). Tan Sitong and the ether reconsidered. *Bulletin of the School of Oriental and African Studies. University of London. School of Oriental and African Studies, 57*(3), 551–575. doi:10.1017/S0041977X00008909

Xu, L. (2007). Bayesian Ying Yang learning. *Scholarpedia, 2*(3), 1809. doi:10.4249/scholarpedia.1809

Zadeh, L. A. (1965). Fuzzy sets. *Information and Control, 8*, 338–353. doi:10.1016/S0019-9958(65)90241-X

Zadeh, L. A. (2001). Causality is undefinable–toward a theory of hierarchical definability. *Proceedings of FUZZ-IEEE*, 67-68.

Zhang, W.-.-R. (2003b). Equilibrium energy and stability measures for bipolar decision and global regulation. *International Journal of Fuzzy Systems, 5*(2), 114–122.

Zhang, W.-R. (1996). NPN fuzzy sets and NPN qualitative algebra: A computational framework for bipolar cognitive modeling and multi-agent decision analysis. *IEEE Transactions on SMC, 16*, 561–574.

Zhang, W.-R. (1998). YinYang bipolar fuzzy sets. *Proceedings of IEEE World Congress on Computational Intelligence – Fuzz-IEEE*, (pp. 835-840). Anchorage, AK.

Zhang, W.-R. (2005a). YinYang bipolar lattices and l-sets for bipolar knowledge fusion, visualization, and decision making. *International Journal of Information Technology and Decision Making, 4*(4), 621–645. doi:10.1142/S0219622005001763

Zhang, W.-R. (2005b). YinYang bipolar cognition and bipolar cognitive mapping. *International Journal of Computational Cognition, 3*(3), 53–65.

Zhang, W.-R. (2006a). YinYang bipolar fuzzy sets and fuzzy equilibrium relations for bipolar clustering, optimization, and global regulation. *International Journal of Information Technology and Decision Making, 5*(1), 19–46. doi:10.1142/S0219622006001885

Zhang, W.-R. (2006b). YinYang Bipolar T-norms and T-conorms as granular neurological operators. *Proceedings of IEEE International Conference on Granular Computing,* (pp. 91-96). Atlanta, GA.

Zhang, W.-R. (2007). YinYang bipolar universal modus ponens (bump)–a fundamental law of non-linear brain dynamics for emotional intelligence and mental health. *Walter J. Freeman Workshop on Nonlinear Brain Dynamics, Proc. of the 10th Joint Conf. of Information Sciences,* (pp. 89-95). Salt Lake City, Utah, USA.

Zhang, W.-R. (2009a). Six conjectures in quantum physics and computational neuroscience. *Proceedings of 3rd International Conference on Quantum, Nano and Micro Technologies (ICQNM 2009),* (pp. 67-72). Cancun, Mexico.

Zhang, W.-R. (2009b). YinYang Bipolar Dynamic Logic (BDL) and equilibrium-based computational neuroscience. *Proceedings of International Joint Conference on Neural Networks (IJCNN 2009),* (pp. 3534-3541). Atlanta, GA.

Zhang, W.-R. (2009c). YinYang bipolar relativity–a unifying theory of nature, agents, and life science. *Proceedings of International Joint Conference on Bioinformatics, Systems Biology and Intelligent Computing (IJCBS).* (pp. 377-383). Shanghai, China.

Zhang, W.-R. (2010). YinYang bipolar quantum entanglement–toward a complete quantum theory. *Proceedings of the 4th Int'l Conference on Quantum, Nano and Micro Technologies (ICQNM 2010),* (pp. 77-82). February, St. Maarten, Netherlands Antilles.

Zhang, W.-R., Chen, S., & Bezdek, J. C. (1989). POOL2: A generic system for cognitive map development and decision analysis. *IEEE Transactions on SMC, 19*(1), 31–39.

Zhang, W.-R., Chen, S., Wang, W., & King, R. (1992). A cognitive map based approach to the coordination of distributed cooperative agents. *IEEE Transactions on SMC, 22*(1), 103–114.

Zhang, W.-R., Chen, S., & Zang, H. (2009). A multitier YinYang-N-element cellular architecture for the Chinese meridian system. [IJFIPM]. *International Journal of Functional Informatics and Personal Medicine, 2*(3), 292–302. doi:10.1504/IJFIPM.2009.030828

Zhang, W.-R., & Chen, S. S. (2009). Equilibrium and non-equilibrium modeling of YinYang Wuxing for diagnostic decision analysis in traditional Chinese medicine. *International Journal of Information Technology and Decision Making, 8*(3), 529–548. doi:10.1142/S0219622009003521

Zhang, W.-R., Pandurangi, A., & Peace, K. (2007). YinYang dynamic neurobiological modeling and diagnostic analysis of major depressive and bipolar disorders. *IEEE Transactions on Bio-Medical Engineering, 54*(10), 1729–1739. doi:10.1109/TBME.2007.894832

Zhang, W.-R., & Peace, K. E. (2007). YinYang MentalSquares–an equilibrium-based system for bipolar neurobiological pattern classification and analysis. *Proceedings of IEEE BIBE,* (pp. 1240-1244). Boston.

Zhang, W.-R., Wang, P., Peace, K., Zhan, J., & Zhang, Y. (2008). On truth, uncertainty, equilibriumequilibrium, and harmony–a taxonomy for YinYang scientific computing. *International Journal of New Mathematics and Natural Computing*, *4*(2), 207–229. doi:10.1142/S1793005708001033

Zhang, W.-R., Wang, W., & King, R. (1994). An agent-oriented open system shell for distributed decision process modeling. *Journal of Organizational Computing*, *4*(2), 127–154. doi:10.1080/10919399409540220

Zhang, W.-R., Zhang, H. J., Shi, Y., & Chen, S. S. (2009). Bipolar linear algebra and YinYang-N-element cellular networks for equilibrium-based biosystem simulation and regulation. *Journal of Biological System*, *17*(4), 547–576. doi:10.1142/S0218339009002958

Zhang, W.-R., & Zhang, L. (2003). Soundness and completeness of a 4-valued bipolar logic. *International Journal on Multiple-Valued Logic*, *9*, 241–256.

Zhang, W.-R., & Zhang, L. (2004). YinYang bipolar logic and bipolar fuzzy logic. *Information Sciences*, *165*(3-4), 265–287. doi:10.1016/j.ins.2003.05.010

ADDITIONAL READING

Aristotle, Posterior Analytics, Book II, Part 2.

Aristotle, Posterior Analytics, Book II, Part 11.

Aristotle, Metaphysics, Book V, Part 1.

Aristotle, Metaphysics, Book XI, Part 9.

KEY TERMS AND DEFINITIONS

Classical Element: Elements that are assumed the simplest essential parts upon which the constitution and fundamental powers of anything are based (Strathern, 2000).

Ether Theory: According to ancient and medieval science, aether, also spelled ether, is the material that fills the region of the universe above the terrestrial sphere.

String Theory: It is a developing theory in theoretical physics which attempts to reconcile quantum mechanics and general relativity into a unifying theory of quantum gravity. String theory originally posits that the most fundamental elements of the universe are 1-dimensional oscillating lines ("strings") (Gefter, 2005). Later it was developed to many dimensional theories of multiple universes up to 11-dimensions. The theory was criticized by Smolin (2006) and Woit (2006).

Platonic Realism: Platonic realism is a philosophical term usually used to refer to the idea of realism regarding the existence of universals after the Greek philosopher Plato (c. 427–c. 347 BC), a student of Socrates, and the teacher of Aristotle. As universals were considered ideal forms by Plato, this stance is confusingly also called Platonic idealism (Theosophy, 1939).

Aristotelian Science: Aristotelian science is the science founded by Greek philosopher Aristotle (384 BC – 322 BC). It is a widely accepted view that Western science is evolved from Aristotelian science.

Aristotle's causality principle has been considered the doctrine of all sciences by many. The principle, however, has not been logically definable by classical truth-based logic (Zadeh, 2001).

Aristotle-Plato Dispute: It is well-known that Aristotle refused to recognize supersensible cognition as the source of knowledge and he refuted his teacher Plato's doctrines (Platonic realism), maintaining that the Platonic method is fatal to science. "But, is there, in the universe or outside of it, an underlying reality which is eternal, immovable, unchanging as claimed by Platonic realism?" (Theosophy, 1939) The dispute has lasted for more than 2300 years without final resolution until now. For instance, Einstein once sided with Plato and bluntly claimed that *"In a certain sense, therefore I hold it true that pure thought can grasp reality, as the ancients dreamed."* (Einstein, 1934)

Causality: Causality denotes a necessary relationship between one event called *cause* and another event called *effect* where the effect is the direct consequence of the cause. However, Aristotle makes it very clear that all his predecessors merely touched upon these causes. That is to say, they did not engage in their causal investigation with a firm grasp of these four causes. They lacked a complete understanding of the range of possible causes and their systematic interrelations. Put differently, and more boldly, their *use* of causality was not supported by an adequate *theory* of causality. According to Aristotle, this explains why their investigation, even when it resulted in important insights, was not entirely successful. (Falcon 2008).

Causality is Undefinable: The title of an article published in 2001 by Lotfi A. Zadeh – founder of fuzzy logic. (Zadeh, 2001)

YinYang Bipolar Causality: Bipolar equilibrium-based logically definable causality with bipolar dynamic logic (BDL) (Zhang, 2009c, 2009d; Ch. 3).

YinYang Bipolar Equilibrium: A dynamic balance between the two reciprocal opposite poles of an agent.

Ultimate Dilemma of Science: The dilemma of logically definable causality.

Constructionism: Constructionists claim that reality is socially constructed. Social constructionism established itself in sociology in 1970s-1980s and advanced to the front of science and technology studies. It is argued that, despite its common perception as objective, science and mathematics are not immune to social constructivist accounts. The "provocative" arguments "incited" heated debates between the scientific community and the social constructionism community.

Part 2
Set Theoretic Foundation

Chapter 3
Bipolar Sets and YinYang Bipolar Dynamic Logic (BDL)

ABSTRACT

In this chapter an equilibrium-based set-theoretic approach to mathematical abstraction and axiomatization is presented for resolving the LAFIP paradox (Ch. 1) and for enabling logically definable causality (Ch. 2). Bipolar set theory is formally presented, which leads to YinYang bipolar dynamic logic (BDL). BDL in zeroth-order, 1st-order, and modal forms are presented with four pairs of dynamic De Morgan's laws and a bipolar universal modus ponens (BUMP). BUMP as a key element of BDL enables logically definable causality and quantum computing. Soundness and completeness of a bipolar axiomatization are asserted; computability is proved; computational complexity is analyzed. BDL can be considered a non-linear bipolar dynamic generalization of Boolean logic plus quantum entanglement. Despite its non-linear bipolar dynamic quantum property, it does not compromise the basic law of excluded middle. The recovery of BDL to Boolean logic is axiomatically proved through depolarization and the computability of BDL is proved. A redress on the ancient paradox of the liar is presented with a few observations on Gödel's incompleteness theorem. Based on BDL, bipolar relations, bipolar transitivity, and equilibrium relations are introduced. It is shown that a bipolar equilibrium relation can be a non-linear bipolar fusion of many equivalence relations. Thus, BDL provides a logical basis for YinYang bipolar relativity–an equilibrium-based axiomatization of social and physical sciences.

INTRODUCTION

In mathematics, a set is any collection of objects (elements) considered as a whole. Although this appears to be a simple idea, set is actually one of the most fundamental concepts in modern logical systems and mathematics that forms the foundation of information and computation.

DOI: 10.4018/978-1-60960-525-4.ch003

The principle of mathematical abstraction in classical set theory was rarely challenged throughout history. The principle claims that the concept of an element is self-evident without the need for proof of any kind and the properties of a set are independent of the nature of its elements. Classical logic is based on this principle. Thus, a classical (crisp) set X in a universe U can be defined in the form of its characterization function $\mu_X:U\rightarrow\{0,1\}$ which yields the value "1" or "true" for the elements belonging to the set X and "0" or "false" for the elements excluded from the set X. Evidently, classical set theory is based on Aristotle's universe of truth objects as defined in the unipolar bivalent space $\{0,1\}$. Since the principle has been rarely questioned in history it is commonly considered "unquestionable".

The LAFIP paradox (Zhang 2009a) (Ch. 1) presents a fundamental challenge to the principle of classical mathematical abstraction. The challenge is aimed at searching for a new mathematical abstraction that leads to logically definable causality, axiomatization of physics, axiomatization of agent interaction, and a logical foundation for quantum gravity. As discussed in early chapters the four goals are interrelated or fundamentally equivalent.

An equilibrium-based approach to mathematical abstraction and axiomatization is presented in this chapter for resolving the LAFIP paradox. YinYang *bipolar set* and bipolar dynamic logic (*BDL*) are introduced. Soundness and completeness of the bipolar axiomatization is asserted. The recovery of BDL to Boolean logic is proved. A redress on the ancient paradox of the liar is presented with a few observations on Gödel's incompleteness theorem. Based on BDL, bipolar relations, bipolar transitivity, and equilibrium relations are introduced. BDL and bipolar equilibrium relations provide a logical basis for YinYang bipolar relativity and bipolar quantum computing.

The remaining presentations and discussions of this chapter are organized in the following sections:

- **Bipolar Sets and Bipolar Dynamic Logic (BDL).** This section lays out the groundwork for this chapter. The concepts of *bipolar element*, *bipolar set*, *bipolar poset*, and *bipolar lattice* are introduced.
- **Laws of Equilibrium and Bipolar Universal Modus Ponens (BUMP).** This section presents equilibrium-based laws including the *bipolar laws of excluded middle*, four pairs of *bipolar dynamic DeMorgan's laws*, and the key element of BDL – *bipolar universal modus ponens* (*BUMP*).
- **Bipolar Axiomatization and Computation**. This section presents a bipolar axiomatization with soundness and completeness theorems. The computability of BDL is axiomatically proved. Computational complexity is analyzed.
- **Bipolar Modality.** This section presents a transformation of BDL to its modal form.
- **Bipolar Relations and Equilibrium Relations.** This section presents the concepts of *bipolar relation*, bipolar transitivity, and *equilibrium relation*.
- **On Gödel's Incompleteness Theorem.** This section gives a redress on the liar's case *"This sentence is not true"* and makes a few observations on Gödel's incompleteness theorem.
- **Research Topics**. This section lists a few research topics.
- **Summary**. This section draws a few conclusions.

BIPOLAR SETS AND BIPOLAR DYNAMIC LOGIC (BDL)

Bipolar Sets vs. Unipolar Sets

From a unipolar perspective, the two opposing poles of a physical equilibrium can be considered negative and positive elements (Zhang, 2005a). Then, *positive elements*, such as positive electric charges, action forces, positive ions, cooperative relations, activation, positive causalities, self-assertion, positive energy, or positive numbers (including 0 as the bottom), etc, can form a *positive set*; and *negative elements*, such as negative electric charges, reaction forces, negative ions, competitive relations, repression, negative causalities, self-negation, negative energy, or negative numbers (including 0 as the bottom), etc, can form a *negative set*. We call positive and negative sets *unipolar sets*. (Note: In truth-based convention, 0 is neither negative nor positive but neutral; in the equilibrium-based approach 0 is used as the bottom of both negative and positive numbers such that (-0,+0) or (0,0) can denote bipolar false or an eternal bipolar equilibrium later.)

A positive set X^+ in a unipolar universe U can be defined as a mapping f_X^+:U→{0,1} which yields the value 1 (or true) for elements belonging to X^+ and 0 (or false) for elements excluded from X^+. Similarly, a negative set X^- in a unipolar universe U can be defined as a mapping f_X^-: U → {-1,0} which yields the value -1 (or -true) for elements belonging to X^- and 0 (-false) for elements excluded from X^-. The bivalent operators ∧, ∨, ¬, and → in the *positive bivalent lattice* L^+ = {0,1} can be extended to the *negative bivalent lattice* L^-={-1,0} and, ∀a,b∈{-1,0}, we can have

$a \wedge b \equiv \max(a,b) \equiv -\min(|a|,|b|);$

$a \vee b \equiv \min(a,b) \equiv -\max(|a|,|b|);$

$\neg a \equiv -1 - a;$

$\neg b \equiv -1 - b;$ and

$a \to b \equiv \neg a \vee b.$

(Note: The use of absolute value |x| through this book is for explicit bipolarity and readability only.)

Observation: *Bipolar equilibrium* is a generic form of multidimensional equilibria (Ch. 1, Figure 1) because electron-positron pair production denoted (e^-, e^+) as an example of the materialization of energy predicted by special relativity has been accurately described by quantum electrodynamics (QED) (Dirac, 1927; Feynman, 1962, 1985). Although the two unipolar lattices (L^+,∧,∨, ¬,→) and (L^-,∧,∨,¬,→) are deemed *isomorphic* and *logically equivalent* to each other from a classical mathematical point of view (Mendelson, 1987), they are not isomorphic from an equilibrium-based holistic mathematical physics perspective because a symmetrical bipolar product lattice $B_1 = L^- \times L^+$ or $L^+ \times L^-$ provides a *natural non-linear dynamic logical structure* that can carry both truth-based logical semantics and some fundamental "illogical" physical semantics such as bipolar fusion, interaction, oscillation, and quantum entanglement (Zhang, 2009a, 2009b, 2009c, 2009d, 2010).

Bipolar Reification. Since the two poles of a physical bipolar equilibrium coexist in a fusion, coupling, or binding with opposite polarities, we say that *they are not isomorphic to each other*. This dis-

Figure 1. Bipolar abstraction and interaction: (a) Linear interaction; (b) Cross-pole non-linear interaction; (d) Oscillation; (e) Two entangled bipolar interactive variables

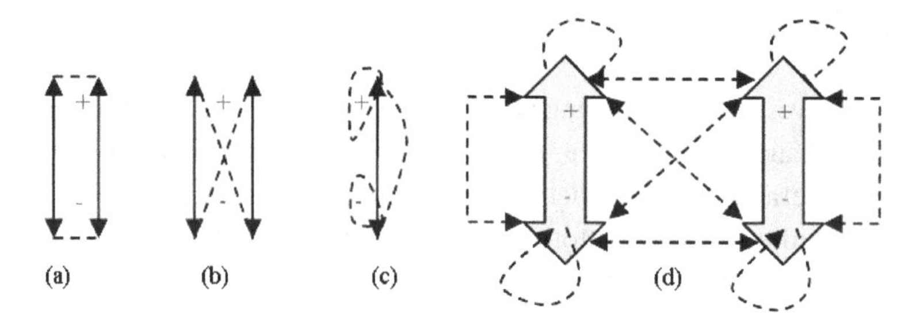

tinction enables (-,+) bipolar semantics/syntax be elicited and distinguished from (+,+) such that bipolar equilibrium can be defined as bipolar holistic truth.

Definition 3.1a. A *bipolar set* is a collection of bipolar equilibria or non-equilibria each of which as a self-evident *bipolar element* has two poles that are unipolar elements with opposite polarities in a bipolar fusion, coupling, or binding.

We refer to Definition 3.1a as *the principle of equilibrium-based bipolar mathematical abstraction.* With this principle, negative and positive poles as unipolar elements are partial concepts derived from self-evident holistic bipolar elements, not vice versa. Thus, bipolar sets leave the syntactic structures of classical sets and their operations intact for syntactic independence between sets and their elements. The semantic property of a bipolar set and the nature of its elements, however, could be interdependent.

A bipolar element can be in singleton or bipolar form such as $e = (e^-,e^+)$ or (e^+,e^-). We say $e = (e^-,e^+)$ is an *equilibrium* if $e^+ = -e^-$, where the negation operator $(-)$ changes the polarity of a pole. For instance, a pair of electromagnetic particles (-q,+q) may form an equilibrium but (-q,0) and (0,+q) form two non-equilibria. When $|e^-| \neq e^+$ we call the binding, coupling, or fusion $e = (e^-,e^+)$ *non-equilibrium* (Note: fuzzy or quasi-equilibrium is discussed in next chapter). The word "equilibrium-based" is to equilibrium and non-equilibrium as the word "truth-based" is to truth and falsity but the two assume fundamentally different syntax and semantics. We call a negative-positive ordered pair (n,p) a *NP equilibrium* and (p,n) a *PN equilibrium*, respectively.

Based on Postulates 1.1 and 1.2 (Ch. 1) YinYang bipolarity is the most fundamental property of nature and bipolarity is *background independent*. However, NP, PN, LR (left-right), and RL (right-left) ordering are background dependent concepts relative to the observer. Thus, without further background specification, in this work, we consider NP and PN equilibria equivalent provided that they exhibit the same polarity and value. For instance, with background independency we have $(0,+1) \equiv (+1,0)$; $(-1,0) \equiv (0,-1)$; $(-1,+1) \equiv (+1,-1)$. Thus, all bipolar variables can be sorted into either NP or PN types. We use NP notation through this book without losing generality. In case background information is given, we provide additional specification. Thus, a bipolar set can be denoted as $\{(x,y)\}$.

Definition 3.1b. A *bipolar poset* is a bipolar set denoted $(B_p,\geq\geq)$ (Zhang, 2005a) where $\geq\geq$ is a bipolar partial order relation and, $\forall(x,y),(u,v)\in B_p$, we have

Bipolar Partial Ordering: $(x,y) \geq\geq (u,v)$, iff $|x| \geq |u|$ and $y \geq v$. (3.1)

Zeroth-Order Bipolar Dynamic Logic (BDL)

Definition 3.2. A *bipolar lattice* (Zhang, 2005a) is a quadruplet (B, ⊕, &, ⊗), where B is a bipolar poset and, $\forall(x,y),(u,v) \in B$, there is a bipolar least upper bound (blub ⊕), a bipolar greatest lower bound (bglb &), and a cross-pole greatest lower bound (cglb ⊗):

$$blub((x,y),(u,v)) \equiv (x,y) \oplus (u,v) \equiv (x \vee u, y \vee v) \equiv (-\max(|x|,|u|), \max(y,v)); \quad (3.2)$$

$$bglb((x,y),(u,v)) \equiv (x,y) \& (u,v) \equiv (x \wedge u, y \wedge v) \equiv (-\min(|x|,|u|), \min(y,v)); \quad (3.3)$$

$$cglb((x,y),(u,v)) \equiv (x,y) \otimes (u,v) \equiv (-(|x| \wedge |v| \vee |y| \wedge |u|), (|x| \wedge |u| \vee |y| \wedge |v|)). \quad (3.4)$$

Definition 3.3. A bipolar operator is said to be *linearly bipolar equivalent* to its unipolar counterpart if it doesn't account for cross-pole interaction and its operation on each pole is equivalent to its unipolar counterpart defined on {-1,0} and {0,1}, respectively. Otherwise, a bipolar operator is said to be *non-linear*.

Based on the above definition, & is *a bipolar conjunctive* linearly bipolar equivalent to ∧; ⊕ is *a bipolar disjunctive* dual to & and linearly bipolar equivalent to ∨; and ⊗ is a *non-linear bipolar interactive conjunctive* in general or an equilibrium-based *bipolar oscillator w.r.t.* the fact that it is a cross-pole operator with the entangled bipolar interactive semantics $- - = +$, $- + = -$, $+ - = -$, and $+ + = +$ defined on each pole such that (-1,0) ⊗ (-1,0) = (0,1); (-1,0) ⊗ (0,1) = (0,1) ⊗ (-1,0)= (-1,0); (0,1) ⊗ (0,1) = (0,1), (-1,1) ⊗ (0,1) = (-1,1) ⊗ (-1,0) = (-1,1), and, $\forall(a,b) \in B_1$, (a,b) ⊗ (0,0)=(0,0). Thus, the operator ⊗, has no equivalent in classical logic. The linear, cross-pole, and oscillatory bipolar interactions are abstracted and depicted, respectively, in Figure 1(a-f) based on Ch. 1, Figure 1. (It should be remarked that ⊗ is *non-linear* in general. That does not mean all its operations are non-linear. It can also be linear in specific cases. For instance, (0,1) ⊗ (x,y) = (x,y) is a linear operation.)

The three operators ⊕, &, and ⊗ are said to be bipolar *equilibrium-based* because they are logical or physical and static or dynamic from an equilibrium-based perspective. The operator ⊕ is a balancer/fusion operator with (-1,0) ⊕ (0,1) = (-1,1); & is a minimizing operator with (-1,0) & (0,1) = (0,0); ⊗ is an intuitive bipolar interactive oscillator as $(-1,0) \otimes .. \otimes (-1,0) = (-1,0)^n$ results in (-1,0) or (0,1), respectively, depending on whether n is odd or even. *Since all closed dynamic systems tend to reach an equilibrium state and all open systems are subject to external disturbance through bipolar interaction and oscillation, ⊗ provides a key for equilibrium-based open-world reasoning.* Furthermore, *BDL* supports a holistic combination of inductive and deductive reasoning. Although ⊗ is not a free operator, it can be used in quantum entanglement and non-linear bipolar dynamic inference discussed later. While the bipolar operators could be deemed illogical from a truth-based perspective, the bipolar equilibrium-based syntax and semantics provide a crucial basis for resolving the LAFIP paradox. With bipolarity, equilibrium-based non-linear dynamics is introduced into logical reasoning – a crucial step toward a probable axiomatization for physics and quantum mechanics.

Definition 3.4a. A bipolar lattice B is *bounded* if it has both a unique minimal element denoted (0,0) and a unique maximal element denoted (-1,1).

Table. 1. Truth Table for – and ¬

(n,p)	(0,0)	(-1,0)	(0,1)	(-1,1)
–(n,p)	(0,0)	(0,1)	(-1,0)	(-1,1)
¬(n,p)	(-1,1)	(0,1)	(-1,0)	(0,0)

Figure 2. Hasse diagram of B_1

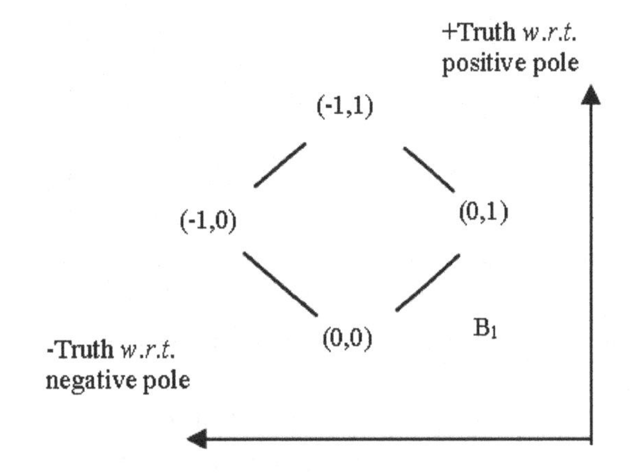

Table 2. Truth Table for (x,y)⇒(u.v)≡¬(x,y)⊕(u,v) ≡ (x→u, y→v)

(x,y)	¬(x,y) ≡ (¬x, ¬y)	(u,v)	(x,y)⇒(u.v)≡¬(x,y)⊕(u,v) ≡(x→u, y→v)
(0,0)	(-1,1)	(0,0) (-1,0) (0,1) (-1,1)	(-1,1) (-1,1) (-1,1) (-1,1)
(-1,0)	(0,1)	(0,0) (-1,0) (0,1) (-1,1)	(0,1) (-1,1) (0,1) (-1,1)
(0,1)	(-1,0)	(0,0) (-1,0) (0,1) (-1,1)	(-1,0) (-1,0) (-1,1) (-1,1)
(-1,1)	(0,0)	(0,0) (-1,0) (0,1) (-1,1)	(0,0) (-1,0) (0,1) (-1,1)
Note: [(x,y) ≤≤ (u,v)] ⇔ [(x,y)⇒(u,v)]			

In the unipolar case a bounded lattice is also complemented, but that is not necessarily true in the bipolar case (Zhang 2005a).

Definition 3.4b. A bounded bipolar lattice B is *complemented* if, ∀(x,y)∈B, we have the *bipolar complement* ¬(x,y) ∈ B, where ¬ is defined in Eq. 3.5.

Bipolar Complement: ¬(x,y) ≡ (¬x,¬y) ≡(-1,1) – (x,y) ≡ (-1-x,1-y); (3.5)

Bipolar complement induces bipolar implication. A bounded and complemented bipolar lattice B can be denoted as a zeroth-order BDL (B,≡,⊕, ⊗, &, –, ¬, ⇒) (Zhang 2005a) where – and ⇒ are defined in Eq. 3.6 and 3.7, respectively.

Bipolar Implication: (x,y)⇒(u,v)≡¬(x,y)⊕(u,v)≡(x→u,y→v)≡(¬x∨u,¬y∨v); (3.6)

Bipolar Negation: –(x,y)≡(-1,0)⊗(x,y) ≡ (–y,–x). (3.7)

Evidently, bipolar complement (¬) and implication (⇒) are linearly bipolar equivalent to their unipolar counterparts, respectively, as we have (x,y) ⇒ (u.v) ≡ ¬(x,y) ⊕ (u,v) ≡ (x→u, y→v) ≡ (¬x∨u, ¬y∨v) and we must have [(x,y)⇒(u,v)] ⇔ [(x,y) ≤≤ (u,v)], where ⇔ stands for bipolar equivalent. In Eq. 3.7 the "–" operator is for arithmetic negation that changes the polarity of the two poles in a switch. In BDL, arithmetic negation and logical negation (¬) are different operations, e.g. –(-1,1)=(-1,1) but ¬(-1,1) = (0,0).

Truth-table for the implication operator $-$, $\neg$, $\Rightarrow$ is shown in table 1 and table 2.

The Hasse diagram of the bounded and complemented unit square bipolar lattice $B_1 = \{-1,0\} \times \{0,1\}$ is shown in Figure 2, where (0,0) stands for logically bipolar false or an eternal equilibrium; (-1,0) stands for negative pole true but positive pole false or an imbalance; (0,1) stands for negative pole false but positive pole true or another imbalance; (-1,1) stands for bipolar true or a bipolar equilibrium. Equilibrium is not a contradiction but a healthy dynamic balance. The four lattice values can be typically illustrated with mood states (Zhang, Pandurangi & Peace, 2007). Let p be a person; $\forall p$, $\varphi(p)$ = (self-negation, self-assertion):$p \Rightarrow B_1$ defines a bipolar mapping from p to B_1. Then, $\varphi(p)$ = (0,1), (-1,0), (-1,1), and (0,0) indicates mania, depression; mental equilibrium; brain death, respectively (Zhang, Pandurangi & Peace, 2007).

Remarks: (1) The equilibrium-based non-linear bipolar dynamic properties and the *bipolar law of excluded middle* (LEM) (see next section) distinguish B_1 from any other 4-valued truth-based logic. (2) While 4-valued logical structures resemble each other's forms, all resemble YinYang-4-Images (Ch. 1, Figure 2).

First-Order Bipolar Syntax and Semantics

A 1st-order BDL doesn't exclude classical logic syntax. It extends Boolean logic from the bivalent lattice {0,1} to B_1 with the addition of non-linear bipolar dynamic properties.

1. User defined primitives:
   - Constant symbols (individual bipolar equilibria, e.g. bipolar disorder patient A).
   - Bipolar function symbols (mapping individual bipolar equilibria to another equilibria, e.g. bipolar function such as a bipolar interactive transitive closure function f maps a bipolar relation to its closure denoted $f:R \Rightarrow \check{R}$.)
   - Bipolar predicate symbols (mapping individual equilibria to B_1, e.g. If $\psi = (\psi^-, \psi^+)$ = (self-negation, self-assertion) abilities of a person, we have $\psi:A \Rightarrow B_1$.
2. BDL supplied primitives:
   - Bipolar variable symbols: e.g., singleton form: v; bipolar form: (v^-, v^+).
   - Bipolar connectives: e.g. $\oplus$, $\otimes$, &, ..., $-$, $\neg$, $\Rightarrow$, $\Leftrightarrow$.
   - Quantifiers: universal ($\forall$) and existential ($\exists$).
3. Bipolar Sentences are built up from bipolar terms and atoms:
   - A bipolar term (denoting real-world equilibrium) is a constant symbol, a variable symbol, or an n-place function of n terms.
   - A bipolar atom (which has value $\in B_1$) is either an n-place predicate of n terms, or, if P and Q are atoms, then $-P$, $\neg P$, $P*Q$, $P \Rightarrow Q$, $P \Leftrightarrow Q$ are atoms. We use * as a universal bipolar binary operator that can be bound to a specific bipolar binary operator such as $\oplus$, $\otimes$, or &. We use $\rightarrow$ and $\leftrightarrow$ for classical (unipolar) implication; we use $\Rightarrow$ and $\Leftrightarrow$ for bipolar implication.
   - A bipolar sentence is a bipolar atom, or, if P is a bipolar sentence and x is a bipolar variable, then $(\forall x)P$ and $(\exists x)P$ are bipolar sentences
   - A bipolar well-formed formula (bwf) is a bipolar sentence containing no "free" variables. i.e., all variables are "bound" by universal or existential quantifiers. E.g., $(\forall x)P[x,y]$, $x=(x^-,x^+)$

Table 3. Truth tables

&	**(0,0)**	**(-1,0)**	**(0,1)**	**(-1,1)**
(0,0)	(0,0)	(0,0)	(0,0)	(0,0)
(-1,0)	(0,0)	(-1,0)	(0,0)	(-1,0)
(0,1)	(0,0)	(0,0)	(0,1)	(0,1)
(-1,1)	(0,0)	(-1,0)	(0,1)	(-1,1)

(a)

$\oplus$	**(0,0)**	**(-1,0)**	**(0,1)**	**(-1,1)**
(0,0)	(0,0)	(-1,0)	(0,1)	(-1,1)
(-1,0)	(-1,0)	(-1,0)	(-1,1)	(-1,1)
(0,1)	(0,1)	(-1,1)	(0,1)	(-1,1)
(-1,1)	(-1,1)	(-1,1)	(-1,1)	(-1,1)

(b)

$\otimes$	**(0,0)**	**(-1,0)**	**(0,1)**	**(-1,1)**
(0,0)	(0,0)	(0,0)	(0,0)	(0,0)
(-1,0)	(0,0)	(0,1)	(-1,0)	(-1,1)
(0,1)	(0,0)	(-1,0)	(0,1)	(-1,1)
(-1,1)	(0,0)	(-1,1)	(-1,1)	(-1,1)

(c)

$\oslash$	**(0,0)**	**(-1,0)**	**(0,1)**	**(-1,1)**
(0,0)	(0,0)	(0,0)	(0,0)	(-1,1)
(-1,0)	(0,0)	(-1,0)	(0,1)	(-1,1)
(0,1)	(0,0)	(0,1)	(-1,0)	(-1,1)
(-1,1)	(-1,1)	(-1,1)	(-1,1)	(-1,1)

(d)

$\&^-$	**(0,0)**	**(-1,0)**	**(0,1)**	**(-1,1)**
(0,0)	(0,0)	(0,0)	(0,0)	(0,0)
(-1,0)	(0,0)	(0,1)	(0,0)	(0,1)
(0,1)	(0,0)	(0,0)	(-1,0)	(-1,0)
(-1,1)	(0,0)	(0,1)	(-1,0)	(-1,1)

(e)

$\oplus^-$	**(0,0)**	**(-1,0)**	**(0,1)**	**(-1,1)**
(0,0)	(0,0)	(0,1)	(-1,0)	(-1,1)
(-1,0)	(0,1)	(0,1)	(-1,1)	(-1,1)
(0,1)	(-1,0)	(-1,1)	(-1,0)	(-1,1)
(-1,1)	(-1,1)	(-1,1)	(-1,1)	(-1,1)

(f)

continued on the following page

Table 3. continued

$\otimes^-$	(0,0)	(-1,0)	(0,1)	(-1,1)
(0,0)	(0,0)	(0,0)	(0,0)	(0,0)
(-1,0)	(0,0)	(-1,0)	(0,1)	(-1,1)
(0,1)	(0,0)	(0,1)	(-1,0)	(-1,1)
(-1,1)	(0,0)	(-1,1)	(-1,1)	(-1,1)
(g)				
$\oslash^-$	(0,0)	(-1,0)	(0,1)	(-1,1)
(0,0)	(0,0)	(0,0)	(0,0)	(-1,1)
(-1,0)	(0,0)	(0,1)	(-1,0)	(-1,1)
(0,1)	(0,0)	(-1,0)	(0,1)	(-1,1)
(-1,1)	(-1,1)	(-1,1)	(-1,1)	(-1,1)
(h)				

and $y = (y^-,y^+)$, is not a bwf because it has x bound as a universally quantified bipolar variable, but y is free.

Definition 3.5a. A *unipolar theorem* is a statement or well-formed formula regarding unipolar truth that can be derived by applying a fixed set of unipolar deduction rules and axioms without any additional assumptions. A *bipolar theorem* is a statement or bwf regarding equilibrium-based bipolar holistic truth that can be derived by applying a fixed set of bipolar deduction rules and axioms without any additional assumptions. In what follows we use the word *"theorem"* for either or both.

For simplicity and without losing generality, we assume in what follows that all bipolar implication in the form $A \Rightarrow B$ is designated bipolar true such that $(A \Rightarrow B) \equiv [(A \Rightarrow B) \equiv (-1,1)]$; any other bwf is not designated without further specification.

Definition 3.5b. *A bipolar tautology* is a bwf whose form makes it always bipolar true (-1,1) regardless of the truth values of its undesignated variables. A bwf is *linear* if only linear bipolar connectives are involved in it. It is *non-linear* if any non-linear bipolar connective is involved.

Based on the above, a bipolar tautology is also a theorem or axiom for bipolar inference.

Two-fold instantiation can be involved in bipolar inference: *operator instantiation* and *variable instantiation*. We use *"IF* $bwf_1 \equiv (-1,1)$*; THEN* $bwf_2 \equiv (-1,1)$*"* when necessary to indicate that if the premise is *designated bipolar true* the consequent is also bipolar true. We use *"IF* (bwf_1*bwf_2)*; THEN* (bwf_3*bwf_4)*"* to indicate that if bipolar interaction * occurs in the premise between bwf_1 and bwf_2, the same interaction * also occurs in the consequent between bwf_3 and bwf_4, however, (bwf_1*bwf_2) and (bwf_3*bwf_4) are not designated. The variables in a designated bwf are ***free but not completely free***

because they must collectively satisfy the designation. Bipolar designation leads to ***equilibrium-based bipolar holistic reasoning.***

LAWS OF EQUILIBRIUM AND BIPOLAR UNIVERSAL MODUS PONENS

In (Zhang, 2005a) logical formulation is limited to the operators ⊕, &, and ⊗ only. In this section we extend these operators to eight linear and non-linear bipolar operators on B_1 that lead to four pairs of dual *bipolar dynamic DeMorgan's laws* and an extended *BUMP*.

Let $\oplus^-$ and $\&^-$ be the arithmetic negations of ⊕ and &, respectively, we have

Negation of blub: $blub^-((x,y),(u,v)) \equiv (x,y)\oplus^-(u,v) \equiv -((x,y)\oplus(u,v)) \equiv (-(y\vee v), (|x|\vee|u|))$; (3.8)

Negation of bglb: $bglb^-((x,y),(u,v)) \equiv (x,y)\&^-(u,v) \equiv -((x,y)\&(u,v)) \equiv (-(y\wedge v), (|x|\wedge|u|))$. (3.9)

A cross-pole least upper bound (club) operator ∅ is defined in Eq. 3.10 as a generally non-linear bipolar disjunctive or cross-pole least upper bound dual to ⊗.

club: $club((x,y),(u,v)) \equiv (x,y)\varnothing(u,v) \equiv \neg(\neg(x,y)\otimes\neg(u,v))$. (3.10)

Further, we define the arithmetic negation of ⊗ as $\otimes^-$ and the arithmetic negation of ∅ as $\varnothing^-$:
Negation of cglb:

Negation of cglb: $cglb-((x,y),(u,v)) \equiv (x,y)\otimes-(u,v) \equiv -((x,y)\otimes(u,v))$. (3.11)

Negation of club: $club^-((x,y),(u,v)) \equiv (x,y)\varnothing^-(u,v) \equiv -((x,y)\varnothing(u,v))$. (3.12)

Truth tables of the eight binary operators &, ⊕, $\&^-$, $\oplus^-$, ⊗, ∅, $\otimes^-$, and $\varnothing^-$are shown in Table 3.

Theorem 3.1. The laws of excluded middle (LEMs) and non-contradiction in Table 4 hold on B_1.

Proof. It follows directly from that ⊕ is linearly bipolar equivalent to ∨ (Def. 3.2) and that $\oplus^-$ (Eq. 3.8) is simply an arithmetic negation of ⊕ (see Eq. 3.8). □

Theorem 3.2. The dual linear *bipolar DeMorgan's laws* in Table 4 hold on B_1.

Proof. It follows directly from that the bipolar operators ⊕, &, and ¬ are linearly bipolar equivalent to the unipolar operators ∨, ∧, and ¬ (Def. 3.2), respectively, and that $\oplus^-$ and $\&^-$ are both arithmetic negations of ⊕ and & Eq. 3.8 and 3.9, respectively. □

The duality does not hold between ⊗ and ⊕ because of the non-linear cross-pole property of ⊗. With {⊗,∅ } and {$\otimes^-$,$\varnothing^-$ }, we have

Theorem 3.3. The dual non-linear bipolar DeMorgan's laws in Table 4 hold on B_1.

Table 4. Bipolar laws

Excluded Middle (LEMs)	$(x,y)\oplus \neg(x,y)$; $(x,y)\oplus^{-} \neg(x,y)$;
Non-Contradiction	$\neg((x,y)\&\neg(x,y))$; $\neg((x,y)\&^{-}\neg(x,y))$;
Linear Bipolar DeMorgan's Laws	$\neg((a,b)\&(c,d)) \equiv \neg(a,b)\oplus\neg(c,d)$; $\neg((a,b)\oplus(c,d)) \equiv \neg(a,b)\&\neg(c,d)$; $\neg((a,b)\&^{-} (c,d)) \equiv \neg(a,b)\oplus^{-}\neg(c,d)$; $\neg((a,b)\oplus^{-} (c,d)) \equiv \neg(a,b)\&^{-}\neg(c,d)$;
Non-Linear Bipolar DeMorgan's Laws	$\neg((a,b)\otimes (c,d)) \equiv \neg(a,b) \varnothing \neg(c,d)$; $\neg((a,b) \varnothing (c,d)) \equiv \neg(a,b) \otimes \neg(c,d)$; $\neg((a,b)\otimes^{-}(c,d)) \equiv \neg(a,b) \varnothing^{-} \neg(c,d)$; $\neg((a,b)\varnothing^{-} (c,d)) \equiv \neg(a,b) \otimes^{-} \neg(c,d)$

Proof. It follows directly from Eq. 3.10-3.12. □

It is interesting to notice that $\otimes^{-}$ implements the effects of $- - = -$, $- + = +$, $+ - = +$, $+ + = -$. These effects may give the impression of absolute nonsense at first glance. A careful examination, however, reveals just the opposite. That is, $\otimes^{-}$ is a counterintuitive oscillator that is the resistance or negation of the intuitive oscillator $\otimes$ such as resistance to a drug therapy.

While $\varnothing$ and $\varnothing^{-}$ are mutual arithmetic negations, the key is to understand the difference between $\varnothing$ and $\otimes$. $\otimes$ is a non-linear cross-pole conjunctive; $\varnothing$ is a non-linear cross-pole disjunctive that provides a least upper bound for an oscillation. For instance, $(-1,0)\otimes(0,1)=(-1,0)$ but $(-1,0)\varnothing(0,1) = (0,1)$; $(0,0)\otimes(-1,1) = (0,0)$ is an annihilation of an equilibrium but $(0,0) \varnothing (-1,1) = (-1,1)$ characterizes an undisturbed equilibrium by $(0,0)$.

Theorem 3.4. The eight binary operators $\&$, $\oplus$, $\&^{-}$,$\oplus^{-}$, $\otimes$, $\varnothing$, $\otimes^{-}$, and $\varnothing^{-}$ are commutative and bipolar monotonic *w.r.t.* to $\geq\geq$ (Eq. 3.1).

Proof. (1) The commutativity and bipolar monotonicity of $\&$ *and* $\oplus$ *follow directly from that they are linearly bipolar equivalent to* $\wedge$ *and* $\vee$*, respectively (Eq. 3.2,3.3); (2)* $\&^{-}$ *and* $\oplus^{-}$ *are both arithmetic negations of* $\&$ *and* $\oplus$*, respectively (Eq. 3.8 & 3.9); (3) the commutativity and bipolar monotonicity of* $\otimes$ *follow from that the definition of* $\otimes$ *is symmetrical,* $\wedge$-$\vee$ *based, absolute value based (Eq. 3.4), and* $\geq\geq$ is absolute value based *(Eq. 3.1); (4) the commutativity and bipolar monotonicity of* $\varnothing$ *follow from the dual non-linear bipolar DeMorgan's laws on* $\otimes$ *and* $\varnothing$ *(Table 4); (5) the commutativity and bipolar monotonicity of* $\otimes^{-}$ *and* $\varnothing^{-}$ *follow fromEq. 3.1 & 3.12) and the dual non-linear bipolar DeMorgan's laws (Table 4) on* $\otimes^{-}$ *and* $\varnothing^{-}$. □

With the four pairs of bipolar DeMorgan's laws, bipolar universal modus ponens (BUMP) (Zhang, 2005a, 2007) can be extended to include $\&$, $\oplus$, $\&^{-}$,$\oplus^{-}$, $\otimes$, $\varnothing$, $\otimes^{-}$, and $\varnothing^{-}$ or any other commutative and bipolar monotonic operators as shown in Table 5. Variables in BUMP can be bound to any equilibrium-based bipolar interactive agents including biochemical or biological agents (Zhang, 2007; Zhang, Pandurangi & Peace, 2007). The 2-fold universal instantiation enables a holistic combination of top-down inductive and bottom-up deductive reasoning.

Theorem 3.5. (a) BUMP (see Table 5) is a bipolar tautology. (b) BUMP is an equilibrium-based non-linear bipolar dynamic generalization of classical modus ponens (MP) with equilibrium-based bipolar quantum entanglement property.

Proof.

(a) *Since,* $\forall \varphi, \varphi \in B_1$, $(\varphi \Rightarrow \varphi) \Leftrightarrow (\varphi \leq\leq \varphi)$, *BUMP is equivalent toEq. 3.14, the theorem follows directly from the commutativity and bipolar monotonicity of* * *(Theorem 3.4).*

$$[(\phi-,\phi+)\Rightarrow(\varphi-,\varphi+)\&(\psi-,\psi+)\Rightarrow(\chi-,\chi+)]\&[(\phi-,\phi+)*(\psi-,\psi+)] \leq\leq [(\varphi-,\varphi+)*(\chi-,\chi+)]; \quad (3.13)$$

$$[(\phi^-,\phi^+)\Rightarrow(\varphi^-,\varphi^+)]\&[(\psi^-,\psi^+)\Rightarrow(\chi^-,\chi^+)] \Rightarrow[((\phi^-,\phi^+)*(\psi^-,\psi^+))\leq\leq((\varphi^-,\varphi^+)*(\chi^-,\chi^+))]. \quad (3.14)$$

(b) *(i) The binding of* $\&$, $\oplus$, $\&^-$, *or* $\oplus^-$ *to the universal operator* * *makes BUMP linearly bipolar equivalent to MP because* & *and* $\oplus$ *are linearly bipolar equivalent to* $\wedge$ *and* $\vee$, *respectively;* $\&^-$ *and* $\oplus^-$ *are simply negations of* & *and* $\oplus$, *respectively;* $\Rightarrow$ *is linearly bipolar equivalent to* $\rightarrow$. *However,* $\otimes$, $\varnothing$, $\otimes^-$, *and* $\varnothing^-$ *are non-linear operators with cross-pole bipolar interaction or oscillation semantics and syntax that find no equivalent in* $\wedge$ *or* $\vee$ *(Eq. 3.4, 3.10, 3.11, 3.12). Therefore, classical MP is part of BUMP but not vice versa. (i) Classical MP is truth-based and static in nature that states "IF A implies B and A is true, THEN B is true." BUMP is bipolar equilibrium-based and dynamic in nature that states "IF the first bipolar equilibrium implies the 2nd and the 3rd implies the 4th, THEN bipolar dynamic interaction between the 1st and the 3rd implies the same type of bipolar dynamic interaction between the 2nd and the 4th. Such non-linear bipolar dynamic physical semantics and syntax find no equivalent in classical MP. These semantics and syntax enable "illogical" or disordered natural properties and behaviors of a physical system (including biological and quantum systems) to be represented for logical solutions. Thus, the binding of the universal operator* * *does not have to be restricted to the so-called "free" operators (such as* & *and* $\oplus$*). (iii) From a classical logic perspective, without non-linear bipolar interaction BUMP would be equivalent to MP; from a quantum mechanics perspective, BUMP exhibits equilibrium-*

Table 5. Bipolar Universal Modus Ponens (BUMP)

BUMP in IF-THEN form:
IF $[(\phi^-,\phi^+)\Rightarrow(\varphi^-,\varphi^+)]\&[(\psi^-,\psi^+)\Rightarrow(\chi^-,\chi^+)]\&[(\phi^-,\phi^+)*(\psi^-,\psi^+)]$;
THEN $[(\varphi^-,\varphi^+)*(\chi^-,\chi^+)]$.
BUMP in Tautological form:
$[(\phi^-,\phi^+)\Rightarrow(\varphi^-,\varphi^+)]\ \&\ [(\psi^-,\psi^+)\Rightarrow(\chi^-,\chi^+)]$
$\Rightarrow \{[((\phi^-,\phi^+)*(\psi^-,\psi^+)) \Rightarrow ((\varphi^-,\varphi^+)*(\chi^-,\chi^+))]\}$.
BUMP in Singleton Form:
IF $[(\phi \Rightarrow \varphi)\ \&\ (\psi \Rightarrow \chi)]$ and $(\phi * \psi)$, THEN $(\varphi * \chi)$.
$(\phi \Rightarrow \varphi)\ \&\ (\psi \Rightarrow \chi) \Rightarrow [(\phi * \psi) \Rightarrow (\varphi * \chi)]$.
Two-fold universal instantiation:
1) Operator instantiation: * as a universal operator can be bound to any commutative and bipolar monotonic (*w.r.t.* $\geq\geq$) operator including $\&$, $\oplus$, $\&^-$, $\oplus^-$, $\otimes$, $\varnothing$, $\otimes^-$, and $\varnothing^-$.
2) $(\phi \Rightarrow \varphi)$ is designated bipolar true; $((\phi^-,\phi^+)*(\psi^-,\psi^+))$ is not designated.
3) Bipolar variable instantiation: $\forall x$, $(\phi^-,\phi^+)(x) \Rightarrow (\varphi^-,\varphi^+)(x)$; $(\phi^-,\phi^+)(A)$; $\therefore (\varphi^-,\varphi^+)(A)$.

based bipolar quantum physics property. That is, with the premise $[(\varphi \Rightarrow \varphi) \,\&\, (\psi \Rightarrow \chi)]$, we have the quantum entanglement $[(\varphi * \psi) \Rightarrow (\varphi * \chi)]$ (Table 5). □

A Unified Architecture for Holistic Reasoning. Figure 3 shows a unified architecture for holistic top-down and bottom-up reasoning in an open-world of bipolar equilibria. The unification is made possible with BUMP through operator instantiation on the one hand and variable instantiation on the other. Operator instantiation uses holistic data and knowledge in a top-down reasoning process and variable instantiation uses specific knowledge and evidence in a bottom-up reasoning process. Two key differences between BUMP and classical modus ponens (MP) are: (1) While MP is unipolar and static in nature, BUMP is bipolar and dynamic; (2) While MP uses variable instantiation only, BUMP enables both operator instantiation and variable instantiation. Without bipolarity and operator instantiation, classical MP can only be applied with static bivalent truth objects in a closed world. With operator instantiation, BUMP bridges a gap from a classical static closed-world to a dynamic open-world of bipolar equilibria, quasi- or non-equilibria. Top-down and bottom-up reasoning with BUMP are scientific processes because BUMP does not compromise the law of excluded middle in its non-linear dynamic non-monotonic extension.

The above is further supported with the following mathematical analysis:

1. If $\&$, $\oplus$, $\&^-$, or $\oplus^-$ is bound to the universal operator * in BUMP, BUMP degenerates to classical modus ponens for truth-based reasoning and linear bipolar fusion, coupling, or binding.
2. If $\otimes$, $\varnothing$, $\otimes^-$, or $\varnothing^-$ is bound to the universal operator * in BUMP, BUMP can accommodate dynamic interaction, oscillation, and quantum entanglement.
3. No matter how the bipolar equilibrium or non-equilibrium states are dynamically changed, BUMP guarantees the bipolar partial ordering $((\phi^-,\phi^+) * (\psi^-,\psi^+)) \leq\leq ((\varphi^-,\varphi^+) * (\chi^-,\chi^+))$ between the premise and the consequence as long as we have $[((\phi^-,\phi^+) \Rightarrow (\varphi^-,\varphi^+))] \,\&\, [((\psi^-,\psi^+) \Rightarrow (\chi^-,\chi^+))]$.

BIPOLAR AXIOMATIZATION AND COMPUTABILITY

Axiomatization and Computability

Now we have a zeroth-order BDL $(B, \equiv, \&, \oplus, \&^-, \oplus^-, \otimes, \varnothing, \otimes^-, \varnothing^-, -, \neg, \Rightarrow)$ and its 1st-order lift. As in classical logic, truth tables enable us to answer many questions concerning bipolar truth-functional connectives, such as whether a given bwf is a tautology or whether it logically implies or is logically

Figure 3. A unified architecture for holistic reasoning

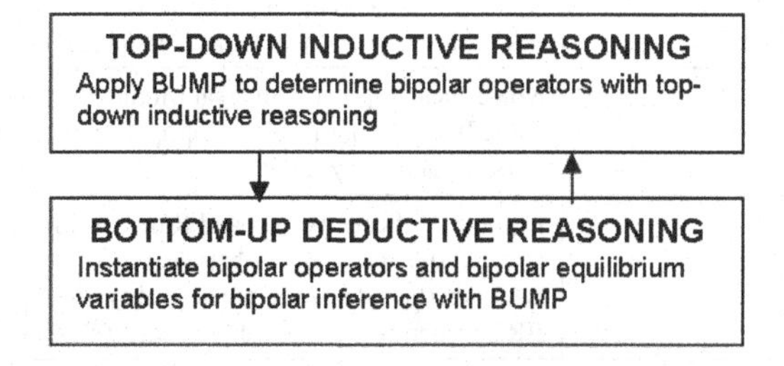

equivalent to some other given bwf. Although a propositional logic may surrender to the truth table method, it is instructive to have an axiomatic method. We define a set of *bipolar axioms (BAs) and rules of inference (BRs)* as a basic set of bipolar tautologies or bwfs from which all bipolar tautologies can be derived.

Definition 3.6. A set Γ of bwfs in zeroth-order is *sound* iff every bwf derived from Γ is a bipolar tautology. In another words, if $\Gamma \vdash A$, then $\Gamma \models A$. Γ is *zeroth-order complete* iff any other bipolar tautology must be a derivation from Γ.

Definition 3.7. A set Γ of bwfs in first-order is *sound* iff there is no bwf A such that $\Gamma \vdash A$ and $\Gamma \vdash \neg A$, $\Gamma \vdash (\neg A^-, A^+)$, or $\Gamma \vdash (A^-, \neg A^+)$. It is *first-order complete* iff any other bwf $A \equiv (-1,1)$ must be either a designation or a derivation from Γ.

Since BDL is a non-linear dynamic polarization of classical logic, bipolar soundness and completeness can be proved based on linear bipolar logical equivalence plus the proof of non-linear dynamic generalization in Hilbert style without a formal proof theory from scratch. On the left side of Table 6 we list a nine-set axiomatization on classical logic that has been proven sound and complete (Bridge, 1977); on the right side we list a nine-set bipolar axiomatization to be proven sound and complete. Some derived bipolar tautologies are listed in Table 7. To "lift" the axiomatization from zeroth-order to 1st-order we introduced two bipolar predicate axioms BA6-7 and one generalization BR2 in addition to BA1-5 and BR1 (BUMP). In the "lift" bipolar quantification assumes the tradition that a bipolar variable x occurring in a formula $\phi=(\phi^-,\phi^+)$ is *free* if it is not within the scope of ∀x or ∃x, otherwise, it is *bound.*

Theorem 3.6. The set {BA1-5, BR1 (BUMP)} (see Table 6) is zeroth-order sound and complete with respect to ¬, &, and ⊗.

Proof. Since the zeroth-order bipolar axioms and inference rules {BA1-BA5, BR1 (BUMP)}(Table 6) are bipolar tautologies because (a) except BUMP all others are linearly bipolar equivalent to their unipolar counterpart and (b) BUMP has been proven a bipolar tautology in Theorem 3.5, soundness

Table 6. Unipolar and bipolar axioms and inference rules

Unipolar Axioms (UAs): UA1: $\phi \rightarrow (\varphi \rightarrow \phi)$; UA2: $(\phi \rightarrow (\varphi \rightarrow \chi)) \rightarrow ((\phi \rightarrow \varphi) \rightarrow (\phi \rightarrow \chi))$; UA3: $\neg\phi \rightarrow \varphi) \rightarrow ((\neg\phi \rightarrow \neg\varphi) \rightarrow \phi)$; UA4: (a) $\phi \wedge \varphi \rightarrow \phi$; (b) $\phi \wedge \varphi \rightarrow \varphi$; UA5: $\phi \rightarrow (\varphi \rightarrow \phi \wedge \varphi)$;	Bipolar Linear Axioms: BA1: $(\phi^-,\phi^+) \Rightarrow ((\varphi^-,\varphi^+) \Rightarrow (\phi^-,\phi^+))$; BA2: $((\phi^-,\phi^+) \Rightarrow ((\varphi^-,\varphi^+) \Rightarrow (\chi^-,\chi^+))) \Rightarrow (((\phi^-,\phi^+) \Rightarrow (\varphi^-,\varphi^+)) \Rightarrow ((\phi^-,\phi^+) \Rightarrow (\chi^-,\chi^+)))$; BA3: $(\neg(\phi^-,\phi^+) \Rightarrow (\varphi^-,\varphi^+)) \Rightarrow ((\neg(\phi^-,\phi^+) \Rightarrow \neg(\varphi^-,\varphi^+)) \Rightarrow (\phi^-,\phi^+))$; BA4: (a) $(\phi^-,\phi^+)\&(\varphi^-,\varphi^+) \Rightarrow (\phi^-,\phi^+)$; (b) $(\phi^-,\phi^+)\&(\varphi^-,\varphi^+) \Rightarrow (\varphi^-,\varphi^+)$; BA5: $(\phi^-,\phi^+) \Rightarrow ((\varphi^-,\varphi^+) \Rightarrow ((\phi^-,\phi^+)\&(\varphi^-,\varphi^+)))$;
Inference Rule – Modus Ponens (MP): UR1: $(\phi \wedge (\phi \rightarrow \varphi)) \rightarrow \varphi$.	Non-Linear Bipolar Universal Modus Ponens (BUMP) BR1: IF $((\phi^-,\phi^+)*(\psi^-,\psi^+))$, $[((\phi^-,\phi^+) \Rightarrow (\varphi^-,\varphi^+))\&((\psi^-,\psi^+) \Rightarrow (\chi^-,\chi^+))]$, THEN $[(\varphi^-,\varphi^+)*(\chi^-,\chi^+)]$;
Predicate axioms and rules UA6: $\forall x,\phi(x) \rightarrow \phi(t)$; UA7: $\forall x, (\phi \rightarrow \varphi) \rightarrow (\phi \rightarrow \forall x,\varphi)$; UR2–Generalization: $\phi \rightarrow \forall x,\phi(x)$	Bipolar Predicate axioms and Rules of inference BA6: $\forall x,(\phi^-(x),\phi^+(x)) \Rightarrow (\phi^-(t),\phi^+(t))$; BA7: $\forall x,((\phi^-,\phi^+) \Rightarrow (\varphi^-,\varphi^+)) \Rightarrow ((\phi^-,\phi^+) \Rightarrow \forall x,(\varphi^-,\varphi^+)$; BR2-Generalization: $(\phi^-,\phi^+) \Rightarrow \forall x,(\phi^-(x),\phi^+(x))$

follows directly. For bipolar completeness, we first recall the proven unipolar completeness of classical propositional logic:

1. *With $\neg$ and $\rightarrow$ as primitive connectives in the unipolar case, UA1-3 alone, together with classical MP, are sufficient to generate all tautologies (Mendelson, 1987).*
2. *With $\neg$ and $\wedge$ as primitive connectives and notice that $p \rightarrow q \equiv \neg p \vee q \equiv \neg(p \wedge \neg q)$, UA4 and UA5 are necessary for zeroth-order completeness (Bridge, 1977).*

Bipolar completeness then follows from that, $\forall(x,y)$, (u,v), P, $Q \in B_1$ and $\forall p,q \in \{-1,0\}$ or $\{0,1\}$,

1. *The bipolar operators $\neg$ and & are linearly bipolar equivalent to their unipolar counterpart $\neg$ and $\wedge$, respectively (See Eq. 3.3 and 3.5).*
2. *Since $P \Rightarrow Q \equiv \neg P \oplus Q \equiv \neg(P \& \neg Q)$ is linearly bipolar equivalent to $p \rightarrow q \equiv \neg p \vee q \equiv \neg(p \wedge \neg q)$ (see Eq. 3.6), $\Rightarrow$ and $\oplus$ can be represented with $\neg$ and &.*
3. *BA1 to BA5 are linearly bipolar equivalent to UA1-5, respectively.*
4. *Classical MP is derivable from BUMP for bipolar linear inference because if we instantiate * with $\otimes$ and substitute $((\psi^-,\psi^+) \Rightarrow (\chi^-,\chi^+))$ with $((0,1) \Rightarrow (0,1))$ we have the linear form $((\phi^-,\phi^+) \& [((\phi^-,\phi^+) \Rightarrow (\varphi^-, \varphi^+))] \Rightarrow (\varphi^-,\varphi^+)$ which is equivalent to the classical MP $\phi \& (\phi \rightarrow \varphi) \rightarrow \varphi$.*
5. *Since $-(x,y) \equiv (-1,0) \otimes (x,y)$ (See Eq. 3.7), $-$ can be represented with $\otimes$.*
6. *Since the bipolar DeMorgan's laws (Table 4) provide the duality between $\oplus$ and &, $\oplus^-$ and $\&^-$, $\varnothing$ and $\otimes$, $\varnothing^-$ and $\otimes^-$, every binary bipolar operator can be represented with $\neg$ and it's dual.*
7. *Since $(x,y) \oplus^- (u,v) \equiv -((x,y) \oplus (u,v))$ (See Eq. 3.8), $\oplus^-$ can be represented with $-$ and $\oplus$. Since $-$ is derived from $\otimes$ and $\oplus$ is dual to &, $\oplus^-$ can be represented with $\otimes$, $\neg$ and &.*
8. *Since $(x,y) \&^- (u,v) \equiv -((x,y) \& (u,v))$ (See Eq. 3.9), $\&^-$ can be represented with $-$ and & and, therefore, can be represented with $\otimes$ and &.*
9. *Since $(x,y) \otimes^- (u,v) \equiv -((x,y) \otimes (u,v))$ (See Eq. 3.11), $\otimes^-$ can be represented with $\otimes$.*

Table 7. Some derived bipolar tautologies

Double negation	$\neg(\neg(\varphi^-,\varphi^+)) \Rightarrow (\varphi^-,\varphi^+)$; $-(-(\varphi^-,\varphi^+)) \Rightarrow (\varphi^-,\varphi^+)$;
Modus Tollens	$(\neg(\varphi^-,\varphi^+) \& ((\phi^-,\phi^+) \Rightarrow (\varphi^-,\varphi^+))) \Rightarrow \neg(\phi^-,\phi^+)$;
Fusion Law	$(\varphi^-,\varphi^+) \Leftrightarrow ((\varphi^-,0) \oplus (0,\varphi^+))$;
Bipolar Separation Law	$\%(\varphi^-,\varphi^+) \Leftrightarrow \{ (\varphi^-,0), (0,\varphi^+) \} \Leftrightarrow ((\varphi^-,\varphi^+) \& (\varphi^-,0))^n \Leftrightarrow (\varphi^-,0)^n$, Where $(\varphi^-,0)^n = (\varphi^-,0) \otimes (\varphi^-,0) \otimes .. \otimes (\varphi^-,0)$, n = 1,2,3,..., with oscillation
Derived linear form of BUMP	$(\phi^-,\phi^+) \& ((\phi^-,\phi^+) \Rightarrow (\varphi^-,\varphi^+)) \Rightarrow (\varphi^-,\varphi^+)$ or in singleton form $(\phi$ and $\phi \rightarrow \varphi) \rightarrow \varphi$. (Note: The classical MP is part of BUMP; but BUMP is not part of MP.)
Derived non-linear forms of BUMP	$[(\phi^-,\phi^+) \Rightarrow (\varphi^-,\varphi^+)] \Rightarrow [((\phi^-,\phi^+)*(\chi^-,\chi^+)) \Rightarrow ((\varphi^-,\varphi^+)*(\chi^-,\chi^+))]$; $[(\phi^-,\phi^+) \Rightarrow (\varphi^-,\varphi^+)] \Leftrightarrow [((\phi^-,\phi^+)*(\phi^-,\phi^+)) \Rightarrow ((\varphi^-,\varphi^+)*(\varphi^-,\varphi^+))]$; $((\varphi^-,\varphi^+)*(\psi^-,\psi^+)) \& [((\varphi^-,\varphi^+)*(\psi^-,\psi^+)) \Rightarrow ((\phi^-,\phi^+)*(\chi^-,\chi^+))] \Rightarrow ((\phi^-,\phi^+)*(\chi^-,\chi^+))$;
BUMP in cascade form:	$[(\phi^-,\phi^+) \Rightarrow (\varphi^-,\varphi^+)] \& [(\psi^-,\psi^+) \Rightarrow (\chi^-,\chi^+)] \& .. \& [(\alpha^-,\alpha^+) \Rightarrow (\beta^-,\beta^+)]$ $\Rightarrow \{ [(\phi^-,\phi^+)*_1(\psi^-,\psi^+)*_2 .. *_i .. *_n (\alpha^-,\alpha^+)]$ $\Rightarrow [(\varphi^-,\varphi^+)*_1(\chi^-,\chi^+)*_2 .. *_i .. *_n(\beta^-,\beta^+)] \}$ where $*_i$, i =1 to n, is the universal operator * that can be instantiated to the same or different bipolar operators.

10. *Since $(x,y) \oslash^- (u,v) \equiv -((x,y) \oslash (u,v))$ (See Eq. 3.12), $\oslash^-$ can be represented with $-$ and $\oslash$ and, therefore, can be represented with $\otimes$ and $\neg$.*
11. *All bipolar tautologies with $\Rightarrow$ and $* \in \{\oplus, \&, \oslash, \otimes, \oplus^-, \&^-, \oslash^-, \otimes^-\}$ must be a derivation from BUMP because BUMP is the only inference rule (Note: a few derived forms of BUMP are listed in Table 7. While classical MP is for deductive reasoning and is generic in nature, BUMP is for a holistic top-down inductive and bottom-up deductive reasoning.)*
12. *Based on (1)-(11), we can conclude that the set of axioms and rules {BA1-BA5, BR1 (BUMP)} is zeroth-order sound and complete with respect to $\neg$, &, and $\otimes$.* □

Theorem 3.7. The set {BA1-7, BR1 (BUMP), BR2} (see Table 6) is first-order sound and complete with respect to $\neg$, &, and $\otimes$.

Proof. Note that UA6, UA7, and UR2 are necessary for the unipolar "lift" from zeroth-order to 1st-order (Bridge, 1977). Since BA6, BA7, and BR2 are linearly bipolar equivalent to their unipolar counterparts UA6, UA7, and UR2, respectively, 1st-order bipolar soundness and completeness with respect to $\neg$, &, and $\otimes$ follow directly from zeroth-order bipolar soundness and completeness. □

Theorem 3.8. (1) The non-linear bipolar dynamic operators $\otimes, \oslash, \otimes^-$, and $\oslash^-$ are recoverable to Boolean operators $\vee$ or $\wedge$ through depolarization; (2) BDL is computational.

Proof.

1. *$\forall(a,b),(c,d),(u,v) \in B_F$, let $(u,v) = (a,b) \otimes (c,d)$, $(u,v) = (a,b) \otimes^- (c,d)$, $(u,v) = (a,b) \oslash (c,d)$, or $(u,v) = (a,b) \oslash^- (c,d)$, we must have $|u| \vee |v| \equiv (|a| \vee |b|) \wedge (|c| \vee |d|)$. Thus, $(|a| \vee |b|) \wedge (|c| \vee |d|)$ is a depolarization filter that recovers $\otimes, \otimes^-, \oslash$ and $\oslash^-$ to $\wedge$. $\forall(a,b),(c,d),(u,v) \in B_F$, let $(u,v) = (a,b) \otimes (c,d)$, $(u,v) = (a,b) \otimes^- (c,d)$, $(u,v) = (a,b) \oslash (c,d)$, or $(u,v) = (a,b) \oslash^- (c,d)$, we must have $|u| \wedge |v| \equiv (|a| \wedge |b|) \vee (|c| \wedge |d|)$. Thus, $(|a| \wedge |b|) \vee (|c| \wedge |d|)$ is a depolarization filter that recovers $\otimes, \otimes^-, \oslash$ and $\oslash^-$ to $\vee$.*
2. *With the three conditions: (i) $\neg$, &, and $\oplus$ are linearly bipolar equivalent to $\neg, \wedge$, and $\vee$, respectively; (ii) $\oplus^-$ and $\&^-$ are both arithmetic negations of $\oplus$ and &, respectively (Eq. 3.8, 3.9); (iii) $\otimes, \otimes^-, \oslash, \oslash^-$ are recoverable to $\wedge$ or $\vee$ through depolarization; we can conclude that BDL is computational.*
□

Theoretically, BUMP established the condition for bipolar quantum entanglement such that $((\phi^-, \phi^+) * (\psi^-, \psi^+))$ in one world implies $((\varphi^-,\varphi^+) * (\chi^-,\chi^+))$ in another world (Theorem 3.5). Since the universal operator can be bound to linear or non-linear bipolar operators such as $\otimes, \oslash, \otimes^-$, and $\oslash^-$ that can be recovered to $\wedge$ or $\vee$ (Theorem 3.8), BUMP bridges a gap between bipolar quantum mechanics and digital computation and communication.

Bipolar Computational Complexity

Based on Figure 3, bipolar causal reasoning is supported with a combination of (1) top-down inductive reasoning to determine the bipolar universal operator and (2) bottom-up deductive reasoning to determine the outcome. Moreover, the following cascading form of BUMP can be used (See Table 7):

$$[(\phi^-,\phi^+)\Rightarrow(\varphi^-,\varphi^+)]\&[(\psi^-,\psi^+)\Rightarrow(\chi^-,\chi^+)]\&..\&[(\alpha^-,\alpha^+)\Rightarrow(\beta^-,\beta^+)]$$

$$\Rightarrow \{[(\phi^-,\phi^+)*_1(\psi^-,\psi^+)*_2 .. *_i .. *_n (\alpha^-,\alpha^+)]\Rightarrow [(\varphi^-,\varphi^+)*_1(\chi^-,\chi^+)*_2 .. *_i .. *_n(\beta^-,\beta^+)]\}$$

where $*_i$, $i = 1$ to n, is the universal operator * that can be instantiated to the same or different bipolar operators.

The complexity of both top-down and bottom-up reasoning can be analyzed as follows:

Complexity of top-down inductive causal reasoning. Notice that instantiating the universal operator * requires the firing of BUMP eight times corresponding to the four pairs of dual bipolar operators &, ⊕, $\&^-$, $\oplus^-$, ⊗, ∅, $\otimes^-$, and $\varnothing^-$. Since the firing of BUMP each time requires a constant number of *c* computational steps, the computational cost of top-down causal reasoning is determined by the number of firings. Let the number be N; the total computational steps would be 8cN which result in *O(N)* complexity.

Complexity of bottom-up deductive causal reasoning. It is interesting to analyze the complexity of bipolar logical (causal) programming in case a Prolog-like bipolar interpreter is involved with bipolar belief revision and dynamic truth maintenance. Intuitively, bipolar logical programming seems to lead to a much high computational complexity. Surprisingly, however, bipolar bottom-up causal reasoning with a Prolog-like bipolar interpreter entails the same computational complexity as required for the unipolar case except algorithms with exponential time or worse. This conclusion is reached based on the following observations and analysis:

Each of the four linear or parallel operators $\{\oplus,\&,\oplus^-,\&^-\}$ doubles the computational cost as for $\vee$ or $\wedge$ based on the definitions of the four operators in Eq. 3.2, 3.3, 3.8, and 3.9 with a constant factor 2 that does not increase computational complexity except for algorithms with exponential time or worse.

Each of the four non-linear operators $\{\varnothing,\otimes,\varnothing^-,\otimes^-\}$ requires 8 times the computational cost as for $\vee$ or $\wedge$ based on their definitions in Eq. 3.4), 3.10, 3.11, and 3.12. However, a larger constant does not increase complexity except for algorithms with exponential time or worse.

Each firing of a bipolar inference rule in forward and backward chaining with the bipolar implication operator ⇒ doubles the computational cost of firing a classical unipolar rule with the unipolar implication operator → (see Eq. 3.6) with a constant 2 that does not increase complexity except for algorithms with exponential time or worse.

Based on the above analysis, it can be concluded that bipolar bottom-up reasoning entails the same computational complexity as that in classical (unipolar) logical programming, albeit algorithms with exponential time or worse are exceptions.

Although the computational complexity may not increase in terms of bipolar logical programming, storage requirements will increase significantly to support bipolar belief revision and dynamic truth maintenance. This will not be further discussed.

BIPOLAR MODALITY

The 1st-order BDL can be "transformed" from a quantificational language to a bipolar modal logic. According to Leibniz, necessity is what is true at every possible world and possibility is what is true at some possible world. Evidently, bipolar equilibrium is a necessity in many micro- or macrocosms and the possibility to reach equilibrium or non-equilibrium is essential for many "worlds". Equilibrium-based bipolar modality is clearly an interesting topic. We show in this section that a *bipolar dynamic modal logic* (*BDML*) is possible.

Definition 3.8. A bipolar sentence of the form □A (necessarily A) is bipolar true if and only if A itself is bipolar true at every possible world; a bipolar sentence of the form ◇A (possible A) is bipolar true just in case A is bipolar true at some possible world. Thus, bipolar necessity is defined as bipolar true; bipolar possibility is defined as possibly bipolar true.

$\forall(x,y)\in B_1$, let the bipolar necessity modality $\Box(x,y) = (x,y)$ and possibility modality $\diamond(x,y) = \neg\Box\neg(x,y)$. Let A,B,C,D be bipolar modal sentences, P_n be one of denumerably many one-place bipolar predicates and $\alpha=(\alpha^-,\alpha^+)$ be a bipolar variable defined on B_1, so that $P_n(\alpha)$ is an atomic formula, and * be the universal operator in BUMP. We define the mapping $\mathfrak{I}$ from the language of bipolar modal logic to the quantificational language of bipolar predicate logic as shown in Table 8, where $\mathfrak{I}$ associates with each bipolar sentence A in the bipolar modal language a unique formula $\mathfrak{I}(A)$ in the quantification language BDL by replacing each atomic bipolar sentence P_n by $P_n(\alpha)$ and putting $\forall\alpha$ and $\exists\alpha$ for occurrences of □ and ◇, respectively. Bipolar modality supports logical reasoning with the form $\mathfrak{I}(A*B)=\mathfrak{I}(A)*\mathfrak{I}(B)$, which enables non-linear dynamic modeling of bipolar equilibria as a form of possibility, necessity, or bipolar truth in an open world. *(see Table 8).*

BIPOLAR RELATIONS AND EQUILIBRIUM RELATIONS

Binary relations are extended to bipolar (binary) relations in (Zhang, 2003).

Definition 3.9. Let X be a bipolar set. A *bipolar relation* in X is a mapping denoted $R{:}X\times X\Rightarrow B_1$.

Definition 3.10. A bipolar relation R in X, where $X = \{x_i\}$, $0<i\leq n$, is said **⊕-⊗ or max-⊗ bipolar transitive** if, $\forall i,j,k$, $0<i,j,k\leq n$, we have

$$\mu_R(x_i,x_k) \geq\geq \max_{x_j}(\mu_R(x_i,x_j)\otimes\mu_R(x_j,x_k)). \quad (3.15)$$

Table 8. Transformation from quantificational language to bipolar modal logic

1) $\mathfrak{I}(P_n) = P_n(\alpha)$, for n = 0,1,2,...;
2) $\mathfrak{I}(x,y) = (x,y)$, $\forall(x,y)\in B_1$;
3) $\mathfrak{I}(-(x,y)) = -(x,y)$, $\forall(x,y)\in B_1$;
4) $\mathfrak{I}(\neg A) = \neg\mathfrak{I}(A)$;
5) $\mathfrak{I}(A*B)=\mathfrak{I}(A)*\mathfrak{I}(B)$;
6) $\mathfrak{I}(A\Rightarrow B) = \mathfrak{I}(A)\Rightarrow\mathfrak{I}(B)$;
7) $\mathfrak{I}(A\Leftrightarrow B)= \mathfrak{I}(A)\Leftrightarrow\mathfrak{I}(B)$;
8) $\mathfrak{I}(\Box A)=\forall\alpha\, \mathfrak{I}(A)$;
9) $\mathfrak{I}(\diamond A)=\exists\alpha\mathfrak{I}(A)$;
10) BUMP in modal form: $\forall A,B,C,D\in B_1$, $\Box(A\Rightarrow B)\&\Box(C\Rightarrow D) \Rightarrow \Box[(A*C) \Rightarrow (B*D)] \Rightarrow [\Box(A*C) \Rightarrow \Box(B*D)]$

The $\oplus$-$\otimes$ **or max-$\otimes$ composition** of an m×n bipolar relation R in X×Y and an n×k bipolar relation Q in Y×Z, denoted by $M=R\otimes Q$, is defined as

$$M(x,z) = \max_y (\mu_R(x,y) \otimes \mu_Q(y,z)), \forall x,y,z,\ x \in X, y \in Y, z \in Z,$$

where max is equivalent to $\oplus$ and the n-fold composition of R with itself is denoted by R^n throughout this book. The bipolar disjunction operator $\oplus$ on m×n bipolar relations R and Q denoted by $R\oplus Q$, is defined by

$$R \oplus Q = [(r_{ij} \oplus q_{ij})], \forall i,j,\ 1 \leq i \leq m,\ 1 \leq j \leq n,$$

where $r_{ij,}$ q_{ij} are the ijth element of R and Q, respectively.

If Ω is used as a binding or coupling operator for a m×n bipolar relation R, $\forall i,j$, $1 \leq i \leq m$, $1 \leq j \leq n$, we have

$$R \equiv (R^-,R^+) = R^-\Omega R^+ = [r_{ij}] = [r_{ij}^-]\Omega[r_{ij}^+] = R^+\Omega R^- = [r_{ij}^+]\Omega[r_{ij}^-] = [(r_{ij}^-,r_{ij}^+)];$$

$$(\neg R^-,R^+) = [(\neg r_{ij}^-,r_{ij}^+)],\ (R^-,\neg R^+) = [(r_{ij}^-,\neg r_{ij}^+)],$$

$$\neg R = (\neg R^-,\neg R^+) = [(\neg r_{ij}^-,\neg r_{ij}^+)].$$

Then, the identities in Table 9 hold.

Definition 3.11. The **bipolar transitive closure** of a bipolar relation R is the smallest transitive bipolar relation containing R, denoted by $\mathfrak{R}$ and

Table 9. Some identities for bipolar relations (Zhang, 2003)

Laws	⊗	⊕
Identity	[I]⊗R≡R; //[I] and R are both n×n	[(0,0)]⊕R≡R; //R is m×n
Null	[(0,0)]⊗R≡ [(0,0)]; //[(0,0)] and R are both n×n	[(-1,1)]⊕R≡ [(-1,1)]; //R is m×n
Idempotency	R⊕R≡R; //R is m×n;	
Commutativity	R⊗Q≡Q⊗R; //R and Q are both n×n	R⊕Q≡Q⊕R; // R,Q are both m×n;
Associativity	(R⊗Q)⊗W≡R⊗(Q⊗W); //R is m×n, Q is n×w, W is w×x;	(R⊕Q)⊕W≡R⊕(Q⊕W); //R,Q,W are all m×n
Distributivity	R⊗(Q⊕W) ≡R⊗Q⊕R⊗W; //R is m×n; N and W are both n×x	
Negation	-[I]⊗R ≡ -R; //[I] and R are both n×n	
Augmentation	$(-1,1)\otimes R \equiv [((-1,1)\otimes r_{ij})]$ $\equiv [((-\max(\lvert r_{ij}^-\rvert,r_{ij}^+),\max(\lvert r_{ij}^-\rvert, r_{ij}^+))]$, $\forall i,j$, $1\leq i\leq m$, $1\leq j\leq n$; // R is m×n	$[(-1,0)]\oplus R\equiv [(-1, r_{ij}^+)]$, $\forall i,j$, $1\leq i\leq m$, $1\leq j\leq n$; $[(0,1)]\oplus R\equiv [(r_{ij}^-,1)]$, $\forall i,j$, $1\leq i\leq m$, $1\leq j\leq n$; //[(-1,0)], [(0,1)],R are all m×n
Inverse	$(R^-,R^+)\oplus(\neg R^-,\neg R^+)\equiv [(-1,1)]$; $(R^-,R^+)\equiv\neg(\neg R)$ // all matrices are m×n	

$$\mathfrak{R} = R^1 \oplus R^2 \oplus R^3 \oplus .... \quad (3.16)$$

An immediate question about (6) is whether the "summation" always terminates, and if so, how soon. The answers are given by

Theorem 3.9. Let $X=\{x_1,x_2,...,x_n\}$ be a finite set, the bipolar transitive closure $\mathfrak{R}$ of R in X exists, is unique, and

$$\mathfrak{R} = R^1 \oplus R^2 \oplus R^3 \oplus ... \oplus R^{2n}. \quad (3.17)$$

Proof. The existence and uniqueness of R follow (1) the 4-valued bipolar space S specifies a closed domain and ≥≥ defines a bipolar partial ordering; (2) the ⊕ operator is idempotent and it does not oscillate; and (3) X is a finite set.

Note that R^i means i-fold composition of R with ⊗. While R is considered as a bigraph, i-fold composition is equivalent to travel i edges through every existing chain between every pair of nodes and each traveled edge is composed to the strength of the traveled chain. A disjunction operation $R^i \oplus R^{i+1}$ enforces the ≥≥ condition of transitivity in Eq. 3.15.

If R were a unipolar relation in X, R would close after applying ⊕ on the first n terms of (6) (Warshall, 1962). This is because n would be the maximum possible length of a nonzero chain from x_i to x_j in the graph. The longest chain would be a circular chain under the conditions (1) i=j and (2) every node appears on the chain exactly once. For instance, the chain (x_1— x_2 —...— x_n—x_1) with length n would be such a circular chain. Any cyclic chain except such a circular chain makes no difference because a cycle on a chain cannot strengthen the chain. After applying ⊕ on the first n terms in (6) for a unipolar closure, every existing chain with length≤ n between every pair of nodes would have been traveled and the strengths of all these chains would have been taken into the "summation" (disjunction) with the idempotent operator ⊕.

In the bipolar case, a bipolar circular chain with length n under the two conditions as above may have a composed strength (-|x|,0). Continuing chaining the same cycle a second time will result in (-|x|,|x|) because (-|x|,0)⊕(-|x|,0)⊗(-|x|,0)=(-|x|,|x|) assuming⊕ -⊗ ∧ transitivity. Thus, the maximum possible length of any effective bipolar chain is 2n in the bipolar case. Cyclic chaining can make a difference as long as travel distance is less than the maximum length 2n. Travel a distance greater than 2n makes no difference because it is already closed on both poles.

Note that all existing bipolar chains with length≤ 2n between every pair of nodes will have been traveled after 2n terms in Eq. 3.16 and the composed strengths of all these chains have been taken into the "summation" (disjunction) with the idempotent bipolar operator ⊕. Therefore, R will close after applying ⊕ on the first 2n terms of Eq. 3.16. □

Theorem 3.10. The computational complexity of (7) can be reduced to $O(n^3)$ using the following equations:

$$\mathfrak{R} \equiv R^1 \oplus R^2 \oplus R^3 \oplus ... \oplus R^{2n}.$$

$$\equiv (R^1 \oplus ... \oplus R^n) \oplus R^n \otimes (R^1 \oplus ... \oplus R^n) \quad (3.18a)$$

$$\equiv (R^1 \oplus ... \oplus R^n) \oplus (R^1 \oplus ... \oplus R^n)^2 \tag{3.18b}$$

$$\equiv (R^1 \oplus ... \oplus R^n) \otimes ([I] \oplus (R^1 \oplus ... \oplus R^n)) \tag{3.18c}$$

$$\equiv (\sum_1^n R^i) \otimes ([(0,1)] \oplus (\sum_1^n R^i)). \tag{3.18d}$$

where [I] in (8c) is an n × n identity matrix.

Proof. Given the identities in Table 9,

(1) *(8a) follows the associative and distributive laws.*

(2) *(8b) follows the associative, idempotent, and commutative laws as we have*

$(R^1 \oplus ... \oplus R^n) \oplus (R^1 \oplus ... \oplus R^n)^2$

$= (R^1 \oplus ... \oplus R^n) \oplus R^1 \otimes (R^1 \oplus ... \oplus R^n) \oplus R^2 \otimes (R^1 \oplus ... \oplus R^n) \oplus ... \oplus R^n \otimes (R^1 \oplus ... \oplus R^n)$

$= R^1 \oplus .. \oplus R^n \oplus R^{n+1} \oplus R^{n+2} ... \oplus R^{2n}$

$= (R^1 \oplus ... \oplus R^n) \oplus R^n \otimes (R^1 \oplus ... \oplus R^n)$

(3) *(8c) follows the distributive law.*
(4) *It is clear from (8a), (8b), or (8c) that R can be computed based on the "summation" of the first n terms in Eq. 3.16.*
(5) *Since the transitive closure C of a classical unipolar binary relation U can be computed based on the first n terms with an $O(n^3)$ algorithm (Warshall, 1962), that is, $C = R^1 \oplus R^2 \oplus ... \oplus R^n$, the computational complexity for $\sum_1^n R^i$ can be reduced to the same order $O(n^3)$.*
(6) *The two operations ⊗ and ⊕ in Eq. 3.18d entail $O(n^3)$ and $O(n^2)$ complexity, respectively.R can be computed with an $O(n^3)$ algorithm.* □

Recall that the subsets involved in a bipolar relation can be bipolar related. Bipolar relations lead to the notions of **bipolar sets including coalition sets, harmony sets, and conflict sets**.

Definition 3.12a. Let $C = \{c_i\}$ and $C \subseteq X$, $0<i \leq n$, C is a **coalition (sub)set** and a relation $\mu(c_j,c_k)$ is **a coalition relation** if, $\forall j,k$, $0<j,k\leq n$, $\mu(c_j,c_k) = (0,1)$.

Definition 3.12b. Let $H=\{h_i\}$ and $H \subseteq X$, $0<i \leq n$, H is **a harmony (sub)set** and $\mu(h_j,h_k)$ is **a harmony relation** if, $\forall j,k$, $0<j,k\leq n$, $\mu(h_j,h_k) = (-1,1)$.

Definition 3.12c. Let C_1 and C_2 be coalition sets, $C_1 = \{c_{1i}\}$ and $C_2 = \{c_{2j}\}$, $0<i\leq n$ and $0<j \leq p$; let $F = C_1 \cup C_2$ and $F \subseteq X$; F is **a conflict (sub)set** and $\mu(c_{1k},c_{2z})$ is **a conflict relation** if, $\forall k,z$, $0<k\leq n$ and $0<z\leq p$, $\mu(c_{1k},c_{2z})=\{-1,0\}$.

Definition 3.13a. Let $C = \{c_i\}$ and $C \subseteq X$, $0<i \leq n$, C is a **coalition subset of X** and a relation $\mu(c_j,c_k)$ is **a coalition relation** if, $\forall j,k$, $0<j,k\leq n$, $\mu(c_j,c_k) = (0,1)$. Let $H=\{h_i\}$ and $H \subseteq X$, $0<i \leq n$, H is **a harmony subset of X** and $\mu(h_j,h_k)$ is **a harmony relation** if, $\forall j,k$, $0<j,k\leq n$, $\mu(h_j,h_k) = (-1,1)$. Let C_1 and C_2 be coalition sets, $C_1 = \{c_{1i}\}$ and $C_2 = \{c_{2j}\}$, $0<i\leq n$ and $0<j \leq p$; let $F = C_1 \cup C_2$ and $F \subseteq X$; F is **a conflict subset of X** and $\mu(c_{1k},c_{2z})$ is **a conflict relation** if, $\forall k,z$, $0<k\leq n$ and $0<z\leq p$, $\mu(c_{1k},c_{2z})=\{-1,0\}$.

Definition 3.13b. A bipolar relation R in X, where $X = \{x_i\}$, $1\leq i\leq n$, is an *equilibrium relation* (Zhang 2003) if it is (1) *bipolar symmetric*; (2) *positive pole or bipolar reflexive*, and (3) *bipolar transitive*. More specifically, a positive pole-reflexive equilibrium relation is a *P-type equilibrium relation*; a negative pole reflexive equilibrium relations is a *N-type equilibrium relation*; and a bipolar reflexive equilibrium relation is an *NP-type equilibrium relation.*

Theorem 3.11. The transitive closure of any negative pole reflexive and symmetric bipolar relation R is an N-type equilibrium relation. The transitive closure of any positive pole reflexive and symmetrical bipolar relation R is a P-type equilibrium relation. The transitive closure of any bipolar reflexive and symmetric bipolar relation R is an NP-type equilibrium relation.

Proof. The theorem follows from that (1) the transitive closure of any symmetrical bipolar relation is still symmetrical because any symmetrical relation is a bigraph; (2) the closure does not change the original reflexivity since (-1,x), (y,1), and (-1,1) are already closed, respectively, on the negative pole, positive pole, or both poles. □

Now we are ready to prove the harmony laws (HLaws):

Theorem 3.12. The following laws hold.

- ***HLaw1***: The bipolar transitive closure $\mathfrak{R}$ of any negative pole reflexive bipolar relation R must be bipolar reflexive.
- ***HLaw2***: Any N-type equilibrium relation must also be an NP-type equilibrium relation and any equilibrium relation is positive-pole reflexive or a P-type equilibrium relation.
- ***HLaw3***: The negative pole R^- and the positive pole R^+ of any NP-type equilibrium relation R in X, where $X = \{x_i\}$, $1\leq i\leq n$, must meet the condition $R^- = -R^+$ or $R^+ = -R^-$, and therefore, all clusters from R must be in a harmonic state.
- ***HLaw4***: A harmony subset must be disjoint by neutral (0,0) relationships with any other types of clusters if any other clusters coexist with a harmony subset in an equilibrium relation.
- ***HLaw5***: The transitive closure of any bipolar relation in a harmony set must be an NP-type equilibrium relation.

Proof.

HLaw1. *It follows from that* $\mu_R(x_i,x_i) \oplus \mu_R(x_i,x_i) \otimes \mu_R(x_i,x_i) = (-1,v) \oplus (-1,v) \otimes (-1,v) = (-1,v) \oplus (-1,1) = (-1,1)$*, which is derived from cycling the negative pole reflexive edge (-1,v) twice (2n) using Eq. 3.18.*

HLaw2. *Since an N-type equilibrium relation must also be NP-type following HLaw1, any equilibrium relation is either: (1) P-type but not N-type, or (2) NP-type. Since an NP-type is both N-type and P-type by definition, any equilibrium relation must be P-type.*

HLaw3. *It follows from that, given any* $\mu_R(x_i,x_j) = (u,v)$*, if* $|u| > v$*, (u,v) must be (-1,0), that would violate the bipolar transitivity requirement* $(u,v) \geq\geq \mu_R(x_i,x_j) \otimes \mu_R(x_j,x_j) = (-1,0) \otimes (-1,1) = (-1,1)$ *because the reflexive edge* $\mu_R(x_j,x_j)$ *must be harmonic for any NP-type equilibrium relation. Similarly, if* $|u| < v$*, (u,v) must be (0,1), that would violate the bipolar transitivity requirement* $(u,v) \geq\geq \mu_R(x_i,x_j) \otimes \mu_R(x_j,x_j) = (0,1) \otimes (-1,1) = (-1,1)$*. Since each relationship in R is either (0,0) or (-1,1), all clusters induced from* R^+ *must be in a harmonic state.*

HLaw4. *According to the augmentation law we have,* $(-1,1) \otimes (-1,0) = (-1,1)$ *and* $(-1,1) \otimes (0,1) = (-1,1)$*. Therefore, if agent A and B are in a harmonic relation, B and C have any non-neutral relationship will necessarily imply A, B, and C are all in a harmonic relation based on the transitivity of an equilibrium relation. Since only when (a,b)=(0,0) we have* $(-1,1) \otimes (a,b) = (0,0)$ *(null law), the only way for other type of clusters to coexist with a harmony cluster is to have a (0,0) relation with it. Otherwise, other clusters have to join the harmony subset.*

HLaw5. *First, given any two nodes x and y with a harmonic relationship (-1,1), the transitive cycle* $x \rightarrow y \rightarrow x$ *always results in bipolar reflexivity. Secondly, harmony implies symmetry.* □

Theorem 3.13. The following laws hold.

- ***ELaw1***. R^+ of any crisp bipolar transitive relation R in X, where $X = \{x_i\}$, $1 \leq i \leq n$, is unipolar transitive.
- ***ELaw2***. R^+ of any crisp equilibrium relation R in X, where $X = \{x_i\}$, $1 \leq i \leq n$, is an equivalence relation.
- ***ELaw3***. For any n×n crisp bipolar transitive relation R, $|R^-| \cup R^+$ results in a transitive unipolar relation.
- ***ELaw4***. For any crisp equilibrium relation R, $|R^-| \cup R^+$ results in an equivalence relation.

Proof.

ELaw1. *Based on the bipolar transitivity definition (Eq. 3.17), if* R^+ *were not unipolar transitive, R would not be bipolar transitive.*

ELaw2. *First, any crisp equilibrium relation is positive pole reflexive. Secondly, equilibrium implies bipolar symmetry. Thirdly, positive pole transitivity* R^+ *follows from bipolar transitivity.*

ELaw3. *Since R is bipolar transitive, let* μ_R *be a mapping function of R in set X with size n, we must have,* $\forall i,j,k$, $1 \leq i,j,k \leq n$, $\mu_R(x_i,x_k) \geq\geq \max_{x_j}(\mu_R(x_i,x_j) \otimes \mu_R(x_j,x_k))$*. Then, let* $R1 = |R^-| \vee R^+$ *be a unipolar relation and, let* ρ_{R1} *be the mapping function of R1, we must have* $\rho_{R1}(x_i,x_k) \geq \max_{x_j}(\rho_{R1}(x_i,x_j) \wedge \rho_{R1}(x_j,x_k))$*, therefore* $|R^-| \vee R^+$ *must be unipolar transitive.*

ELaw4. *First, positive (unipolar) reflexivity follows from the definition of an equilibrium relation. Secondly, unipolar symmetry follows from the fact that the* ∨ *operation does not change the property of symmetry. Thirdly, its transitivity follows from bipolar transitivity.* □

On the bipolar partitioning laws (BPLaws) we have:

Theorem 3.14. The following laws hold.

- ***BPLaw1.*** $|R^-|\cup R^+$ of any equilibrium relation R in X, where $X = \{x_i\}$, $1 \le i \le n$, induces disjoint subsets including coalition subset(s) not in conflict, harmony subset(s), and conflict subset(s) (if any) in X.
- ***BPLaw2.*** Given any equilibrium relation R in X, where $X = \{x_i\}$, $1 \le i \le n$, R^+ induces disjoint subsets including coalition subset(s) that are in or not in conflicts and harmony subset(s) (if any) in X.
- ***BPLaw3.*** Given any equilibrium relation R in X, where $X = \{x_i\}$, $1 \le i \le n$, let the set of disjoint subsets induced from ($|R^-|\cup R^+$) be S1; let the set of disjoint subsets induced from R^+ be S2.

1. S1∩S2 results in a set S3 of disjoint harmony subset(s) and coalition subset(s) that are not in any conflict in X.
2. S1∪S2 results in a set S4 of disjoint conflict subset(s), coalition subsets in the conflict subset(s), coalition subset(s) not in conflict, and harmony subset(s).
3. S1-S2 results in a set S5 of disjoint conflict subset(s), each of which contains two coalition subsets.
4. S2-S1 results in a set S6 of disjoint coalition subset(s) that are involved in a conflict.

Proof.

BPLaw1. *First, $|R^-|\cup R^+$ is an equivalence relation(ELaw4). Non-neutral relationships (-1,0), (0,1), and (-1,1) in R become related in $|R^-|\cup R^+$ and neutral (0,0) relationships in R become. Second, coalition subsets, conflict subsets, and harmony subsets neutral to each other in R become unrelated to each other. Third, based on HLaw4, a harmony subset must be disjoint with other clusters. Fourth, two and at most two coalition subsets can join a conflict set because (1) if coalition A is in conflict with B and B is in conflict with C implies A and C is a coalition due to the fact that (-1,0)⊗(-1,0)=(0,1) ("an enemy's enemy is a friend"); and (2) if any element in a coalition is in conflict with an element in another coalition, all elements in the first coalition must be in conflict with everyone in the second coalition due to the fact that (0,1)⊗(-1,0)=(-1,0) ("a friend's enemy is an enemy").*

BPLaw2. *First R^+ is an equivalence relation (ELaw2) that induces disjoint partitions. Second, (0,1) and (-1,1) relationships in R become 1 in R^+, the (-1,0) relationships are excluded. Therefore, the partitions induced from R^+ can only be coalition or harmony subsets.*

BPLaw3. *It follows from BPLaw1 and 2 directly.* □

Theorem 3.15. Given a bipolar relation R in X, where $X = \{x_i\}$, $1 \le i \le n$, the following conditions are necessary and sufficient for R to be an equilibrium relation:

1. R^+ is an equivalence relation;
2. $|R^-|\vee R^+$ is an equivalence relation;
3. if ($R^+\wedge|R^-|$) is not null it must be a local equivalence;

4. if $(|R^-|\vee R^+)-(R^+\wedge|R^-|)$ is not null it must be a local equivalence;
5. if $(|R^-|\vee R^+)-|R^-|$ is not null it must be a local equivalence; and
6. $R^+-(R^+\wedge|R^-|)\equiv(|R^-|\vee R^+)-|R^-|$, null or not null.

Proof. It follows from

1. *Positive pole reflexivity of R follows from that R^+ is an equivalence relation;*
2. *Bipolar transitivity of R follows from its reflexivity and the fact $(R^+\wedge|R^-|)$, $(|R^-|\vee R^+)-(R^+\wedge|R^-|)$, and $R^+-(R^+\wedge|R^-|)\equiv(|R^-|\vee R^+-|R^-|)$ are all local equivalence if not null. That is, the disjoint harmony sub-relations in X are bipolar transitive; the disjoint coalition sub-relations are bipolar transitive, and the disjoint conflict subrelations are bipolar transitive. Since there are no other types of subrelations besides the three types, R must be bipolar transitive.*
3. *Symmetry of R follows from the fact that R is transitive and both R^+ and $|R^-|\vee R^+$ are equivalence relations.* □

On the local equivalence laws (LELaws), we have

Theorem 3.16. The following laws hold.

- ***LELaw1***: Given any equilibrium relation R in X, where $X = \{x_i\}$, $1\leq i\leq n$, if $R^+\wedge|R^-| = [|r^-_{ij}|\wedge r^+_{ij}]$, $\forall i,j,\ 1\leq i,j\leq n$, is not null it must be a local equivalence, which induces disjoint harmony subsets.
- ***LELaw2***: Given any equilibrium relation R in X, where $X = \{x_i\}$, $1\leq i\leq n$, if $(|R^-|\vee R^+)-(R^+\wedge|R^-|)$ is not null, it must be a local equivalence in X, which induces either disjoint conflict-free coalition subsets, or disjoint conflict subsets, or disjoint subsets of both types
- ***LELaw3***: Given any equilibrium relation R in X, where $X = \{x_i\}$, $1\leq i\leq n$, if $(|R^-|\vee R^+)-|R^-|$ is not null it must be a local equivalence in X that induces one or more coalition subsets that are either involved or not involved in a conflict.
- ***LELaw4***: Given any equilibrium relation R in X, where $X = \{x_i\}$, $1\leq i\leq n$, we have $R^+-(R^+\wedge|R^-|) \equiv (|R^-| \vee R^+)-|R^-|$.

Proof.

LELaw1*. First, following the 5th harmony law a transitive binary relation on a harmony set must be an NP-type equilibrium relation. Secondly, only a harmonic relationship (-1,1) in R results in a nonzero value in $R^+\wedge|R^-|$. Thirdly, following the 4th harmony law, harmonic clusters must be disjoint with each other and any other clusters. Thus, if $R^+\wedge|R^-|$ is not null, it must be a local equivalence that identifies one or more disjoint harmony subsets.*

LELaw2. *We have*

1. *Following last theorem $(|R^-|\vee R^+)$ is an equivalence relation;*
2. *Following last theorem if $(R^+\wedge|R^-|)$ is not null it is an local equivalence that identifies harmonic relationships and induces one or more disjoint harmony subsets; and*
3. *the subtraction $(|R^-|\vee R^+)-(R^+\wedge|R^-|)$ simply removes the harmonic type relationships without affecting the unipolar reflexivity, symmetry, and transitivity among other types of true relationships that are positively related or negatively related, because Truth(-1,0)=1 and Truth(0,1)=1.*

LELaw3. *It follows from*

1. *$|R^-|\vee R^+$ is an equivalence relation;*
2. *(-1,0) and (-1,1) values in R lead to 1 values in $|R^-|$; and*
3. *The subtraction removes disjoint harmonic relationships and conflict relationships without affecting the reflexivity, symmetry, and transitivity among coalition subsets.*

LELaw4. *It follows from*

1. *(0,1) and (-1,1) in R lead to 1 in R^+; (-1,1) values in R lead to 1 in $(R^+\wedge|R^-|)$;*
2. *(-1,0),(0,1), and (-1,1) in R all lead to 1 in $(|R^-|\vee R^+)$; (-1,0) and (-1,1) in R lead to 1 in $|R^-|$; and*
3. *R^+-$(R^+\wedge|R^-|)$ and $(|R^-|\vee R^+)$-$|R^-|$ both identify the (0,1)-values in R.* □

On the necessary and sufficient conditions (besides reflexivity, symmetry, and bipolar transitivity) for a bipolar relation R to be an equilibrium relation, we have

Corresponding to equivalence classes, an **equilibrium class** of an equilibrium relation E can be defined as a subequilibrium of E associated with $x_i \in X$ that identifies a row of the n×n matrix of a bipolar equilibrium relation (Zhang, 2003).

Thus, an equilibrium relation is a non-linear bipolar fusion of many equivalence relations. It is shown (Zhang, 2003, 2005a, 2005b) that $(R^+\cap|R^-|)$ induces *harmony sets*; R^+-$(R^+\cap|R^-|)$ induces *coalition sets*; R^--$(R^+\cap|R^-|)$ induces *bipolar conflict sets* each of which consists of two coalition subsets.

From an application perspective, equilibrium relations can be used for causal reasoning and multiagent coordination and global regulation (Wellman, 1994; Wellman & Hu, 1998; Chaib-draa, 2002; Zhang *et al.*, 1989, 1992; Zhang, 2003, 2005a, 2005b, 2006). Theoretically, as bipolar sets equilibrium relations keep the existing syntactic structures of classical sets intact. However, elementary bipolarity leads to bipolar interaction and enables the interdependence between a set and its elements such as coalition, harmony, and conflict sets.

ON GÖDEL'S INCOMPLETENESS THEOREM

Bipolar inference leads to a brand new redress on the liar's case in the ancient paradox (Rucker, 1987, p. 221). The paradox, which was known even to Aristotle, consists of a single sentence L "This sentence is not true." If L is true, then L is not true. Thus, L is both true and false. But that is against the fundamental law of non-contradiction. So the sentence has been considered a "liar" for more than 2300 years. Now with bipolar representation, the "liar" can be proven an honest being with the virtue of self-negation ability (Note: A sentence is not a being, but it was made a being when it is called a liar.)

Let (ϕ^-,ϕ^+) = (self-negation, self-assertion) and (φ^-,φ^+) = (self-consciousness, self-assurance), we then have the bipolar inference rule $(\phi^-,\phi^+)\Rightarrow(\varphi^-,\varphi^+)$. Following BUMP we have

$$[((\phi^-,\phi^+)\Rightarrow(\varphi^-,\varphi^+))] \Rightarrow [(\phi^-,\phi^+)*(\phi^-,\phi^+) \Rightarrow (\varphi^-,\varphi^+)*(\varphi^-,\varphi^+)].$$

Since L has self-negation ability, we have $(\phi^-,\phi^+)(L) = (-1,0)$ and $(\varphi^-,\varphi^+)(L) = (-1,0)$. Then, we have

$$[(\phi^-,\phi^+)(L) \Rightarrow (\varphi^-,\varphi^+)\ (L)]; \quad (1)$$

$$[(\phi^-,\phi^+)*(\phi^-,\phi^+)](L) \Rightarrow [(\varphi^-,\varphi^+)*(\varphi^-,\varphi^+)]\ (L)]; \quad (2)$$

$$[(\phi^-,\phi^+)\otimes(\phi^-,\phi^+)](L) \Rightarrow [(\varphi^-,\varphi^+)\otimes(\varphi^-,\varphi^+)]\ (L)]; \quad (3)$$

From (1) to (3) we have

$$\{(\phi^-,\phi^+)\oplus[(\phi^-,\phi^+)\otimes(\phi^-,\phi^+)]\}(L) \Rightarrow \{(\varphi^-,\varphi^+)\oplus[(\varphi^-,\varphi^+)\otimes(\varphi^-,\varphi^+)]\}(L). \quad (4)$$

If we substitute the values of (-1,0) into $(\phi^-,\phi^+)(L)$ and (φ^-,φ^+) in (4) we have

$$(-1,0)\oplus(-1,0)^2 = (-1,0)\oplus[(-1,0)\otimes(-1,0)] = (-1,0) \oplus (0,1) = (-1,1). \quad (5)$$

As commonsense, (5) shows that self-negation can lead to self-correction. If we define self-correction as negative reflexivity, we have $(-1,0)^2 = (0,1)$ and the bipolar fusion $(-1,0)\oplus[(-1,0)\otimes(-1,0)] = (-1,0) \oplus (0,1) = (-1,1)$. Therefore, the "liar" L could be diagnosed as having perfect bipolar mental balance after being mistreated for 2300 years, provided L's bipolar fusion and oscillation functionalities $\oplus$ and $\otimes$ are intact.

After all, the so-called "liar" was just like an honest person saying "Sorry, I was wrong" or like a computer system that detects its own error (e.g., parity checking). Evidently, the paradox was caused by the representational and computational limitation of unipolar truth-based logic in modeling an agent not by the agent itself.

It should be remarked that the above redress is different from any other possible redress with an existing 4-valued logic. For instance, the paradox can be proven a contradiction with a 4-valued paraconsistency logic (Belnap, 1977) or bilattice (Ginsberg, 1990) or an intuitionistic logic (Brouwer, 1912). Although a proof of contradiction can lead to a "not guilty verdict", it cannot prove the liar's "self-consciousness" and "honesty."

According to Rucker (1987, p. 221), the ancient paradox of the liar was the original inspiration for Gödel's incompleteness theorem (Gödel, 1930). The redress leads to the following observations on Gödel's incompleteness theorem:

1. The "universal truth machine" questioned by Gödel in his proof was mistreated like the "liar" because the machine was only programmed to determine bivalent truth while Gödel's question required the machine to reason about itself with self-negation, self-consciousness, or mental equilibrium that is beyond bivalent logic.
2. When the machine is called a "liar" it is entitled to act as a being, to agree or disagree. However, Gödel's questioning of the universal machine is like "Even if I give you a contradiction you must tell me whether it is true or false. No other answer."
3. With Belnap's 4-valued logic or Brower's intuitionistic logic an answer to Gödel's questioning would simply be "Your sentence is a contradiction." With BDL the answer could be "I am sorry. Your sentence is a contradiction in bivalent logic. I might be called a liar if I say 'I am false'. How about I say *'As an agent I am either in equilibrium or in disorder but I cannot be true or false.'*"

4. It is clear that bipolar dynamic logic is needed for modeling an equilibrium world of bipolar agents. Gödel's incompleteness theorem proved the incompleteness of classical bivalent logic, which does not necessarily hold in the same way for BDL. BDL may fail but not necessarily in the same way as Gödel proved.

Many scientists believe that Einstein's views in physics were related to the mathematical views of Hilbert and Gödel. It is well-known that Einstein was a friend and colleague of Gödel at Princeton University. Einstein once visited German mathematician Hilbert by invitation and was aware of Hilbert's effort in axiomatizing physics (Hilbert 1901). Einstein asserted that *"Physics constitutes a logical system of thought which is in a state of evolution, whose basis (principles) cannot be distilled, as it were, from experience by an inductive method, but can only be arrived at by free invention"* (Einstein, 1916). Clearly, Einstein believed that Hilbert's effort in axiomatizing physics (Hilbert, 1901) was hopeful. In 1930, Gödel published his incompleteness theorem (Gödel, 1930) and shattered the hope for an axiomatization of physics. Four years later, Einstein reaffirmed his view (Einstein, 1934) that *"pure thought can grasp reality"* and *"nature is the realization of the simplest conceivable mathematical ideas."* In 1940, Einstein asserted that the grand unification of general relativity and quantum mechanics needs a new logical foundation: *"For the time being we have to admit that we do not possess any general theoretical basis for physics which can be regarded as its logical foundation"* (Einstein, 1940).

Evidently, Einstein never wavered on a possible logical foundation for physics. Gödel's incompleteness theorem, Hilbert's effort in axiomatizing physics, and Einstein's assertion on a new logical foundation for physics were all giant steps. To be honest, however, we have to say that the three giants stopped short of pointing out the inevitable:

1. The incompleteness of classical logic is due to its lack of syntax and semantics for the fundamental physical concept *"equilibrium"* or *"symmetry"*.
2. A logical foundation for physics requires a philosophically deeper cosmology beyond spacetime and a different mathematical abstraction beyond classical truth-based unipolar cognition.

RESEARCH TOPICS

Bipolar sets and BDL presents an equilibrium-based mathematical abstraction for the first time ever that transcends truth-based tradition. More theoretical formulations are yet to come on the integration of unipolar and bipolar systems for complex applications. Some suitable application domains are outlined in the following:

Bipolar Causality and Bipolar Relativity. Bipolar Sets and BDL provide the mathematical abstraction for logically definable causality and YinYang bipolar relativity (Zhang 2009a, 2009b, 2009c, 2009d) (Ch. 6).

Bipolar Quantum Linear Algebra and Bipolar Quantum Cellular Automata. Bipolar sets and BDL provide a logical basis for bipolar fuzzy sets, bipolar dynamic fuzzy logic, bipolar quantum linear algebra and bipolar quantum cellular automata (Chs. 8,9).

Bipolar Quantum Computing. While classical logics use a bottom-up approach for computing and quantum logics use a holistic top-down approach for computing, both use bit patterns for information encoding. Therefore, bivalent and quantum logics are both unipolar systems that cannot be directly used for the representation and visualization of bipolar equilibria in micro- or macroscopic agent worlds like those in nuclear magnetic resonance (Khitrin & Fung, 2000; Fung, 2001) and quantum cellular automata (Lent *et al.,* 1993). YinYang bipolar logic and sets present an alternative holistic approach to information encoding in quantum computing (Zhang, 2009a, 2009b, 2009c, 2009d, 2010) (Chs. 6,7).

Axiomatization of Physics. Bipolar sets and BDL provide a basis for resolving the LAFIP paradox and for a minimal but most general axiomatization of all physics beyond spacetime geometry (Zhang, 2009c, 2009d) (Ch. 6).

Biomedicine. A direct application of bipolar sets and BDL is YinYang bipolar dynamic neurobiological modeling, diagnostic analysis, and psychopharmacology for mental depression and mental disorders (Zhang, 2007; Zhang, Pandurangi & Peace, 2007). Evidently, given a set of bipolar disorder patients P and a set of biomedical treatment T, both P and T can be characterized as bipolar sets. Preliminary results show that bipolar inference with BUMP can unravel mental or neurophysiologic intrinsics of bipolar disorders from certain counter-intuitive symptoms (Zhang, 2007; Zhang, Pandurangi & Peace, 2007) (Ch. 10).

Bipolar Fuzzy Sets. Bipolar sets and BDL are extended to bipolar fuzzy sets and bipolar dynamic fuzzy logic (BDFL) (Zhang, 2005a, 2006a) (Chs. 4,5).

Bipolar Cognitive Mapping. With bipolar sets, fuzzy sets and equilibrium relations, bipolar cognitive mapping can be effectively used for equilibrium analysis, decision, and multiagent coordination in global regulation (Zhang, 2003a, 2003b, 2005b, 2006) (Ch. 11).

Bipolar Partitioning and Clustering. While an equivalence relation leads to hard partitioning and a crisp equilibrium relation induces disjoint ***coalition sets, harmony sets, and conflict sets,*** it is shown that a bipolar fuzzy equilibrium relation induces disjoint or joint ***quasi-coalition sets, harmony sets, and conflict sets in X*** (Zhang, 2003b, 2006) (Chs. 5, 11).

Equilibrium Energy and Stability Analysis for Global Regulation. Equilibrium energy can be defined on equilibrium relations for stability analysis in both social and natural sciences, especially in macroeconomics (Ch. 5) and international relations (Ch. 11). From equilibrium relations and equilibrium classes bipolar laws can be derived for optimization, decision, coordination, and global regulation (Zhang 2003a,b, 2005b,2006) (Chs. 5, 11).

SUMMARY

An equilibrium-based approach to mathematical abstraction and axiomatization has been presented. To avoid LAFIP, YinYang bipolar set theory has been introduced. Bipolar sets lead to YinYang bipolar dynamic logic (BDL). Soundness and completeness of a bipolar axiomatization has been asserted. The computability of BDL has been axiomatically proved. The recovery of BDL to Boolean logic has been proved through depolarization. Computational complexity has been analyzed. A redress on the ancient paradox of the liar has been presented that leads to a few observations on Gödel's incompleteness theorem.

Bipolar relations, bipolar transitivity, and equilibrium relations have been introduced. It has been shown that a bipolar equilibrium relation enables non-linear bipolar fusion of many equivalence relations. Bipolar partitioning has been discussed with equilibrium relations.

A key element of BDL is bipolar universal modus ponens (BUMP). For thousands of years, classical modus (MP) has been the only generic inference rule in truth-based logical deduction. BUMP as a non-linear bipolar dynamic generalization of MP exhibits both classical logical property and bipolar quantum entanglement. Thus, BDL and bipolar equilibrium relations provide a mathematical basis for bipolar causality, bipolar relativity, and an equilibrium-based axiomatization for bipolar quantum computing.

REFERENCES

Belnap, N. (1977). A useful 4-valued logic . In Epstein, G., & Dunn, J. M. (Eds.), *Modern uses of multiple-valued logic* (pp. 8–37). Reidel.

Bridge, J. (1977). *Beginning model theory–the completeness theorem and some consequence*. Oxford University Press.

Brouwer, L. E. (1912). Intuitionism and formalism. *Bulletin of the American Mathematical Society*, *20*, 81–96. doi:10.1090/S0002-9904-1913-02440-6

Chaib-draa, B. (2002). Causal maps: Theory, implementation, and practical applications in multi-agent environments. *IEEE Transactions on KDE*, *14*(6), 1201–1217.

Dirac, P. A. M. (1927). The quantum theory of the emission and absorption of radiation. *Proceedings of the Royal Society of London. Series A*, *114*, 243–265. doi:10.1098/rspa.1927.0039

Feynman, R. P. (1985). *QED: The strange theory of light and matter*. Princeton University Press.

Ginsberg, M. (1990). Bilattices and modal operators. [Oxford University Press.]. *Journal of Logic and Computation*, *1*(1), 41–69. doi:10.1093/logcom/1.1.41

Gödel, K. (1930). Uber formal unentscheidbare Sätze der Principia Mathematica und verwandter Systeme. *Monats hefte fur Math. und Phys.*, *37*, 349-360.

Hilbert, D. (1901). Mathematical problems. *Bulletin of the American Mathematical Society*, *8*, 437–479. doi:10.1090/S0002-9904-1902-00923-3

Mendelson, E. (1987). *Introduction to mathematical logic* (3rd ed.). Monterey, CA: Wadsworth & Brooks/Cole Adv. Books & Software.

Warshall, S. (1962). A theorem on Boolean matrices. *Journal of the ACM*, *9*(1), 11–12. doi:10.1145/321105.321107

Wellman, M. P. (1994). Inference in cognitive maps. *Mathematics and Computers in Simulation*, *36*, 137–148. doi:10.1016/0378-4754(94)90028-0

Wellman, M. P., & Hu, J. (1998). Conjectural equilibrium in multiagent learning. *Machine Learning*, *33*, 179–200. doi:10.1023/A:1007514623589

Zhang, W.-R. (2003). Equilibrium relations and bipolar cognitive mapping for online analytical processing with applications in international relations and strategic decision support. *IEEE Transactions on SMC . Part B*, *33*(2), 295–307.

Zhang, W.-R. (2005a). YinYang bipolar lattices and L-sets for bipolar knowledge fusion, visualization, and decision making. *International Journal of Information Technology and Decision Making*, *4*(4), 621–645. doi:10.1142/S0219622005001763

Zhang, W.-R. (2005b). YinYang bipolar cognition and bipolar cognitive mapping. *International Journal of Computational Cognition*, *3*(3), 53–65.

Zhang, W.-R. (2006). YinYang bipolar fuzzy sets and fuzzy equilibrium relations for bipolar clustering, optimization, and global regulation. *International Journal of Information Technology and Decision Making*, *5*(1), 19–46. doi:10.1142/S0219622006001885

Zhang, W.-R. (2007). YinYang bipolar universal modus ponens (bump)–a fundamental law of non-linear brain dynamics for emotional intelligence and mental health. *Walter J. Freeman Workshop on Nonlinear Brain Dynamics, Proceedings of the 10th Joint Conference of Information Sciences*, (pp. 89-95). Salt Lake City, Utah, USA.

Zhang, W.-R. (2009a). Six conjectures in quantum physics and computational neuroscience. *Proceedings of 3rd International Conference on Quantum, Nano and Micro Technologies (ICQNM 2009)*, (pp. 67-72). Cancun, Mexico.

Zhang, W.-R. (2009b). YinYang Bipolar Dynamic Logic (BDL) and equilibrium-based computational neuroscience. *Proceedings of International Joint Conference on Neural Networks (IJCNN 2009)*, (pp. 3534-3541). Atlanta, GA.

Zhang, W.-R. (2009c). YinYang bipolar relativity–a unifying theory of nature, agents, and life science. *Proceedings of International Joint Conference on Bioinformatics, Systems Biology and Intelligent Computing (IJCBS)*. (pp. 377-383). Shanghai, China.

Zhang, W.-R. (2009d). The logic of YinYang and the science of TCM–an Eastern road to the unification of nature, agents, and medicine. [IJFIPM]. *International Journal of Functional Informatics and Personal Medicine*, *2*(3), 261–291. doi:10.1504/IJFIPM.2009.030827

Zhang, W.-R. (2010). YinYang bipolar quantum entanglement–toward a complete quantum theory. *Proceedings of the 4th International Conference on Quantum, Nano and Micro Technologies (ICQNM 2010)*, (pp. 77-82). St. Maarten, Netherlands Antilles.

Zhang, W.-R., Chen, S., & Bezdek, J. C. (1989). POOL2: A generic system for cognitive map development and decision analysis. *IEEE Transactions on SMC*, *19*(1), 31–39.

Zhang, W.-R., Chen, S., Wang, W., & King, R. (1992). A cognitive map based approach to the coordination of distributed cooperative agents. *IEEE Transactions on SMC*, *22*(1), 103–114.

Zhang, W.-R., Pandurangi, A., & Peace, K. (2007). YinYang dynamic neurobiological modeling and diagnostic analysis of major depressive and bipolar disorders. *IEEE Transactions on Bio-Medical Engineering*, *54*(10), 1729–1739. doi:10.1109/TBME.2007.894832

ADDITIONAL READING

Zhang, W.-R., & Zhang, L. (2003). Soundness and Completeness of a 4-Valued Bipolar Logic. *Int'l Journal on Multiple-Valued Logic, Vol.*, *9*, 241–256.

Zhang, W.-R., & Zhang, L. (2004). YinYang Bipolar Logic and Bipolar Fuzzy Logic. *Information Sciences*, *165*(3-4), 265–287. doi:10.1016/j.ins.2003.05.010

KEY TERMS AND DEFINITIONS

Bipolar Element: A bipolar element is an element or agent with a negative pole and a positive pole.

Bipolar Set: A bipolar set is a set of bipolar elements.

Bipolar Poset: A bipolar poset is a bipolar set satisfying the bipolar partial ordering relation $\geq\geq$.

Bipolar Lattice: A bipolar lattice is a bipolar poset with the three operators $\oplus$, &, and $\otimes$ of BDL defined on every pair of its elements.

BDL: It stands for "bipolar dynamic logic." BDL is defined on a bipolar lattice typically on the 4-valued bipolar lattice $B_1 = \{-1,0\}\times\{0,1\}$.

BDML: It stands for "bipolar dynamic modal logic."

BUMP: It stands for "bipolar universal modus ponens" – a non-linear bipolar dynamic and symmetrical generalization of classical modus ponens from the bivalent lattice $\{0,1\}$ to $B_1 = \{-1,0\}\times\{0,1\}$ or any other bipolar lattice.

Bipolar LEM: It stands for "bipolar laws of excluded middle."

Bipolar Dynamic DeMorgan's Laws: They are non-linear bipolar dynamic symmetrical generalizations of the classical De Morgan's laws from the bivalent lattice $\{0,1\}$ to $B_1 = \{-1,0\}\times\{0,1\}$ or to any other bipolar lattice.

Bipolar Relation: It is a binary relation characterized by bipolar dynamic logic (BDL) values (Zhang 2003a).

Equilibrium Relation: A given bipolar relation R on a set, *A* is an equilibrium relation if and only if it is bipolar reflexive, symmetric and bipolar $\oplus$-$\otimes$ transitive or interactive.

Chapter 4
Bipolar Quantum Lattice and Dynamic Triangular Norms

ABSTRACT

Bipolar quantum lattice (BQL) and dynamic triangular norms (t-norms) are presented in this chapter. BQLs are defined as special types of bipolar partially ordered sets or posets. It is shown that bipolar quantum entanglement is definable on BQLs. With the addition of fuzziness, BDL is extended to a bipolar dynamic fuzzy logic (BDFL). The essential part of BDFL consists of bipolar dynamic triangular norms (t-norms) and their co-norms which extend their truth-based counterparts from a static unipolar fuzzy lattice to a bipolar dynamic quantum lattice. BDFL has the advantage in dealing with uncertainties in bipolar dynamic environments. With bipolar quantum lattices (crisp or fuzzy), the concepts of bipolar symmetry and quasi-symmetry are defined which form a basis toward a logically complete quantum theory. The concepts of strict bipolarity, linearity, and integrity of BQLs are introduced. A recovery theorem is presented for the depolarization of any strict BQL to Boolean logic. The recovery theorem reinforces the computability of BDL or BDFL.

INTRODUCTION

The concept *lattice* is central in logic and set theory. A lattice is a partially ordered set (poset) in which any two elements have a least upper bound or supremum ($\vee$) and a greatest lower bound or infimum ($\wedge$). Boolean logic (Boole, 1854) is defined on the bivalent lattice $\{0,1\}$; Zadeh's fuzzy logic (Zadeh, 1965) is defined on the fuzzy or real-valued lattice $[0,1]$; YinYang bipolar dynamic logic (BDL) is defined on the YinYang bipolar lattice $\{-1,0\} \times \{0,1\}$ (Ch. 3).

DOI: 10.4018/978-1-60960-525-4.ch004

Bipolar lattice introduces equilibrium or non-equilibrium-based physical syntax and semantics into lattice theory. Such physical syntax and semantics lead to bipolar quantum entanglement as defined in bipolar universal modus ponens (BUMP) (Ch. 3). In this chapter we reference bipolar lattice as bipolar quantum lattice (*BQL*) due to its salient feature of bipolar quantum entanglement. This should be distinguished from *lattice quantum chromodynamics* (lattice QCD) in physics, which is a theory of quarks and gluons formulated on a spacetime lattice.

The properties of BQLs are analyzed in this chapter. Notably, the quantum nature of BQLs is distinguished from truth-based unipolar lattice. Energy equilibrium, balance, and symmetry are formally defined and axiomatically formulated.

Although lattice-ordered *triangular norms* or *t-norms* and *t-conorms* are the fundamental operators in probability, logic, and fuzzy set theory, which play essential roles in computational intelligence, artificial intelligence (AI), cognitive informatics, and decision making, they have a number of truth-based fundamental limitations:

1. The static truth-based nature of classical mathematical abstraction in set theory leads to the well-known closed-world assumption in logical computation that forms a bottleneck in automated reasoning especially in equilibrium-based open-world open-ended dynamic reasoning.
2. Without bipolarity any truth-based logical value (crisp or fuzzy) can't directly carry bipolar equilibrium-based holistic physical semantics. Therefore, truth-based norms are too "logical" for reasoning on the "illogical" but nevertheless natural or physical aspects such as bipolar interaction, oscillation, disorders, and quantum entanglement (see Ch. 3, Figure 1). This limitation leads to the LAFIP or LAFIB paradox (Zhang, 2009a) (see Ch. 1).

The above two fundamental limitations prevent any unipolar truth-based operator from being directly used for holistic knowledge representation and computation in a world of nonlinear bipolar dynamic equilibria especially for bipolar quantum entanglement. To overcome the limitations, equilibrium-based YinYang bipolar quantum lattices and lattice-ordered bipolar dynamic norms are proposed in this chapter that generalize the classical logical connectives $\wedge$ and $\vee$ from $\{0,1\}$ and $[0,1]$ to $B_1 = \{-1,0\} \times \{0,1\}$, $B_F = [-1,0] \times [0,1]$, and any other BQLs, respectively. Thus, we have the basic classification of lattices and sets:

1. A classical (unipolar) crisp set X in a universe U is defined in the form of its characteristic function μ_X:U$\rightarrow\{0,1\}$ which yields the value 1 for elements belonging to the set X and 0 for elements excluded from the set X. Evidently, classical set theory is based on Aristotle's universe of truth objects defined in the unipolar bivalent lattice $\{0,1\}$, where the concept of element in a set is claimed self-evident without the need of proof.
2. A classical (unipolar) fuzzy set (Zadeh, 1965) X in a universe U is defined in the form of its characteristic function μ_X:U$\rightarrow[0,1]$. The fuzzification preserves the unipolar truth-based property of classical set theory with the addition of an infinite number of levels of membership degrees which can be defuzzified to 0 or 1.
3. A YinYang bipolar crisp set X (Ch. 3) in a universe U of bipolar equilibria is defined in the form of its characteristic function μ_X:U $\Rightarrow B_1$, where $B_1=\{-1,0\}\times\{0,1\}=\{(-1,0),(0,0),(0,1),(-1,1)\}$ is a bounded complemented crisp BQL.

4. A YinYang bipolar fuzzy set X (Zhang, 1998, 2006a) in a universe U of bipolar fuzzy equilibria can be defined in the form of its characteristic function μ_X:U⇒B_F, where $B_F = [-1,0] \times [0,1]$ is a bounded complemented *bipolar quantum fuzzy lattice* (*BQFL*). Evidently, bipolar fuzzy sets are to bipolar crisp sets as fuzzy sets are to classical crisp sets; bipolar fuzzy sets are to unipolar fuzzy sets as bipolar crisp sets are to classical crisp sets. Physically speaking, bipolar fuzzy sets are to bipolar crisp sets as quasi-equilibria are to full equilibria to a certain extent.
5. An arbitrary YinYang BQL is not necessarily bounded and complemented but satisfies the definition of a BQL with bipolar quantum entanglement features, which finds applications in quantum mechanics and computing.

Since an *L-fuzzy set* (Goguen, 1967) is defined as a function φ:X→L where L can be any lattice in the bipolar interval (-∞,∞) and the lattices B_1 and B_F are both bipolar product lattices, bipolar crisp and fuzzy sets can be characterized as *bipolar L-sets* (Zhang, 2005a). Following Goguen, ∀x∈X, we have:

1. If L = {0,1}, φ:X ⇒ L, or alternatively, φ(x) ∈ L is a unipolar Boolean predicate that yields either 0 or 1;
2. if L = [0,1], φ:X ⇒ L, or alternatively, φ(x) ∈ L is a unipolar fuzzy function that yields a membership degree in [0,1];
3. If L = B, where B is any BQL, we have φ:X ⇒ B or alternatively, φ(x) ∈ B, is a bipolar L-set. Furthermore, following Goguen, given bipolar L-sets ϕ:X⇒ B and φ:Y⇒ B, the expression ϕ(x)*φ(y) denotes a bipolar binary operation where * is any valid binary operator.

To further clarify the essential difference between bipolar and unipolar systems, please see the YinYang classification (Ch. 2, Fig. 1).

Through this chapter, ***linearity*** is referred to as no cross-pole bipolar interaction and no quantum entanglement; ***non-Linearity*** is referred to as having such bipolar interaction or quantum entanglement (see Ch. 3, Fig. 1). We use the word ***bipolar norm*** for bipolar t-norm, t-conorm, p-norm, or p-conorm to be defined on a bipolar lattice and ***unipolar norm*** for truth-based t-norms and t-conorms defined on a unipolar lattice. We follow the unipolar notations:

a. $T_1(x, y) = \min(x, y) = x \wedge y$; $\perp_1(x, y) = \max(x, y) = x \vee y$;
b. $T_2(x, y) = x \times y$; $\perp_2(x, y) = x + y - x \times y$;
c. $T_3(x, y) = \max(0, x + y - 1) = \Delta(x,y)$; $\perp_3(x, y) = \min(1, x + y)$.

In this chapter, formal definitions and classifications of BQLs, t-norms, p-norms, and their conorms are presented with related theorems and proofs. The symmetrical nature of BQLs is examined following the definition of symmetry in particle physics. The norms can be used as quantum operators in different applications. Bipolar L-relations are introduced as L-sets based BQL and bipolar norms. As a new family of fundamentally different operators the bipolar dynamic norms on B_F together with a fuzzified version of BUMP form an equilibrium-based *bipolar dynamic fuzzy logic* (*BDFL*).

The remaining presentations and discussions in this chapter are organized in the following sections:

- **Bipolar Quantum Lattices and L-Sets.** In this section we introduce the classification and axiomatization of BQLs and lay out the background for bipolar norms. The concept of *bipolar energy symmetry* is introduced.
- **Bipolar Dynamic T-norms and P-norms**. In this section we introduce different bipolar norms.
- **Norm-Based Bipolar Universal Modus Ponens.** In this section we present a norm-based granular version of BUMP.
- **Comparison and Discussion.** In this section we present a comparison between bipolar norms and unipolar norms with a discussion.
- **Lattice-Ordered Bipolar Dynamic Logic**. In this section we present a lattice-ordered bipolar dynamic logic.
- **Bipolarity, Linearity, Integrity, and Recovery Theorem.** This section presents the concept of strict bipolarity, linearity, and integrity of BQLs. A *recovery theorem* is presented for the recovery of any strict BQL to Boolean logic.
- **Research Topics.** A few research topics are listed in this section.
- **Summary.** The chapter is summarized in this section.

BIPOLAR QUANTUM LATTICES AND L-SETS

Bipolar Crisp and Fuzzy Sets

Based on the concepts ***positive elements***, ***positive set***, ***negative elements***, ***negative set,*** and ***unipolar sets*** as defined in Chapter 2, a positive fuzzy set X^+ in universe U can be defined in the form of its characteristic function μ_X^+:U→[0,1] which yields the value 1 for elements belonging to the set X^+, 0 for elements excluded from the set X^+, and a real value in [0, 1] for elements partially belonging to the set X^+. Evidently, Zadeh's fuzzy set theory (Zadeh, 1965) is a unipolar real-valued truth-based extension of classical set theory based on a universe of unipolar truth objects. Similarly, following Goguen (1967) a negative fuzzy set X^- in a universe U can be defined in the form of its characteristic function μ_X^-:U→[-1,0] which yields the value -1 for elements belonging to the set X^-, 0 for elements excluded from the set X^-, and a real value in [-1, 0] for elements partially belonging to the set X^-. The logical operators ∧,∨, ¬, and → in the **unipolar lattice** L^+ = [0,1] can be symmetrically extended to the **negative lattice** L^- = [-1,0] such that, ∀a,b∈ [-1,0], we have **a∧b = -min(|a|,|b|); a∨b = -max(|a|,|b|); ¬a = -1 – a; and a→b = ¬a∨b,** where |x| denotes the absolute value of x. *(Note: The use of absolute value |x| through this book is for explicit bipolarity only.)*

Although the two fuzzy **lattices** (L^+, ∧,∨, ¬, →) and (L^-, ∧,∨, ¬, →) can be considered **isomorphic or logically equivalent** from a unipolar perspective, in what follows it is shown that the product lattice $L = L^- \times L^+$ leads to equilibrium-based bipolar fuzzy norms with real-valued non-linear bipolar dynamic fusion, interaction, oscillation, and quantum entanglement properties. The bipolar norms together with a real-valued version of bipolar universal modus ponens (BUMP) form a bipolar dynamic fuzzy logic (BDFL). Without bipolar syntax and semantics, these would be virtually impossible due to the loss of negative identity.

Based on *background independence* of bipolarity as discussed in Chapter 3, NP, PN, LR (left-right), and RL (right-left) ordering are background dependent concepts relative to the observer, and we follow the background independence convention where (0,+1) ≡ (+1,0), (-1,0) ≡ (0,-1), and (-1,+1) ≡ (+1,-1).

Given the bipolar equilibrium ***(e^-,e^+) or*** equilibrium ***(e^+,e^-)***, if *(e^-,e^+)* or ***(e^+,e^-)*** is defined on a discrete domain it is called a **crisp equilibrium**; if it is defined on a continuous domain it is called a **fuzzy equilibrium.** When $|e^-| \neq e^+$ we call the coupling ***(e_x^-,e_y^+)* a quasi- or non-equilibrium.** We use the forms *(n,p) and (x,y)* for NP **equilibrium variables**. We call a set of bipolar equilibria {(x,y)} a **bipolar set.** A bipolar set is said **crisp** if it is a collection of crisp equilibria or non-equilibria; it is said **fuzzy** if it is a collection of fuzzy equilibria. *(Note: In this work we continue to use NP notation without further specification.)*

Based on bipolar poset (Ch. 3), a bipolar fuzzy poset (B_p,≥≥) is defined as a bipolar fuzzy set *{(x,y)}*, where *(x,y)* as a bipolar element characterizes a bipolar fuzzy equilibrium, ≥≥ is a bipolar partial order relation and, $\forall (x,y),(u,v) \in B_p$,

$$(x,y) \geq\geq (u,v), \text{ iff } |x| \geq |u| \text{ and } y \geq v. \quad (4.1)$$

Bipolar Quantum Lattice and L-Sets

The concept of bipolar lattice (Ch. 3) can be naturally extended to an arbitrary bipolar lattice, crisp or fuzzy. Since bipolar lattice leads to bipolar quantum entanglement (Ch. 3), it is quantum in nature. Thus, we adopt the name *Bipolar Quantum Lattice* or BQL.

Definition 4.1. A *bipolar quantum lattice* is a quadruplet *(B, ⊕, &, ⊗)*, where *B* is any crisp or fuzzy bipolar poset, and, $\forall (x,y),(u,v) \in B$, there is a bipolar least upper bound (blub), a bipolar greatest lower bound (bglb), and a cross-pole greatest lower bound (cglb) as defined in Eq. 4.2 – 4.4 in NP order. *B* is a ***bipolar crisp lattice*** if it is a bipolar crisp poset; *B* is a ***bipolar fuzzy lattice*** if it is bipolar fuzzy poset.

$$blub((x,y),(u,v)) \equiv (x,y) \oplus (u,v) \equiv (x \vee u, y \vee v) \equiv (-max(|x|,|u|), max(y,v)); \quad (4.2)$$

$$bglb((x,y),(u,v)) \equiv (x,y) \& (u,v) \equiv (x \wedge u, y \wedge v)) \equiv (-min(|x|,|u|), min(y,v)); \quad (4.3)$$

$$cglb((x,y),(u,v)) \equiv (x,y) \otimes (u,v) \equiv (-(|x| \wedge |v| \vee |y| \wedge |u|), (|x| \wedge |u| \vee |y| \wedge |v|)). \quad (4.4)$$

Similar to crisp cases, & is **a linear bipolar conjunctive**; ⊕ is **a linear bipolar disjunctive dual to &**; and ⊗ is a **non-linear bipolar conjunctive operator.** We say a bipolar operator is **bipolar equivalent** to its unipolar counterpart if its operation on each pole is equivalent to its unipolar counterpart.

While the two linear operators ⊕ and & are bipolar equivalent to ∨ and ∧ in unipolar fuzzy logic, respectively, the non-linear operator ⊗ has no equivalent in classical fuzzy logic. All three bipolar operators are dynamic in nature regardless of their linearity or non-linearity as justified in the following:

1. ⊕ is a dynamic balancer. For instance, (-0.9,0) ⊕ (0,0.7)=(-0.9,0.7).
2. & is a dynamic minimizer or life distinguisher, e.g. (-0.9,0) & (0,0.7) = (0,0).
3. ⊗ is an intuitive bipolar interactive oscillator with the bipolar infused semantics − − = +, − + = −, + − = −, and + + = + on each pole. Since all closed dynamic systems tend to reach equilibrium through oscillation and all open dynamic systems tend to be disturbed to an oscillatory state, ⊗ is a necessary operator and important part for dynamic reasoning with bipolar equilibria. An oscil-

lation sequence such as $(-0.9,0) \otimes (-0.9,0) \otimes .. \otimes (-0.9,0) = (-0.9,0)^n$ can be used in non-linear brain dynamics or bipolar strings (Zhang, 2007, 2009a, 2009b, 2009c, 2009d). Thus, $\otimes$ introduces non-linear dynamics into logical reasoning – a key step toward a bipolar dynamic fuzzy logic (BDFL) (Zhang & Zhang, 2004; Zhang 2005a, 2009b).

4. Bipolar dynamic fuzzy logic (BDFL) supports holistic top-down inductive and bottom-up deductive reasoning with different gray levels. It also leads to bipolar quantum entanglement with a real-valued bipolar universal modus ponens.

Although any chain in a unipolar lattice is a lattice, a chain C in a bipolar lattice can be verified a linear lattice $(C, \oplus, \&)$ that is isomorphic to $(C,\vee,\wedge)$ but not necessarily a bipolar lattice because it may not have a cglb ($\otimes$) for some pair of elements. On the other hand, some subsets of a bipolar lattice may form a linear lattice or another bipolar lattice. To distinguish from classical unipolar chains, we call any subset C of a bipolar lattice L a *bipolar chain* of L if and only if C forms a linear lattice $(C, \oplus, \&)$ or another bipolar lattice $(C, \oplus, \&, \otimes)$. Thus, a bipolar chain must be part of a bipolar lattice while a classical chain can be part of any poset. Two elements in a bipolar chain do not have to be comparable by the partial order relation.

Theorem 4.1. Any bipolar lattice B is a distributive structure, That is, $\forall(x,y),(u,v)\in B$, we must have

Parallel Distributivity:

$$(a,b)\&((c,d)\oplus(e,f))\equiv((a,b)\&(c,d))\oplus((a,b)\&(e,f)); \tag{4.5a}$$

$$(a,b)\oplus((c,d) \& (e,f)) \equiv ((a,b)\oplus(c,d)) \&((b) \oplus (e,f)); \tag{4.5b}$$

Serial Distributivity:

$$(a,b)\otimes((c,d)\oplus(e,f)) \equiv ((a,b)\otimes(c,d))\oplus((a,b)\otimes(e,f)). \tag{4.6}$$

Proof. It follows from the definitions of blub, bglb, and cglb and their existences in B. □

Theorem 4.2. Let $B_n^- =(-n,-(n-1),..,-1,0)$ and $B_n^+ = - B_n^- = (0,1,..,n+1,n)$, $\forall n$, $0\leq n \leq\infty$; we have: (1) the Cartesian product or discrete space $B_n = B_n^- \times B_n^+$ forms a NP bipolar crisp lattice; and (2) the Cartesian product or continuous space $B_X = [-x,0] \times [0,x]$, $\forall x$, $0\leq x \leq\infty$, forms a NP bipolar fuzzy lattice.

Proof. Since B_n^- and B_n^+ are bipolar symmetrically ordered sets and B_X is a bipolar symmetrical continuous domain, B_n and B_X must have a blub, a bglb, and a cglb for every pair of its elements. □

Definition 4.2. An NP bipolar lattice B (crisp or fuzzy) is ***bounded*** if it has both a unique minimal element denoted (0,0) and a unique maximal element denoted (-1,1). A bounded bipolar lattice B is ***complemented*** if, $\forall(x,y)\in B$, we have the ***bipolar complement*** $\neg(x,y) \in B$.

Note that bipolar complement induces bipolar implication. Thus, a bounded and complemented bipolar lattice B (crisp or fuzzy) can be denoted as $(B,\equiv,\oplus, \otimes, \&, -, \neg, \Rightarrow)$ with $-$, $\neg$ and $\Rightarrow$ defined as follows.

Complement: $\neg(x,y) \equiv (\neg x,\neg y) \equiv (-1-x,1-y)$; (4.7)

Implication: $(x,y)\Rightarrow(u,v) \equiv (x\Rightarrow u,y\Rightarrow v) \equiv (\neg x \vee u, \neg y \vee v)$; and (4.8)

Negation: $-(x,y) \equiv (-1,0)\otimes(x,y)=(-y,-x)$; (4.9)

where the negation operator is derived from $\otimes$.

It is evident that every bounded and complemented bipolar lattice B must be a bipolar chain of $B_F = [-1,0] \times [0,1]$.

Based on the notion of bipolar lattice, we extend L-fuzzy sets to bipolar L-sets:

Definition 4.3. A bipolar L-set $B=(B^-,B^+)$ in a bipolar set X to a bipolar lattice B_L is a bipolar equilibrium function or variable $B:X\Rightarrow B_L$. If B_L is crisp we call B a *bipolar L-crisp set*; if B_L is fuzzy we call B a *bipolar L-fuzzy set*.

Since a bipolar L-crisp set is a special case of a bipolar L-fuzzy set, we use the terms *bipolar L-set, bipolar L-fuzzy set*, *equilibrium function* or *equilibrium variable* alternatively which can be denoted as $B = (B^-, B^+)$, (x,y), or (ϕ^-,ϕ^+).

Remarks: The necessity for the notions of bipolar sets and bipolar lattice can be clarified with the following fundamental differences:

(1) A usual lattice is for modeling a universe of unipolar sets; a bipolar lattice is for modeling a world of YinYang bipolar equilibria with bipolar quantum entanglement.
(2) A unipolar lattice is not necessarily distributive, but a bipolar lattice must be distributive.
(3) A chain of a bipolar lattice is not necessarily a bipolar lattice because it may not be distributive.
(4) A usual complete lattice is defined as a poset with a lub and glb for each non-empty subset. Bipolar completeness, however, is not a natural property because we cannot expect a bipolar lattice to have a blub, bglb, and a cglb for each non-empty subset.
(5) Completeness plays a major role in L-fuzzy sets; balance and strictness (discussed in latter sections) play a major role in bipolar L-sets.
(6) While classical (unipolar) fuzzy logic is defined on the bounded and complemented unipolar fuzzy lattice [0,1] for reasoning with unipolar fuzzy sets, a bipolar dynamic fuzzy logic (BDFL) can be defined on the bounded and complemented bipolar fuzzy lattice $B_F = [-1,0] \times [0,1]$ for reasoning with bipolar fuzzy sets of equilibria.
(7) Modus Ponens is the only generic rule in classical logic for closed-world reasoning, bipolar lattices and L-sets lead to dynamic bipolar universal modus ponens (BUMP) for open-world reasoning.
(8) Although a bounded unipolar bivalent or fuzzy lattice is also complemented, a bounded bipolar lattice does not necessarily lead to bipolar complementarity (see examples in next section).

Illustrations and Observations

A bipolar lattice is different from a traditional lattice. A bipolar lattice B can be considered a combination of two lattices – a linear bipolar lattice $(B, \oplus, \&)$ and a non-linear bipolar lattice $(B, \oplus, \otimes)$. The linear part is an isomorphic structure of a traditional lattice $(B, \vee, \wedge)$. The non-linear part provides a crucial extension for modeling bipolar dynamic interactions. Although we have blub $\geq\geq$ bglb for any pair of elements in a bipolar lattice, we do not have the relation blub $\geq\geq$ cglb.

For instance, given (x,y) = (u,v) = (-1, 0), we have blub((x,y),(u,v)) = (-1,0); bglb((x,y),(u,v)) = (-1,0); and cglb((x,y),(u,v)) = (0,1). Evidently, a bipolar lattice is not simply a traditional lattice due to the non-linear bipolar conjunctive $\otimes$.

To illustrate, the Hasse diagrams of three NP bipolar lattices are shown in Figure 1. Where B_1 and B_2 are crisp bipolar lattices and B_F is a bipolar fuzzy lattice. While B_1 is a polarization of the bivalent space {0,1} that leads to BDL; B_2 is a polarization of a 3-valued logical space {0,1,2}, and B_F is a polarization of Zadeh's fuzzy space [0,1].

A bipolar chain in a bipolar quantum lattice might be a lattice but not a bipolar lattice. The Hasse diagrams of two bipolar chains from the bounded NP bipolar lattices B_2 and B_F are shown in Figure 2. The bipolar chains C_1 and C_2 in Figure 2 are perfect lattices as every pair of elements has both a lub and glb, but C_1 is not bipolar lattices because cglb((-2,0),(-2,0)) = (0,2) is undefined in C_1. Interestingly, C_2 of Figure 2 can be verified as bipolar lattice based on Equations 4.1 – 4.4.

Some bipolar chains of a bipolar lattice are also bipolar lattices. In Figure 3, we show that the Hasse diagrams of S_{2a},S_{2b},S_{2c},$\subseteq B_2$. S_{2a},S_{2b} can both be verified as bipolar lattices. S_{2c} is a usual lattice but not a bipolar lattice because cglb((-2, 0), (-2,0)) = (0,2) does not exist in S_{2c}. These features support the observations that ***non-equilibrium may exist in equilibrium and equilibrium may exist within non-equilibrium.***

A bipolar lattice does not have to be balanced. It is interesting to examine the bipolar chains S_{2d} in Figure 4. It can be verified a perfect bipolar lattice but it is not balanced. Therefore, a bipolar lattice

Figure 1. Hasse diagrams of three bipolar lattices B_1, B_2, and B_F (Zhang, 2005a)

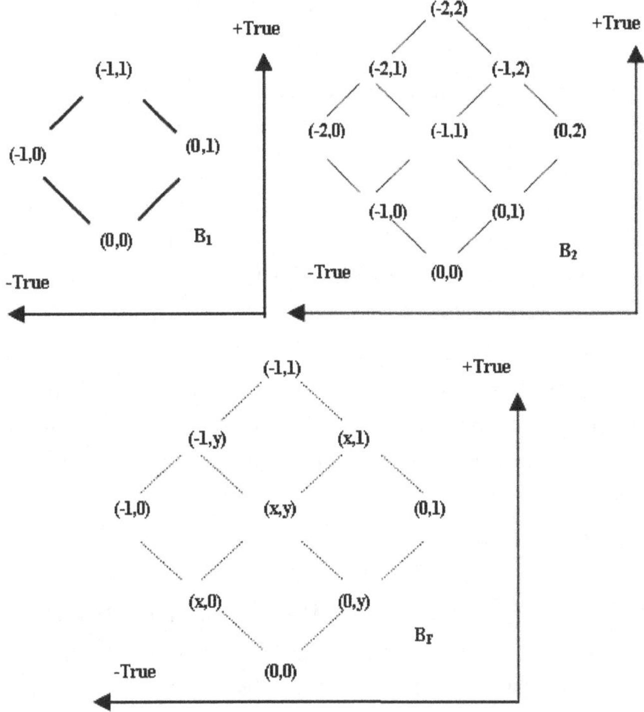

does not have to be balanced. The three chains B_{1a},B_{1b}, $B_{1c} \subseteq B_1$ in Figure 4 can also be verified as bipolar lattices. B_{1a} is bounded but not balanced; B_{1b} is bounded and balanced; but B_{1c} is not balanced and not upper bounded. Although B_{1a} is bounded it is not complemented due to the missing ¬(0,1).

Equilibrium, Balance, Energy, Stability, and Bipolar Symmetry

Recall that a bipolar L-set ϕ is a bipolar crisp or fuzzy equilibrium function that maps a set of objects X onto a crisp or fuzzy bipolar lattice. For instance, (self-negation, self-assertion), (competition, coopera-

Figure 2. Hasse diagrams of two chains C_1 and C_2 (adapted from Zhang, 2005a)

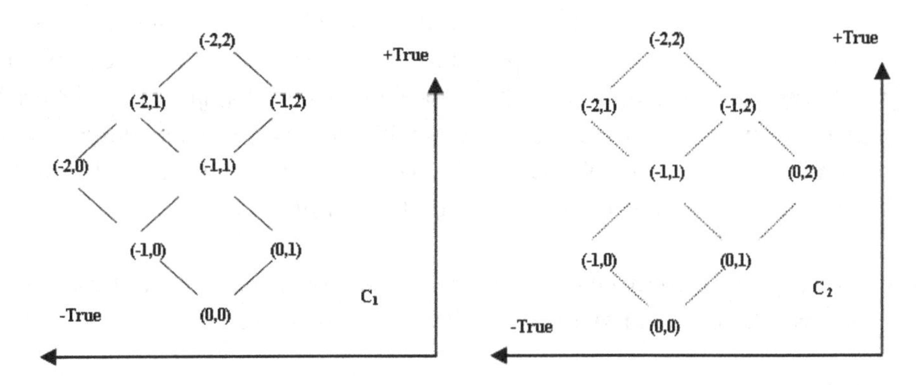

Figure 3. Hasse diagrams of three bipolar chains of B_2 (Zhang, 2005a)

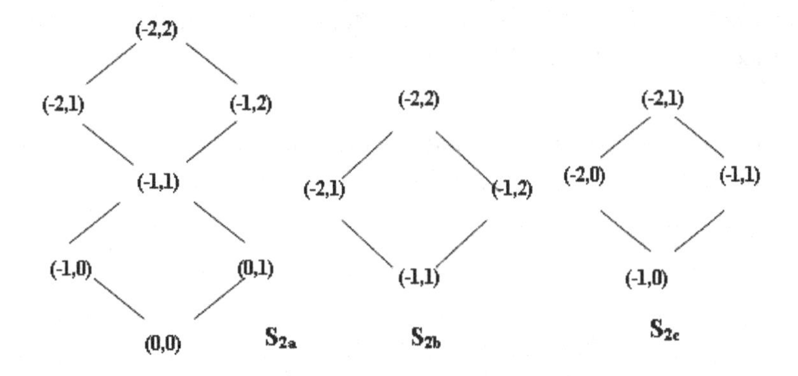

Figure 4. Hasse diagrams of four bipolar lattices (Zhang, 2005a)

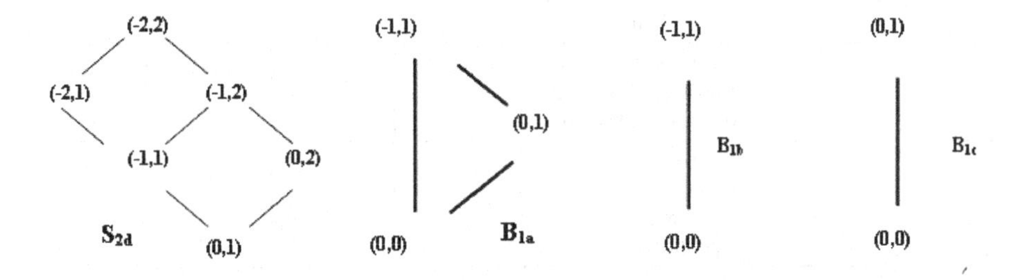

tion), (conflict-interest, common-interest), (negative-charge, positive-charge), (reaction-force, action-force), (cost, profit), (output, input), (left, right), and (Yin, Yang), etc, are all equilibrium functions or bipolar L-sets. Since a bipolar lattice has balanced and unbalanced elements, the concept of equilibrium has quasi- or non-equilibrium aspects self-contained.

Definition 4.4. Let W = $\{\phi_i\}$ be a set of bipolar equilibria (including quasi- or non-equilibria) or bipolar L-sets; a truth object A *exists in W* iff $\exists\phi \in$ W such that **the equilibrium energy of A** in ϕ is greater than zero. Formally, we have: *A exists* iff $\{\exists\phi\in W \mid (\phi^-(A),\phi^+(A)) \neq (0,0)\}$ or $|\phi^-(A)| + \phi^+(A) > 0$, where $(\phi^-(A),\phi^+(A))$ and $|\phi^-(A)| + \phi^+(A)$ identify the equilibrium energy of A in bipolar and unipolar forms, respectively.

The equilibrium energy definition finds its root in quantum physics. The annihilation of a particle (such as electron) with its antiparticle (such as positron) will release energy. It is well-known that electron-positron pair production denoted (e^-,e^+) is the most fundamental example of the materialization of energy that was predicted in Einstein's special theory of relativity and accurately described by quantum electrodynamics (QED) – the jewel of physics (Dirac, 1927, 1928; Feynman, 1962, 1985). Thus, the name of bipolar quantum lattice (BQL) is justified with its *equilibrium energy*.

Definition 4.5. Let bipolar reflexivity be a bipolar equilibrium $\phi\in$W and $\phi = (\phi^-,\phi^+)$ = (self-negation, self-assertion), **an agent A has life in W** iff (self-negation(A), self-assertion(A)) >> (0,0).

The above definitions follow the commonsense that anything has to exist in at least one type of equilibrium or non-equilibrium and any agent has to have some self-negation and/or self-assertion ability to adapt itself to a world. These definitions provide a basis for equilibrium energy and stability measurement in accordance with the ancient Chinese Yin-Yang theory. *Negative and positive energy* and stability functions are specified in Table 1 for an NP-bipolar lattice.

Table 1. $Dplr^{\vee}(\varphi^-(A),\varphi^+(A))=\varphi^-(A)\vee\varphi^+(A)$ defines an existence or life function (Dplr: depolarization function) (Zhang, 2005a)

$(\phi^-(A), \phi^+(A))$	$Dplr^{\vee}(\phi^-(A),\phi^+(A)) = \phi^-(A)\vee\phi^+(A)$	Existence/life	Negative Energy	Positive Energy	Energy Total (T)	Energy Imbalance (I)	Stability S=(T-I)/T
(0,0)	0	No existence/No Life	0	0	0	0	Undefined
(0,y)	y y>0	Existence/Life with yin-deficiency (non-equilibrium)	0	y	y	y	0
(x,0)	$\|x\|$ $\|x\|>0$	Existence/Life with yang-deficiency (non-equilibrium)	x	0	$\|x\|$	$\|x\|$	0
(x,y)	$\|x\|\vee y$ $\|x\|\neq y$	Existence/Life with Unbalanced yin-yang (quasi-equilibrium)	x	y	$\|x\|+y$	$\|x+y\|$	$1-\|x+y\|/(\|x\|+y)$
(-1,1)	1	Existence/ Life with balanced, yin-yang in harmony and full equilibrium	-1	1	2	0	1

It is noteworthy to mention that the gauge theory in quantum electrodynamics (QED) is best understood in terms of something physicists refer to as *symmetry*. *Symmetry is an operation that doesn't change how something behaves relative to the outside world.* Based on the concept of equilibrium, equilibrium energy, and stability we have

Definition 4.6. A bipolar lattice B is balanced if, $\forall(x,y)\in B$, we also have $(-y,-x) \in B$. A bipolar operation on a lattice B is ***a symmetry*** if it does not change any property of B. A bipolar operation on a lattice B is ***an energy symmetry*** if it does not change the energy total of B.

Definition 4.7. The negative energy E^- of a bipolar lattice B is the sum of the negative energies of its elements. **The positive energy E^+ of B** is the sum of the positive energies of its elements. The **energy equilibrium** (including quasi- or non-equilibrium) is denoted (E^-,E^+). **The energy total** is the sum $T=|E^-|+E^+$. **The energy imbalance I** is $|E^-+ E^+|$ and **the stability of B** is $S=1-I/T$.

Theorem 4.3. (1) Any NP bipolar lattice with element (-1,0) must be balanced. (2) Given any balanced bipolar quantum lattice B, the negation operation –B is a symmetry. (3) Given any bipolar quantum lattice B, balanced or not, the negation operation –B is an energy symmetry.

Proof. (1) If (-1,0) is an element of B, $\forall(x,y)\in B$, $cglb((-1,0),(x,y)) = ((-1,0)\otimes(x,y)) = (-y,-x) \in B$. *(2) If B is balanced, -B=B from a complete background independent perspective. (3) Given any bipolar quantum lattice B, T(B)=T(-B).* □

Theorem 4.4. Any balanced NP bipolar lattice B (crisp or fuzzy) must be the Cartesian product $B^-\times B^+$ where B^+ is a unipolar lattice and $B^- = - B^+$.

Proof. If B is not the Cartesian product $B^-\times B^+$, *B must have at least one unbalanced element.* □

Evidently, the energy imbalance of a balanced bipolar lattice B must be zero and its stability must 1.

Definition 4.8. Given the energy equilibrium $E_x=(E_x^-,E_x^+)$ of any NP bipolar lattice B_x (balanced or unbalanced) and the energy equilibrium $E_y=(E_y^-, E_y^+)$ of the PN bipolar lattice $B_y = -B_x$, the **energy equilibrium** of both lattices is defined as $E = (E^-,E^+)= (E_1^-+E_2^-, E_1^++E_2^+)$. **The energy total of both lattices** is defined as $T=|E^-|+E^+$. **The energy imbalance of both lattices** is defined as $I = | E^- + E^+|$ and **the stability of B and -B** is defined as $S=1-I/T$.

Theorem 4.5. (1) The energy imbalance of any bipolar lattice B and –B must be zero and their stability must be 1. (2) If the coupling (-B, B) is a bipolar variable, the negation –(-B,B) is a symmetry.

Proof. (1) Since for every element (x,y) in B there must be a corresponding element (-x,-y) in –B, it follows directly from the definition of bipolar negation and NP/PN bipolar poset. (2) –(-B,B) = (-B,B). □

The significance of energy balance and symmetry is 3-fold:

1. A local unbalanced structure can be globally balanced by another unbalanced structure. Two bipolar lattices can be two such structures. This property can be used to model real world equilibria, e.g. the equilibrium of two magnetic or quantum fields.
2. Given any bipolar quantum lattice B the negation operation –B is a symmetry in terms of total energy reservation.
3. Given any coupling of two bipolar quantum lattices (-B,B) the negation operation –(-B,B) = (-B,B) is a symmetry.

Balancer, Extinguisher, Oscillator, and Equilibrant

Based on the notions of life and energy, we call the blub operator $\oplus$ a life/energy **maximizer** or **balancer** due to its bipolar fusion functionality. We call the bglb operator & a life/energy **minimizer or extinguisher** because it leads to the bipolar minimum of two bipolar variables. We call the cglb operator $\otimes$ a **life/energy oscillator** due to its embedded semantics of $- - = +$; $- + = -$; and $+ - = -$. We call the bipolar element (-1, 0) in a bounded bipolar lattice an **equilibrant** due to the property $(-1,0) \oplus (-1,0)^2 = (-1,1)$, which indicates an equilibrium process with a balancer, oscillator and the equilibrant.

Bipolar Isomorphism and Strict Bipolar Lattices

Definition 4.9. **A bipolar isomorphism** is an order-preserving bipolar bijection from one bipolar lattice to another that also preserves blubs, bglb, and cglbs.

To support bipolar isomorphism we define bipolar arithmetic operations on bipolar functions or variables (x,y) and (u,v) as

Bipolar addition: $(x,y) + (u,v) \equiv (x+u,y+v)$; (4.10a)

Bipolar subtraction: $(x,y) - (u,v) \equiv (x-u,y-v)$; (4.10b)

Multiplication: $(x,y) \times a \equiv (x\times a,y\times a)$; (4.10c)

Division: $(x,y)/a \equiv (x/a,y/a)$. (4.10d)

We define the bipolar isomorphic arithmetic operations on a bipolar lattice B as,

Bipolar addition: B + (u,v) maps (x,y) to $(x,y) + (u,v)$, $\forall(x,y)\in B$; (4.11a)

Bipolar subtraction: B – (u,v) maps (x,y) to $(x,y) - (u,v)$, $\forall(x,y)\in B$; (4.11b)

Multiplication: B * a maps (x,y) to (x,y) * a, $\forall(x,y)\in B$; (4.11c)

Division: B/a maps (x,y) to (x,y)/a, $\forall(x,y)\in B$. (4.11d)

Evidently, two bipolar lattices are bipolar isomorphic if one can be converted to the other using Eq. (4.10) and (4.11). For instance, the bipolar lattice $B = \{-3,-2\}\times\{2,3\}$ is isomorphic to $B_1 = \{-1,0\}\times\{0,1\}$

because we have $B_1 = B - (-2,2)$. It should be noted that bipolar isomorphism is defined on bipolar lattices. A bipolar poset cannot be isomorphic to a bipolar lattice if it does not have a cglb. For instance, the bipolar poset S_{2c} in Figure 3 is a usual lattice but not a bipolar lattice because it does not have a cglb. Although $S_{2c}-(-1,0) = B_1$ it is not isomorphic to B_1.

Definition 4.10. Each smallest square in the Hasse diagram of a bipolar lattice B is called **an inner square**. Each grid on the border of a bipolar lattice B is called **an outer grid**. A lattice formed by the four corners of an outer grid of B is called **a corner lattice of the grid.** A lattice formed by the corners of all outer grids of B is called **a corner lattice of B.**

Definition 4.11. A balanced bipolar lattice B is called **a strict bipolar lattice** if the corner lattice of B is isomorphic to $B_1 = \{-1,0\} \times \{0,1\}$;

Theorem 4.6. (1) The Cartesian product $B_n=(-n,-(n-1),..,-1,0) \times (0,1,..,n+1,n)$, $\forall n$, $0 \leq n \leq \infty$, is a strict bipolar lattice; and (2) the Cartesian product $B_X = [-x,0] \times [0,x]$, $\forall x$, $0 \leq x \leq \infty$. is a strict bipolar lattice.

Proof. It follows from: (1) B_n and B_x are bipolar lattices; (2) Let the corner lattice of B_n be B, we have $B/n= B_1$; and (3) Let the corner lattice of B_x be B, we have $B/x= B_1$. □

Definition 4.12. A bipolar lattice B is called **a strict bipolar lattice chain** if every outer grid of B is a strict bipolar lattice.

A strict bipolar lattice chain and its corner lattice are illustrated in Figure 5. On the left, the bottom square is equal to B_1 and the top outer grid is isomorphic to B_2. The corner lattice of the lattice chain on the right is clearly a strict bipolar lattice chain as each square in the lattice is isomorphic to B_1.

Figure 5. A strict bipolar lattice chain and its corner lattice (Zhang, 2005a)

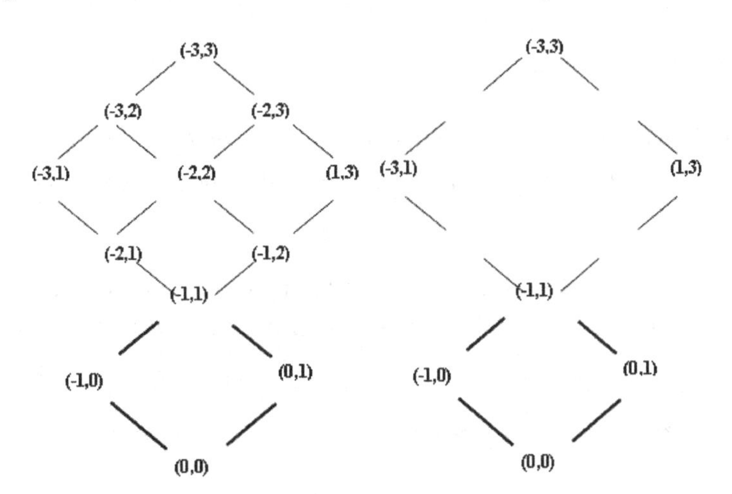

A Generalized Bipolar Axiomatization on Strict BQLs

Theorem 4.7. A sound and complete axiomatization on B1 (see Ch. 3, Table 6) is also sound and complete on the corner lattice of any strict bipolar lattice B.

Proof. It follows directly from the fact that the corner lattice of a strict bipolar lattice is isomorphic to B_I. □

The significance of the above theorem lies in the fact that it generalizes the soundness and completeness theorem on B_1 onto any bipolar lattice isomorphic to B_1 which can be the corner lattice of any strict bipolar quantum lattice either crisp or fuzzy. Since any fuzzy logic does not satisfy the law of excluded middle (LEM), soundness and completeness does not apply to any fuzzy lattice. The above generalization, however, has excluded bipolar fuzzy lattice because the corner lattice of any strict bipolar fuzzy lattice has only four discrete bipolar values.

BIPOLAR DYNAMIC T-NORMS AND P-NORMS

Linear and Non-Linear Bipolar DeMorgan's Laws on B_F

If we define $\oplus^-$ and $\&^-$ as the negation of $\oplus$ and & on B_F, respectively, $\forall (x,y),(u,v)\in B_F$ we have

$$\text{blub}^-((x,y),(u,v)) \equiv (x,y)\oplus^-(u,v) \equiv -((x,y)\oplus(u,v)) \equiv (-(y\vee v), (|x|\vee|u|)); \tag{4.12}$$

$$\text{bglb}^-((x,y),(u,v)) \equiv (x,y)\ \&^-(u,v) \equiv -((x,y)\&(u,v)) \equiv (-(y\wedge v), (|x|\wedge|u|))). \tag{4.13}$$

With $\{\oplus,\&\}$ and $\{\oplus^-,\&^-\}$, the laws of excluded middle (LEMs) and the laws of non-contradiction hold on B_1 (Ch. 1) but do not hold on B_F due to the fact $\neg 0.5 \equiv 0.5$. However, bipolar DeMorgan's laws still hold on B_F.

Theorem 4.8. The two pairs of linear bipolar DeMorgan's laws in Table 2 hold on B_F.

Proof. Since the pair of bipolar t-norm and t-conorm $\{\oplus, \&\}$ *on* B_F *are bipolar equivalent to the pair of unipolar t-norm and t-conorm* $\{\vee, \wedge\}$ *on [0,1], the bipolar laws are bipolar equivalent to their unipolar counterpart.* □

The duality does not hold on $\otimes$ and $\oplus$ because of the non-linear cross-pole property of $\otimes$. In Eq. 4.14, a cross-pole least upper bound (club) operator or $\varnothing$ is extended from B_1 to B_F as a disjunctive oscillator dual to $\otimes$.

$$\text{club}((x,y),(u,v)) \equiv (x,y)\varnothing(u,v) \equiv \neg\,(\neg(x,y)\otimes\neg(u,v)) \equiv (-1,1)-(\neg(x,y)\otimes\neg(u,v)), \tag{4.14}$$

where $\neg(x,y) = (-1,1) - (x,y) = (-1-x,1-y)$.

Table 2. Linear Bipolar DeMorgan's Laws

Linear or Parallel Bipolar DeMorgan's Laws $\forall (a,b),(c,d) \in B_F$	$\neg((a,b)\&(c,d)) \equiv \neg(a,b)\oplus\neg(c,d)$; $\neg((a,b)\oplus(c,d)) \equiv \neg(a,b)\&\neg(c,d)$;
	$\neg((a,b)\&^-(c,d)) \equiv \neg(a,b)\oplus^-\neg(c,d)$; $\neg((a,b)\oplus^-(c,d)) \equiv \neg(a,b)\&^-\neg(c,d)$.

Table 3. Nonlinear Bipolar DeMorgan's Laws

Non-Linear Bipolar DeMorgan's Laws $\forall (a,b),(c,d) \in B_F$	$\neg((a,b) \otimes (c,d)) \equiv \neg(a,b) \varnothing \neg(c,d)$; $\neg((a,b) \varnothing (c,d)) \equiv \neg(a,b) \otimes \neg(c,d)$;
	$\neg((a,b) \otimes^- (c,d)) \equiv \neg(a,b) \varnothing^- \neg(c,d)$. $\neg((a,b) \varnothing^- (c,d)) \equiv \neg(a,b) \otimes^- \neg(c,d)$.

Further, we extend $\otimes^-$ and $\varnothing^-$ from B_1 to B_F as shown in of Eq. 4.15 and Eq. 4.16, respectively.

$$cglb^-((x,y),(u,v)) \equiv (x,y) \otimes^- (u,v) \equiv -((x,y) \otimes (u,v)). \quad (4.15)$$

$$club^-((x,y),(u,v)) \equiv (x,y) \varnothing^- (u,v) \equiv -((x,y) \varnothing (u,v)). \quad (4.16)$$

Theorem 4.9. The non-linear bipolar DeMorgan's laws in Table 3 hold on B_F.

Proof. The dual DeMorgan's laws on $\otimes$ and $\varnothing$ follow directly from Eq. 4.14. The dual DeMorgan's laws on $\otimes^-$ and $\varnothing^-$ follow directly from Eq. 4.15 – 4.16. □

Now we have two pairs of linear connectives $\{\&, \oplus\}$ and $\{\&^-, \oplus^-\}$, two pairs of non-linear connectives $\{\otimes,\varnothing\}$ and $\{\otimes^-,\varnothing^-\}$, and four pairs of bipolar DeMorgan's laws. Now we have the questions: What are the differences between unipolar and bipolar DeMorgan's laws especially in the fuzzy case? What does each bipolar operator mean in B_F?

First, as in B_1 the bipolar DeMorgan's laws are dynamic in B_F. Such dynamic properties are easy to see for the non-linear operators $\{\otimes,\varnothing\}$ and $\{\otimes^-,\varnothing^-\}$ with bipolar oscillation. For the linear operators $\{\&, \oplus\}$ and $\{\&^-, \oplus^-\}$ the dynamic nature can be analyzed with an example. For instance, if we use (0, 0.8) for characterizing a mental mania and (-0.7,0) for mental depression, $(0,0.8) \oplus (-0.7,0) = (-0.7,0.8)$ describes a dynamic binding, coupling, or fusion of the two unbalanced mental states into a nearly healthy fuzzy mental equilibrium. Thus, $\oplus$ can be used as a mental fusion operator. On the other hand, the bipolar union $\{(0,0.8)\} \cup \{(-0.7,0)\} = \{(-0.7,0),(0,0.8)\}$ can be used to characterize a bipolar psychiatric disorder known as a mixed state with both mania and depression symptoms (Zhang & Peace 2007). Note that a bipolar set consists of bipolar elements that are inseparable for representing bipolar equilibria, quasi-equilibria or non-equilibria. The advantage of using bipolar fuzzy values in B_F leads to more precise knowledge representation. With B_1 it would be impossible to represent a slightly manic mental state without quasi-equilibrium such as (-0.7,0.8).

It is interesting to notice that $\otimes^-$ combines the semantics of $- - = -$, $- + = +$, $+ - = +$, $+ + = -$ on each pole. These semantics may give the impression of absolutely nonsense at first glance. A careful examination, however, reveals just the opposite. That is, $\otimes^-$ is a counterintuitive oscillator which is the resistance or negation of the intuitive oscillator $\otimes$. Such a negation could be useful in modeling drug reaction (see later chapter).

While the semantics of $\varnothing$ and $\varnothing^-$ are mutual negations, the key is to understand the difference between $\varnothing$ and $\otimes$. While $\otimes$ is a non-linear cross-pole conjunctive, $\varnothing$ is a non-linear cross-pole disjunctive that provides a least upper bound for an oscillation derived from the complement or resistance of the oscillator ⊗ as shown in Eq. (4.14). This non-linear least upper bound is more evident with B_F than with B_1. For instances,

- (-0.6,0.3) ⊗ (-0.3,0.6) = (-0.6,0.3) is one end of an oscillation; (-0.6,0.3) ∅ (-0.3,0.6) = (-0.3,0.6) is another end in symmetry;
- (-0.3,0.3) ⊗ (-0.6,0.6) = (-0.3,0.3) is a cglb; (-0.3,0.3) ∅ (-0.6,0.6) = (-0.6,0.6) is a *club*;
- (0,0) ⊗ (-0.3,0.6) = (0,0) characterizes a disturbed or eliminated quasi-equilibrium; (0,0) ∅ (-0.3,0.6) = (-0.3, 0.6) shows an undisturbed quasi-equilibrium.

Bipolar P-Norms, P-Conorms, T-Norms and T-Conorms

Definition 4.13a. Given the bipolar fuzzy lattice $B_F = [-1,0] \times [0,1]$, ***a linear bipolar minimizer t-norm*** is a function $T: B_F \times B_F \Rightarrow B_F$ and, $\forall (x,y),(u,v)(a,b),(c,d) \in B_F$, T satisfies the following properties:

- *Commutativity:* $T((x,y), (u,v)) \equiv T((u,v), (x,y))$;
- *Monotonicity:* $T((x,y),(u,v)) \geq\geq T((a,b),(c,d))$, *if* $(x,y) \geq\geq (a,b)$ *and* $(u,v) \geq\geq (c,d)$;
- *Associativity:* $T((x,y),T((u,v),(a,b))) \equiv T(T((x,y),(u,v)),(a,b))$; *and*
- *Positive Identity:* $T((a,b),(-1,1)) \equiv (a,b)$ *or* $-T((a,b),(-1,1)) \equiv (a,b)$ *or*
- *Negative Identity:* $-T((a,b),(-1,1)) \equiv (a,b)$.

It is interesting to notice that in the bipolar case we have positive and negative identity and the null law $T((a,b),(0,0)) \equiv (0,0)$ is derivable from the above definition. The following theorem is obvious and it proof is omitted.

Theorem 4.10. $\forall (x,y),(u,v) \in B_F$, $T_{\&i}$ as defined in Eq. 4.17 is a linear bipolar t-norm.

$$T_{\&i}((x,y),(u,v)) \equiv (x,y)\&_i(u,v) \equiv (-T_i(|x|,|u|),T_i(|y|,|v|)). \quad (4.17)$$

It is evident that on the bipolar lattice $B_F = [-1,0] \times [0,1]$, the linear bipolar t-norm & can be any $T_{\&i.}$ Three popular ones $T_{\&1} = \&_\wedge$, $T_{\&2} = \&_\times$, or $T_{\&3} = \&_\Delta$ are defined in Eq. 4.18 which show different granularities similar to those of their unipolar counterparts.

$$T_{\&1}((x,y),(u,v)) \equiv (x,y)\ \&_1\ (u,v) \equiv (x,y)\ \&_\wedge\ (u,v) \equiv (-(|x|\wedge|u|),y\wedge v)); \quad (4.18a)$$

$$T_{\&2}((x,y),(u,v)) \equiv (x,y)\ \&_2\ (u,v) \equiv (x,y)\ \&_\times\ (u,v) \equiv (-(|x|\times|u|),y\times v)); \quad (4.18b)$$

$$T_{\&3}((x,y),(u,v)) \equiv (x,y)\ \&_3\ (u,v) \equiv (x,y)\ \&_\Delta\ (u,v) \equiv (-(|x|\Delta|u|),y\Delta v)). \quad (4.18c)$$

Definition 4.13b. Given the bipolar fuzzy lattice $B_F = [-1,0] \times [0,1]$, ***a non-linear cross-pole bipolar interactive or oscillatory t-norm*** is a function $T: B_F \times B_F \Rightarrow B_F$ and, $\forall (x,y),(u,v)(a,b),(c,d) \in B_F$, T satisfies the following properties:

- *Commutativity:* $T((x,y), (u,v))\ \ T((u,v), (x,y))$;
- *Monotonicity:* $T((x,y), (u,v)) \geq\geq T((a,b), (c,d))$, *if* $(x,y) \geq\geq (a,b)$ *and* $(u,v) \geq\geq (c,d)$;

- *Associativity:* $T((x,y),T((u,v),(a,b)))\ \ T(T((x,y),(u,v)),(a,b))$; *and*
- *Positive Identity:* $T((a,b),\ (0,1))\ \ (a,b)$ *or Negative Identity:* $T((a,b),\ (-1,0))\ \ (a,b)$.

It should be noted that the null law $T((a,b),(0,0)) \equiv (0,0)$ is derivable from the above definition.

Definition 4.13c. Given the bipolar fuzzy lattice $B_F = [-1,0] \times [0,1]$, ***a non-linear cross-pole bipolar interactive or oscillatory p-norm*** is a function $P\colon B_F \times B_F \Rightarrow B_F$ and, $\forall (x,y),(u,v)(a,b),(c,d) \in B_F$, P satisfies the following properties:

- *Commutativity:* $P((x,y),\ (u,v)) \equiv P((u,v),\ (x,y))$;
- *Monotonicity:* $P((x,y),\ (u,v)) \geq\geq P((a,b),\ (c,d))$, *if* $(x,y) \geq\geq (a,b)$ *and* $(u,v) \geq\geq (c,d)$; *and*
- *Positive Identity:* $P((a,b),\ (0,1)) \equiv (a,b)$ *or Negative Identity:* $P((a,b),(-1,0)) \equiv (a,b)$.

Based on the Definitions 4.13(b-c) all non-linear *bipolar t-norms* are also p-norms but not vice versa due to the lack of associatively. The definition of p-norms can be justified with the following arguments:

1. The associative law is too strict for dynamic sequential data mining especially in psychopharmacology and mental health data mining where the order of giving different medication to patients at different times does make a difference and hence the effects of different medications do not have to be associative.
2. BUMP can be applied to determine p-norms as well as t-norms where associativity is not necessary.

Intuitively, we assume m_1 is a positive antidepressant drug characterized with $\phi(m_1) = (0,1)$, and m_2 is a negative drug for treating mania characterized with $\phi(m_2) = (-1,0)$. Let p be a patient, m_1 and m_2 be two drugs, T be a t-norm for patient-drug reaction, ϕ be a bipolar L-set for mental equilibrium, and φ be a bipolar L-set for drug effect, and, $\phi(p),\varphi(m_1),\varphi(m_2) \in B_F$, it is easy to see that commutativity is natural as we have $T(\phi(p),\varphi(m_1)) = T(\varphi(m_1),\phi(p))$ that stands for "p takes m_1 is the same as m_1 is used to treat p." Alternatively, $T(\varphi(m_1),\varphi(m_2)) = T(\varphi(m_2),\varphi(m_1))$ stands for "the combo treatment using m_1 and m_2 is the same as using m_2 and m_1." But the associative law $T(\phi(p),\varphi(m_1),\varphi(m_2)) = T(T(\phi(p),\varphi(m_1)),\varphi(m_2)) = T(T(\phi(p),\varphi(m_2)),\varphi(m_1)) = T(\phi(p),(\varphi(m_1),\varphi(m_2)))$ would mean "p takes m_1 first and m_2 later is the same as p takes m_2 first and m_1 later or p takes m_1 and m_2 in a combo." Evidently, the sequencing of drugs may lead to different biological or clinical outcomes, where associativity is not a natural property to impose and p-norms is meaningful.

Theorem 4.11a. $\forall (x,y),(u,v) \in B_F$, any P_{ij} as defined in Eq. 4.19 is a non-linear bipolar p-norm.

$$P_{ij}((x,y),(u,v)) \equiv (x,y)\otimes_{ij}(u,v) \equiv (-\bot_i(T_j(|x|,|v|),T_j(|y|,|u|)),\ \bot_i(T_j(|x|,|u|),T_j(|y|,\ |v|))). \tag{4.19}$$

Proof. The commutativity and monotonicity of P_{ij} *follow from that of* $\bot_i$ *and* T_j *in Eq. 4.19. The positive and negative identity law for* P_{ij} *can be strictly verified with Eq. (4.19) and the neutral element (0,1).* □

Theorem 4.11b. When i = 1 any non-linear bipolar p-norm P_{ij} is a non-linear bipolar t-norm T_{1j}.

Proof. Let $\bullet$ *be any unipolar t-norm* T_j*. Note that when i=1 we have* $\bot_i = \vee$ *and,* $\forall (x,y),(u,v),(a,b) \in B_F$*, we have the associative law:*

$P_{1j}((x,y),(u,v),(a,b)) \equiv [(x,y) \otimes_{1j} (u,v)] \otimes_{1j} (a,b) \equiv (x,y) \otimes_{1j} [(u,v) \otimes_{1j} (a,b)]$

$\equiv (-(|x|\bullet|v|\vee|y|\bullet|u|), (|x|\bullet|u|\vee y\bullet v)) \otimes_{1j} (a,b)$

$\equiv (-(|x|\bullet|v|\bullet b \vee |y|\bullet|u|\bullet b \vee |x|\bullet|u|\bullet|a|\vee y\bullet v\bullet|a|), (|x|\bullet|v|\bullet|a| \vee |y|\bullet|u|\bullet|a| \vee |x|\bullet|u|\bullet b\vee y\bullet v\bullet b))$. □

A non-linear bipolar p-norm can be any $\otimes_{ij}$. Based on the three most popular and distinctive classical t-norms $\wedge$, $\times$, and Δ (Bezdek & Harris, 1978), nine of the popular bipolar p-norms with distinct granularities are defined in Eq. (4.20a-i). The bipolar distinction in granularity is derived from their unipolar counterparts. The first three of them have been proven bipolar t-norms (see last Theorem); the other six can be verified not associative.

$P_{\otimes 11}((x,y),(u,v)) \equiv (x,y) \otimes_{11} (u,v) \equiv (-\bot_1(|x|\wedge|v|,|y|\wedge|u|), \bot_1(|x|\wedge|u|,y\wedge v));$ (4.20a)

$P_{\otimes 12}((x,y),(u,v)) \equiv (x,y) \otimes_{12} (u,v) \equiv (-\bot_1(|x|\times|v|,|y|\times|u|), \bot_1(|x|\times|u|,y\times v));$ (4.20b)

$P_{\otimes 13}((x,y),(u,v)) \equiv (x,y) \otimes_{13} (u,v) \equiv (-\bot_1(|x|\Delta|v|,|y|\Delta|u|), \bot_1(|x|\Delta|u|,y\Delta v));$ (4.20c)

$P_{\otimes 21}((x,y),(u,v)) \equiv (x,y) \otimes_{21} (u,v) \equiv (-\bot_2(|x|\wedge|v|,|y|\wedge|u|), \bot_2(|x|\wedge|u|,y\wedge v));$ (4.20d)

$P_{\otimes 22}((x,y),(u,v)) \equiv (x,y) \otimes_{22} (u,v) \equiv (-\bot_2(|x|\times|v|,|y|\times|u|), \bot_2(|x|\times|u|,y\times v));$ (4.20e)

$P_{\otimes 23}((x,y),(u,v)) \equiv (x,y) \otimes_{23} (u,v) \equiv (-\bot_2(|x|\Delta|v|,|y|\Delta|u|), \bot_2(|x|\Delta|u|,y\Delta v));$ (4.20f)

$P_{\otimes 31}((x,y),(u,v)) \equiv (x,y) \otimes_{31} (u,v) \equiv (-\bot_3(|x|\wedge|v|,|y|\wedge|u|), \bot_3(|x|\wedge|u|,y\wedge v));$ (4.20g)

$P_{\otimes 32}((x,y),(u,v)) \equiv (x,y) \otimes_{32} (u,v) \equiv (-\bot_3(|x|\times|v|,|y|\times|u|), \bot_3(|x|\times|u|,y\times v));$ (4.20h)

$P_{\otimes 33}((x,y),(u,v)) \equiv (x,y) \otimes_{33} (u,v) \equiv (-\bot_3(|x|\Delta|v|,|y|\Delta|u|),\bot_3(|x|\Delta|u|,y\Delta v)).$ (4.20i)

Definition 4.14a. A linear bipolar balancer t-conorm $\bot$ satisfies the following properties:

- *Commutativity:* $\bot((x,y), (u,v)) \equiv \bot((u,v), (x,y));$
- *Monotonicity:* $\bot((x,y), (u,v)) \gg \bot((a,b), (c,d))$, *if* $(x,y) \gg (a,b)$ *and* $(u,v) \gg (c,d);$
- *Associativity:* $\bot((x,y), \bot((u,v), (a,b))) \equiv \bot(\bot((x,y), (u,v)),(a,b))$; *and*
- *Positive Identity:* $\bot((a,b),(0,0)) \equiv (a,b)$ *or Negative Identity:* $-\bot((a,b),(0,0)) \equiv (a,b)$.

Note that the null law $\bot((a,b),(-1,1)) \equiv (-1,1)$ is a direct derivation of the above definition.

Definition 4.14b. A cross-pole non-linear bipolar oscillatory t-conorm $\bot$ satisfies the following relations:

- *Commutativity:* $\perp((x,y), (u,v)) \equiv \perp((u,v), (x,y))$;
- *Monotonicity:* $\perp((x,y), (u,v)) \gg \perp((a,b), (c,d))$, *if* $(x,y) \gg (a,b)$ *and* $(u,v) \gg (c,d)$;
- *Associativity:* $\perp((x,y), \perp((u,v), (a,b))) \equiv \perp(\perp((x,y), (u,v)),(a,b))$; *and*
- *Positive Identity:* $\perp((a,b),(0,1)) \equiv (a,b)$ *or Negative Identity:* $\perp((a,b),(-1,0)) \equiv (a,b)$.

Note that the null law $\perp((a,b),(-1,1)) \equiv (-1,1)$ is a direct derivation of the above definition.

Definition 4.14c. Given the bipolar fuzzy lattice $B_F = [-1,0] \times [0,1]$, ***a cross-pole non-linear bipolar interactive or oscillatory p-conorm*** is a function $P': B_F \times B_F \Rightarrow B_F$ and, $\forall (x,y),(u,v)(a,b) \in B_F$ we have the following properties:

- *Commutativity:* $P'((x,y), (u,v)) \equiv P'((u,v), (x,y))$;
- *Monotonicity:* $P'((x,y), (u,v)) \gg P'((a,b), (c,d))$, *if* $(x,y) \gg (a,b)$ *and* $(u,v) \gg (c,d)$; *and*
- *Positive Identity:* $P'((a,b), (0,1)) \equiv (a,b)$ *or Negative Identity:* $\boldsymbol{P'}((a,b), (-1,0)) \equiv (a,b)$.

Based on the above definitions non-linear *bipolar t-conorms* are also p-conorms but not vice versa.

Theorem 4.12. Any bipolar t-norm T and p-norm P has a dual bipolar t-conorm and p-conorm as defined by Eq. 4.21a and.21b, respectively. Formally, given a bipolar t-norm T or p-norm P, we have

$$\perp((x,y), (u,v)) \equiv (-1,1) - T(\neg(x,y), \neg(u,v)). \tag{4.21a}$$

$$P'_{ij}((x,y), (u,v)) \equiv (-1,1) - P_{ij}(\neg(x,y), \neg(u,v)). \tag{4.21b}$$

Proof. The commutativity, monotonicity, and associativity follow from that of its duals, respectively. The identity laws can be strictly verified with the neutral element (0,0) for linear cases and (-1,0) or (0,1) for non-linear cases. □

It should be remarked that the identity law for any linear bipolar t-conorm $\oplus_i$ is obviously $(a,b) \oplus_i (0,0) \equiv (a,b)$. The identity law for a non-linear bipolar p-conorm $\varnothing_{ij}$ can be verified as $((a,b)\varnothing_{ij}(-1,0)) \equiv (a,b)$. It is interesting to notice that $((a,b)\varnothing_{ij}(0,0)) \neq (a,b)$; instead, $((a,b)\varnothing_{ij}(0,0)) \equiv (-min(|a|, b), min(|a|,b))$, which assumes the semantics: ***Any closed bipolar dynamic system (a,b) with zero disturbance (0,0) will at least reach the least upper bound equilibrium (-min(|a|, b), min(|a|,b)).***

Similarly we have $\perp\oplus_1 = \oplus_1$, $\perp\oplus_2 = \oplus_2$, $\perp\oplus_3 = \oplus_3$, respectively, as defined in Eq. (4.22a-c).

$$\perp_{\oplus 1}((x,y),(u,v)) \equiv (x,y) \oplus_1 (u,v) \equiv (-\perp_1(|x|,|u|), \perp_1(y,v)) \equiv (-(|x| \vee |u|), y \vee v); \tag{4.22a}$$

$$\perp_{\oplus 2}((x,y),(u,v)) \equiv (x,y) \oplus_2 (u,v) \equiv (-\perp_2(|x|,|u|), \perp_2(y,v)) \equiv (-(|x|+|u|-|x|\times|u|), y + v - y\times v); \tag{4.22b}$$

$$\perp_{\oplus 3}((x,y),(u,v)) \equiv (x,y) \oplus_3 (u,v) \equiv (-\perp_3(|x|,|u|), \perp_3(y,v)) \equiv (-(|x|+|u|-|x|\Delta|u|), y + v - y\Delta v). \tag{4.22c}$$

Based on the non-linear bipolar DeMorgan's laws (see Table 4.3), a non-linear bipolar p-conorm $\varnothing$ can be any $\varnothing_{ij}$ dual to $\otimes_{ij}$ as defined in Eq. 4.23:

Table 4. Bipolar norms and their neutral elements

Norm	Identity Law	Neutral Element
$\&_i$	$(x,y)\&_i(-1,1) \equiv (x,y)$	(-1,1)
$\oplus_i$	$(x,y)\oplus_i(0,0) \equiv (x,y)$	(0,0)
$\&_i^-$	$-[(x,y)\&_i^-(-1,1)] \equiv (x,y)$	(-1,1), –
$\oplus_i^-$	$-[(x,y)\oplus_i^-(0,0)] \equiv (x,y)$	(0,0), –
$\otimes_{ij}$	$(x,y) \otimes_{ij}(0,1) \equiv (x,y)$	(0,1)
$\otimes_{ij}^-$	$(x,y)\otimes_{ij}^-(-1,0) \equiv (x,y)$	(-1,0)
$\varnothing_{ij}$	$(x,y)\varnothing_{ij}(-1,0) \equiv (x,y)$	(-1,0)
$\varnothing_{ij}^-$	$(x,y)\varnothing_{ij}^-(0,1) \equiv (x,y)$	(0,1)

$$P'_{ij}((x,y),(u,v)) \equiv (x,y)\varnothing_{ij} (u,v) \equiv \neg(\neg(x,y)\otimes_{ij} \neg(u,v)) \equiv (-1,1)-(\neg(x,y)\otimes_{ij} \neg(u,v)). \quad (4.23)$$

Theorem 4.13. Let i = 1, any non-linear bipolar p-conorm $P'_{ij} = \varnothing_{ij}$ is a bipolar t-conorm $\perp_{ij}$.

Proof. It follows from Theorem 4.11 and the duality condition in Eq. 4.23 between $\varnothing_{1j}$ and $\otimes_{1j}$. □

The negations of $\oplus_i$, $\&_i$, $\otimes_{ij}$ and $\varnothing_{ij}$ are defined in Eq. (4.24a-d).

$$T_{\&i}^-((x,y),(u,v)) \equiv (x,y)\&_i^-(u,v) \equiv -((x,y)\&_i(u,v)); \quad (4.24a)$$

$$\perp_{\oplus i}^-((x,y),(u,v)) \equiv (x,y)\oplus_i^-(u,v) \equiv -((x,y)\oplus_i(u,v)); \quad (4.24b)$$

$$P_{\otimes ij}^- ((x,y),(u,v)) \equiv (x,y) \otimes_{ij}^- (u,v) \equiv -((x,y)\otimes_{ij}(u,v)); \quad (4.24c)$$

$$P'_{\varnothing ij}^- ((x,y),(u,v)) \equiv (x,y) \varnothing_{ij}^- (u,v) \equiv -((x,y) \varnothing_{ij} (u,v)). \quad (4.24d)$$

It can be proven that the negations of all t-norms, t-conorms, p-norms, p-conorms (linear or non-linear) are also t-norms, t-conorms, p-norms, and p-conorms, respectively. The identity laws and neutral elements of different norms are listed in Table 4.

NORM-BASED BIPOLAR UNIVERSAL MODUS PONENS

Norm-Based BUMP

With bipolar norms, we generalize bipolar universal modus ponens (BUMP) to a norm-based granular or fuzzy version as shown in Table 5. The new BUMP achieves practical universality in terms of universal bipolar operator binding to any t-norm, p-norm, and their co-norms.

Theorem 4.14. The norm-based BUMP in (Table 5) is a bipolar tautology.

Table 5. Norm-based BUMP

BUMP in IF-THEN form:

IF $[((\phi^-,\phi^+)\Rightarrow(\varphi^-,\varphi^+)) \equiv(-1,1)]$,

$[((\psi^-,\psi^+)\Rightarrow(\chi^-,\chi^+)) \equiv(-1,1)]$, $((\phi^-,\phi^+)*(\psi^-,\psi^+))\}$;

THEN $((\varphi^-,\varphi^+)*(\chi^-,\chi^+))$.

BUMP in Tautological form:

$\{ [((\phi^-,\phi^+)\Rightarrow(\varphi^-,\varphi^+)) \equiv(-1,1)] \,\&\, [((\psi^-,\psi^+)\Rightarrow(\chi^-,\chi^+)) \equiv(-1,1)] \}$

$\Rightarrow \{[((\phi^-,\phi^+)*(\psi^-,\psi^+)) \Rightarrow ((\varphi^-,\varphi^+)*(\chi^-,\chi^+))] \equiv (-1,1) \}$.

Two-fold universal instantiation:

(1) Operator instantiation: Bipolar operator instantiation: * is a universal operator that can be bound to any t-norm, p-norm, t-conorm, or p-conorm including $\&_i$, $\oplus_i$, $\&_i^-$, $\oplus_i^-$, $\otimes_{ij}$, $\varnothing_{ij}$, $\otimes_{ij}^-$, or $\varnothing_{ij}^-$.

(2) $(\phi \Rightarrow \varphi) \equiv (-1,1)$ is designated (bipolar true); $((\phi^-,\phi^+)*(\psi^-,\psi^+))$ is not designated.

(3) Bipolar variable instantiation: $\forall x$, $(\phi^-,\phi^+)(x) \Rightarrow (\varphi^-,\varphi^+)(x)$; $(\phi^-,\phi^+)(A)$; $\therefore$ $(\varphi^-,\varphi^+)(A)$.

Proof. $\forall(\phi^-,\phi^+),(\psi^-,\psi^+),(\varphi^-,\varphi^+),(\chi^-,\chi^+)\in B_F$ *any binding of * to a bipolar norm including* $\&_i$, $\oplus_i$, $\&_i^-$, $\oplus_i^-$, $\otimes_{ij}$, $\varnothing_{ij}$, $\otimes_{ij}^-$, *or* $\varnothing_{ij}^-$, *BUMP satisfies Eq. (4.25) provided that* $(a^-,a^+) \Rightarrow (b^-,b^+)\equiv (-1,1)$ *satisfies* $(a^-,a^+)\leq\leq(b^-,b^+)$ *as defined in Eq. (4.6).*

$$\{ [((\phi^-,\phi^+)\Rightarrow(\varphi^-,\varphi^+)) \equiv (-1,1)]\&[((\psi^-,\psi^+)\Rightarrow(\chi^-,\chi^+)) \equiv(-1,1)]\& [((\phi^-,\phi^+)*(\psi^-,\psi^+))] \}$$

$$\leq\leq [((\varphi^-,\varphi^+)*(\chi^-,\chi^+))]; \tag{4.25a}$$

$$\{ [((\phi^-,\phi^+)\Rightarrow(\varphi^-,\varphi^+)) \equiv(-1,1)] \,\&\, [((\psi^-,\psi^+)\Rightarrow(\chi^-,\chi^+)) \equiv(-1,1)] \}$$

$$\Rightarrow [((\phi^-,\phi^+)*(\psi^-,\psi^+))] \leq\leq [((\varphi^-,\varphi^+)*(\chi^-,\chi^+))]. \tag{4.25b}$$

The theorem follows directly from the commutativity and bipolar monotonicity of all bipolar p-norms, p-conorms, t-norms, and t-conorms. □

Theorem 4.15. The norm-based BUMP in (Table 5) is a real-valued non-linear bipolar fuzzy dynamic generalization of classical modus ponens (MP) with quantum entanglement property.

Proof.

1. *The binding of* $\&_i$, $\oplus_i$, $\&^-_i$, or $\oplus^-_i$ *to the universal operator * makes BUMP linearly bipolar equivalent to MP because* $\&_i$, $\oplus_i$, $\&^-_i$, and $\oplus^-_i$ *are linearly bipolar equivalent to* $\wedge$ *or* $\vee$, *respectively, and* $\Rightarrow$ *is linearly bipolar equivalent to* $\rightarrow$. *However,* $\otimes_{ij}$, $\varnothing_{ij}$, $\otimes_{ij}^-$, and $\varnothing_{ij}^-$ *are non-linear operators with cross-pole bipolar interaction or oscillation semantics and syntax that find no equivalent in* $\wedge$ *or* $\vee$ *(Eq. 4, 10, 11, and 12). Therefore, classical MP is part of BUMP but not vice versa.*
2. *Classical MP is truth-based and static in nature that states "IF A implies B and A is true, THEN B is true." BUMP is bipolar equilibrium-based and dynamic in nature that states "IF the first bipolar equilibrium implies the 2nd and the 3rd implies the 4th, THEN bipolar dynamic interaction between the 1st and the 3rd implies the same type of bipolar dynamic interaction between the 2nd and the 4th. Such non-linear bipolar dynamic physical semantics and syntax find no equivalent in classical MP. These bipolar quantum entanglement semantics and syntax enable "illogical" or disordered natural properties and behaviors of a physical system (including biological and quantum systems)*

*to be represented for logical solutions. Thus, the binding of the universal operator * does not have to be restricted to the so-called "free" linear logical operators.* □

Applicability of Bipolar Norms and BUMP

Bipolar sets and logic have been used in different research areas including quantum teleportation and computing (Zhang, 2010), bipolar cognitive mapping and multiagent decision/coordination (e.g. Zhang *et al.,* 1989, 1992, 2003a, 2003b), business decision analysis (e.g. Kwahk & Kim, 1999; Lee & Kwon, 2008), knowledge fusion and global regulation (e.g. Zhang, 2005a, 2005b, 2006), bipolar disorder diagnostic analysis (e.g. Zhang, Pandurangi & Peace, 2007). While bipolar reasoning is generally applicable in an open-world of equilibria, they are particularly suitable in bipolar neurobiological modeling and pattern analysis. According to NIMH (US National Institute of Mental Health) millions of children and adults are affected by major depression or bipolar disorders in the US alone. Understanding the differences in how people respond to medication is one of the most important goals of psychiatric drug therapy and robot pet therapy (Tamura *et al.,* 2004). Progress in this area has been slow because of the complexity of human emotion problem. To the author's knowledge, there is no commonly accepted logical system for modeling human and robot emotions in AI. Since all closed dynamic systems tend to reach a form of equilibrium and all open dynamic systems tend to be disturbed to non-equilibrium, bipolar norms and BUMP provide a computational basis for exploratory knowledge discovery in neurobiological data mining and agent emotion modeling.

COMPARISON AND DISCUSSION

Bipolar Norms vs. Unipolar Norms

Classical unipolar norms have been intensively investigated in fuzzy set research with thousands of publications. Unipolar norms provide the logical basis for computation with truth and falsity in different forms of uncertainties. Unipolar norms are truth-based and not bipolar equilibrium-based.

Uninorms presents a generalization of both t-norms and t-conorms (Yager & Rebalov, 2000). While uninorms allow for an identity element lying anywhere in the unit interval rather than at one or zero as in the case of t-norms and t-conorms, respectively, bipolar norms allow for an identity element to be (-1,0), (0,0), (0,1) or (-1,1) (see Table 4). Since equilibrium-based (-,+) bipolar syntax and semantics are not observed in uninorms, they are fundamentally truth-based and unipolar in nature.

Bipolar Norms vs. Balanced Norms

Balanced norms (Homanda, 2006) extend the negative truth value 0 in [0,1] to [-1,0]. The "negative" and "positive" concepts in this extension are used for "false" and "true", respectively, which find applications in neural networks. The extension is, therefore, motivated by the balance of truth and falsity *w.r.t.* individual truth objects, where the reification and binding of the negative and positive poles of a physical equilibrium (Zhang & Zhang, 2003, 2004, 2005a) are not observed.

Bipolarity vs. Paraconsistency and Intuitionism

Paraconsistency and intuitionistic models have been defined, respectively, on the unit square lattices $\{0,1\}^2$ (Belnap, 1977; Ginsberg, 1990) and $[0,1]^2$ (Atanassov, 1986). These structures are truth-based in nature and fundamentally different from equilibrium-based YinYang bipolar lattice (Zhang, 2005a). Firstly, they do not satisfy the law of excluded middle (LEM) which is satisfied by the crisp bipolar lattice B_1. Secondly, they are static and traditionally used for reasoning on truth and falsity with contradiction or indetermination as permissible truth values, where dynamic physical syntax and semantics pertaining to two poles are not observed.

Bipolar Norms vs. Unipolar Norms on Product Lattices

It is interesting to compare bipolar norms with unipolar norms on product lattices. A proposition on product lattices is stated as *"Any lattice-ordered t-norm * on a product lattice $L = L_1 \times L_2$ is the direct product of two lattice ordered t-norms on L_1 and L_2, respectively"* (Jenei & De Baets, 2003) which can be formally described as Eq. 4.26.

$$T_1 \times T_2((x_1,y_1),(x_2,y_2)) \equiv (T_1\,(x_1,x_2),\, T_2\,(y_1,y_2)). \qquad (4.26)$$

Evidently, the product lattice $L = L_1 \times L_2$ in Eq. (4.26) doesn't account for YinYang bipolar lattices such $B_F = [-1,0]\times[0,1]$ and the non-linear bipolar oscillatory t-norms defined in Definition 4.14. The equilibrium-based bipolar oscillator norms $\otimes_{ij}$, $\varnothing_{ij}$, $\otimes_{ij}^{-}$, and $\varnothing_{ij}^{-}$ have built-in bipolar syntax and semantics such as $-- = +$, $-+ = -$, $+- = -$, and $++ = +$ on each pole. Without bipolarity, however, the product lattice $[0,1]^2$ and any of its unipolar extension provides no representation for the nonlinear bipolar semantics. Evidently, the linearity in Eq. 4.26 would not hold for $\otimes_{ij}$, $\varnothing_{ij}$, $\otimes_{ij}^{-}$, and $\varnothing_{ij}^{-}$.

For instance, $(-0.3, 0)\otimes_{\times}(-0.7, 0) = (0, +0.21)$, where -0.3 and -0.7 belong to the negative lattice [-1, 0] but the result +0.21 swings to the positive lattice [0,1]. Such bipolar equilibrium-based oscillatory property wasn't taken into consideration in Eq. 4.26. Without such non-linear bipolar dynamic property BUMP would degenerate to classical MP (Theorem 4.15).

From the above analysis, we can conclude:

1. If YinYang bipolar lattice and the family of non-linear bipolar t-norms were taken into consideration, the proposition *"Any lattice-ordered t-norm * on a product lattice $L = L_1 \times L_2$ is the direct product of two lattice ordered t-norms on L_1 and L_2, respectively"* would not hold.
2. The t-norms on the product lattice $[0,1]^2$ are unipolar static truth-based in nature that is inadequate for non-linear bipolar knowledge representation and equilibrium-based dynamic reasoning.
3. Equilibrium-based bipolar norms and truth-based unipolar norms should be complementary to each other in problem solving.

A Distinguishing Factor

Among all differences between unipolar and bipolar systems, BUMP is only observed in a bipolar dynamic crisp or fuzzy logic. As a fundamentally different inference rule from classical MP and a cornerstone of equilibrium-based bipolar dynamic reasoning, it can be considered a distinguishing factor for this work.

Lattice-Ordered Bipolar Dynamic Logic

Although the norm-based BUMP in Table 5 holds on B_F for all t-norms and p-norms and their co-norms, it does not hold on an arbitrary strict bipolar lattice B. This is because that, if B is not normalized, the bipolar norms that are not defined solely based on $\wedge$ and $\vee$ are not bounded within B. For instance, $(-1,1) \otimes_{12} (-1,1) = (-1,1)$ but $(-2,2) \otimes_{12} (-2,2) >> (-2,2)$. This can be resolved if we restrict all bipolar operators to $\wedge$-$\vee$-based norms on an arbitrary strict bipolar quantum lattice $B = [N,0]\times[0,P]$. In addition we redefine bipolar complement, implication, and negation as in Eq. 4.27 – 4.29.

Complement: $\neg(x,y)\equiv(N,P)-(x,y)\equiv(\neg x,\neg y)\equiv(N-x,P-y)$. (4.27)

Logical Negation (Implication): $(x,y)\Rightarrow(u,v)\equiv(x\rightarrow u,y\rightarrow v)\equiv(\neg x\vee u,\ \neg y\vee v)$. (4.28)

Arithmetic Negation: $-(x,y) \equiv (-y,-x)$. (4.29)

Theorem 4.16. The lattice-based BUMP in Table 6 is a bipolar tautology.

Proof. It follows directly from (1) B is strict, and (2) the norms are $\wedge$-$\vee$-*based.* □

BIPOLARITY, LINEARITY, INTEGRITY AND RECOVERY THEOREM

Strict ***bipolarity*** meets the following conditions:

1. ***Bipolar Coexistence***: The existence of one pole conceptually depends on another in one concept regardless of the truth value of each pole. Generally speaking, let Yin and Yang be the negative and positive poles of any agent; without the concept Yin there would be no the concept Yang or vice versa.
2. ***Bipolar Equilibrium***: The two poles form a bipolar equilibrium or non-equilibrium that are bipolar interactive and reciprocal through the universal non-linear bipolar dynamic operator *.
3. ***Bipolar Symmetry***: Negation converts one pole to another in symmetry such that -(yin,yang)=(-yang,-yin).

Table 6. Lattice-ordered BUMP

BUMP in IF-THEN form:

IF $[((\phi^-,\phi^+)\Rightarrow(\varphi^-,\varphi^+)) \equiv(N,P)]$,
$[((\psi^-,\psi^+)\Rightarrow(\chi^-,\chi^+)) \equiv(N,P)],\ ((\phi^-,\phi^+)*(\psi^-,\psi^+))\}$;
THEN $((\varphi^-,\varphi^+)*(\chi^-,\chi^+))$.

BUMP in Tautological form:

$\{ [((\phi^-,\phi^+)\Rightarrow(\varphi^-,\varphi^+)) \equiv(N,P)]\ \&\ [((\psi^-,\psi^+)\Rightarrow(\chi^-,\chi^+)) \equiv(N,P)] \}$
$\Rightarrow \{[((\phi^-,\phi^+)*(\psi^-,\psi^+)) \Rightarrow ((\varphi^-,\varphi^+)*(\chi^-,\chi^+))] \equiv (N,P) \}$.

<u>Two-fold universal instantiation:</u>

(1) Operator instantiation: Bipolar operator instantiation: * is a universal operator that can be bound to any $\wedge$-$\vee$-based $\&_1$, $\oplus_1$, $\&_1^-$, $\oplus_1^-$, $\otimes_{11}$, $\varnothing_{11}$, $\otimes_{11}^-$, or $\varnothing_{11}^-$ or simply $\&$, $\oplus$, $\&^-$, $\oplus^-$, $\otimes$, $\varnothing$, $\otimes^-$, or $\varnothing^-$ on any strict bipolar lattice $B = [N,0]\times[0,P]$.
(2) $(\phi \Rightarrow \varphi) \equiv (N,P)$ is designated (bipolar true); $((\phi^-,\phi^+)*(\psi^-,\psi^+))$ is not designated.
(3) Bipolar variable instantiation: $\forall x,\ (\phi^-,\phi^+)(x) \Rightarrow (\varphi^-,\varphi^+)(x);\ (\phi^-,\phi^+)(A);\ \therefore\ (\varphi^-,\varphi^+)(A)$.

4. ***Bipolar Linearity***: Each pole of a strict bipolar lattice can be recovered to bivalent lattice and Boolean logic linearly and separately.
5. ***Bipolar Integrity***: Both poles of a strict bipolar lattice as a non-linear dynamic combination can be recovered to bivalent lattice and Boolean logic.

Fuzzy bipolarity should meet the additional conditions:

6. ***Bipolar Fuzziness:*** Each pole can be characterized with different degrees of gray levels.
7. ***Bipolar Fuzzy Equilibrium***: The two poles can form a fuzzy or quasi-equilibrium.
8. ***Bipolar Fuzzy Symmetry***: Negation converts one pole to another in fuzzy symmetry such that $-(x,y) = (-y, -x)$.
9. ***Bipolar Fuzzy Linearity***: Each pole can be recovered to Zadeh's fuzzy sets and fuzzy logic (Zadeh, 1965) linearly and separately.
10. ***Bipolar Fuzzy Integrity***: Both poles as a non-linear dynamic binding, coupling, or fusion can be recovered to Zadeh's fuzzy lattice [0,1] and fuzzy logic defined on [0,1].

While the four conditions of ***bipolar coexistence, equilibrium, negation, and fuzziness*** for strict bipolarity are apparently inherent in the definition of strict bipolar lattices. ***Linearity and integrity*** are proved in the following that establish the recovery of strict lattice, BDL, and BDFL to classical logic.

Theorem 4.17 (Linearity Theorem). Without bipolar interaction $\otimes,\otimes^-,\varnothing$, and $\varnothing^-$, any strict bipolar lattice $B = (B,\equiv,\&, \oplus, \&^-, \oplus^-, -, \neg, \Rightarrow)$ can be recovered to the unipolar lattice $(B^+,\equiv,\vee,\wedge,\neg,\rightarrow)$ or $(|B^-|, \equiv,\vee,\wedge\neg,\rightarrow)$ linearly through one of the two depolarization functions (or filters) $Dplr^-(x,y)=|x|$ and $Dplr^+(x,y)=y$, where $(x,y)\in B$, which, in turn, can be recovered to the bivalent lattice $\{0,1\}$ through defuzzification.

Proof. Without bipolar interaction, the two depolarization functions (or filters) $Dplr^-(x,y)=|x|$ and $Dplr^+(x,y)=y$ ***linearly degenerates*** *B to two unipolar lattices $(B^+,\equiv,\vee,\wedge,\neg,\rightarrow)$ and $(|B^-|, \equiv,\vee,\wedge\neg,\rightarrow)$ on its two poles, respectively.* □

Theorem 4.18 (Integrity Theorem). With bipolar interaction $\otimes,\otimes^-,\varnothing$, and $\varnothing^-$, any strict bipolar lattice $B = (B,\equiv,\&,\oplus, \&^-,\oplus^-,\otimes,\otimes^-,\varnothing, \varnothing^-,-, \neg, \Rightarrow)$ can be recovered to a unipolar lattice through the depolarization function (or filter) $Dplr^\vee(x,y)=|x|\vee y$, which, in turn, can be recovered to a Boolean lattice.

Proof. With $\otimes-,\varnothing$, and $\varnothing^-$, B is bipolar interactive. Since $\otimes-,\varnothing$, and $\varnothing^-$ can be defined by $\otimes$, let $U = |x|\vee y$, $\forall(x,y)\in B$, the integrity of B will follow if $Dplr\vee(x,y) = |x|\vee y$ is a homomorphism that degenerates B to $\{U,\equiv,\vee,\wedge,\neg,\rightarrow\}$. First, notice that $-(x,y) \equiv (x,y)\otimes(-1,0)$, the recovery of $-(x,y)$ follows from that of $\otimes$. Secondly, the recovery of $\otimes-,\varnothing$, and $\varnothing^-$ follows from that of $\otimes$. Thirdly, the recovery of $\oplus^-$ follows from that of $\oplus$. Fourthly, the recovery of & and $\&^-$ can be integrated with that of $\otimes$ because the definition of & and $\&^-$ is part of that of $\otimes$ and $\otimes^-$, respectively (see Eq. 4.3, Eq. 4.4, and Theorem 5.14).

Thus, if $\oplus$ and $\otimes$ in B can be recovered to unipolar $\vee$ and $\wedge$, respectively, B will recover to the unipolar lattice [0,1] and then to {0,1} as both $\neg$ and $\rightarrow$ follow $\vee$ and $\wedge$ directly in unipolar case. This is proved as follows. $\forall(a,b),(c,d)\in B$,

1. $Dplr^{\vee}((a,b)\oplus(c,d)) \equiv Dplr^{\vee}(-(|a|\vee|c|),b\vee d) \equiv ((|a|\vee|c|)\vee(b\vee d)) \equiv ((|a|\vee b)\vee(c\vee d))$;
2. $Dplr^{\vee}((a,b) \otimes_{\wedge} (c,d))\equiv Dplr^{\vee}(-\max(|a|\wedge|d|,|b|\wedge|c|),\max(|a|\wedge|c|,b\wedge d))$
   $\equiv (|a|\wedge|d|\vee b\wedge|c|)\vee(|a|\wedge|c|\vee b\wedge d) \equiv ((|a|\vee b)\wedge(c\vee d))$. □

Theorem 4.19 (No-Contradiction Theorem). The integrity theorem does not contradict the linearity theorem.

Proof. It follows from that $Dplr\vee((a,b) \otimes\wedge (c,d)) \equiv (|a|\vee b)\wedge(c\vee d) \geq Dplr\vee((a,b)\&(c,d)) \equiv Dplr\vee(-|a|\wedge|c|,b\wedge d) \equiv |a|\wedge|c|\vee b\wedge d$, *where* $\geq$ *indicates that bipolar interaction is an augmentation of linear conjunction through bipolar interaction with the additional semantics of serial conjunction* $+\wedge+=+$, $-\wedge-=+$, $-+\wedge=-$, *and* $+\wedge-=-$. *That is, bipolar interaction is a truth augmentation without contradicting the linear bipolar operator & and linear truth.* □

Theorem 4.20 (Recovery Theorem). Any strict BQL $B=(B_F,\equiv,\oplus, \oplus^-, \&, \&^-, \otimes, \otimes^-,\varnothing, \varnothing^-,-,\neg,\Rightarrow)$ is a nonlinear dynamic generalization of Boolean logic and B can be recovered to Boolean logic.

Proof. It follows from the proofs of Theorems 4.17-4.19 directly. □

Now an interesting question can be raised: Can the depolarization function (or filter) $Dplr^-(x,y)=|x|$ or $Dplr^+(x,y)=y$, where $(x,y)\in B$, be used to recover the non-linear BQL $B = (B,\equiv,\&,\oplus, \&^-,\oplus^-,\otimes,\otimes^-,\varnothing, \varnothing^-,-, \neg, \Rightarrow)$ to a unipolar lattice? The answer is negative (Zhang & Zhang 2004).

Thus, based on Theorems 4.17 – 4.20, it can be asserted:

1. Any strict BQL $B = (B_F,\equiv,\oplus, \oplus^-, \&, \&^-, \otimes, \otimes^-,\varnothing, \varnothing^-,-,\neg,\Rightarrow)$ is a non-linear bipolar dynamic generalization of Boolean logic or fuzzy logic with quantum entanglement features;
2. Bipolar set theory is a bipolar fusion or generalization of classical set theory from a static truth-based world to a bipolar dynamic quantum world;
3. The coexistence of a negative and a positive pole cannot be directly represented in a unipolar logic; and
4. The recovery function $Dplr^{\vee}(x,y)=|x|\vee y$ conforms to the basic semantics of YinYang bipolar equilibrium or non-equilibrium.

RESEARCH TOPICS

The strength of bipolar quantum lattice lies in its inclusion and non-linear bipolar augmentation of its unipolar counterpart. More theoretical formulations are yet to come on the integration of unipolar and bipolar systems for complex applications. The unifying features of bipolar sets are enumerated in the following:

1. Unipolar L-fuzzy sets are based on unipolar lattices; bipolar L-sets are based on bipolar lattices. The object set X in a unipolar set assumes unipolar truth; the object set X in a bipolar set assumes equilibrium-based bipolar holistic truth.

2. Classical (unipolar) lattices are originated from bivalent logic; bipolar lattices can be considered as mathematical structures inspired by the ancient Chinese YinYang theory. Bipolar lattices provide a basis for holistic top-down cognition with the presumption of an equilibrium or non-equilibrium bipolar world.
3. Unipolar sets do not support non-linear bipolar interaction; bipolar sets support bipolar interaction.
4. Modus ponens is the only generic inference rule for inference with classical sets; bipolar sets lead to bipolar universal modus ponens (BUMP) with non-linear bipolar dynamic quantum entanglement.
5. Unipolar sets are linear, static, and assume a closed world; bipolar L-sets are non-linear, dynamic, and work in an open world due to bipolar augmentation.
6. Unipolar sets are for modeling a universe of truth objects without polarity; bipolar L-sets are for modeling a universe of bipolar equilibria (including quasi or non-equilibria).
7. Unipolar sets do not support equilibrium energy and stability analysis; energy and stability in bipolar L-sets lead to commonsense laws for bipolar optimization and decision.
8. Unipolar lattices do not support equilibrium-based symmetry; bipolar quantum lattice presents a unification of equilibrium and symmetry that leads to an explanation to quantum non-local connection or quantum entanglement to be discussed in chapter 7.

Some suitable application domains are identified in the following:

- **Biomedicine.** A direct application of bipolar sets and bipolar modus ponens is YinYang bipolar dynamic neurological modeling, diagnostic analysis, and psychopharmacology for mental depression and neurological disorder (Zhang, 2007; Zhang, Pandurangi, & Peace, 2007) (Ch. 10). Evidently, given a set of bipolar disorder patients P and a set of biomedical treatment T, both P and T can be characterized as bipolar sets. Preliminary results show that bipolar inference with BUMP can unravel mental or neurophysiological intrinsics of bipolar disorders from certain counter-intuitive symptoms.
- **Bipolar Cognitive Mapping.** With bipolar L-sets and L-fuzzy sets, incomplete, unipolar, or partial cognitive maps can be pooled or mined from text data into bipolar relations for visualization. Equilibrium relations can then be computed for equilibrium analysis, decision, and multiagent coordination (Zhang, 2003a, 2003b, 2005b, 2006) (Ch. 11).
- **Bipolar Partitioning and Clustering.** While an equivalence relation leads to hard partitioning and a crisp equilibrium relation R in X to B_1 induces disjoint ***coalition sets, harmony sets, and conflict sets in X,*** it is shown that a bipolar fuzzy equilibrium relation R in X to $B_F = [-1,0]\times[0,1]$ induces disjoint or joint ***quasi-coalition sets, harmony sets, and conflict sets in X*** (Zhang, 2003b, 2006) (Chs. 5, 11).
- **Equilibrium Energy and Stability Analysis for Global Regulation.** Energy analysis can be extended from bipolar lattices to bipolar relations and applied to both social and natural sciences, especially to international relations and macroeconomics. From equilibrium relations and equilibrium classes bipolar laws can be derived for optimization, decision, coordination, and global regulation (Zhang, 2003a, 2003b, 2005b, 2006) (Chs. 5, 11).
- **Bipolar Evaluation and Decision.** With bipolar cognition, a decision making process can be viewed as a weighing or depolarization process on the desired effects and the side effects of available options. Based on this view, a decision making process can be considered a bipolar equilibrium. Bipolar information/knowledge fusion can be used for bipolar evaluation and decision on

a wide variety of matters. Bipolar evaluations of stock performance, academic publications, job performance, gymnastic performance,.., etc, are a few examples. Bipolar evaluation provides both positive and negative perspectives for decision support with added information gain to unipolar evaluations (Chs. 5, 11).

- **Bipolar Fuzzy Relativity.** BDFL enables real-valued bipolarity that provides a basis for handling uncertainty in bipolar relativity (Ch. 7).
- **Bipolar Quantum Linear Algebra and Bipolar Quantum Cellular Automata.** Bipolar sets, BDL and BDFL provide the logical basis for a bipolar quantum linear algebra and bipolar quantum cellular automata (Chs. 8, 9).

SUMMARY

Bipolar quantum lattice, equilibrium energy, balance, and symmetry have been defined and axiomatically characterized; bipolar t-norms, t-conorms, p-norms and p-conorms have been defined on the fuzzy lattice $B_F = [-1,0] \times [0,1]$; bipolar universal modus ponens (BUMP) has been extended to a norm-based granular or fuzzy version. The concept of strict bipolarity, linearity, and integrity of BQLs have been presented and discussed. A recovery theorem has been presented for the depolarization of any strict BQL to Boolean logic. The recovery theorem reinforces the computability of BDL and BDFL. The originality and significance of bipolar quantum lattices and bipolar dynamic norms are 3-fold: (1) they can carry direct equilibrium-based physical syntax and semantics such as bipolar fusion, interaction, oscillation, and quantum entanglement properties; (2) they enable logical reasoning with BUMP on both logical and "illogical" but nevertheless physical phenomena such as bipolar disorders, symmetry, and quantum entanglement; (3) they enable hypothesis-driven theory validation and exploratory knowledge discovery.

Since equilibrium is central in all dynamic systems, equilibrium-based open-world reasoning and data mining are theoretically imperative in quantum mechanics as well as computational intelligence. Bipolar norms are expected to pave the way for: (1) further theoretical investigation on bipolar continuous norms and bipolar uninorms; (2) the development of bipolar neurofuzzy systems for different applications such as computational neuroscience, agent emotion modeling, and medical robots (Tamura *et al.,* 2004); (3) gene network simulation and regulation (Shi *et al.,* 1991; Ai *et al.,* 2000; Kim *et al.,* 2007); (4) bipolar axiomatization for exploratory scientific knowledge discovery in uncertain or uncharted territories especially in bipolar quantum physics and biophysics (Zhang, 2009a, 2009b, 2009c, 2009d, 2010a).

ACKNOWLEDGMENT

This chapter reuses part of published material in Zhang, W. –R. (2005a) YinYang Bipolar Lattices and L-Sets for Bipolar Knowledge Fusion, Visualization, and Decision. *World Scientific Publishing (WSP): Int'l J. of Inf. Technology and Decision Making*, Vol. 4, No. 4: 621-645, Dec. 2005. Permission to reuse is acknowledged.

REFERENCES

Atanassov, K. T. (1986). Intuitionistic fuzzy sets. *Fuzzy Sets and Systems, 20*, 87–96. doi:10.1016/S0165-0114(86)80034-3

Belnap, N. (1977). A useful 4-valued logic. In Epstein, G., & Dunn, J. M. (Eds.), *Modern uses of multiple-valued logic* (pp. 8–37). Reidel.

Bezdek, J. C., & Harris, J. D. (1978). Fuzzy partitions and relations: An axiomatic basis for clustering. *Fuzzy Sets and Systems, 1*, 111–127. doi:10.1016/0165-0114(78)90012-X

Dirac, P. A. M. (1927). The quantum theory of the emission and absorption of radiation. *Proceedings of the Royal Society of London. Series A, 114*, 243–265. doi:10.1098/rspa.1927.0039

Dirac, P. A. M. (1928). The quantum theory of the electron. *Proceedings of the Royal Society of London. Series A, 117*(778), 610–624. doi:10.1098/rspa.1928.0023

Feynman, R. P. (1962). *Quantum electrodynamics*. Addison Wesley.

Feynman, R. P. (1985). *QED: The strange theory of light and matter*. Princeton University Press.

Fung, B. M. (2001). Use of pairs of pseudopure states for NMR quantum computing. *Physical Review A., 63*(2). doi:10.1103/PhysRevA.63.022304

Ginsberg, M. (1990). Bilattices and modal operators. [Oxford University Press.]. *Journal of Logic and Computation, 1*(1), 41–69. doi:10.1093/logcom/1.1.41

Goguen, J. (1967). L-fuzzy sets. *Journal of Mathematical Analysis and Applications, 18*, 145–174. doi:10.1016/0022-247X(67)90189-8

Jenei, S., & De. Baets, B. (2003). On the direct decomposability of t-norms on product lattices. *Fuzzy Sets and Systems, 139*, 699–707. doi:10.1016/S0165-0114(03)00125-8

Khitrin, A. K., & Fung, B. M. (2000). Nuclear magnetic resonance quantum logic gates using quadrupolar nuclei. *The Journal of Chemical Physics, 112*(16), 6963–6965. doi:10.1063/1.481293

Kwahk, K.-Y., & Kim, Y.-G. (1999). Supporting business process redesign using cognitive maps. *Decision Support Systems, 25*(2), 155–178. doi:10.1016/S0167-9236(99)00003-2

Lee, K. C., & Kwon, S. (2008). A cognitive map-driven avatar design recommendation DSS and its empirical validity. *Decision Support Systems, 45*(3), 461–472. doi:10.1016/j.dss.2007.06.008

Lent, C. S., Tougaw, P. D., Porod, W., & Bernstein, G. H. (1993). Quantum cellular automata. *Nanotechnology, 4*, 49–57. doi:10.1088/0957-4484/4/1/004

Navara M. (2007). Triangular norms and conorms. *Scholarpedia*, 2(3):2398. Created: 12 November 2006, reviewed: 12 March 2007, accepted: 15 March 2007.

Shi, Y., Seto, E., Chang, L.-S., & Shenk, T. (1991). Transcriptional repression by YY1, a human GLI-Kruppel-related protein, and relief of repression by adenovirus E1A protein. *Cell, 67*(2), 377–388. doi:10.1016/0092-8674(91)90189-6

Tamura, T., Satomi, Y., Akiko, I., Daisuke, O., Akiko, K., & Yuji, H. (2004). Is an entertainment robot useful in the care of elderly people with severe dementia? *Journals of Gerontology, A. Biological Sciences and Medical Sciences*, *59*(1), M83–M85.

Yager, R. R., & Rybalov, A. (1996). Uninorm aggregation operators. *Fuzzy Sets and Systems*, *80*(1), 111–120. doi:10.1016/0165-0114(95)00133-6

Zadeh, L. (1965). Fuzzy sets. *Information and Control*, *8*, 338–353. doi:10.1016/S0019-9958(65)90241-X

Zhang, W.-R. (1998). YinYang Bipolar Fuzzy Sets. *Proc. of IEEE World Congress on Computational Intelligence – Fuzz-IEEE*, Anchorage, AK, May 1998, 835-840.

Zhang, W.-R. (2003a). Equilibrium relations and bipolar cognitive mapping for online analytical processing with applications in international relations and strategic decision support. *IEEE Transactions on SMC. Part B*, *33*(2), 295–307.

Zhang, W.-R. (2003b). Equilibrium energy and stability measures for bipolar decision and global regulation. *International Journal of Fuzzy Systems*, *5*(2), 114–122.

Zhang, W.-R. (2005a). YinYang bipolar lattices and L-sets for bipolar knowledge fusion, visualization, and decision making. *International Journal of Information Technology and Decision Making*, *4*(4), 621–645. doi:10.1142/S0219622005001763

Zhang, W.-R. (2005b). YinYang bipolar cognition and bipolar cognitive mapping. *International Journal of Computational Cognition*, *3*(3), 53–65.

Zhang, W.-R. (2006). YinYang bipolar fuzzy sets and fuzzy equilibrium relations for bipolar clustering, optimization, and global regulation. *International Journal of Information Technology and Decision Making*, *5*(1), 19–46. doi:10.1142/S0219622006001885

Zhang, W.-R. (2007). YinYang bipolar universal modus ponens (bump) – a fundamental law of non-linear brain dynamics for emotional intelligence and mental health. *Walter J. Freeman Workshop on Nonlinear Brain Dynamics, Proc. of the 10th Joint Conference of Information Sciences*, (pp. 89-95). Salt Lake City, Utah.

Zhang, W.-R. (2009a). Six conjectures in quantum physics and computational neuroscience. *Proceedings of 3rd International Conference on Quantum, Nano and Micro Technologies (ICQNM 2009)*, (pp. 67-72). Cancun, Mexico.

Zhang, W.-R. (2009b). YinYang Bipolar Dynamic Logic (BDL) and equilibrium-based computational neuroscience. *Proceedings of International Joint Conference on Neural Networks (IJCNN 2009)*, (pp. 3534-3541). Atlanta, GA.

Zhang, W.-R. (2009c). YinYang bipolar relativity–a unifying theory of nature, agents, and life science. *Proceedings of International Joint Conference on Bioinformatics, Systems Biology and Intelligent Computing (IJCBS)*. (pp. 377-383). Shanghai, China.

Zhang, W.-R. (2009d). The logic of YinYang and the science of TCM–an Eastern road to the unification of nature, agents, and medicine. [IJFIPM]. *International Journal of Functional Informatics and Personal Medicine*, *2*(3), 261–291. doi:10.1504/IJFIPM.2009.030827

Zhang, W.-R. (2010). YinYang bipolar quantum entanglement–toward a logically complete quantum theory. *Proceedings of the 4th International Conference on Quantum, Nano and Micro Technologies (ICQNM 2010)*, (pp. 77-82). St. Maarten, Netherlands Antilles. Additional Readings Zhang, W. -R. (1996). NPN Fuzzy Sets and NPN Qualitative Algebra: A Computational Framework for Bipolar Cognitive Modeling and Multiagent Decision Analysis. *IEEE Trans. on SMC., Vol. 16*, 561-574.

Zhang, W.-R., Chen, S., & Bezdek, J. C. (1989). POOL2: A generic system for cognitive map development and decision analysis. *IEEE Transactions on SMC*, *19*(1), 31–39.

Zhang, W.-R., Chen, S., Wang, W., & King, R. (1992). A cognitive map based approach to the coordination of distributed cooperative agents. *IEEE Transactions on SMC*, *22*(1), 103–114.

Zhang, W.-R., Pandurangi, A., & Peace, K. (2007). YinYang dynamic neurobiological modeling and diagnostic analysis of major depressive and bipolar disorders. *IEEE Transactions on Bio-Medical Engineering*, *54*(10), 1729–1739. doi:10.1109/TBME.2007.894832

Zhang, W.-R., & Peace, K. E. (2007). YinYang MentalSquares–An equilibrium-based system for bipolar neurobiological pattern classification and analysis. *Proceedings of IEEE BIBE*, (pp. 1240-1244). Boston.

Zhang, W.-R., & Zhang, L. (2004). YinYang bipolar logic and bipolar fuzzy logic. *Information Sciences*, *165*(3-4), 265–287. doi:10.1016/j.ins.2003.05.010

KEY TERMS AND DEFINITIONS

BQL: Bipolar quantum lattice – an equilibrium-based non-linear bipolar symmetrical or dynamic generalization of truth-based lattice.

BDFL: Bipolar dynamic fuzzy logic – an equilibrium-based non-linear bipolar symmetrical or dynamic generalization of truth-based fuzzy logic.

Bipolar L-Set (or Bipolar L-Fuzzy Set): A mapping ϕ of a bipolar set S to a BQL or BQFL B denoted ϕ:S→B.

Triangular Norms (t-norms): In mathematics, a t-norm (also T-norm or, unabbreviated, triangular norm) is a binary operation used in the framework of probabilistic metric spaces and in multi-valued logic, specifically in fuzzy logic. A t-norm generalizes intersection in a lattice and conjunction in logic. The name *triangular norm* refers to the fact that in the framework of probabilistic metric spaces t-norms are used to generalize triangle inequality of ordinary metric spaces. Technically, a t-norm is a function T: [0, 1] × [0, 1] → [0, 1] which satisfies the following properties (Navara, 2007): (1) Commutativity: $T(a, b) = T(b, a)$; (2) Monotonicity: $T(a, b) \leq T(c, d)$ if $a \leq c$ and $b \leq d$; (3) Associativity: $T(a, T(b, c)) = T(T(a, b), c)$; (4) The number 1 acts as identity element: $T(a, 1) = a$

Triangular Co-Norms (t-conorms): T-conorms (also called S-norms) are dual to t-norms under the order-reversing operation which assigns $1 - x$ to x on [0, 1]. Given a t-norm, the complementary conorm is defined by $\perp(a,b) = 1 - T(1-a,1-b)$.

Bipolar Triangular Norms (Bipolar t-norms): A non-linear bipolar symmetrical or dynamic generalization of t-norms from [0,1] to the BQL [-1,0]×[0,+1].

Bipolar Triangular T-Conorms (Bipolar t-conorms): A non-linear bipolar symmetrical or dynamic generalization of t-conorms from [0,1] to the BQL [-1,0]×[0,+1].

Recovery Theorem: It proves the recovery of BDL to Boolean logic through depolarization and therefore proves the computability of BDL.

Energy Symmetry: A bipolar operation on a lattice B that may change the polarity of the energy of B but does not change the energy total of B.

Chapter 5
Bipolar Fuzzy Sets and Equilibrium Relations

ABSTRACT

Based on bipolar sets and quantum lattices, the concepts of bipolar fuzzy sets and equilibrium relations are presented in this chapter for bipolar fuzzy clustering, coordination, and global regulation. Related theorems are proved. Simulated application examples in multiagent macroeconomics are illustrated. Bipolar fuzzy sets and equilibrium relations provide a theoretical basis for cognitive-map-based bipolar decision, coordination, and global regulation.

INTRODUCTION

Based on the concepts of set, binary *relation*, *equivalence relation*, fuzzy set, *fuzzy relation*, *similarity relation*, we introduce the concepts of *bipolar fuzzy relation*, *fuzzy equilibrium relation*, *bipolar partitioning*, and *bipolar clustering* in this chapter. Decision analysis in agent-oriented macroeconomics is illustrated using equilibrium-based bipolar clustering.

While set theory provides the doctrine for classical mathematical abstraction, Zadeh's fuzzy set theory (Zadeh 1965) presents a serious challenge to the doctrine. The fuzzy set research community successfully fought the decisive battles in seeking recognition in the last quarter century before the new millennium with a number of heated debates and dramatic stories. Now fuzzy set theory has been widely accepted and is commonly considered an established field in both academic and applied terms.

It can be observed that fuzzification in Zadeh's fuzzy set theory preserves the truth-based property of classical set theory with the addition of an infinite number of truth gray levels which can be defuzzified to true or false (1 or 0). Therefore, fuzzy sets can be deemed an extension of classical set theory

DOI: 10.4018/978-1-60960-525-4.ch005

that is based largely on Aristotle's philosophy of science where the universe consists of a set of truth objects. Fundamentally, the fuzzy extension is realized by fuzzifying the bivalent lattice {0,1} to the fuzzy lattice [0,1].

As stated in Postulate 1.1 (Ch. 1), truth and fuzziness cannot serve as the most fundamental property of nature because they are not physical but logical properties subjected to the LAFIP paradox. Three decades after inventing his fuzzy set theory, the founder of fuzzy sets Lotfi Zadeh concluded that *"Causality Is Undefinable"* (Zadeh, 2001). Evidently, the purpose of fuzzy set theory has not been for providing logically definable causality or providing the logical foundation for physics as envisioned by Einstein in his quote: *"For the time being we have to admit that we do not possess any general theoretical basis for physics which can be regarded as its logical foundation."*

Although it conveys no physical information if we say that the universe is fuzzy, it does make sense if we say the universe is a quasi- or fuzzy equilibrium. Thus, based on bipolar sets, bipolar dynamic logic (BDL), and bipolar quantum lattice as introduced in earlier chapters, we discuss bipolar fuzzy sets and fuzzy equilibrium relations in this chapter. The word "bipolar fuzzy sets" was first coined in 1994 (Zhang, 1994). Further development of the theory is documented in (Zhang, 1996, 1998, 2003b; Zhang & Zhang, 2004). The formal presentation of the theory is documented in (Zhang, 2005a, 2006). (The author acknowledges Lotfi Zadeh (2006) for publically recognizing bipolar fuzzy sets.)

From Chapters 3 – 4, it is clear that:

1. The bipolar crisp lattice $\mathbf{B}_1 = \{-1,0\} \times \{0,+1\}$ (Ch. 3) is the corner lattice of $\mathbf{B}_F = [-1,0] \times [0,+1]$;
2. $\mathbf{B}_1$ is a polarization of the bivalent lattice {0,1};
3. $\mathbf{B}_F$ is a polarization of the fuzzy lattice [0,1]; and
4. Both $\mathbf{B}_1$ and $\mathbf{B}_F$ are strict bipolar lattices (Ch. 4) (Zhang, 2005a).

In this chapter, we use the terms *bipolar crisp sets* and *bipolar fuzzy sets* to distinguish their mapping to $\mathbf{B}_1$ and $\mathbf{B}_F$, respectively, and we follow the basic concepts and notations of bipolar quantum lattice, bipolar sets, bipolar relation, bipolar transitivity, and bipolar equilibrium relation as defined in Chapters 3-4.

Strict bipolarity provides a basis for modeling bipolar systems and relations as sets of equilibria and quasi- or non-equilibria. This chapter is focused on the unit square bipolar fuzzy lattice $B_F = [-1,0] \times [0,1]$ and fuzzy equilibrium relations mapped to B_F for bipolar clustering, coordination, and global regulation. Simulated examples in multiagent macroeconomics are used for illustration.

The remaining presentations and discussions of this chapter are organized in the following sections:

- **Bipolar Fuzzy Relations.** In this section bipolar fuzzy relations are formally defined; the existence of bipolar transitive closure (BTC) is proved; a BTC algorithm is presented and its $O(N^3)$ complexity is axiomatically proved.
- **Bipolar α-Level Sets.** In this section bipolar α-level sets, bipolar defuzzification, depolarization, and focus generation are formulated.
- **Fuzzy Equilibrium Relations.** In this section we formally define fuzzy equilibrium relations and examine their properties with simulations in multiagent macroeconomics.
- **Bipolar Fuzzy Clustering**. Bipolar fuzzy clustering theorems and algorithms are discussed in this section with illustrations in multiagent macroeconomics.

- **Equilibrium Energy and Stability for Multiagent Coordination and Global Regulation.** This section presents the notions of *equilibrium energy* and stability that provide a mathematical basis for *multiagent coordination and global regulation* as discussed in Chapter 11.
- **Research Topics**. This section lists a few research topics.
- **Summary**. This section draws a few conclusions.

BIPOLAR FUZZY RELATIONS

Based on bipolar lattices (Ch. 4), bipolar relations (crisp or fuzzy) can be generalized to bipolar L-relations (Zhang, 2005a, 2006).

Definition 5.1. A **bipolar (binary) L-relation** R from set X to set Y, where $X = \{x_i\}$, $0<i\leq m$, and $Y = \{y_j\}$, $0<j\leq n$, is a collection of ordered pairs or subsets of $X \times Y$ characterized by a bipolar L-set or membership function $\mu_R(x_i,y_j)$ which maps each ordered pair (x_i,y_j) to a bounded or unbounded bipolar lattice B. Formally, R is the set $\{\mu_R(x_i,y_j) | \forall i,j,\ 0<i\leq m,\ 0<j\leq n,\ \mu_R:(x_i,y_j)\Rightarrow B\}$. When B is finite-valued R is called **a bipolar L-crisp relation**; when B is infinite-valued R is called **a bipolar L-fuzzy relation**. When B is bounded, it is said **bounded**. When $B = B_F$ we say a bipolar L-fuzzy relation is a **bipolar fuzzy relation**.

Evidently, a bipolar L-crisp relation is a special case of a bipolar L-fuzzy relation. Among bipolar fuzzy lattices, B_F can be considered the normalized fuzzy lattice which is both bounded and complemented. In this chapter we assume all bipolar relations under consideration are mapped to the same bipolar lattice B_F without further specification.

Given two bipolar fuzzy relations R_1 and R_2 in X, where $X = \{x_i\}$ with size n, if $\forall i,j,\ 0<i,j\leq n$, $\mu_{R1}(x_i,x_j) \oplus \mu_{R2}(x_i,x_j) = \mu_{R2}(x_i,x_j)$, we say that R_1 is **contained in (smaller than or equal to)** R_2 denoted by $R_1 \subset R_2$ or $R_1 \leq\leq R_2$.

The bipolar disjunction operator $\oplus$ on m×n bipolar fuzzy relations R and Q, denoted by $R\oplus Q$, is defined by

$R\oplus Q= [(r_{ij}\oplus q_{ij})]$, $\forall i,j$, $1\leq i\leq m$, $1\leq j\leq n$, where r_{ij} and q_{ij} are the ijth element of R and Q, respectively.

If Ω is used as a commutative NP coupling operator for a m×n bipolar fuzzy relation R, $\forall i,j$, $1\leq i\leq m$, $1\leq j\leq n$, we have

$$R \equiv (R^-,R^+) = R^-\Omega R^+ = [r_{ij}] = [r_{ij}^-]\Omega[r_{ij}^+] = R^+\Omega R^- = [r_{ij}^+]\Omega[r_{ij}^-] = [(r_{ij}^-,r_{ij}^+)].$$

The $\oplus$-$\otimes$ **or max-**$\otimes$ **composition** of an m×n bipolar fuzzy relation R in X×Y and an n×k bipolar Fuzzy relation Q in Y×Z, denoted by $R\otimes Q$, is defined as

$$\mu_{R\otimes Q}(x,z)= \max_y(\mu_R(x,y) \otimes \mu_Q(y,z)),\ \forall x,y,z,\ x \in X, y \in Y, z \in Z; \tag{5.1}$$

where max is equivalent to $\oplus$. The n-fold composition of R in X with itself is denoted by R^n.

- **Definition 5.2.** A bipolar Fuzzy relation R in X, where X = $\{x_i\}$, 0<i≤n, is said **⊕-⊗ or max-⊗ bipolar interactive or transitive** if, ∀i,j,k, 0<i,j,k≤n, we have

$$\mu_R(x_i, x_k) \geq\geq \max_{x_j}(\mu_R(x_i, x_j) \otimes \mu_R(x_j, x_k)), \tag{5.2a}$$

where (a,b) ≥≥ (c,d) iff blub((a,b),(c,d))=(a,b), and the ordering defined by ≥≥ is **a bipolar crisp or fuzzy partial ordering.**

Similarly, a bipolar Fuzzy relation R in X, where X = $\{x_i\}$, 0<i≤n, is said **⊕-& or max-& linear transitive** if, ∀i,j,k, 0<i,j,k≤n, we have

$$\mu_R(x_i, x_k) \geq\geq \max_{x_j}(\mu_R(x_i, x_j) \,\&\, \mu_R(x_j, x_k)). \tag{5.2b}$$

While the linear ⊕-& transitive closure of R is isomorphic to a unipolar transitive relation on each pole, the non-linear ⊕-⊗ bipolar transitive closure of R in X is defined in the following:

Definition 5.3. The **bipolar transitive closure** of a bipolar fuzzy relation R is the smallest transitive bipolar fuzzy relation containing R.

We denote the bipolar transitive closure of R by $\mathfrak{R}$ and the *i-fold composition of R with ⊗ by* R^i, we would have

$$\mathfrak{R} = R^1 \oplus R^2 \oplus R^3 \oplus \ldots. \tag{5.3}$$

An immediate question about (5.3) is whether the "summation" always terminates, and if so, how soon. The answers are given by

Theorem 5.1. Let X = $\{x_1,x_2,\ldots,x_n\}$ be a finite set, the bipolar transitive closure $\mathfrak{R}$ of any bipolar fuzzy relation R in X exists, is unique, and

$$\mathfrak{R} = R^1 \oplus R^2 \oplus R^3 \oplus \ldots \oplus R^{2n}. \tag{5.4}$$

Proof. The existence and uniqueness of R follow (1) ≥≥ defines a bipolar partial ordering; (2) the ⊕ operator is idempotent and it does not oscillate; and (3) X is a finite set.

Note that R^i means i-fold composition of R with ⊗. While R is considered as a bigraph, i-fold composition is equivalent to travel i edges through every existing chain between every pair of nodes and each traveled edge is composed to the strength of the traveled chain. A disjunction operation $R^i \oplus R^{i+1}$ enforces the bipolar partial ordering (≥≥) condition of transitivity in Eq. 5.4.

If R were a unipolar relation in X, R would close after applying ⊕ on the first n terms of Eq. 5.4 (Warshall, 1962). This is because n would be the maximum possible length of a nonzero chain from x_i to x_j in the graph. The longest chain would be a circular chain under the conditions (1) i=j and (2) every node appears on the chain exactly once. For instance, the chain (x_1— x_2 —…— x_n—x_1) with length n would be such a circular chain. Any cyclic chain except such a circular chain makes no difference be-

cause a cycle on a chain cannot strengthen the chain. After applying ⊕ on the first n terms inEq. (5.4) for a unipolar closure, every existing chain with length ≤ n between every pair of nodes would have been traveled and the strengths of all these chains would have been taken into the "summation" (disjunction) with the idempotent operator ⊕.

In the bipolar case, a bipolar circular chain with length n under the two conditions as above may have a composed strength (-|x|,0). Since (-|x|,0)⊕[(-|x|,0)⊗(-|x|,0)]=(-|x|,|x|) assuming ⊕-⊗ transitivity, continuing chaining the same cycle a second time will result in (-|x|,|x|). Thus, the maximum possible length of any effective bipolar chain is 2n in the bipolar case. Cyclic chaining can make a difference as long as travel distance is less than the maximum length 2n. Travel a distance greater than 2n makes no difference because it is already closed on both poles.

Note that all existing bipolar chains with length≤ 2n between every pair of nodes will have been traveled after 2n terms inEq. 5.4and the composed strengths of all these chains have been taken into the "summation" (disjunction) with the idempotent bipolar operator ⊕. Therefore, R will close after applying ⊕ on the first 2n terms ofEq. 5.4. □

Theorem 5.2. The computational complexity of 5.4 can be reduced to $O(n^3)$ using the following equations:

$$\mathfrak{R} \equiv R^1 \oplus R^2 \oplus R^3 \oplus \ldots \oplus R^{2n}.$$

$$\equiv (R^1\oplus\ldots\oplus R^n)\oplus R^n\otimes(R^1\oplus\ldots\oplus R^n) \quad (5.5a)$$

$$\equiv (R^1\oplus\ldots\oplus R^n)\oplus(R^1\oplus\ldots\oplus R^n)^2 \quad (5.5b)$$

$$\equiv (R^1\oplus\ldots\oplus R^n)\otimes([I]\oplus(R^1\oplus\ldots\oplus R^n)) \quad (5.5c)$$

$$\equiv (\sum_1^n R^i)\otimes([(0,1)]\oplus(\sum_1^n R^i)). \quad (5.5d)$$

where [I] in (5.5c) is an n × n identity matrix.

Proof. It can be verified that the identities in Ch. 3, Table 8 hold on bipolar fuzzy relations as well. Thus, we have

1. *Eq. 5.5a follows the associative and distributive laws (Ch. 3, Table 8).*
2. *Eq. 5.5b follows the associative, idempotent, and commutative laws (Ch. 3, Table 8) as we have*

$$(R^1\oplus\ldots\oplus R^n)\oplus(R^1\oplus\ldots\oplus R^n)^2$$

$$= (R^1\oplus\ldots\oplus R^n)\oplus R^1\otimes(R^1\oplus\ldots\oplus R^n)\oplus R^2\otimes(R^1\oplus\ldots\oplus R^n)\oplus\ldots\oplus R^n\otimes(R^1\oplus\ldots\oplus R^n)$$

$$= R^1 \oplus .. \oplus R^n \oplus R^{n+1} \oplus R^{n+2} \ldots \oplus R^{2n}$$

$$= (R^1\oplus\ldots\oplus R^n)\oplus R^n\otimes(R^1\oplus\ldots\oplus R^n)$$

3. *Eq. 5.5c follows the distributive law (Ch. 3, Table 9).*
4. *It is clear from Eq. 5.5a, 5.5b, or 5.5c that R can be computed based on the "summation" of the first n terms in Eq. 5.4.*
5. *Since the transitive closure C of a classical unipolar binary relation U can be computed based on the first n terms with an $O(n^3)$ algorithm (Warshall, 1962), that is, $C = R^1 \oplus R^2 \oplus \ldots \oplus R^n$, the computational complexity for $\sum_{1}^{n} R^i$ can be reduced to the same order $O(n^3)$.*
6. *The two operations ⊗ and ⊕ inEq. 5.5dentail $O(n^3)$ and $O(n^2)$ complexity, respectively. R can be computed with an $O(n^3)$ algorithm.* □

It should be remarked that bipolar transitivity takes bipolar interaction, oscillation, and quantum entanglement with ⊗ and balancing with ⊕ into consideration that is fundamentally different from unipolar transitivity. A bipolar transitive closure can also be called **a bipolar interactive closure** or **a bipolar quantum closure**. Bipolar interaction or transitivity leads to the notion of **equilibrium relations** to be discussed later.

BIPOLAR α-LEVEL SETS

While fuzzy truth can be recovered to bivalent truth through defuzzification (Zadeh, 1965), bipolar truth can be coerced and recovered to unipolar truth through depolarization (Zhang & Zhang, 2004; Zhang, 2006). A decision making process can be viewed as a weighing or depolarization of a bipolar equilibrium that consists of considerations on both negative and positive effects of available options. If the desired effect exceeds the undesired or side effect to a threshold, an option is identified as good. Such depolarization, however, must necessarily depend on bipolar representation. Therefore, bipolar representation and depolarization as well as defuzzification are all important steps in a decision process.

In our discussion, all classical set operations can be naturally applied to bipolar sets where each element is inherently bipolar. We follow the convention for unipolar and bipolar partial ordering: (1) if Θ is used as **a unipolar comparison operator,** Θ∈{=, ≤, ≥, <,>}; and (2) if Θ is used as **a bipolar comparison operator,** Θ∈{=, ≤≤, ≥≥, ≤≥, ≥≤, <<, >>, <>, ><, <≤, >≥, <≥, >≤, ≤<, ≥>, ≤>, ≥>}, where Θ consists of a left side and a right side comparison except the equal sign.

Bipolar Qualitative Equality

Bipolar qualitative equality (BQ-Equality) can be defined as

$$a \cong b, \text{ if } \exists\alpha\in S \text{ such that } a\Theta\alpha \text{ and } b\Theta\alpha, \tag{5.6}$$

where α is a criterion or reference in a bipolar logical space S; ≅ reads "Qualitatively Equals", and Θ reads "Satisfies". Θ is any bipolar ordering or bipolar comparison operator. For instance, given the (side_effect, effect) of decision x and y as (side_effect, effect)(x) = (-0.1,+0.8) and (side_effect, effect)(y) = (-0.2,+0.9), and good(d) is defined as (side_effect, effect)(d) ≤≥ (-0.2,0.8), we have [((-0.1,+0.8)≤≥(-0.2,0.8)) ⇒ good(x) = True)] ∧ [((-0.2,+0.9) ≤≥(-0.2,0.8)) ⇒ good(y) is true] ⇒ [x ≅ y], which reads "x and y both satisfy the good criterion implies x and y are qualitatively equal."

A-Type Bipolar Level Sets: $R_{\Theta\alpha}$ Form

The Q-equality definition in Eq. (5.6) is actually an extension of Zadeh's α-level-set (Zadeh, 1971) to the bipolar space [-1,0]×[0,1]. In Zadeh (1971), the α-level-set of a (unipolar) fuzzy relation R is denoted by R_α, which is a non-fuzzy relation from the set X to set Y or in X × Y defined by

$$R_\alpha = \{(x,y) | \mu_R(x,y) \geq \alpha\}. \tag{5.7}$$

R_α forms a nested sequence of non-fuzzy relations with $\alpha 1 \geq \alpha 2 \Rightarrow R_{\alpha 1} \subset R_{\alpha 3.}$ Zadeh proved that any (unipolar) fuzzy relation from X to Y admits of the resolution

$$R = \sum_{\alpha} \alpha R_\alpha, 0 < a \leq 1. \tag{5.8}$$

Based on Zadeh's α-level-sets of a (unipolar) fuzzy relation, **A-type bipolar α-level-sets** of a bipolar fuzzy relation is proposed as

$$R_{\Theta\alpha} = \{(x,y) | \mu_R(x,y)\ \Theta\alpha\}, \tag{5.9}$$

where α = (u,v) is a bipolar fuzzy threshold variable; Θ is any bipolar comparison operator.

To illustrate, we show different $R_{\Theta\alpha}$ of R in the following:

$$\mathrm{R} = \begin{bmatrix} (-0.6 & 0.7) & (-0.9 & 0) & (-0.6 & 0.7) \\ (-0.2 & 0) & (-0.7 & 0.8) & (-0.5 & 0.6) \\ (-0.3 & 0.8) & (-0.2 & 0.1) & (-0.4 & 0.9) \end{bmatrix};$$

$$\mathrm{R}_{\geq\geq(-0.5,\ 0.5)} = \begin{bmatrix} 1 & 0 & 1 \\ 0 & 1 & 1 \\ 0 & 0 & 0 \end{bmatrix}; \mathrm{R}_{\geq\leq(-0.5,\ 0.5)} = \begin{bmatrix} 0 & 1 & 0 \\ 0 & 0 & 0 \\ 0 & 0 & 0 \end{bmatrix};$$

$$\mathrm{R}_{\leq\leq(-0.5,\ 0.5)} = \begin{bmatrix} 0 & 0 & 0 \\ 1 & 0 & 0 \\ 0 & 1 & 0 \end{bmatrix}; \mathrm{R}_{\leq\geq(-0.5,\ 0.5)} = \begin{bmatrix} 0 & 0 & 0 \\ 0 & 0 & 0 \\ 1 & 0 & 1 \end{bmatrix};$$

$$\mathrm{R}_{<\geq(-0.5,\ 0.6)} = \begin{bmatrix} 0 & 0 & 0 \\ 0 & 0 & 0 \\ 1 & 0 & 1 \end{bmatrix}; \mathrm{R}_{\geq<(-0.5,\ 0.6)} = \begin{bmatrix} 0 & 1 & 0 \\ 0 & 0 & 0 \\ 0 & 0 & 0 \end{bmatrix}.$$

It is quite interesting to examine the properties of bipolar α-level-sets. First, in the bipolar case α is a bipolar variable and Θ can be any bipolar comparison; second, the negative and positive relationships are interrelated in a transitive bipolar relation. The reasoning power of a bipolar relation lies in the interplay of bipolar relationships.

We use the notion $R_{\Theta(u,v)} = R^-_{\Theta|u|} \cap R^+_{\Theta v} = \left[r^-_{ij\Theta|u|} \wedge r^+_{ij\Theta v}\right], \forall r_{ij} \in R$, to indicate that an A-type bipolar level set of R is determined by applying Θα, α=(u,v), to the absolute values of the two poles of R, respectively. An immediate and yet important consequence of the bipolar α-level-set definition is stated in the following theorems:

Theorem 5.3. Given $\alpha = \forall(u,v) \in R$, any bipolar fuzzy relation from set X to set Y admits of the resolution identity

$$R \equiv \sum_{\alpha} \alpha R_{\Theta\alpha} \equiv \sum_{(u,v)} (uR^{-}_{\Theta|u|} \cap vR^{+}_{\Theta v}); \tag{5.10}$$

and we have $\alpha 1 \geq\geq \alpha 2 \Rightarrow R_{\Theta\alpha 1} \subset R_{\Theta\alpha 2}$, which forms a nested sequence of non-fuzzy relations from X to Y.

Proof. Note that bipolar resolution is defined on the 2-dimensional domain $[-1,0]\times[0,1]$. *For any given* Θ *and* (x,y), $\mu_R(x,y)\Theta(u,v)$ *specifies a condition on the negative (Yin) dimension and a condition on the positive (Yang) dimension. Both conditions should be met. Since any bipolar relation R is a binding or coupling of its negative pole* R^- *with its positive pole* R^+, *Zadeh's unipolar resolution (Zadeh, 1971) can be applied to the magnitude values of* R^- *and* R^+, *respectively. We must have*

$$R^{-} \equiv \sum_{u} uR_{\Theta|u|}, \text{ where } 0 \leq |u| \leq 1; \text{ and}$$

$$R^{+} \equiv \sum_{v} vR_{\Theta v}, \text{ where } 0 \leq v \leq 1.$$

Evidently, when $\mu_R(x,y)\Theta(u,v)$ *we have*

$$R \equiv \sum_{(u,v)} (u,v) R_{\Theta(u,v)} \equiv \sum_{(u,v)} (uR^{-}_{\Theta|u|} \cap vR^{+}_{\Theta v}) \equiv \sum_{(u,v)} (u,v)(R^{-}_{\Theta|u|} \cap R^{+}_{\Theta v}))$$

and $\alpha 1 >> \alpha 2 \Rightarrow R_{\Theta\alpha 1} \subset R_{\Theta\alpha 3.}$ □

As an illustration, assume X = Y = {x1, x2}, with the bipolar fuzzy relation matrix μ_R as given by

$$R = \begin{bmatrix} (-0.6, 0.9) & (-0.8, 0.9) \\ (-0.9, 0.5) & (0.0, 0.5) \end{bmatrix};$$

$$R^{-} = \begin{bmatrix} -0.6 & -0.8 \\ -0.9 & 0.0 \end{bmatrix}; R^{+} = \begin{bmatrix} 0.9 & 0.9 \\ 0.5 & 0.5 \end{bmatrix};$$

if Θ is $\underline{\diamond}$, the resolutions of R^-, R^+, and R read, respectively, as

R^- = (-0.0){(x_2,x_2)} + (-0.4){(x_2,x_2)} + (-0.6){$(x_1,x_1)(x_2,x_2)$} + (-0.8){$(x_1,x_1)(x_1,x_2)(x_2,x_2)$} + (-0.9) {$(x_1,x_1)(x_1,x_2)(x_2,x_1)(x_2,x_2)$}

$R^+ = 0.1\{(x_1,x_1)(x_1,x_2)(x_2,x_1)(x_2,x_2)\} + 0.3\{(x_1,x_1)(x_1,x_2)(x_2,x_1)(x_2,x_2)\} + 0.5\{(x_1,x_1)(x_1,x_2)(x_2,x_1)(x_2,x_2)\} + 0.9\{(x_1,x_1)(x_1,x_2)\}$

$R = (-0.6,0.9)\{(x_1,x_1)\} + (-0.8,0.9)\{(x_1,x_1)(x_1,x_2)\} +(-0.0,0.5)\{(x_2,x_2)\} + (-0.9,0.5)\{(x_1,x_1)(x_1,x_2)(x_2,x_1)(x_2,x_2)\}+ (-0.6,0.3)\{(x_1,x_1)(x_2,x_2)\} + (-0.0,0.3)\{(x_2,x_2)\}+ (-0.7,0.9)\{(x_1,x_1)\} + (-0.4,0.1)\{(x_2,x_1)(x_2,x_2)\}$

In the bipolar case each bipolar comparison operator Θ leads to a different resolution. This is quite different from a unipolar case. In the unipolar case, there is only one dimension involved where all α values can be ordered into a linear list within [0,1]. In the bipolar case, two dimensions (negative and positive) are involved where α values are distributed within the 2-D space $B_F = [-1,0]\times[0,1]$. This is depicted in Figure 1, where each of the four operators is headed toward a corner of the 2-D fuzzy space (Note that we did not place Figure 1 in the 2nd quadrant. It is converted to the 4th quadrant following the convention in graphical user interface (GUI) design. The conversion is possible because YinYang geometry is completely background independent).

The four corners of Figure 1 have well defined semantics. The (0,0) corner can be best described as "Negative Small and Positive Small" (NS,PS); the (-1,1) corner can be best described as "Negative Large and Positive Large" (NL,PL); the (-1,0) corner can be best described as "Negative Large and Positive Small" (NL,PS); the (0,+1) corner can be best described as "Negative Small and Positive Large" (NS,PL); the center (-0.5,+0.5) can be best described as "Negative Medium and Positive Medium" (NM,PM); where [NL,NM,NS,0,PS,PM,PL] form a bipolar linguistic fuzzy set (Zhang 1996).

Any bipolar threshold α divides the 2-D fuzzy geometry into four regions, each of which identifies an A-type bipolar α-level set with respect to the bipolar comparison operator located in its corner. Each bipolar value in Figure 1 can be fuzzified with a bipolar linguistic fuzzy set that generalize Zadeh's extension principle from [0,1] to B_F. For instance (-0.6,+0.35) can be described as (NL(0.2)-NM(0.8), PS(0.3)-PM(0.7)) in Figure 1.

It should be remarked that the A-type bipolar α-level sets of a bipolar fuzzy relation R are the results of conjunctive operation (∧) of the α-cut of the two poles. This type of α-cut of a bipolar fuzzy relation R results in a unipolar crisp binary relation. Therefore A-type α-cut is a conversion from bipolar fuzzy relation back to unipolar crisp binary relation. If we consider Zadeh's α-cut as a defuzzification opera-

Figure 1. Bipolar comparison

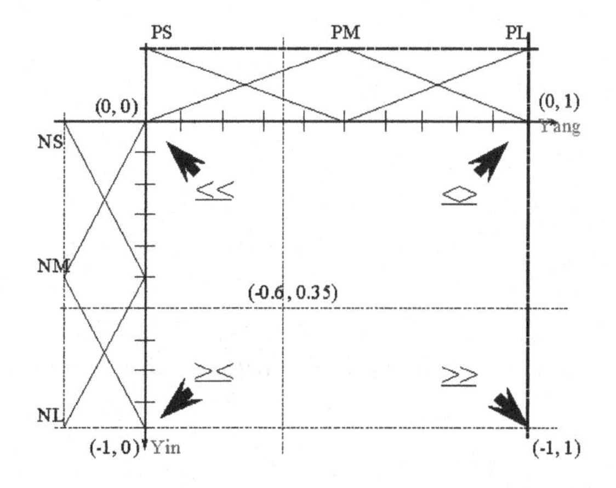

tion from a unipolar fuzzy relation to a unipolar crisp relation, A-type α-cut combines *defuzzification* with *depolarization* and converts a bipolar fuzzy relation to a unipolar crisp relation.

B-Type Bipolar α-Level Sets: $R_{(u,v)}$ Form

Another approach to bipolar α-cut of R is to bind the α-cut of the two poles and form a crisp bipolar relation. This approach defuzzifies R without depolarizing it. In this case, comparison on each pole is limited to bipolar greater or equal (≥≥). For instance, (1) if the comparisons on both poles result in 0, we have a bipolar level (0,0); (2) if the comparison on the negative pole results in 0 and on the positive pole results in 1, we have a bipolar level (0,1); (3) if the comparison on the negative pole results in 1 and that on the positive pole results in 0, we have a bipolar level (-1,0); (4) if the comparisons on both poles result in 1, we have a harmony level (-1,1). This can be realized using the NP coupling operator Ω on the unipolar α-level sets resulted from the two poles of R. We call this type of α-level sets ***B-type bipolar α-level-sets*** denoted

$$R_{(u,v)} = (R^{-}\mathrm{W}R^{+})_{(u,v)} = -\mid R^{-}\mid_{|u|} \Omega R^{+}_{v}. \tag{5.11a}$$

The following may be used for notational simplicity,

$$R_{(u,v)} = (R^{-}\mathrm{W}R^{+})_{(u,v)} = -\mid R^{-}\mid_{|u|} \Omega R^{+}_{v} = R^{-}_{|u|}\Omega R^{+}_{v}. \tag{5.11b}$$

To illustrate, $R_{(-0.5,0.5)}$ of R is shown in the following:

$$\mathrm{R} = \begin{bmatrix} (-0.6, & 0.9) & (-0.8, & 0.9) & (-0.0, & 0.5) \\ (-0.9, & 0.5) & (-0.0, & 0.5) & (-0.6, & 0.3) \\ (-0.0, & 0.3) & (-0.7, & 0.9) & (-0.4, & 0.1) \end{bmatrix};$$

$$\mathrm{R}_{(-0.5,0.5)} = \begin{bmatrix} (-1, & 1) & (-1, & 1) & (-0, & 1) \\ (-1, & 1) & (-0, & 0) & (-1, & 0) \\ (-0, & 0) & (-1, & 1) & (-0, & 0) \end{bmatrix}$$

Theorem 5.4. Given α = ∀(u,v)∈R, any bipolar fuzzy relation from X to Y admits of the bipolar resolution identity

$$R \equiv \sum_{\alpha} \alpha R_{\alpha} \equiv \sum_{(u,v)} (uR^{-}_{|u|}\Omega vR^{+}_{v}), \tag{5.12}$$

and we have α1≥≥α2⇒$R_{\alpha 1}$⊂$R_{\alpha 2}$, which form a nested sequence of crisp bipolar relations from X to Y.

Proof. The proof is the same as for Theorem 5.3 except that in the B-type case the resolution on the two poles are bound with Ω and R_α is a crisp bipolar relation, while in the A-type case $R_{\theta\alpha}$ is a non-fuzzy unipolar relation. □

To illustrate, we show the resolution of R in X×Y, where X={x1,x2} and Y = {y}, in the following:

$$\mathrm{R} \;=\; \begin{matrix} & y \\ x1 & \\ x2 & \end{matrix} \begin{bmatrix} (-0.5 & 0.5) \\ (-0.8 & 0.4) \end{bmatrix};$$

$$\mathrm{R}_{(-0.5,0.5)} = \begin{bmatrix} (-1 & 1) \\ (-1 & 0) \end{bmatrix}; \mathrm{R}_{(-0.8,0.4)} = \begin{bmatrix} (0 & 1) \\ (-1 & 1) \end{bmatrix};$$

R = (-0.5,0.5){[(x1,y),(x2,y)]Ω[(x1,y)]} + (-0.8,0.4){[(x2,y)]Ω[(x1,y),(x2,y)]}

Semantics of A-Type and B-Type Bipolar Level Sets

From the above, we can see that A-type uses different bipolar comparison operators; B-type only applies "greater than or equal to" on each pole. A-type α-cut of a bipolar fuzzy relation R defuzzifies R to a crisp unipolar relation, a B-type α-cut defuzzifies R to a crisp bipolar relation. Evidently, the two types should be semantically consistent with each other. The consistency is proved in

Theorem 5.5. Given a bipolar fuzzy relation R, we have

$$(R_{(u,v)})^{-} \equiv -\,|\, R_{\geq<(u,v)} \cup R_{\geq\geq(u,v)} \,|; \tag{5.13a}$$

$$(R_{(u,v)})^{+} \equiv R_{<\geq(u,v)} \cup R_{\geq\geq(u,v)}; \tag{5.13b}$$

$$|\,(R_{(u,v)})^{-}\,| \cap (R_{(u,v)})^{+} \equiv R_{\geq\geq(u,v)}. \tag{5.13c}$$

Proof: Based on their definitions, the B-type α-cut $R_{(u,v)}$ *is the binding or coupling of the unipolar α-cuts on two poles; A-type α-cut* $R_{\Theta(u,v)}$ *is the result of "ANDing" the unipolar α-cuts on the two poles. The theorem follows from the definitions.* □

Corollary: Given a bipolar fuzzy relation R, we have

$$|\,(R_{(u,v)})^{-}\,| \cup (R_{(u,v)})^{+} \equiv R_{<\geq(u,v)} \cup R_{\geq<(u,v)} \cup R_{\geq\geq(u,v)}; \tag{5.13d}$$

$$|\,(R_{(u,v)})^{-}\,| \cup (R_{(u,v)})^{+} - R_{\geq\geq(u,v)} \equiv R_{<\geq(u,v)} \cup R_{\geq<(u,v)}; \tag{5.13e}$$

$$(R_{(u,v)})^{+} - R_{\geq\geq(u,v)} \equiv R_{<\geq(u,v)}. \tag{5.13f}$$

Proof. The corollary follows from Theorem 5.5. □

Eqs. 5.13 a – f are illustrated in Figure 2. Note that B-type level sets use the inclusive bipolar comparison operator $\geq\geq$. The subtraction operator in Eq. 5.13e is exclusive. The $<\geq$ and $\geq<$ operators are necessary for the subtraction operation that makes the 0 regions and 1 regions in Figure 2e to be disjoint on the dotted border and at the joint point (u,v). Otherwise, the regions will overlap on the dotted lines and the joint point. This technical point is crucial for bipolar partitioning.

From the above discussions it is clear that A-type and B-type α-cuts of R can complement each other. A-type α-cut combines defuzzification and **depolarization** into one step and results in a unipolar crisp relation; B-type α-cut retains the polarity of R in a defuzzification process and results in a crisp bipolar relation. As noted in (Zhang *et al.*, 1992) that a physical world may be represented with a conceptual cognitive map (CCM). Now we have shown that a CCM represented as a bipolar fuzzy relation can be transformed to a (spatially) visual CM (VCM) (Zhang *et al.*, 1992) as in Figure 2 to enhance the understanding of a CCM.

Figure 2. (a) Illustration of Eq. 5.13a; (b) Illustration of Eq. 5.13b; Illustration of Eq. 5.13c; (d) Illustration of Eq. 5.13d; (e) Illustration of Eq. 5.13e; and (f) Illustration of Eq. 5.13f (Zhang, 2006)

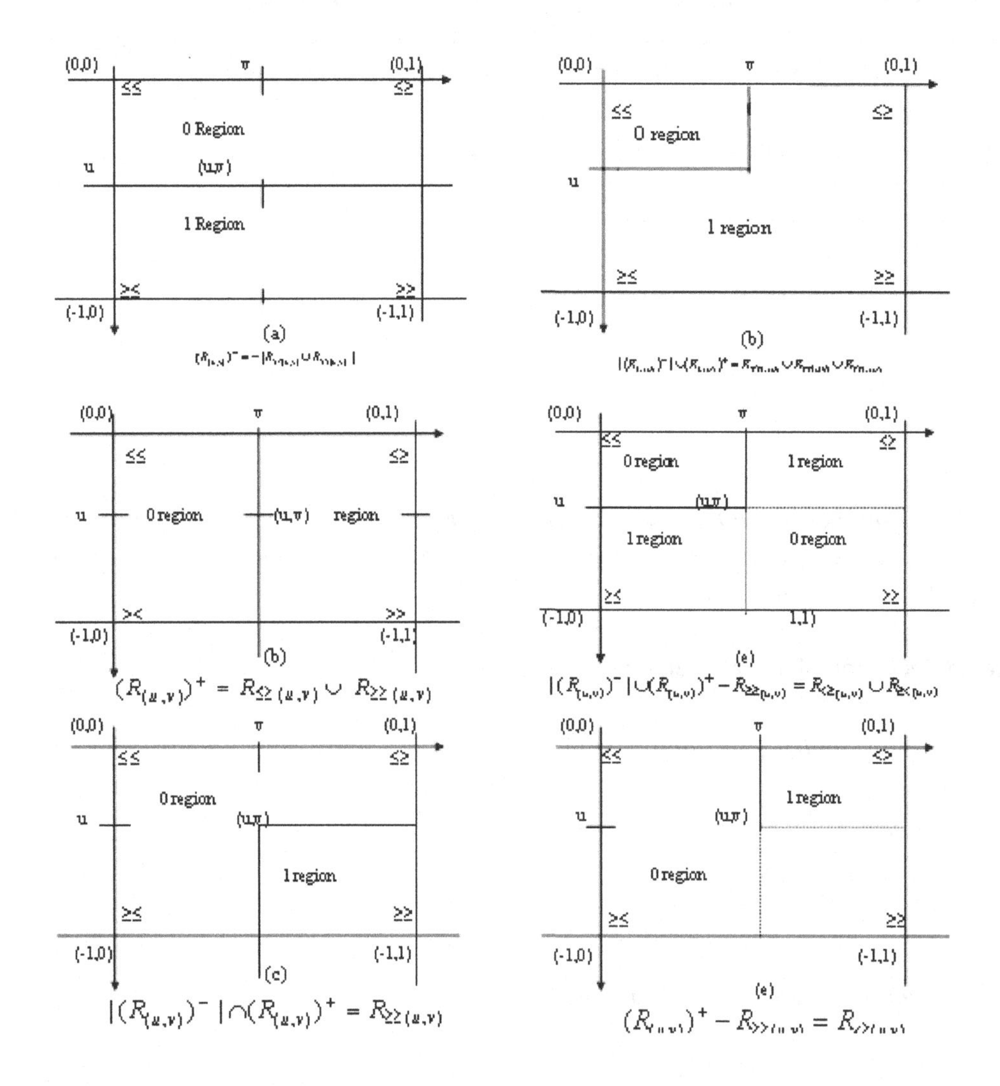

C-Type Bipolar Level Sets: $R_{©\partial}$ Forms

Besides the A-type and B-type level sets, there is a 3rd type of level set. The third type is based on the difference in magnitudes of the two poles of a bipolar variable instead of using a fixed level. This type of level set is named as C-type. **C-type bipolar α-level-sets**[7] of a bipolar fuzzy relation is proposed as

$$R_{©\partial} = \{(x,y) \mid \mu_R(x,y) = (\alpha^- + \alpha^+), (\alpha^- + \alpha^+) © \partial\}, \tag{5.14}$$

where ∂ is a unipolar threshold variable; © is any unipolar comparison operator =, <, >, ≤, or ≥..

Evidently, the C-type bipolar α-cut $R_{©\partial}$ is equivalent to the unipolar α-cut $(R^- + R^+)_\partial$. If we call $R^- + R^+$ the **depolarization** of R, $R_{©\partial}$ combines defuzzification and depolarization into one step which is similar to A-type α-cut but not equivalent.

To illustrate, we show different $R_{©\partial}$ of R in the following:

$$R = \begin{bmatrix} (-0.6 & 0.8) & (-0.9 & 0) & (-0.6 & 0.7) \\ (-0.2 & 0) & (-0.7 & 0.8) & (-0.5 & 0.8) \\ (-0.3 & 0.8) & (-0.2 & 0.1) & (-0.4 & 0.9) \end{bmatrix};$$

$$\left(R^- + R^+\right) = \begin{bmatrix} 0.2 & -0.9 & 0.1 \\ -0.2 & 0.1 & 0.3 \\ 0.5 & -0.1 & 0.5 \end{bmatrix};$$

$$R_{\geq 0.1} = \begin{bmatrix} 1 & 0 & 1 \\ 0 & 1 & 1 \\ 1 & 0 & 1 \end{bmatrix}; R_{\leq -0.5} = \begin{bmatrix} 0 & 1 & 0 \\ 0 & 0 & 0 \\ 0 & 0 & 0 \end{bmatrix}; R_{\geq 0.5} = \begin{bmatrix} 0 & 0 & 0 \\ 0 & 0 & 0 \\ 1 & 0 & 1 \end{bmatrix}.$$

Defuzzification, Depolarization, and Focus Generation with Bipolar α-Cuts

Depolarization is referred to as an operation to coerce and convert a bipolar set to a unipolar set. While defuzzification has been used frequently in fuzzy decision and control, depolarization can be as important as defuzzification. Both defuzzification and depolarization are essential in focus generation and visualization. Bipolar level sets provide a basis for both depolarization and defuzzification.

It should be remarked, however, depolarization with A-type and C-type α-cuts will necessarily lead to the loss of polarity information. This loss is a natural one in some cases but unacceptable in many other cases. For instance, let (side_effect, effect)=(-0.1, 0.9) in a decision process, net_effect = side_effect + effect = 0.8. This depolarization is a natural one due to the overwhelming positive effect. In another case, let (competition, cooperation) = (-1,1) characterize the relationship between the US and China, we have competition + cooperation = 0. This depolarization apparently leads to unacceptable information loss and clearly illustrates the importance of bipolar information fusion and visualization.

FUZZY EQUILIBRIUM RELATIONS

Bipolarity and fuzziness opened a door for the generalization of Zadeh's similarity relation (Zadeh, 1971) to nonlinear bipolar fuzzy equilibrium relation (Zhang, 2006). We extend the notion of L-fuzzy equilibrium relation (Zhang, 2005a) to include strong reflexivity, weak reflexivity, and non-reflexivity as in the follows.

Definition 5.4. A (binary) bipolar fuzzy relation R in X to B_F, where $X = \{x_i\}$, $1<i\leq n$, is **bipolar symmetric** if, $\forall i,k$, $0<i,k\leq n$, we have

$$\mu_R(x_i,x_k) = \mu_R(x_k,x_i); \quad (5.15)$$

it is **positive pole reflexive** if, $\forall i$, $0<i\leq n$, we have

$$\mu_R(x_i,x_i) = (x,1); \quad (5.16a)$$

it is **negative pole reflexive** if, $\forall i$, $0<i\leq n$, we have

$$\mu_R(x_i,x_i) = (-1,y); \quad (5.16b)$$

it is **bipolar reflexive** if, $\forall i$, $0<i\leq n$, we have

$$\mu_R(x_i,x_i) = (-1,1); \quad (5.16c)$$

it is **weakly reflexive** or has **weak reflexivity** if, $\exists i$, $0<i\leq n$, we have

$$\mu_R(x_i,x_i) = (u_i,v_i),\ (0,0) << (u_i,v_i) << (-1,1); \quad (5.16d)$$

it is **irreflexive** or has **no reflexivity** if, $\forall i$, $0<i\leq n$, we have

$$\mu_R(x_i,x_i) = (u_i,v_i) =(0,0); \quad (5.16e)$$

it is **⊕-⊗ or max-⊗ bipolar transitive** if, $\forall i,j,k$, $0<i,j,k\leq n$, we have

$$\mu_R\left(x_i,x_k\right) \geq\geq \max_{x_j}\left(\mu_R\left(x_i,x_j\right) \otimes \mu_R\left(x_j,x_k\right)\right). \quad (5.16f)$$

Definition 5.5. A bipolar fuzzy relation R in X, where $X = \{x_i\}$, $0 \leq I \leq n$, is **a (strong) fuzzy equilibrium relation** if it is (1) **bipolar symmetric**; (2) **positive pole, negative pole, or bipolar reflexive**, and (3) **bipolar transitive**. A **fuzzy equilibrium class** is a subequilibrium of R associated with $x_i \in X$.

A fuzzy equilibrium relation represents an equilibrium, quasi- or non-equilibrium state of a bipolar relation. Semantically, negative and positive reflexivity represents the equilibrium of (self-negation, self-assertion). Self-negation provides a measure for self-adjustability to equilibrium and harmony. Bipolar transitivity allows both negative and positive relationships to propagate with the cross-pole conjunctive ⊗ to reach equilibrium. Bipolar transitivity and transitive closure computation is discussed in (Zhang, Chen & Bezdek, 1989). Bipolar symmetry makes a bipolar graph bidirectional. Bipolar symmetry and transitivity are defined here the same way as for bipolar crisp relations; bipolar reflexivity is quite different in the fuzzy case. Here, the concept of weak reflexivity is introduced. As pointed out by Zadeh that in many applications of the concept of a fuzzy partial ordering, the condition of (strong) reflexivity is not a natural one to impose (Zadeh, 1971). While weak unipolar reflexivity is usually referred to as "irreflexive", the concept of weak reflexivity is a natural one in the bipolar case and it plays an important role in bipolar fuzzy partial ordering.

Definition 5.6. A bipolar fuzzy relation R in X, where $X = \{x_i\}, 0 \leq I \leq n$, is a **weak fuzzy equilibrium relation** if it is **bipolar symmetric**, **bipolar transitive**, and **weakly reflexive**. A weak equilibrium relation R1 is **weaker** than R2 if R1<<R3.

A fuzzy equilibrium relation is a fuzzification of a bipolar equilibrium relation (Zhang, 2003a) and a non-linear generalization or fusion of Zadeh's similarity relations (Zadeh, 1971) through polarization.

Theorem 5.6. The following fuzzy equilibrium laws (E_FLaws) hold:

1. **E_FLaw1**: R^+ of a bipolar transitive fuzzy relation R is unipolar transitive.
2. **E_FLaw2**: R^+ of a (strong) fuzzy equilibrium relation R is a similarity relation.
3. **E_FLaw3**: $|R^-| \cup R^+$ of a bipolar transitive fuzzy relation is a unipolar transitive relation.
4. **E_FLaw4**: $|R^-| \cup R^+$ of a (strong) fuzzy equilibrium relation is a similarity relation.
5. **E_FLaw5**: The B-type α-level sets $R_{(-\alpha,\alpha)}$, $\forall\alpha$, $0<\alpha\leq 1$, of a ⊕-∧ transitive (strong) fuzzy equilibrium relation R is a crisp equilibrium relation, where R^+_α and $(|R^-| \cup R^+)_\alpha$ are both equivalence relations.

Proof.

1. *For E_FLaw1: Based on the bipolar transitivity definition (Eq. 5.16f), if R^+ were not unipolar transitive, R would not be bipolar transitive.*
2. *For E_FLaw2: The reflexivity, symmetry, and transitivity of R^+ follow from positive-pole reflexivity, bipolar symmetry and transitivity of R. (Note: Due to bipolar transitivity, negative-pole reflexivity leads to positive-pole reflexivity and all (strong) fuzzy equilibrium relation must be positive-pole reflexive.)*
3. *For E_FLaw3: Since R is bipolar transitive, we must have,*
   $\forall i, j, k,\ 0 < i, j, k \leq n,\ \mu_R(x_i, x_k) \geq\geq \max_{x_j}(\mu_R(x_i, x_j) \otimes \mu_R(x_j, x_k))$
   and $\mu_{|R^-| \cup |R^+|}(x_i, x_k) \geq \max_{x_j}(\mu_{|R^-| \cup |R^+|}(x_i, x_j) \otimes \mu_{|R^-| \cup |R^+|}(x_j, x_k))$
   therefore $|R^-| \cup R^+$ must be unipolar transitive.
4. *For E_FLaw4: First, the unipolar reflexivity of $|R^-| \cup R^+$ follows from the second harmony law. Secondly, its symmetry follows from that the ∪ operation does not change the property of symmetry. Thirdly, its transitivity follows from last theorem.*
5. *For E_FLaw5: Positive pole reflexivity of $R_{(-\alpha,\alpha)}$ follows from that R is a (strong) fuzzy equilibrium relation. Symmetry of $R_{(-\alpha,\alpha)}$ follows from bipolar fuzzy symmetry. Bipolar transitivity of $R_{(-\alpha,\alpha)}$ follows from that, if R is ⊕-$\otimes_\wedge$ bipolar transitive, we must have,*

$$\forall i, j, k,\ 0 < i, j, k \le n, \mu_R(x_i, x_k) \ge\ge \max_{x_j}(\mu_R(x_i, x_j) \wedge \mu_R(x_j, x_k))$$

$$\textit{and}\ \mu_{(R^-_{|\alpha|} \wedge R^+_\alpha)}(x_i, x_k) \ge \max_{x_j}(\mu_{(R^-_{|\alpha|} \wedge R^+_\alpha)}(x_i, x_j) \wedge \mu_{(R^-_{|\alpha|} \wedge R^+_\alpha)}(x_j, x_k)).$$

R^+_α and $(|R^-| \cup R^+)_\alpha$ are both equivalence relations because $R_{(-\alpha,\alpha)}$ is a crisp equilibrium relation (Zhang, 2003a). □

Similar to unipolar cases, a symmetric bipolar relation can be represented as a bi-directional graph (bigraph) and a non-symmetrical bipolar relation can be represented as a directed graph (digraph). The edges in the graphs are marked with bipolar relationship strengths. Based on bipolar crisp relations and its transitive closure computation (Zhang, Chen & Bezdek, 1989), Warshall's algorithm (Warshall, 1962) has been extended for computing the bipolar transitive closure with $O(n^3)$ complexity. The shortest bipolar transitive or causal (strongest) paths can be identified as **critical paths** for bipolar decision and coordination. These paths are also called heuristic paths (Zhang, Chen & Bezdek, 1989) for focus generation and heuristic decision.

In macroeconomics (Cossette & Lapointe, 1997) the global economy can be modeled as a dynamic multiagent system in which major players form an equilibrium or quasi-equilibrium relation of common and conflict interests. Global regulation can be used in coordinating the major players for the health of a market economy. Figure 3 shows an example bipolar fuzzy cognitive map (FCM) of six major players in a macroeconomy for illustrating bipolar equilibrium and bipolar clustering. Figure 4 shows the positive pole reflexive and symmetric bipolar fuzzy relation of the FCM. The bipolar relationship strengths in Figure 3 are measures of the negative and positive ties between each other (or common interests and conflict interests). A CM can be the result of text mining from the Web that is beyond the scope of this chapter. Here, we focus on bipolar clustering and visualization as a post mining process in OLAP/OLAM. Its $\oplus$-$\otimes_\wedge$, $\oplus$-$\otimes_\times$, and $\oplus$-$\otimes_\Delta$ transitive closures or equilibrium relations are computed in Figure 5(a-c). It is easy to verify that $\oplus$-$\otimes_\wedge \ge\ge \oplus$-$\otimes_\times \ge\ge \oplus$-$\otimes_\Delta$. The equilibrium (transitive or causal) paths are listed in Figure 6, which identifies a strongest negative path and a strongest positive path between any pair of elements in the set of C. These paths provide focal points in decision analysis.

Different properties of fuzzy equilibrium relations vs. similarity and equivalence relations, local equilibrium relations, and crisp equilibrium relations are summarized in Figure 7.

Figure 3. A bipolar conceptual graph or cognitive map in macroeconomics (Zhang, 2006)

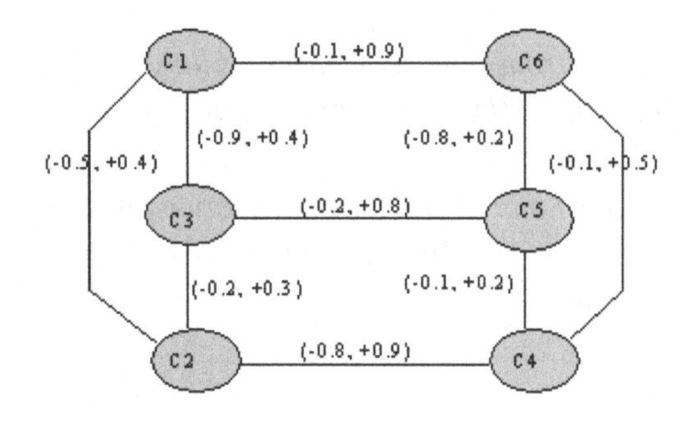

Figure 4. A positive pole reflexive and symmetric bipolar fuzzy relation (Zhang, 2006)

	C1	c2	c3	c4	c5	c6
c1	(-0 1)	(-0.5 0.4)	(-0.9 0.1)	(-0 0)	(-0 0)	(-0.1 0.9)
c2	(-0.5 0.4)	(-0 1)	(-0.2 0.3)	(-0.8 0.9)	(-0 0)	(-0 0)
c3	(-0.9 0.1)	(-0.2 0.3)	(-0 1)	(-0 0)	(-0.2 0.8)	(-0 0)
c4	(-0 0)	(-0.8 0.9)	(-0 0)	(-0 1)	(-0.1 0.2)	(-0.1 0.5)
c5	(-0 0)	(-0 0)	(-0.2 0.8)	(-0.1 0.2)	(-0 1)	(-0.8 0.2)
c6	(-0.1 0.9)	(-0 0)	(-0 0)	(-0.1 0.5)	(-0.8 0.2)	(-0 1)

BIPOLAR FUZZY CLUSTERING

Bipolar Fuzzy Clustering

It is clear that fuzzy equilibrium relation is a fuzzification of crisp equilibrium relation (Ch. 1) or a bipolar fusion or non-linear generalization of equivalence and similarity relations (Zadeh, 1971). With this generalization, we are ready to discuss bipolar fuzzy clustering (Zhang, 2006).

Definition 5.7. A **bipolar fuzzy cluster** is a bipolar fuzzy subset of a bipolar set. Let $\mu(x,y)$ be a fuzzy equilibrium relation on a set S to B_F. (1) A subset C in S is a **fuzzy coalition set or cluster** that satisfies a bipolar coalition condition $\alpha_c = (a,b)$ if, $\forall x,y \in C$, we have $\mu(x,y) \leq\geq \alpha_c$, where $\varepsilon_c = b-|a| > 0$ is the minimum surplus of positive relationship strength for coalition. (2) A subset H in S is **a fuzzy harmony**

Figure 5. (a) $\oplus$-$\otimes_\wedge$ transitive closure of Figure 4 – a fuzzy equilibrium relation; (b) $\oplus$-$\otimes_\times$ transitive closure of Figure 4 – a fuzzy equilibrium relation; (c) $\oplus$-$\otimes_\Delta$ transitive closure of Figure 4 – a fuzzy equilibrium relation (Adapted from Zhang, 2006)

	c1	c2	c3	c4	c5	c6
c1	(-0.5 1)	(-0.5 0.5)	(-0.9 0.5)	(-0.5 0.5)	(-0.8 0.5)	(-0.5 0.9)
c2	(-0.5 0.5)	(-0.8 1)	(-0.5 0.5)	(-0.8 0.9)	(-0.5 0.5)	(-0.5 0.5)
c3	(-0.9 0.5)	(-0.5 0.5)	(-0.5 1)	(-0.5 0.5)	(-0.5 0.8)	(-0.9 0.5)
c4	(-0.5 0.5)	(-0.8 0.9)	(-0.5 0.5)	(-0.8 1)	(-0.5 0.5)	(-0.5 0.5)
c5	(-0.8 0.5)	(-0.5 0.5)	(-0.5 0.8)	(-0.5 0.5)	(-0.5 1)	(-0.8 0.5)
c6	(-0.5 0.9)	(-0.5 0.5)	(-0.9 0.5)	(-0.5 0.5)	(-0.8 0.5)	(-0.5 1)

(a)

	c1	c2	c3	c4	c5	c6
c1	(-0.2 1)	(-0.5 0.4)	(-0.9 0.18)	(-0.45 0.45)	(-0.72 0.18)	(-0.22 0.9)
c2	(-0.5 0.4)	(-0.72 1)	(-0.36 0.45)	(-0.8 0.9)	(-0.36 0.36)	(-0.45 0.45)
c3	(-0.9 0.18)	(-0.36 0.45)	(-0.16 1)	(-0.4 0.4)	(-0.2 0.8)	(-0.81 0.2)
c4	(-0.45 0.45)	(-0.8 0.9)	(-0.4 0.4)	(-0.72 1)	(-0.4 0.32)	(-0.4 0.5)
c5	(-0.72 0.18)	(-0.36 0.36)	(-0.2 0.8)	(-0.4 0.32)	(-0.16 1)	(-0.8 0.2)
c6	(-0.22 0.9)	(-0.45 0.45)	(-0.81 0.2)	(-0.4 0.5)	(-0.8 0.2)	(-0.2 1)

(b)

	C1	c2	c3	c4	c5	c6
c1	(0 1)	(-0.5 0.4)	(-0.9 0.1)	(-0.4 0.4)	(-0.7 0.1)	(-0.1 0.9)
c2	(-0.5 0.4)	(-0.7 1)	(-0.3 0.4)	(-0.8 0.9)	(-0.2 0.2)	(-0.4 0.4)
c3	(-0.9 0.1)	(-0.3 0.4)	(-0 1)	(-0.3 0.3)	(-0.2 0.8)	(-0.8 0)
c4	(-0.4 0.4)	(-0.8 0.9)	(-0.3 0.3)	(-0.7 1)	(-0.3 0.2)	(-0.3 0.5)
c5	(-0.7 0.1)	(-0.2 0.2)	(-0.2 0.8)	(-0.3 0.2)	(-0 1)	(-0.8 0.2)
c6	(-0.1 0.9)	(-0.4 0.4)	(-0.8 0)	(-0.3 0.5)	(-0.8 0.2)	(-0 1)

(c)

Figure 6. Equilibrium paths (Zhang, 2006)

	c1	c2	c3	c4	c5	c6
c1	((1 6 4 2 1)(1 1))	((1 2)(1 6 4 2))	((1 3)(1 6 4 2 1 3))	((1 2 4)(1 6 4))	((1 6 5)(1 6 5))	((1 2 4 6)(1 6))
c2	((2 1)(2 4 6 1))	((2 4 2)(2 2))	((2 4 6 1 3)(2 1 3))	((2 4)(2 4))	((2 4 6 5)(2 1 6 5))	((2 1 6)(2 4 6))
c3	((3 1)(3 1 6 4 2 1))	((3 1 6 4 2)(3 1 2))	((3 1 2 4 6 1 3)(3 3))	((3 1 6 4)(3 1 2 4))	((3 5)(3 5))	((3 1 6)(3 1 2 4 6))
c4	((4 2 1)(4 6 1))	((4 2)(4 2))	((4 6 1 3)(4 2 1 3))	((4 2 4)(4 4))	((4 6 5)(4 2 1 6 5))	((4 2 1 6)(4 6))
c5	((5 3 1)(5 6 1))	((5 6 4 2)(5 3 1 2))	((5 3)(5 3))	((5 6 4)(5 3 1 2 4))	((5 3 5)(5 5))	((5 6)(5 6))
c6	((6 4 2 1)(6 1))	((6 1 2)(6 4 2))	((6 1 3)(6 4 2 1 3))	((6 1 2 4)(6 4))	((6 5)(6 5))	((6 1 2 4 6)(6 6))

Figure 7. (a) Fuzzy equilibrium relations vs. similarity relations and equivalence relations; (b) Fuzzy equilibrium relations vs. local equivalence relations; (c) Fuzzy equilibrium relations vs. crisp equilibrium relations; (Adapted from Zhang, 2006)

Bipolar Fuzzy Relation R	R^+	$\lvert R^-\rvert \cup R^+$	R^+_α	$(\lvert R^-\rvert \cup R^+)_\alpha$	$\forall(u,v)\in R$, $\lvert R^-\rvert_{\lvert u\rvert} \cup R^+_v$
⊕-⊗ transitive	⊕-⊗ transitive	⊕-⊗ transitive	may be not transitive	may be not transitive	may be not transitive
⊕-⊗ (⊗≠⊗$_\wedge$) transitive strong equilibrium	⊕-⊗ transitive similarity relation	⊕-⊗ transitive similarity relation	reflexive symmetric but may be not transitive	reflexive symmetric but may be not transitive	reflexive symmetric but may be not transitive
⊕-⊗$_\wedge$ transitive strong equilibrium	⊕-∧ transitive similarity relation	⊕-∧ transitive similarity relation	equivalence relation	equivalence relation	equivalence relation

(a)

Fuzzy Equilibrium Relation R	A-type α-cut $R_{\gg(-\alpha,\alpha)}$	A-type α-cut $R_{<\gtrsim(-\alpha,\alpha)}$	$R_{\Omega(-\alpha,\alpha)}$ ∪ $R_{\gg\Omega(-\alpha,\alpha)}$	A-type α-cut $R_{\gg(u,v)}$	A-type α-cut $R_{\Omega(u,v)}$	$R_{\Omega(u,v)}$ ∪ $R_{\gg\Omega(-u,v)}$
⊕-⊗$_\wedge$ transitive strong or weak	local equivalence relations					

(b)

Fuzzy Equilibrium Relation R	B-type α-cut $R_{(-\alpha,\alpha)} = R^-_{\lvert\alpha\rvert} \Omega R^+_\alpha$	B-type α-cut $R_{(u,v)} = R^-_{\lvert u\rvert} \Omega R^+_v$, $\forall(u,v)\in R$
⊕-⊗ (⊗≠⊗$_\wedge$) transitive strong	reflexive, symmetric but may be not transitive bipolar relation	reflexive, symmetric but may be not transitive bipolar relation
⊕-⊗$_\wedge$ transitive strong	strong crisp equilibrium relation	strong crisp equilibrium relation

(c)

set or cluster that satisfies a harmony condition $\alpha_h = (-\alpha,\alpha)$ if, $\forall x,y\in H$, we have $\mu(x,y) \geq\geq \alpha_h$. (3) Let μ be a fuzzy equilibrium relation on a set S, a subset F in S is **a conflict fuzzy set or cluster** that satisfies a conflict condition $\alpha_f = (a,b)$ if F contains two coalition subsets A and B and, $\forall x\in A$ and $\forall y\in B$, we have $\mu(x,y) \geq\leq \alpha_f$, where $\varepsilon_f = |a|-b > 0$ is the minimum discrepancy between the two poles for conflict.

Definition 5.8. Let μ be a fuzzy equilibrium relation on a set S. (1) A subset C in S is a **quasi-coalition set or cluster** that satisfies a bipolar coalition condition $\alpha_c = (a,b)$ if, $\forall x,y\in C$, $x\neq y$, we have $\mu(x,y) \leq\geq \alpha_c$, where $\varepsilon_c = b-|a| > 0$ is the minimum surplus of positive relationship strength for coalition. (2) A subset H in S is a **quasi-harmony set or cluster** that satisfies a harmony condition $\alpha_h = (-\alpha,\alpha)$ if, $\forall x,y\in H$, $x\neq y$, we have $\mu(x,y) \geq\geq \alpha_h$. (3) A subset F in S is **a quasi-conflict fuzzy set or cluster** that satisfies a conflict condition $\alpha_f = (a,b)$ if F contains two quasi-coalition subsets A and B and, $\forall x\in A$ and $\forall y\in B$, we have $\mu(x,y) \geq\leq \alpha_f$, where $\varepsilon_f = |a|-b > 0$ is the minimum discrepancy between the two poles for conflict.

Note that the $x\neq y$ condition in Definition 5.8 eliminates the reflexivity requirement for quasi-coalition, quasi-harmony, and quasi-conflict fuzzy sets. This is necessary because two elements may be in a coalition or harmony or conflict fuzzy cluster while each element to itself may fail the coalition or harmony condition. This treatment allows overlapped clusters to be derived from $\oplus$-$\otimes$ transitive equilibrium relations where $\otimes \neq \otimes_{\wedge}$.

Fuzzy Harmony

Negative reflexivity can be used as a measure of self-adjustability from a non-harmonic state to a harmonic state of an equilibrium relation. While in a crisp equilibrium relation there is only one level of harmony identified by (-1,1), in the fuzzy case, an infinite number of weak reflexivity and harmonic levels can be captured in an equilibrium relation because we may consider any $\mu(x,y) \geq\geq (-\alpha,\alpha) \in [-1,0]\times[0,1]$ or $\mu(x,y) \geq\geq (u,v) \in [-1,0]\times[0,1], |u+v| \leq \varepsilon$, as **a harmony condition** that identifies **a fuzzy harmony cluster**. It can be observed that a fuzzy equilibrium relation in a set X does not have to have strong reflexivity. Therefore, the 5th harmony law in the crisp case is no longer applicable. Instead, we have the following fuzzy harmony laws:

Theorem 5.7. The following fuzzy harmony laws (H_FLaws) hold:

1. **H_FLaw1**: The bipolar transitive closure $\mathfrak{R}$ of any negative pole reflexive bipolar fuzzy relation R must be bipolar reflexive.
2. **H_FLaw2**: Any negative-pole reflexive (strong) fuzzy equilibrium relation must also be bipolar reflexive, and therefore, any (strong) equilibrium fuzzy relation is a positive-pole reflexive (strong) fuzzy equilibrium relation.
3. **H_FLaw3**: The negative pole R^- and the positive pole R^+ of any bipolar (strong) fuzzy equilibrium relation R must meet the condition $-R^- = R^+$, and therefore, all clusters induced from R must be in a **fuzzy harmonic state.**
4. **H_FLaw4**: (i) Given a strong or weak $\oplus$-$\otimes_{\wedge}$ transitive fuzzy equilibrium relation R in X and a fuzzy harmony condition $\mu(x,y) \geq\geq (-\alpha,\alpha)$, $\forall\alpha\in(0,1]$, in the A-type α-cut $R_{\geq\geq(-\alpha,\alpha)}$ the fuzzy harmony subsets that satisfy the condition must be disjoint. (ii) Given a strong or weak $\oplus$-$\otimes_{\wedge}$ transitive fuzzy equilibrium relation R and a fuzzy harmony condition $\mu(x,y) \geq\geq (u,v)$, $\forall(u,v)\in R$, in the A-type α-cut $R_{\geq\geq(u,v)}$ the fuzzy harmony subsets that satisfy the condition must be disjoint.
5. **H_FLaw5**: The stronger the negative pole reflexivity of a symmetric bipolar relation R, the closer is $|E^-|$ to E^+ of the fuzzy equilibrium relation E of R, and therefore, the more fuzzy harmonic the relation E.

Proof.

H_FLaw1 *follows from that* $\mu_R(x_i,x_i) \oplus \mu_R(x_i,x_i) \otimes \mu_R(x_i,x_i) = (-1,v)\oplus(-1,v)\otimes(-1,v) = (-1,v)\oplus(-1,1) = (-1,1)$, *which is derived from cycling the negative pole reflexive edge (-1,v) twice (2n).*

H_FLaw2 *follows from (1) HLaw1; (2) any equilibrium relation is either positive-pole reflexive but not negative-pole reflexive or bipolar reflexive. Since negative-pole reflexivity implies bipolar reflexivity and positive-pole reflexivity by definition, any equilibrium relation must be positive-pole reflexive.*

H_FLaw3 *follows from that given any* $\mu_R(x_i,x_j) = (u,v)$, *if* $|u| > v$, *it would violate the bipolar transitivity requirement* $(u,v) \geq\geq \mu_R(x_i,x_j) \otimes \mu_R(x_j,x_j) = (u,v)\otimes(-1,1) = (-(max(|u|,v), max(|u|,v))$ *because the reflexive edge* $\mu_R(x_j,x_j)$ *must be harmonic for any bipolar reflexivie equilibrium relation. Similarly, if* $|u| < v$, *it would violate the bipolar transitivity requirement* $(u,v) \geq\geq \mu_R(x_i,x_j) \otimes \mu_R(x_j,x_j) = (u,v)\otimes(-1,1) = (-$

(max(|u|,v),max(|u|,v)). Since each relationship in R is (u,v) and |u|=v, all fuzzy clusters induced from R^+ must be in a harmonic state.

H_FLaw4. *Note that $R_{\geq\geq(-\alpha,\alpha)}$ identifies bipolar relationships with both negative and positive strengths greater than α. That is, let μ_L be the member function of $R_{\geq\geq(-\alpha,\alpha)}$, $\mu_L(x_i,x_j) = 1$ iff $|\mu_R(x_i,x_j)^-| \geq \alpha$ and $\mu_R(x_i,x_j)^+ \geq \alpha$, or $Truth(\mu_R(x_i,x_j) \geq \alpha$. Then, due to the symmetry and ⊕-∧ bipolar transitivity of R, if $\mu_L(x_i,x_j) = 1$ and $\mu_L(x_j,x_k) = 1$, we must have $\mu_L(x_i,x_k) = 1$, $\mu_L(x_k,x_j) = 1$, $\mu_L(x_j,x_i) = 1$, $\mu_L(x_k,x_i) = 1$, $\mu_L(x_i,x_i) = 1$, $\mu_L(x_j,x_j) = 1$, and $\mu_L(x_k,x_k) = 1$. That is, two fuzzy harmony clusters H1 and H2 satisfying the same condition must necessarily be disjoint with each other. It can be proved similarly for the condition μ(x,y)≥≥(u,v), ∀(u,v)∈R.*

H_FLaw5. *The theorem follows from that fuzzy harmony in a fuzzy equilibrium relation is the result of negative pole reflexivity of R and bipolar transitivity.* □

Bipolar Fuzzy Clustering with A-Type α-Cut

Theorem 5.8. Let R be a $\oplus$-$\otimes_\wedge$ transitive strong or weak fuzzy equilibrium relation in a bipolar set X, ∀α, 0<α≤1, we have the following:

1. Each A-type α-cut $R_{\geq\geq(-\alpha,\alpha)}$ is a local equivalence relation if it is not null, which induces a set $\{H_i\}$ of disjoint fuzzy harmony subsets (if any) in X, where each subset H_i satisfies the fuzzy harmony condition, ∀x,y∈ H_i, μ(x,y)≥≥(-α,α).
2. Each A-type α-cut $R_{<\geq(-\alpha,\alpha)}$ is an local equivalence relation, if it is not null, which induces a set $\{C_i\}$ of disjoint fuzzy coalition subsets (if any) in X, where each subset C_i satisfies the fuzzy coalition condition, ∀x,y∈ C_i, μ(x,y)=(u,v)<≥(-α,α).
3. $R_{<\geq(-\alpha,\alpha)} \vee R_{\geq<(-\alpha,\alpha)}$ is an local equivalence relation, if it is not null, which induces a set of disjoint fuzzy subsets that can be further classified as fuzzy coalition subsets $\{C_i\}$ {if any} that are not involved in a conflict and fuzzy conflict subsets $\{F_j\}$ (if any), where each coalition satisfies the coalition condition μ(x,y)<≥(-α,α) and each F_j satisfies the conflict condition μ(x,y) ≥<α_f=(-α,α), for all x,y involved in a conflict.
4. Given an harmony condition μ_h(x,y)≥≥(-α,α), a coalition condition μ_c(x,y)<≥(-α,α), and a conflict condition μ_f(x,y) ≥<(-α,α), let H be the set of clusters in X induced from $R_{\geq\geq(-\alpha,\alpha)}$; let C be the set of coalition clusters induced from $R_{<\geq(-\alpha,\alpha)}$; let S be the set of clusters induced from $R_{<\geq(-\alpha,\alpha)} \cup R_{\geq<(-\alpha,\alpha)}$; then
   a. H is a set of disjoint fuzzy harmony clusters;
   b. C is a set of disjoint coalition clusters;
   c. S is a set of disjoint coalition and conflict clusters;
   d. F = S - C results in a set of disjoint fuzzy conflict subsets;
   e. C1 = S - F results in a set of disjoint fuzzy coalition subsets that are not involved in a conflict in F;
   f. C2 = C - C1 results in a set of disjoint fuzzy coalition subsets that are involved in a conflict in F;
   g. Clusters in C and H are disjoint; Clusters in S and H are disjoint.

Proof.

1. *It follows from the 4th fuzzy harmony law H_FLaw4.*

2. *Note that $R_{<\geq(-\alpha,\alpha)}$ identifies bipolar relationships with negative pole strength < |-α| and positive strength ≥ |α|. That is, let μ_L be the member function of $R_{<\geq(-\alpha,\alpha)}$, $\mu_L(x_i,x_j) = 1$ iff $|\mu_R(x_i,x_j)^-| < \alpha$ and $\mu_R(x_i,x_j)^+ \geq \alpha$. Then, due to the symmetry and ⊕-$\otimes_\wedge$ bipolar transitivity of R, if $\mu_L(x_i,x_j) = 1$ and $\mu_L(x_j,x_k) = 1$, we must have $\mu_L(x_i,x_k) = 1$, $\mu_L(x_k,x_j) = 1$, $\mu_L(x_j,x_i) = 1$, $\mu_L(x_k,x_i) = 1$, $\mu_L(x_i,x_i) = 1$, $\mu_L(x_j,x_j) = 1$, and $\mu_L(x_k,x_k) = 1$. Therefore, $R_{<\geq(-\alpha,\alpha)}$ must be a local equivalence relation.*
3. *Following last theorem, $R_{<\geq(-\alpha,\alpha)}$ is a local equivalence relation, then it must induce disjoint clusters. Due to the symmetry and ⊕-$\otimes_\wedge$ bipolar transitivity of R, for any pair of clusters A and B induced from $R_{<\geq(-\alpha,\alpha)}$ and any pair of elements (a,b), a∈A and b∈B, $\mu_R(a,b) \geq <(-\alpha,\alpha) \Rightarrow \mu_R(x,y) \geq <(-\alpha,\alpha)$, ∀x∈A and ∀y∈B. Therefore, $R_{<\geq(-\alpha,\alpha)} \vee R_{\geq<(-\alpha,\alpha)}$ must be a local equivalence relation.*
4. *It follows from (1)-(3) directly.* □

When ⊗≠∧, a bipolar fuzzy relation does not lead to disjoint clusters. The is formalized as

Theorem 5.9. Let R be a ⊕-⊗ (⊗≠$\otimes_\wedge$) transitive strong or weak fuzzy equilibrium relation in a bipolar set X, given an harmony condition $\mu_h(x,y) \geq\geq(-\alpha,\alpha)$, a coalition condition $\mu_c(x,y) <\geq(-\alpha,\alpha)$, and a conflict condition $\mu_f(x,y) \geq<(-\alpha,\alpha)$, let H be the clusters induced from $R_{\geq\geq(-\alpha,\alpha)}$; let C be the set of clusters induced from $R_{<\geq(-\alpha,\alpha)}$; let S be the set of clusters induced from $R_{<\geq(-\alpha,\alpha)} \vee R_{\geq<(-\alpha,\alpha)}$; we have

(1) H is a set of quasi-harmony clusters;
(2) C is a set of quasi-coalition clusters;
(3) S is a set of quasi-coalition and quasi-conflict clusters;
(4) F = S - C results in a set of quasi-conflict clusters;
(5) C1 = S - C results in a set of quasi-coalition clusters that are not involved in a conflict in F;
(6) C2 = C - C1 results in a set of quasi-coalition clusters that are involved in a conflict in F.

Proof. It follows from the proof for the last theorem. □

Note that the criterion (-α,α) used in Theorem 5.9 is not a natural condition to impose because it does not give the flexibility for a user to specify his/her own criterion in OLAP /OLAM. If we replace (-α,α) with (u,v) in R, we have

Theorem 5.10. Let R be a ⊕-$\otimes_\wedge$ transitive strong or weak fuzzy equilibrium relation in a bipolar set X, ∀(u,v)∈R, we have the following **bipolar fuzzy clustering laws**:

1. Each A-type α-cut $R_{\geq\geq(u,v)}$ is an local equivalence relation, provided it is not null, that induces a set $\{H_i\}$of disjoint fuzzy harmony subsets (if any) in X, where each subset H_i satisfies the fuzzy harmony condition: ∀x,y∈ H_i, $\mu(x,y) \geq\geq(u,v)$.
2. When |u|<v, each A-type α-cut $R_{\leq\geq(u,v)}$ is an local equivalence relation, provided it is not null, that induces a set $\{C_i\}$ of disjoint fuzzy coalition subsets (if any) in X, where each subset C_i satisfies the fuzzy coalition condition: ∀x,y∈ C_i, $\mu(x,y) \leq\geq(u,v)$.
3. When |u|<v, $R_{\leq\geq(u,v)} \vee R_{\geq\leq(-v,-u)}$ is an local equivalence relation if it is not null that induces a set S of disjoint fuzzy subsets that can be further classified as fuzzy conflict subsets (if any) and coalition subsets {if any} that are not involved in a conflict, where each coalition subset satisfies the

coalition condition $\mu(x,y)\leq\geq(u,v)$ and each F_j satisfies the conflict condition $\mu(x,y)\geq\leq(-v,-u)$ for all x,y involved in a conflict.

4. Given a coalition condition $\mu_c(x,y)<\geq(u,v)$, a conflict condition $\mu_f(x,y) \geq<(-v,-u)$, and an harmony condition $\mu_h(x,y) \geq\geq(u,-u)$, let H be the set of clusters induced from $R_{\geq\geq(u,-u)}$; let C be the set of clusters induced from $R_{<\geq(u,v)}$; and let S be the set of clusters induced from $R_{<\geq(u,v)} \vee R_{\geq<(-v,-u)}$; we have
    a. H is a set of disjoint fuzzy harmony clusters;
    b. C is a set of disjoint coalition clusters;
    c. S is a set of disjoint coalition and conflict clusters;
    d. F = S - C results in a set of disjoint fuzzy conflict subsets;
    e. C1 = S - F results in a set of disjoint fuzzy coalition subsets that are not involved in a conflict in F;
    f. C2 = C - C1 results in a set of disjoint fuzzy coalition subsets that are involved in a conflict in F;
    g. Clusters in C and H are disjoint; Clusters in S and H are disjoint.

Proof. The proof is similar to that for Theorem 5.8. □

We have a dual to Theorem 5.9 by replacing replace $(-\alpha,\alpha)$ with (u,v) in R.

Theorem 5.11. Let R be a $\oplus$-$\otimes$ ($\otimes\neq\otimes_\wedge$) transitive strong or weak fuzzy equilibrium relation in a bipolar set X, given a coalition condition $\mu_c(x,y)<\geq(u,v)$ ($|u|<v$), a conflict condition $\mu_f(x,y) \geq<(-v,-u)$, and an harmony condition $\mu_h(x,y) \geq\geq(u,-u)$, let H be the set of clusters induced from $R_{\geq\geq(u,-u)}$; let C be the set of clusters induced from $R_{<\geq(u,v)}$; and let S be the set of clusters induced from $R_{<\geq(u,v)} \vee R_{\geq<(-v,-u)}$; we have

1. H is a set of quasi-harmony clusters;
2. C is a set of quasi-coalition clusters;
3. S is a set of quasi-coalition and quasi-conflict clusters;
4. F = S - C results in a set of quasi-conflict clusters;
5. C1 = S - C results in a set of quasi-coalition clusters that are not involved in a conflict in F;
6. C2 = C - C1 results in a set of quasi-coalition clusters that are involved in a conflict in F.

Proof. The proof is similar to that for Theorem 5.9. □

It should be remarked that disjoint coalition, harmony, and conflict relationships fall into three different regions in the 2-D bipolar space. With the set of conditions $\{\mu_h(x,y)\geq\geq(-\alpha,\alpha), \mu_c(x,y)<\geq(-\alpha,\alpha), \mu(x,y) \geq<(-\alpha, \alpha)\}$, different α levels lead to four rectangle regions of neutral, coalition, harmony, and conflict areas that fully occupy the bipolar space $[-1,0] \times [0,1]$. With the set of conditions $\{\mu_c(x,y)<\geq(u,v), \mu_f(x,y)\geq<(-v,-u), \mu_h(x,y)\geq\geq(u,-u)\}$, different (u,v) levels (Note: based on Def. 5.7 we have $|u|<v$) lead to neutral, coalition, harmony, and conflict areas that may leave unoccupied gray regions. This is illustrated in Figure 8(a) and 8(b).

To illustrate Theorems 5.8-5.11, let R be the fuzzy equilibrium relation in Figure 5a, we remove the zero rows and columns from the local equivalence relations $R_{\geq\geq(-0.6,0.6)}$ or $R_{\geq\geq(-0.8,0.9)}$ induced from R and rearrange the remaining rows and columns we have the disjoint fuzzy harmony subset {c2,c4} and fuzzy

Figure 8. (a) Coalition, harmony, and conflict regions with symmetrical (-α,α) partitioning; (b) Coalition, harmony, and conflict regions with (u,v) partitioning

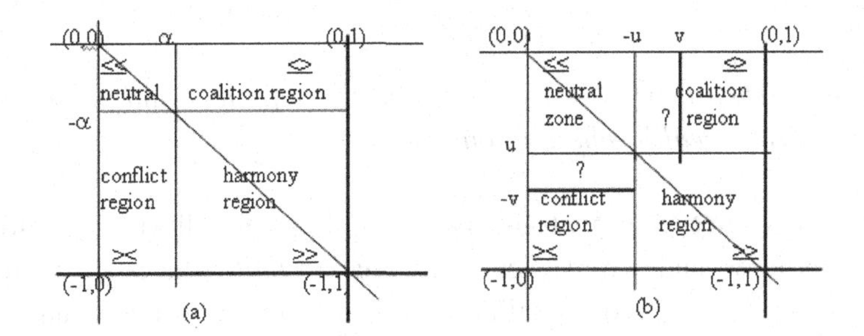

coalition subsets {c1,c6} and {c3,c5} as in Figure 9(a)-(b); the two coalitions are joined into a fuzzy conflict subset {c1,c6,c3,c5} as in Figure 9(c).

Bipolar Fuzzy Clustering with B-Type α-Cut

Theorem 5.12. Let R be a $\oplus$-$\otimes_{\wedge}$ transitive strong fuzzy equilibrium relation,

1. $\forall v$, $0<v\leq 1$, a B-type α-cut $R_{(-v,v)}$ is a crisp equilibrium relation that induces disjoint fuzzy coalition sets and fuzzy harmony sets from R where two fuzzy coalition sets can be joined into a fuzzy conflict set.

Figure 9. (a) Disjoint fuzzy harmony cluster from $R_{\underline{\gg}(-0.6,0.6)} = R_{\underline{\gg}(-0.8,0.9)}$ (R is Figure 5a); (b) Disjoint fuzzy coalition subsets from $R_{\underline{<\geq}(-0.6,0.6)} = R_{\underline{\lessgtr}(-0.8,0.9)}$; (c) Disjoint fuzzy conflict subset from $R_{\underline{<\geq}(-0.6,0.6)} \cup R_{\underline{\geq<}(-0.6,0.6)} = R_{\underline{\lessgtr}(-0.5,0.8)} \cup R_{\underline{\gtrless}(-0.8,0.5)}$

	{c2	c4}
c2	1	1
c4	1	1

(a)

	{c1	c6}	{c3	c5}
c1	1	1	0	0
c6	1	1	0	0
c3	0	0	1	1
c5	0	0	1	1

(b)

	{c1	c6	c3	c5}
c1	1	1	1	1
c6	1	1	1	1
c3	1	1	1	1
c5	1	1	1	1

(c)

2, $\forall(u,v)\in R$, a B-type α-cut $R_{(-u,v)}$, is a crisp equilibrium relation that induces disjoint casual fuzzy coalition sets and casual fuzzy harmony sets from R where two coalition sets can be joined into a conflict set.

Proof. It follows from Theorems 5.8-5.11 above and the sufficient and adequate conditions for a bipolar relation to be a crisp equilibrium relation. □

Theorem 5.13. Let R be a $\oplus$-$\otimes$ ($\otimes \neq \otimes_{\wedge}$) transitive strong fuzzy equilibrium relation, $\forall(u,v)\in R$, a B-type α-cut $R_{(-u,v)}$, is a reflexive and symmetrical crisp relation *that* induces casual quasi-coalition sets and casual quasi-harmony sets where two coalition sets can be joined into a casual quasi-conflict set.

Proof. It follows from Theorem 5.9. □

To illustrate, let R1 be the fuzzy equilibrium relation as in Figure 5(a), three disjoint clusters are derived from $(R1_{(-0.6,0.6)})^{+}$ (see Figure 10(a)), where {c1, c6} and {c3,c5} are coalition sets and {c2,c4} is a harmony set. Two disjoint clusters are derived from $|(R1_{(-0.6,0.6)})^{-}| \cup (R1_{(-0.6,0.6)})^{+}$ (see Figure 10(b)), where {c1,c6,c3,c5} form a conflict set and {c2,c4} forms a harmony set. The types of the clusters can be determined based on R1. As a comparison, let R2 be the equilibrium relation as in Figure 10(c), four clusters are derived from $(R2_{(-0.5,0.5)})^{+}$ (see Figure 10(c)), where {c1,c6},{c6,c4},{c3,c5} are quasi-coalition sets

Figure 10. (a) Clusters from $(R1_{(-0.6,0.6)})^{+}$; (b) Clusters from $|(R1_{(-0.6,0.6)})^{-}| \cup (R1_{(-0.6,0.6)})^{+}$; (c) Quasi-Clusters from $(R2_{(-0.5,0.5)})^{+}$ (Adapted from Zhang, 2006)

	{c1	c6}	{c2	c4}	{c3	c5}
c1	1	1	0	0	0	0
c6	1	1	0	0	0	0
c2	0	0	1	1	0	0
c4	0	0	1	1	0	0
c3	0	0	0	0	1	1
c5	0	0	0	0	1	1

(a)

	{c1	c6	c3	c5}	{c2	c4}
c1	1	1	1	1	0	0
c6	1	1	1	1	0	0
c2	1	1	1	1	0	0
c4	1	1	1	1	0	0
c3	0	0	0	0	1	1
c5	0	0	0	0	1	1

(b)

		{c6	c4}			
	{c1	c6}	{c4	c2}	{c3	c5}
c1	1	1	0	0	0	0
c6	1	1	1	0	0	0
c2	0	1	1	1	0	0
c4	0	0	1	1	0	0
c3	0	0	0	0	1	1
c5	0	0	0	0	1	1

(c)

and {c2,c4} is a quasi-harmony set. It is interesting to note that the clusters are overlapped. The overlap is due to the fact that $(R2_{(-0.5,0.5)})^{+}$ is not an equivalence relations because of the ⊕-× transitivity of R3.

EQUILIBRIUM ENERGY AND STABILITY FOR MULTIAGENT COORDINATION AND GLOBAL REGULATION

The concepts of equilibrium energy and stability of a bipolar lattice (Zhang, 2005a) can be naturally applied to equilibrium relations, local equilibrium, and equilibrium classes in cognitive mapping (Zhang, 2003a, 2003b).

Definition 5.9. Given an equilibrium or quasi-equilibrium relation $R = (R^{-},R^{+})$, the summation of all elements in R^{-}, denoted $\varepsilon^{-}(R)$ is the negative energy of R; the summation of all elements in R^{+}, denoted $\varepsilon^{+}(R)$, is the positive energy of R; the polarized total, denoted $\varepsilon(R) = (\varepsilon^{-}, \varepsilon^{+})$, is the bipolar energy equilibrium of R; the absolute total, denoted $|\varepsilon|(R) = |\varepsilon^{-}| + |\varepsilon^{+}|$, is the total energy of R. The sum $\varepsilon_{imb}(R) = \sum_{i=1}^{m}\sum_{j=1}^{n}(r_{ij}^{+} - |r_{ij}^{-}|)$ is the energy imbalance of R.

While energy values provide unnormalized measures for cooperation and competition among clusters, we now define normalized measures for different clusters or quasi-equilibrium relations in the following definitions:

Definition 5.10. Given $\varepsilon(R) = (\varepsilon^{-}, \varepsilon^{+})$ where R is a quasi-equilibrium relation, (1) if R is a relation in an equilibrium class E(c), the common interest of c with the others is measured as $c^{+} = \varepsilon^{+}/|\varepsilon|(R)$, and the conflict interest is measured as $c^{-} = 1 - c^{+} = |\varepsilon^{-}|/|\varepsilon|(R)$; (2) if R is a relation in a fuzzy coalition, the coalition strength is measured as $c^{+} = \varepsilon^{+}/|\varepsilon|(R)$, and the coalition weakness is measured as $c^{-} = 1 - c^{+} = |\varepsilon^{-}|/|\varepsilon|(R)$; (3) if R is a relation in a fuzzy conflict set, the conflict strength is measured as $c^{-} = |\varepsilon^{-}|/|\varepsilon|(R)$; (4) if R is a relation in a fuzzy harmony set, the harmony level is measured as $h^{\pm} = (\varepsilon^{-}/n^{2}, \varepsilon^{+}/n^{2})$, where n is the size of the harmony set.

Definition 5.11. Given an equilibrium relation or quasi-equilibrium R, the stability of R is defined as *Stability*$(R) = (|\varepsilon|(R) - |\varepsilon_{imb}(R)|)/|\varepsilon|(R)$.

Based on equilibrium energy and stability, the following commonsense laws can be derived for optimization in cognitive-map-based multiagent decision, coordination, and global regulation (Zhang, 2003a, 2003b).

Law 1. Reduce negative energy and increase positive energy among a coalition set for stronger coalition.
Law 2. Reduce negative energy in a conflict set for higher stability.
Law 3. Increase negative energy in a conflict set for more competition.
Law 4. Increase positive energy with neutral agents for stronger coalition.
Law 5. Decrease negative energy and increase positive energy with partially cooperative and partially competitive agents for a stronger coalition.

Law 6. Decrease negative energy and increase positive energy with cooperative and competitive agent sets for more cooperation and less competition.

Law 7. Cooperative agents sometimes need to sacrifice for collective goals.

Law 8. Competitive agent sets sometimes need to compromise for higher stability.

Law 9. Cooperation is needed for competition. Coordination is necessary for cooperation and competition.

Law 10. The equilibrium paths (see Figure 6) identify the critical links that are the most effective for increase or decrease the negative or positive energies among different equilibrium classes for global regulation and coordination.

RESEARCH TOPICS

While BDL and BDFL provide a logical basis for YinYang bipolar relativity, bipolar fuzzy sets and equilibrium relations provide a relational model for equilibrium analysis with a wide spectrum of applications.

Bipolar Cognitive Mapping. Equilibrium relations can be used in cognitive mapping for equilibrium analysis, decision, and multiagent coordination (Zhang, 2003a, 2003b, 2005a, 2005b, 2006) (Ch. 11).

Global Optimization and Regulation. Cognitive-map-based global optimization and regulation can be used in macroeconomics and international relations (Zhang, 2003a, 2003b, 2005b, 2006) (Ch. 11).

Bipolar Quantum Linear Algebra and Bipolar Quantum Cellular Automata. Bipolar fuzzy sets and equilibrium relations provide a relational basis for bipolar quantum linear algebra and bipolar quantum cellular automata that are further discussed in Chapter 8 (Zhang, 1996; Zhang & Chen, 2009; Zhang *et al.,* 2009).

Biosystem Simulation and Regulation. Bipolar sets and equilibrium relations provide a relational basis for biosystem simulation and regulation at the system, subsystem, genomic, and molecular levels that are further discussed in Chapter 9 (Zhang & Chen, 2009; Zhang *et al.,* 2009).

Bipolar quantum Gravity. Bipolar equilibrium relations provide a bipolar relational approach to quantum symmetry, bipolar relativity, and quantum gravity (Zhang, 2009a, 2009b, 2009c, 2009d) (Ch. 6).

SUMMARY

Bipolar fuzzy sets and fuzzy equilibrium relations have been presented for bipolar clustering, optimization, and global regulation with illustrations in multiagent macroeconomics. Bipolar fuzzy set theory provides a non-linear bipolar generalization of unipolar fuzzy sets by adding polarity, an extension of bipolar crisp sets by adding fuzziness, and a generalization of unipolar crisp sets by adding both polarity and fuzziness. Three types of bipolar α-cuts have been proposed for bipolar defuzzification and depolarization. Fuzzy equilibrium relations have been introduced. It is shown that a fuzzy equilibrium relation is a non-linear bipolar fusion of equivalence and/or similarity relations (Zadeh, 1971) with bipolar quantum entanglement that lead to bipolar fuzzy clustering.

The bipolar fuzzy space B_F is a fusion of fuzziness from fuzzy logic and bipolarity from BDL, respectively. Similarity relations and equilibrium relations both inherit reflexivity, symmetry, and transitivity from equivalence relations. Using a multiple inheritance hierarchy as in object-oriented languages, the inheritance relationship is shown in Figure 11.

It should be remarked that, the inheritance relationship in object-oriented language is a logical but not cosmological or biological inheritance. Physically, it is so far not clear whether bipolarity came first or singularity came first, equilibrium came first or the universe came first. Only one thing is sure, we say that the universe is in a kind of equilibrium, quasi- or non-equilibrium but not vice versa.

From a YinYang perspective, everything has two sides. The two poles of nature are the two sides of bipolar dynamic logic (BDL); 0 and 1 are the two sides of Boolean logic. Thus, fuzzy set theory is a fuzzification or generalization of classical set theory; bipolar crisp set theory is a bipolar generalization of classical set theory; bipolar fuzzy set theory is a fuzzification or generalization of bipolar crisp set theory (Figure 11).

Finally, it should be remarked that some researchers in the fuzzy set community attempted to redefine intuitionistic fuzzy sets (Atanassov, 1986) as bipolar fuzzy sets. This effort is misguided and misleading because: (1) bipolar sets and fuzzy sets are originated from the ancient Chinese YinYang philosophy where two opposite poles can form a bipolar equilibrium; (2) intuitionism (Brouwer, 1912) as a truth-based extension to Aristotelian logic has never been intended to be bipolar equilibrium-based; (3) While *"Intuitionistic logic can be succinctly described as classical logic without the Aristotelian law of excluded middle (LEM): ($A \vee \neg A$), but with the law of contradiction ($\neg A \rightarrow (A \rightarrow B)$)"* (Moschovakis, 2010), the 4-valued YinYang bipolar dynamic logic (BDL) satisfies the law of excluded middle and is free from contradiction.

ACKNOWLEDGMENT

This chapter reuses part of published material in Zhang, W. –R. (2006) "YinYang Bipolar Fuzzy Sets and Fuzzy Equilibrium Relations for Bipolar Clustering, Optimization, and Global Regulation", *World Scientific Publishing (WSP): International Journal of Information Technology and Decision Making (IJITDM), Vol. 5 No. 1: 19-46,* March 2006. Permission to reuse is acknowledged.

REFERENCES

Atanassov, K. T. (1986). Intuitionistic fuzzy sets. *Fuzzy Sets and Systems*, *20*, 87–96. doi:10.1016/S0165-0114(86)80034-3

Boole, G. (1854). *An investigation of the laws of thoughts*. London: MacMillan.

Brouwer, L. E. (1912). Intuitionism and formalism. *Bulletin of the American Mathematical Society*, *20*, 81–96. doi:10.1090/S0002-9904-1913-02440-6

Cossette, P., & Lapointe, A. (1997). A mapping approach to conceptual models: The case of macroeconomic theory. *Canadian Journal of Administrative Sciences*, *14*(1), 41–51. doi:10.1111/j.1936-4490.1997.tb00117.x

Goguen, J. (1967). L-fuzzy sets. *Journal of Mathematical Analysis and Applications*, *18*, 145–174. doi:10.1016/0022-247X(67)90189-8

Moschovakis, J. (2010). Intuitionistic logic. *Stanford Encyclopedia of Philosophy*. Retrieved from http://plato.stanford.edu/entries/logic-intuitionistic/

Warshall, S. (1962). A theorem on Boolean matrices. *Journal of the ACM, 9*(1), 11–12. doi:10.1145/321105.321107

Zadeh, L. A. (1965). Fuzzy sets. *Information and Control, 8*, 338–353. doi:10.1016/S0019-9958(65)90241-X

Zadeh, L. A. (1971). Similarity relations and fuzzy orderings. *Information Sciences, 3*, 177–200. doi:10.1016/S0020-0255(71)80005-1

Zadeh, L. A. (2006). Fuzzy logic. *Scholarpedia, 3*(3), 1766. doi:10.4249/scholarpedia.1766

Zadeh, L. A. (2007). *From fuzzy logic to extended fuzzy logic—the concept of f-validity and the impossibility principle*. FUZZ-IEEE Imperial College, London, UK. Retrieved from www.fuzzieee2007.org/ZadehFUZZ-IEEE2007London.pdf

Zadeh, L. A. (2008). Toward human-level machine intelligence–is it achievable? WSEAS AIKED'08, WSEAS SEPADS'08, University of Cambridge, UK. Retrieved from www.wseas.org/wseas-zadeh-2008.pdf

Zhang, W.-R. (1994). Bipolar fuzzy sets and relations: A computational framework for cognitive modeling and multi-agent decision analysis. *Proceedings of 1st International Joint Conference of the North American Fuzzy Information Processing Society,* (pp. 305 – 309).

Zhang, W.-R. (1996). NPN fuzzy sets and NPN qualitative algebra: A computational framework for bipolar cognitive modeling and multi-agent decision analysis. *IEEE Transactions on SMC, 16*, 561–574.

Zhang, W.-R. (1998). YinYang bipolar fuzzy sets. *Proceedings of IEEE World Congress on Computational Intelligence (Fuzz-IEEE)*, (pp. 835-840). Anchorage, AK.

Zhang, W.-R. (2003a). Equilibrium relations and bipolar cognitive mapping for online analytical processing with applications in international relations and strategic decision support. *IEEE Transactions on SMC . Part B, 33*(2), 295–307.

Zhang, W.-R. (2003b). Equilibrium energy and stability measures for bipolar decision and global regulation. *International Journal of Fuzzy Systems, 5*(2), 114–122.

Zhang, W.-R. (2005a). YinYang bipolar lattices and L-sets for bipolar knowledge fusion, visualization, and decision making. *International Journal of Information Technology and Decision Making, 4*(4), 621–645. doi:10.1142/S0219622005001763

Zhang, W.-R. (2005b). YinYang bipolar cognition and bipolar cognitive mapping. *International Journal of Computational Cognition, 3*(3), 53–65.

Zhang, W.-R. (2006). YinYang bipolar fuzzy sets and fuzzy equilibrium relations for bipolar clustering, optimization, and global regulation. *International Journal of Information Technology and Decision Making, 5*(1), 19–46. doi:10.1142/S0219622006001885

Zhang, W.-R. (2009a). Six conjectures in quantum physics and computational neuroscience. *Proceedings of 3rd International Conference on Quantum, Nano and Micro Technologies (ICQNM 2009)*, (pp. 67-72). Cancun, Mexico.

Zhang, W.-R. (2009b). YinYang Bipolar Dynamic Logic (BDL) and equilibrium-based computational neuroscience. *Proceedings of International Joint Conference on Neural Networks (IJCNN 2009)*, (pp. 3534-3541). Atlanta, GA.

Zhang, W.-R. (2009c). YinYang bipolar relativity–a unifying theory of nature, agents, and life science. *Proceedings of International Joint Conference on Bioinformatics, Systems Biology and Intelligent Computing (IJCBS)*. (pp. 377-383). Shanghai, China.

Zhang, W.-R. (2009d). The logic of YinYang and the science of TCM–an Eastern road to the unification of nature, agents, and medicine. [IJFIPM]. *International Journal of Functional Informatics and Personal Medicine*, *2*(3), 261–291. doi:10.1504/IJFIPM.2009.030827

Zhang, W.-R., Chen, S., & Bezdek, J. C. (1989). POOL2: A generic system for cognitive map development and decision analysis. *IEEE Transactions on SMC*, *19*(1), 31–39.

Zhang, W.-R., Chen, S., Wang, W., & King, R. (1992). A cognitive map based approach to the coordination of distributed cooperative agents. *IEEE Transactions on SMC*, *22*(1), 103–114.

Zhang, W.-R., & Chen, S. S. (2009). Equilibrium and non-equilibrium modeling of YinYang Wuxing for diagnostic decision analysis in traditional Chinese medicine. *International Journal of Information Technology and Decision Making*, *8*(3), 529–548. doi:10.1142/S0219622009003521

Zhang, W.-R., Zhang, H. J., Shi, Y., & Chen, S. S. (2009). Bipolar linear algebra and YinYang-n-element cellular networks for equilibrium-based biosystem simulation and regulation. *Journal of Biological System*, *17*(4), 547–576. doi:10.1142/S0218339009002958

Zhang, W.-R., & Zhang, L. (2004). YinYang bipolar logic and bipolar fuzzy logic. *Information Sciences*, *165*(3-4), 265–287. doi:10.1016/j.ins.2003.05.010

ADDITIONAL READING

Any textbook on discrete mathematics.

Any textbook on matrix operations or linear algebra.

KEY TERMS AND DEFINITIONS

Bipolar Clustering: Clustering a set of bipolar fuzzy elements or agents into disjoint or overlapped subsets based on bipolar fuzzy equilibrium relations of the set.

Bipolar Fuzzy Relation: A binary relation characterized by bipolar fuzzy logic (BDFL) values (Zhang & Zhang, 2004).

Bipolar Partitioning: Partitioning a set of bipolar crisp elements or agents into disjoint subsets based on bipolar equilibrium relations of the set.

Equilibrium Energy: Numerical characterization of energy embedded in an equilibrium relation.

Equilibrium-Based Multiagent Coordination and Global Regulation: Coordination and global regulation of distributed agents with equilibrium and non-equilibrium measures of bipolar relations.

Equivalence Relation: In mathematics, an **equivalence relation** is, loosely, a relation that specifies how to partition a set such that every element of the set is in exactly one of the partitions. Two elements of the set are considered equivalent (with respect to the equivalence relation) if and only if they are elements of the same partition. A given binary relation R on a set A is an equivalence relation if and only if it is reflexive, symmetric and transitive.

Fuzzy Equilibrium Relation: A given bipolar fuzzy relation R on a set A is a fuzzy equilibrium relation if and only if it is bipolar reflexive, symmetric and bipolar ($\oplus$-$\otimes$) transitive or interactive.

Fuzzy Relation: A binary relation characterized by fuzzy logic values (Zadeh, 1965).

Relation (Binary): In mathematics, a **binary relation on** a set A is a collection of ordered pairs of elements of A. In other words, it is a subset of the Cartesian product $A^2 = A \times A$. More generally, a binary relation between two sets A and B is a subset of $A \times B$. The terms dyadic relation and 2-place relation are synonyms for binary relations.

Similarity Relation: A given fuzzy relation R on a set A is a similarity relation if and only if it is reflexive, symmetric and transitive (Zadeh, 1971).

Part 3
YinYang Bipolar Relativity and Bipolar Quantum Computing

Chapter 6
Agents, Causality, and YinYang Bipolar Relativity

ABSTRACT

This chapter presents the theory of bipolar relativity–a central theme of this book. The concepts of YinYang bipolar agents, bipolar adaptivity, bipolar causality, bipolar strings, bipolar geometry, and bipolar relativity are logically defined. The unifying property of bipolar relativity is examined. Space and time emergence from YinYang bipolar geometry is proposed. Bipolar relativity provides a number of predictions. Some of them are domain dependent and some are domain independent. In particular, it is conjectured that spacetime relativity, singularity, gravitation, electromagnetism, quantum mechanics, bioinformatics, neurodynamics, and socioeconomics are different phenomena of YinYang bipolar relativity; microscopic and macroscopic agent interactions in physics, socioeconomics, and life science are directly or indirectly caused by bipolar causality and regulated by bipolar relativity; all physical, social, mental, and biological action-reaction forces are fundamentally different forms of bipolar quantum entanglement in large or small scales; gravity is not necessarily limited by the speed of light; graviton does not necessarily exist.

INTRODUCTION

It is postulated in Chapter 1 that the most fundamental property of the universe is YinYang bipolarity. It is pointed out that YinYang was hinted by Einstein in his own theory of relativity. Einstein focused on gravity in spacetime geometry but left open a *YinYang bipolar relativity* theory. Chapter 2 introduces equilibrium-based YinYang bipolarity as a key for logically definable causality. With BDL and BDFL as discussed in Chapters 3 – 5, the Yin and the Yang hinted by Einstein can be logically characterized as bipolar symmetrical variables and systematically processed.

The previous chapters have prompted us to seek answers for the following questions:

1. If bipolarity is the most fundamental property of the universe, should nature's basic forces (which are all bipolar) be also the direct or indirect governing forces for all agents and agent interactions in physics, socioeconomics, cognitive informatics and life sciences?
2. Should all agents in the universe including the universe itself be manifestations of YinYang bipolar relativity?
3. If YinYang bipolar relativity constitutes a logical foundation for physics, how is it related to spacetime relativity, quantum theory, socioeconomics, cognitive informatics, and life sciences?
4. How can agents, agent interactions and adaptivity be defined in the most fundamental way such that agent adaptivity and interactions lead to logically definable causality and bipolar relativity that extends our current understanding of the universe beyond spacetime?
5. How can agents, agent interactions and adaptivity be defined in the most fundamental way such that the ubiquitous effects of quantum entanglement will make sense and will confront us at every turn in our daily lives in simple comprehendible terms?
6. Could all types of action-reaction be fundamentally different forms of bipolar quantum entanglement in large or small scales?
7. Could the speed of gravity exceed the speed of light?

In this chapter we formally present the theory of YinYang bipolar relativity as a logical axiomatization of physics and a partial solution to Hilbert's Problem 6 (Hilbert, 1901). It is postulated that all agents are bipolar in nature; agents can be adaptive to equilibrium; space and time can emerge in *YinYang bipolar geometry* following arrivals of agents. Predictions of the theory are presented as 21 conjectures spanning physical, social, environmental, cognitive, and life sciences. Some of the conjectures are domain dependent and some of them are domain independent. In particular, it is conjectured that spacetime relativity, singularity, gravity, electromagnetism, quantum mechanics, bioinformatics, and socioeconomics are different phenomena of bipolar relativity; microscopic and macroscopic agent interactions are directly or indirectly caused and regulated by bipolar relativity; gravity is not necessarily limited by the speed of light.

The remaining presentations and discussions of this chapter are organized in the following sections:

- **Agents.** This section presents a review on the concepts of agents in AI, social and physical sciences.
- **Bipolar Agents**. This section presents logical definitions of *bipolar agent*, *bipolar adaptivity*, logical and physical bipolar interaction with a discussion on the observability and generality of bipolar agents.
- **Bipolar Causality and Bipolar Causal Reasoning.** This section presents equilibrium-based logically definable bipolar causality.
- **YinYang Bipolar Relativity**. This section presents the central theme of this book. *Background independence* is introduced. An observable *bipolar string* theory is presented as a component of bipolar relativity. *YinYang bipolar geometry* is introduced and examined. It is shown that, with definable causality, the *emergence of space and time* naturally follows the arrivals of agents in YinYang bipolar geometry. Time reversal is briefly discussed. It is shown that scalable bipolar strings play a unifying role for singularity, bipolarity, and particle-wave duality. Predictions from the theory of bipolar relativity are presented as 21 conjectures. The falsifiability of the predictions is briefly discussed. A number of salient fundamental geometric features of bipolar relativity and its difference from spacetime relativity are examined. It is concluded that bipolar relativity forms

a deeper theory beyond spacetime geometry, which can play a regulatory role of all agents and agent interactions in spacetime including the universe itself.

- **Axiomatization of Physics**. It is concluded in this section that bipolar dynamic logic and bipolar relativity form a minimal but most general *equilibrium-based logical axiomatization of physics* in both macroscopic and microscopic scales including quantum scales.
- **Research Topics**. This section lists a few research topics.
- **Summary.** This section summarizes the major points of this chapter and draws a few conclusions.

AGENTS

Philosophically, an ***agent*** is an entity that can act or exert power to produce an effect either intentionally or unintentionally. The agent concept is multifaceted. We classify agents roughly into the following informal categories:

- A most popular view of an agent is the representative view. If you hire a lawyer, the lawyer becomes your representative or agent. In an espionage movie, the major player must be a ***secret agent*** of certain agency or sometimes a free agent.
- If you are engaged in AI research you are aware of the concepts of ***intelligent agent***, ***autonomous agent***, and ***multiagent systems.*** Such agents can be intelligent software running on computers or roaming the World Wide Web (WWW) performing some user designated tasks such as online shopping. If the intelligent agents are autonomous robots, they may perform tasks in people's home or roam their designated territories in the air, on land, or on the ocean floor.
- If you are majoring in physical sciences you are interested in chemical, biological, or physical agents at the molecular, genetic, particle, or quantum level. These agents are the major players of the microscopic worlds.
- In information science, researchers are concerned with the collection, classification, manipulation, storage, retrieval and dissemination of information about certain agents, their interaction, and organization.
- In cognitive information science researchers investigate the natural intelligence and internal information processing mechanisms of the brain as well as the processes involved in perception and cognition in humans or animals.

Despite the tremendous efforts in scientific research on agents, we still don't know how our brain works exactly, how large the largest agent – the universe is, and how small the smallest agent, the most fundamental subatomic particle, is. As a matter of fact we still have not found a unifying mathematical definition for the word *"agent"* that is fundamental for all the aforementioned agents and their interactions. Even though string theory was considered *"theory of everything"*, it is not observable, not experimentally testable, and has so far failed to provide falsifiable predictions (Smolin, 2006; Woit, 2006). Therefore, we don't know whether strings really exist or whether there is actually any smallest fundamental agent at all. It is legitimate to ask: *If we don't know how large the universe is, how could we know how small the most fundamental subatomic particle can be?*

Scientifically speaking, observers and the observed are all agents. The failure of finding a unifying mathematical definition for the concept of agent is mainly due to two limitations: (1) the limited ability of observers to observe and (2) the limited observability of the observed. As a result, all the current agent definitions have failed to provide a basic theory of agent interaction that can lead to logically definable causality for the unification of social and physical sciences as well as general relativity and quantum mechanics.

Just like water and air were considered as basic elements of nature in different ancient civilizations of the East and the West, modern concepts of agents are also influenced by socioeconomical factors besides pure scientific curiosity. That is understandable because all human agents need to consume water and air as well as to earn a living in a socioeconomical environment.

Can scientific research on agents and agent interactions ever be free from socioeconomics? It seems to be difficult if not impossible. Firstly, agents in socioeconomics are the major players on the Earth and scientific researchers themselves cannot escape from socioeconomics. Secondly, any scientific agent theory that is applicable in physical and life sciences can be borrowed for socioeconomics because physical and social sciences share similar equilibrium-based dynamics. After all, a unifying equilibrium-based scientific agent theory is supposed to be applicable in physics, socioeconomics, and life sciences that, after thousands of years, are still directly related to water, air, and the very existence of human societies due to the threatening factor of global warming.

Then, what kind of agent is the most fundamental agent? Without being able to find the smallest generic agent for a *"theory of everything"*, an open-world and open-ended approach seems to be the best choice. In such an approach, we don't have to assume any smallest fundamental agent. Instead, we hypothesize the most fundamental property of the universe based on well-established observation such as (matter, antimatter) particles and then we develop a logical construction with predictions for verification or falsification. This approach actually follows the principle of exploratory scientific knowledge discovery.

Mainstream science is based on *singularity* in which time starts with the so-called *big bang* and ends with a *black hole* (Hawking & Penrose, 1970). Two pitfalls of singularity are gravitational infinity and quantum infinity. One happens to the equations of general relativity at the starting and ending points; another happens in quantum theory in spacetime (Smolin, 2006) (Ch. 2). Another pitfall is the number of ending points of the universe. One big bang from nowhere eventually leading to many black holes would imply one starting point and many ending points for spacetime unless all black holes are synchronized. Otherwise time stops at different spacetime points for different galaxies and the dead galaxies remain part of the living universe until the universe ends completely.

Furthermore, the theory of singularity is incompatible with the bipolar property of nature's basic forces and particles. For any force to exist there must be one agent that exerts force on another agent. That is why electromagnetic force is denoted as (-q,+q) and gravitational force is denoted as (-f,+f) in modern physics textbooks. A big scientific question is then *"Without bipolarity could we have causality and singularity or even spacetime?"*

BIPOLAR AGENTS

Definitions and Axioms

Definition 6.1. An ***agent*** is any physical entity identified symbolically or mathematically.

By definition 6.1, an agent can be physical or mathematical. For instance, an electron e^- or a positron e^+ is an agent because it is an existing entity; the negative number -1 for e^- or +1 for e^+ is an agent if -1 or +1 is used as a mathematical identity for e^- or e^+, respectively. Since any identifiable physical entity must exists in a certain form of bipolar equilibrium or non-equilibrium, one way or another, we have

Definition 6.2a. An agent *a* ***exists*** iff $\exists\varphi$ $\{\varphi(a)\in B_1$ *and* $\varphi(a) \neq (0,0)\}$, *where* φ is a bipolar predicate and $\varphi(a)\in B_1$ ensures $\varphi(a){:}a \Rightarrow B_1$ (Ch. 3) or φ *maps a to the bipolar lattice* B_1.

Definition 6.2b. A fuzzy agent *a* ***exists*** iff $\exists\varphi$ $\{\varphi(a) \in B_F$ *and* $\varphi(a) \neq (0,0)\}$, *where* φ is a bipolar predicate and $\varphi(a)\in B_F$ ensures $\varphi(a){:}a \Rightarrow B_F$ (Ch. 4) or φ *maps a to the bipolar fuzzy lattice* B_F.

Definition 6.2c. A lattice-ordered agent *a* ***exists*** iff $\exists\varphi$ $\{\varphi(a)\in B$ *and* $\varphi(a) \neq (0,0)\}$, *where* φ is a bipolar predicate and $\varphi(a)\in B$ ensures $\varphi(a){:}a \Rightarrow B$ or φ *maps a to any bipolar quantum lattice B* (Ch. 4).

Noticing that any strict bipolar lattice B can be normalized to B_1 or B_F (Ch. 4), without losing generality, we assume $\forall a$, we have $\varphi(a){:}a \Rightarrow B_1$ or B_F. Following Definition 6.2, let $\varphi = (-,+)$ be a bipolar predicate of BDL, ***e*** be an electron, positron, or a bipolar binding or coupling (e^-,e^+), ***e*** exists if $\varphi(e)=(-1,0)$, $(0,+1)$, *or* $(-1,+1)$;***e*** doesn't exist if $\varphi(e)=(0,0)$. If we consider a TV set as an agent, we can have φ = *(reaction, action) such that* φ *(TV)* = $(0,+1)$ that indicates the TV set exerts an action force on its supporting frame or table. Then we have φ *(Table)* = $(-1,0)$; φ *(Table,TV)* = $(-1,+1)$. (Note: we may omit the positive sign for +1 such that +1=1.)

Definition 6.3. A ***primitive agent*** is an agent in a universe of discourse for which there is only a single bipolar equilibrium or non-equilibrium φ under consideration such that $\{\varphi(A) \in B_1$ or B_F & $\varphi(A) \neq (0,0)\}$. Otherwise, an agent is a ***non-primitive agent***.

Following Definition 6.3, we have the postulate:

Postulate 6.1. Any non-primitive agent in a universe of discourse is a multidimensional equilibrium or non-equilibrium, where the number of dimensions ≥ 2.

With Postulate 6.1, we can say that the universe is a multidimensional equilibrium or non-equilibrium because we have gravitational force (-f,+f), electromagnetic force (-q,+q), and particle-antiparticle (-p,+p) symmetry, quasi-symmetry or non-symmetry. With this rationale, the four dimensions of spacetime in general relativity, namely, length, width, height, and time, can also be considered dimensions. Although each of these four dimensions has a negative side and a positive side such as (-x,x) and (past,future), the negative and positive sides of these dimensions are not quantum in nature, not bipolar interactive and, therefore, do not form bipolar equilibria and cannot lead to quantum entanglement. Thus, they may be considered non-interactive bipolar dimensions and less fundamental concepts than bipolarity. That could be the fundamental reason why the supposedly ubiquitous effects of quantum entanglement still do not show up and make sense in our daily life.

On the other hand, it is well known that all opposites are complementary. Without aging there would be no growing; without death there would be no life; with out input there would be no output; and without contraction there would be no expansion. Since any agent must exist in at least one bipolar equilibrium or non-equilibrium state, we have the next postulate:

Postulate 6.2. All agents in the universe are bipolar.

Postulate 6.2 opens an equilibrium-based paradigm of agent interaction in logical, physical, social, mental, and biological terms. Technically, we humans are all bipolar in nature. Physically, we must consume energy in order to give out energy. Biologically, it is a well-known fact that our YinYang 1 genetic regulator protein (Shi *et al.,* 1991) regulates our gene expression with bipolar repression/activation to achieve certain equilibrium or non-equilibrium. Mentally, most of us are in bipolar mental equilibrium with reciprocal self-negation and self-assertion abilities denoted by a bipolar predicate (self-negation, self-assertion) (Zhang, Pandurangi & Peace, 2007). While some get bipolar disorder due to the loss of such abilities, most of us in mental equilibrium are *more or less "depressed" or "manic"* because no one has perfect self-negation and self-assertion abilities. That is why we need bipolar fuzzy sets.

For instances, (-0.9 +0.8) can characterize a quasi-equilibrium with slight "depression"; (-0.8 +0.9) can characterize a quasi-equilibrium with slight "mania". Such slight "depression" or "mania" are not really clinical psychiatric symptoms. These are characteristics of a person's emotional stability. Based on such characteristics, most people can be generally and approximately classified into two categories: Yang deficiency or pessimistic and Yin deficiency or optimistic, respectively.

Someone may argue that if "I am in bipolar equilibrium" how can "I am bound to be true"? Actually, "I am bound to be true" really meant to be "I believe I am on the right side of the history" or "I believe my cause will be justified." It has nothing contrary to "I am in bipolar equilibrium."

While an electron $(e^-,0)$ or positron $(0, e^+)$ or the coupling (e^-, e^+) can be considered a ***primitive agent*** in certain cases, a person, similar to the universe, isn't primitive because there are other bipolar equilibria involved besides (self-negation, self-assertion) such as (reaction, action). Evidently, a multidimensional equilibrium can be deconstructed into a number of bipolar equilibria (see Ch. 1, Figure 1).

Definition 6.4a. An agent A is **bipolar adaptive** iff A has either self-negation ability or self-assertion ability or both and A can regain at least one type of bipolar equilibrium state. Formally, let φ = *(self-negation, self-assertion)* $\in B_1$ and time $t = t_0, t_1, t_2$, A is adaptive iff $\varphi(A) \neq (0,0)$ and A exhibits both bipolar fusion (e.g. $*_1 = \oplus$) and interaction (e.g. $*_2 = \otimes$) functionalities such that,

(1) if $\varphi(A(t_0))=(-1,0)$, A exhibits self-adaptivity:

$$\varphi(A(t_1))=\varphi(A(t_0)) *_1[\varphi(A(t_0)) *_2\varphi(A(t_0))]=(-1,1); \qquad (6.1)$$

(2) if $\varphi(A)=(0,1)$, A exhibits assisted adaptivity with external input (-1,0):

$$\varphi(A(t_1)) *_1[\varphi(A(t_0)) *_2(-1,0)]=(-1,1). \qquad (6.2)$$

Definition 6.4b. A fuzzy agent A is **bipolar adaptive** iff A has either self-negation ability or self-assertion ability or both in any degree and A can regain at least one type of bipolar equilibrium state to certain degree. Formally, let φ=*(self-negation, self-assertion)* $\in B_F$ and time $t = t_0, t_1, t_2$, A is adaptive iff $\varphi(A) \neq (0,0)$ and A exhibits both bipolar fusion (e.g. $*_1 = \oplus_{ij}$) and interaction (e.g. $*_2 = \otimes_{ij}$) functionalities such that:

(1) if $\varphi(A(t_0))=(x,0)$, $-1\leq x<0$, A exhibits self-adaptivity:

$$\varphi(A(t_1))=\varphi(A(t_0)) *_1[\varphi (A(t_0)) *_2\varphi (A(t_0))]=(x,y),\ y>0; \tag{6.3}$$

(2) if $\varphi(A(t_1))=(0,y)$, A exhibits assisted adaptivity with external input (-1,0):

$$\varphi(A(t_1))=\varphi(A(t_0)) *_1[\varphi (A(t_0)) *_2\varphi (A(t_0))]=(-x,y),\ x>0; \tag{6.4}$$

Eq. 6.1 and Eq. 6.3 can be instantiated with Eq. 6.5 in logical form; Eq. 6.2 and Eq. 6.4 can be instantiated with Eq. 6.6 in logical form. (Note: It is shown in Chapter 8 that collective bipolar adaptivity can also be realized with bipolar quantum linear algebra and bipolar quantum cellular automata.)

$$\varphi(A(t_1)) = \varphi(A(t_0))\oplus [\varphi (A(t_0))\otimes\varphi (A(t_0))]$$

$$= (-1,0) \oplus [(-1,0) \otimes (-1,0)]$$

$$= (-1,0) \oplus (0,1)$$

$$= (-1,1); \tag{6.5}$$

$$\varphi(A(t_1)) = \varphi(A(t_0))\oplus [\varphi (A(t_0))\otimes(-1,0)]$$

$$= (0,1) \oplus [(0,1) \otimes (-1,0)]$$

$$= (-1,0) \oplus (-1,0)$$

$$= (-1,1); \tag{6.6}$$

Eq. 6.5 typically characterizes a depressed person who may lose self-assertion ability temporarily but is self-adaptive to mental equilibrium as long as whose bipolar fusion (⊕) and bipolar oscillation (⊗) neurobiological functionalities are intact. Thus, clinical mental depression could be diagnosed as the loss of bipolar fusion (⊕) or bipolar interaction (⊗) *functionalities. This interpretation is also supported by a* Traditional Chinese Medicine principle which states Yin (-) produces Yang (+) but not vice versa.

Eq. 6.6 typically characterizes a manic person who may lose self-negation ability temporarily but is able to recover to mental equilibrium with a medical intervention characterized with (-1,0). The recovery is possible because the patient's bipolar fusion (⊕) and bipolar interaction (⊗) functionalities responded to the medical intervention. Thus, clinical mental mania could also be diagnosed as the loss of bipolar fusion (⊕) or bipolar interaction (⊗) functionalities.

Definition 6.5a. A ***logical agent*** is an agent whose behavior is governed by Boolean logic; a ***bipolar agent*** is an agent whose behavior is governed by BDL.

Definition 6.5b. A ***fuzzy logical agent*** is an agent whose behavior is governed by unipolar fuzzy logic; a ***bipolar fuzzy agent*** is an agent whose behavior is governed by BDFL.

The above definitions lead to another postulate:

Postulate 6.3. A logical agent is not bipolar adaptive to equilibrium; a bipolar adaptive agent is not strictly logical.

Postulate 6.3 may sound ironic but it is actually long overlooked observable reality. For instance, a digital computer is strictly logical but by no means adaptive to mental equilibrium even being programmed because it does not have a biological mind. On the other hand, a normal person is mentally and biologically adaptive to mental equilibrium but can by no means be strictly logical like a digital computer. Thus, BDL can avoid the LAFIP paradox but Boolean logic cannot. Thus, Postulates 6.1 – 6.3 form an axiomatic basis for bipolar agent interaction and axiomatization in a fundamental way.

Definition 6.6. Two bipolar sets X and Y are **physically bipolar interactive** if, $\forall x \in X$ and $\forall y \in Y$, x and y are physically bipolar interactive denoted x♦y. Two bipolar predicates $\varphi(x) \in B_1$ *or* B_F *and* $\phi(y) \in B_1$ *or* B_F are **logically bipolar interactive** denoted $\varphi(x)*\phi(y)$ iff x and y are either physically bipolar interactive (x♦y) or logically bipolar interactive (x*y).

Definition 6.6 establishes the physical basis for bipolar equilibrium-based logical reasoning. For instance, a set of bipolar disorder patients X and a set of bipolar disorder medicines Y can be physically bipolar interactive such that two functions $\varphi(x)$ and $\phi(y)$ are logically bipolar interactive. On the other hand, two sets of bipolar disorder patients X and Y can't be physically bipolar interactive and $\varphi(x)$ and $\phi(y)$ can't be logically bipolar interactive either. Since bipolar variables or functions can form sets X and Y themselves, the recursive nature of Definition 6.6 is necessary such that if $a = \varphi(x)$ and $b = \phi(y)$ the logical interactivity of a and b depends on that of $\varphi(x)$ and $\phi(y)$. In the meantime, physical interactivity can be a default in hypothesis-driven reasoning.

Theorem 6.1. The following hold on physical and logical interactivities:

$\forall x,y,\ [(\varphi^-,\varphi^+)(x)*(\phi^-,\phi^+)(y)] \rightarrow [(x♦y)\vee(x*y)]$; // unipolar logical implication
$\forall x,y$, if x♦y, then $[(\phi^-,\phi^+)(x)*(\phi^-,\phi^+)(y)] \Rightarrow (\phi^-,\phi^+)(x♦y)$; // bipolar implication
$\forall x,y$, if x*y, then $[(\phi^-,\phi^+)(x)*(\phi^-,\phi^+)(y)] \Rightarrow (\phi^-,\phi^+)(x*y)$; // bipolar implication
$\forall x,y \in B_1$ or B_{F},, $(x♦y) \equiv (x*y)$. // logical-physical equivalence

Proof. It follows from Definition 6.6. □

Bipolar physical interaction can be chemical, biological, organizational, or any type. Physical interaction may lead to ***scalability*** with composition or decomposition. It is noteworthy to distinguish bipolar agent interactions from those in multiagent systems (Huhns & Shehory, 2007; Singh & Huhns, 2005; Huhns & Singh, 1997; Huhns, 1987, 2000; Zhang, 1998; Zhang & Zhang, 2004a). The goal of bipolar agent interaction is for logically definable causality that is fundamental in the unification of general relativity and quantum mechanics with application in quantum computing, socioeconomics, and life sciences especially in the basic understanding of the universe.

Observability of Bipolar Agents

It is easy to say that we are all bipolar: in equilibrium or in disorder. The observability of bipolarity, however, has been a long lasting dispute between the Eastern and the Western philosophies. Even though YinYang equilibrium has been the theoretical basis for Traditional Chinese Medicine (TCM) for thousands of years and thermodynamic equilibrium is an undisputed central concept in modern science, equilibrium, as holistic truth, has so far been considered not observable by mainstream science. The dispute is partially due to its lack of formal definition.

As discussed in Chapter 1, every scientist would agree that without bipolar oscillation and interaction there would be no memory scan, no brain, and no equilibrium; without a certain form of dynamic equilibrium or quasi-equilibrium any agent or universe would be in total disorder or chaos and we would have no observables, no observers, and no universe. Similarly, without mental equilibrium mental disorder would be "big bang" from nowhere. Therefore, the concept of equilibrium is unavoidable in agent interaction at the system, molecular, genomic, particle, or quantum level. In addition to nature's basic forces (-q, +q) and (-f,+f) in modern physics textbooks, the bipolar fusion of self-negation and self-assertion abilities denoted (self-negation, self-assertion) in neurobiology and mental health (Zhang, 2007, 2009b; Zhang, Pandurangi & Peace, 2007), the (antioxidation, oxidation) pairs in biochemistry and traditional Chinese medicine (TCM) (Ou *et al.,* 2003), repression-activation abilities denoted (repression, activation) (Vasudevan, Tong & Steitz, 2007) of a YinYang genetic regulator such as YinYang1 (YY1) (Shi *et al.,* 1991; Park & Atchison, 1991) that can act as a transcriptional activator or repressor (Ai, Narahari & Roman, 2000; Kim, Faulk & Kim, 2007; Palko *et al.,* 2004; Zhou & Yik, 2006; Santiago *et al.,* 2007; Liu *et al.,* 2007; Wilkinson *et al.,* 2006), and the subatomic particle, the B-sub-s meson, discovered at the Fermi National Accelerator Laboratory (Fermilab, 2006) that can switch between matter and antimatter three trillion times a second show typical bipolar equilibrium/non-equilibrium properties. Furthermore, it is becoming scientifically evident that brain bioelectromagnetic field is crucial for neurodynamics and different mental states (Carey, 2007) where bipolarity is unavoidable.

The above well observed bipolar equilibria or non-equilibria form the scientific basis of bipolar agents. After all, equilibrium or non-equilibrium is essential in thermodynamics – the most fundamental concept of physical science; as the most fundamental example of the materialization of energy electron-positron (e^-,e^+) pair production predicted in Einstein's special relativity has been accurately described by quantum electrodynamics (QED) – *"the jewel of physics"* (Dirac, 1927, 1928) (Feynman, 1962, 1985).

While the observability of bipolar agents establishes the scientific basis of bipolar relativity, their generality is a key for bipolar agents to be scalable fundamental elements in nature. Such generality has been established with Definitions 6.1-6.6 and Postulates 6.1-6.3. With observability and generality, bipolar causality, bipolar relativity, and falsifiable predictions are within reach and equilibrium-based exploratory scientific knowledge discovery is made possible.

BIPOLAR CAUSALITY AND BIPOLAR CAUSAL REASONING

Equilibrium-Based Logically Definable Causality

As discussed in Chapter 2, Aristotle's causality principle has been considered the doctrine all sciences but not free of controversy. It is not particularly useful in a quantum world because, while we know for sure a bronze statue is made of bronze, we know little about the building blocks of quantum particles. After all, none of the four causes in the principle are purely quantum in nature.

As discussed in Chapter 2, *causality* in science can be considered the matching concept to *motive* of an agent in the court of law. While Aristotle's causality principle is not logically definable with truth-based logic (Zadeh, 2001) and probabilistic definability (Pearl 1988, 2000) is empirical in nature, equilibrium-based bipolar agents provides a unique basis for logically definable causality (Zhang 2005a, 2009a, 2009b, 2009c, 2009d, 2010a) – the first and the only logically definable causality from a mathematical physics or biophysics perspective.

The key element for logically definable causality is bipolar universal modus ponens (BUMP). BUMP is special because (1) it is meaningful in terms of equilibria or non-equilibria even if space, time, agents, and matter are all disappeared and (2) it is a non-linear bipolar dynamic generalization of classical modus ponens (MP). While MP has been the only generic inference rule for thousands of years in logical deduction, BUMP is the only inference rule that ever provides logically definable causality.

BUMP states that, given bipolar equilibria or non-equilibria $A,B,C,D \in B_1$ or B_F, if A implies B and C implies D, we must have $A*C$ implies $B*D$, where the star * is a universal operator for bipolar interaction that satisfy bipolar commutativity and monotonicity. With the arrivals of bipolar agents, BUMP exhibits the potential to reveal the ubiquitous effects of quantum entanglement in simple comprehendible logical terms.

Scenarios

Syntactically and semantically, bipolar causality enables equilibrium-based bipolar agent interaction between different types of equilibria. With bipolar causality we can mathematically represent the following scenarios:

Scenario A: My friends X and Y used to be mentally healthy, but the economic recession E caused their mental depression. Fortunately, X recovered to mental equilibrium through a positive psychiatric medication M. Unfortunately, Y became suicidal after taking M.

Let ϕ = *(self-negation, self-assertion)* be a bipolar predicate for the measure of mental health, φ = *(negative, positive)* be a bipolar predicate for a medication, ψ = *(negative-side, positive-side)* be a bipolar predicate for the economy;

Let time $t = t_0, t_1$, the fact X and Y used to be mental healthy can be mathematically represented as

$$\phi(X(t_0)) = \phi(Y(t_0)) = (-1,1).$$

The economic recession E can be mathematically represented as

$$\psi(E(t_0)) = (-1,0).$$

The fact X and Y became depressed due to the economic recession can be mathematically represented as

$$\phi(X(t_0))*\psi(E(t_0)) = (-1,+1)*(-1,0) = (-1,0);$$

$$\phi(Y(t_0))*\psi(E(t_0)) = (-1,+1)*(-1,0) = (-1,0).$$

At this point the universal operator * for both *X* and *Y* can be determined as &, which can be considered a neurobiological operator reacting to the economic recession.

The fact that *M* is a positive medication can be mathematically represented as

$$\varphi(M) = (0,+1);$$

The fact that M brought mental equilibrium to *X* but suicidal attempt to *Y* can be mathematically represented as

$$\phi(X(t_1))*\psi(M(t_1)) = (-1,0)*_x(0,+1) = (-1,+1);$$

$$\phi(Y(t_1))*\psi(M(t_1)) = (-1,0)*_y(0,+1) = (-1,0).$$

At this time, the neurobiological operator $*_x$ for *X* can be determined as the bipolar fusion operator ⊕. But the neurobiological operator $*_y$ for *Y* can be determined as the bipolar oscillation operator ⊗ because *Y* became suicidal that has to be a deeper state of depression. With the 4-valued bipolar lattice $B_1 = \{-1,0\} \times \{0,+1\}$, deeper depression can't be more precisely represented. With the real-valued bipolar fuzzy lattice $B_F = [-1,0]\times[0,+1]$, lighter and deeper depression can be precisely characterized such as (-0.8, 0) and (-1.0, 0), respectively.

Scenario B: Logically definable causality can also be illustrated with the following bipolar quantum agents *A,B,C,* and *D* in quantum entanglement:

Let $A \Rightarrow B$ and $C \Rightarrow D$, we must have $(A*C) \Rightarrow (B*D)$,

If $A,B,C,D = (-1,0)$ and $(A*C) \Rightarrow (B*D) = (0,+1)$, the universal operator * can be determined as ⊗.

If $A,B,C,D = (0,+1)$ and $(A*C) \Rightarrow (B*D) = (-1,0)$, the universal operator * can be determined as $\otimes^-$.

If $A,C, = (-1,0)$, $B,D =$ *unknown*, and $(A*C) \Rightarrow (B*D) = (0,+1)$, it can be determined that $B,D = (-1,0)$ and the universal operator * is ⊗.

If $A,B, = (-1,0)$, $C,D =$ *unknown*, and $(A*C) \Rightarrow (B*D) = (0,+1)$, it can be estimated that $C,D = (-1,0)$ and the universal operator * can be estimated as ⊗ with probabilistic reasoning.

Consequence of Logically Definable Causality

With logically definable bipolar causality we have the questions: *If the 'big bang' theory is valid, what caused it to happen? Could it be caused by equilibrium or non-equilibrium? Did singularity come first or did bipolarity come first?* Evidently, the impact of logically definable bipolar causality can be fundamentally far reaching. With bipolar causality and bipolar agents, YinYang bipolar relativity is within reach.

YINYANG BIPOLAR RELATIVITY

Emergence of Space and Time

BDL and BDFL so far have not taken time and space dimensions into consideration. Since the universal operator * in BUMP is symmetrical and presents in both the premise and the consequent, *bipolar* *relativity* can be embedded in BUMP and BDL can be naturally extended to a temporal logic with the emergence of space and time dimensions.

Let $\psi = (\psi^-,\psi^+)$, $\phi = (\phi^-,\phi^+)$, $\chi = (\chi^-,\chi^+)$, and $\varphi = (\varphi^-,\varphi^+)$ be any bipolar predicates; let $a(t_1,p_1)$, $b(t_1,p_2)$, $c(t_2,p_3)$, and $d(t_2,p_4)$ be any bipolar agents where $a(t,p)$ stands for *"agent a at time t and space p"* where t_x and t_y can be the same or different points in time and p_x and p_y can be the same or different section or points in space, BUMP with time and space dimensions is shown in Eq. 6.7. Based on Eq. 6.7, an agent without time and space is assumed at any time t and space p that is more general. An agent at time *t* and space *p* is therefore more specific. However, time and/or space can be omitted in some discussion for simplicity without losing generality.

$$\forall a,b,c,d,\ [\psi(a(t_x,p_1))\Rightarrow\chi(c(t_y,p_3))]\&[\phi(b(t_x,p_2))\Rightarrow\varphi(d(t_y,p_4))]$$

$$\Rightarrow[\psi(a(t_x,p_1))*\varphi(b(t_x,p_2))\Rightarrow\chi(c(t_y,p_3))*\varphi(d(t_y,p_4))]. \qquad (6.7)$$

Based on general relativity, gravity "travels" at the speed of light and the effect of a disturbance to the Sun (S) could take 499 seconds to reach the Earth (E). Let $f(S)=f(E)=(-f,f)(S)=(-f,f)(E)$ be the bilateral gravitational *(reaction, action)* forces between S and E; let time *t* be in seconds; let p_1 and p_2 be points for S and E, respectively; let (0,0)(S) be the hypothetical Sun's vanishment or eternal equilibrium; we have

$$[f(S(t,p_1))\Rightarrow f(E(t+499,p_2))] \Rightarrow [f(S(t,p_1))\blacklozenge(0,0))\Rightarrow f(E(t+499,p_2))\blacklozenge(0,0)]. \qquad (6.8a)$$

If $f()$ is normalized to a bipolar predicate, ♦ can be replaced with *, and the binding of &, $\&^-$, ⊗, $\otimes^-$, ∅, or $\varnothing^-$ to * in Eq. 6.8a would lead to the vanishment of the Sun and then the disappearing of the Earth from its orbit after 499 seconds. Thus, bipolar equilibrium/non-equilibrium and general relativity are logically (but not mathematically yet) unified under bipolar relativity.

It should be remarked that YinYang bipolar relativity can accommodate different space and time due to space and time emergence following agents' arrivals. Eq. 6.8a assumes that the speed of gravity equals the speed of light based on general theory of relativity. This assumption is actually questionable. If we assume gravitation is a kind of large scale quantum entanglement it would have a minimum lower bound of 10,000 times the speed of light (Salart et al., 2008), gravity would travel from the sun to the Earth in less than 0.0499 second and Eq. (6.8a) could be revised to Eq. 6.8b.

$$[f(S(t,p_1))\Rightarrow f(E(t+0.0499,p_2)] \Rightarrow [f(S(t,p_1))\blacklozenge(0,0)\Rightarrow f(E(t+0.0499,p_2))\blacklozenge(0,0)]. \qquad (6.8b)$$

A comparison of Eq. 6.8b with Eq. 6.8a reveals an equilibrium-based logical *"bridge"* from spacetime relativity to quantum mechanics or a bridge of quantum gravity. Why cannot other logical systems be used for the above equilibrium-based bipolar unification? The answer is: without bipolarity any truth value 0 or 1 (true or false) is incapable of carrying any shred of direct physical syntax or semantics such as equilibrium (-1,+1), non-equilibrium (-1,0) or (0,+1), and eternal equilibrium (0,0) and, therefore, can not represent non-linear bipolar dynamic interaction such as particle-wave duality, bipolar binding, fusion, coupling, oscillation, decoupling, quantum entanglement, and annihilation.

Time Reversal

Based on Eq. 6.7 bipolar relativity can also support causal reasoning with time reversal because the premise of Eq. 6.7 could be a future event and the consequent a past one. Although time travel in physics and cosmology is highly speculative in nature, time reversal analysis has been proven very useful in many other scientific, technological, and engineering research and development (e.g. Fink, Montaldo & Tanter, 2003). This is not further discussed.

Bipolar Strings

Fundamentally different from string theory or *"theory of everything"*, bipolar relativity provides the logical and physical bindings for the "strings" of the universe but retains the open-world non-linear dynamic property of nature tailored for open-ended exploratory scientific discovery. While string theory is far from observable reality, the non-linear dynamic property of bipolar relativity does not compromise the law of excluded middle (Ch. 3, Table 4) – a unique basis for a scalable and observable bipolar string theory.

Since $(-1,0) \otimes (-1,0) = (-1,0)^2 = (0,1)$ and $(-1,1) \otimes (-1,1) = (-1,1)^2 = (-1,1)$, $(-1,0)^n$ defines an oscillatory non-equilibrium and $(-1,1)^n$ defines a non-linear dynamic equilibrium (Ch. 3, Figure 1). Such properties provide a unifying logical representation for particle-wave duality. For instances, $\phi(P)(f)=(-1,0)^n(3\times 10^{12})$ can denote that "particle P changes polarity three trillion times per second"; $\phi(P)(f)=(-1,1)^n(3\times 10^{12})$ can denote that "The two poles of P interact three trillion times per second."

As strings can be one-dimensional oscillating lines or points (Gefter, 2005), a ***bipolar string*** can be defined as an elementary bipolar variable or agent $e=(-e,+e)$ and characterized as $\phi(e)(f)(m)$ where $\phi(e) \in B_1$, f is the frequency of bipolar interaction or oscillation, and m is mass. If e is massless we have $m = 0$. The two poles of e as ***negative*** and ***positive strings*** are non-exclusive, reciprocal, entangled, and inseparable. Thus, bipolar strings cannot be dichotomous and bipolar string theory is a non-linear dynamic unification of singularity, bipolarity, and particle-wave duality.

YinYang Bipolar Geometry

The renowned theoretical physicist and author Lee Smolin is a strong advocate of background independence and causality in the quest for quantum gravity (Smolin, 2005, 2006). He wrote:

The main unifying idea is simple to state: Don't start with space, or anything moving in space. Start with something that is purely quantum-mechanical and has, instead of space, some kind of purely quantum structure. If the theory is right, then space must emerge, representing some average properties of the structure – in the same sense that temperature as a representation of the average motion of atoms. (Smolin, 2006, p. 240)

Smolin continued: *"Thus, many quantum-gravity theorists believe there is a deeper level of reality, where space does not exist (this is taking background-independence to its logical extreme."*

Smolin emphasized the difficulty to realize *spacetime emergence* from something more fundamental in practice. He wrote (Smolin, 2006, p. 241): *"These days, many of us working on quantum gravity believe that* ***causality*** *itself is fundamental – and is thus meaningful even at a level where the notion of space has disappeared."*

Smolin stated, *"I believe there is something basic we are all missing, some wrong assumption we are all making"* (Smolin, 2006, p. 256).

Smolin noted, *"During a panel discussion on 'The Next Superstring Revolution' at the 2005 Strings Conference, Stephen Shenker, the director of the Stanford Institute for Theoretical Physics, observed that it was likely to come from a topic outside string theory"* (Smolin, 2006, p. 276).

Lee Smolin's view of causality as a deeper level of background-independent fundamental concept is a profound idea. Without logically definable causality, however, Aristotle's original principle of cause and effect are just words that cannot be directly used in formulating mathematical equations in the quest for quantum gravity. Furthermore, most approaches to causal reasoning have so far been truth-based rooted in singularity that cannot avoid the LAFIP paradox. After all, without the pure quantum property such as bipolar interaction and entanglement, any causal model is bound to fail in the quest for quantum gravity, let alone socioeconomics, cognitive informatics, and life sciences.

YinYang bipolar dynamic logic, bipolar lattice, and BUMP enable the definability of equilibrium-based bipolar causality – a deeper level fundamental concept in the quest for quantum gravity as predicted in (Smolin, 2006). With logically definable bipolar causality, the emergence of space and time simply follows the arrivals of bipolar agents including agents as large as the universe and as small as a quantum. Thus, bipolar relativity forms a completely background-independent theory – a theory deeper than spacetime relativity. Its equilibrium-based physical semantics is meaningful even without time, space, and agents. In that case, it simply becomes BUMP.

It is noteworthy to point out the following:

- It is well-known that classical modus ponens (MP) has been the only generic inference rule for truth-based logical deduction for thousands of years.
- Equilibrium-based YinYang bipolar universal modus ponens (BUMP) as a fundamentally different inference rule could not have been brought to light without (-,+) bipolar physical semantics and syntax.
- Classical MP can be derived from BUMP but not vice versa because the non-linear bipolar quantum entanglement of BUMP is "illogical" from a truth-based perspective.
- Classical quantum logical systems and mathematical models are truth-based in nature rooted in singularity which does not claim logical unification of spacetime relativity and quantum mechanics despite its success in quantum computing.
- While this chapter is focused on the theoretical aspect of YinYang bipolar relativity, the practical aspect of bipolar quantum entanglement and bipolar quantum computing will be further discussed in Chapter 7.

Figure 1 is a sketch of YinYang bipolar geometry vs. spacetime geometry, where the 4-valued bipolar lattice $B_1 = \{-1,0\} \times \{0,1\}$ is extended to the real-valued bipolar lattice $B_F = [-1,0] \times [0,1]$ (Ch. 4 and 5) (Zhang, 2006, 2009b) or $B_\infty = [-\infty,0]\times[0,+\infty]$ (Zhang & Chen, 2009; Zhang *et al.*, 2009). As this work is focused on the logical foundation, quantization is postponed for future research. The qualitative features of YinYang bipolar geometry are analyzed as follows.

The acceleration or gravitation curve in spacetime geometry is shown in Figure 1a (Smolin, 2006, p. 42). In spacetime geometry gravitation or acceleration is visualized as a single curve where the effects of acceleration are indistinguishable from that of gravitation. The spacetime curve of a car running in

Figure 1. (a) Geometry of spacetime; (b) YinYang geometry and bipolar equilibrium; (c) Curve of bipolar disorder or deceleration-acceleration switch; (d) Triangle of black hole - big bang - equilibrium - black hole (Adapted from Zhang, 2009d)

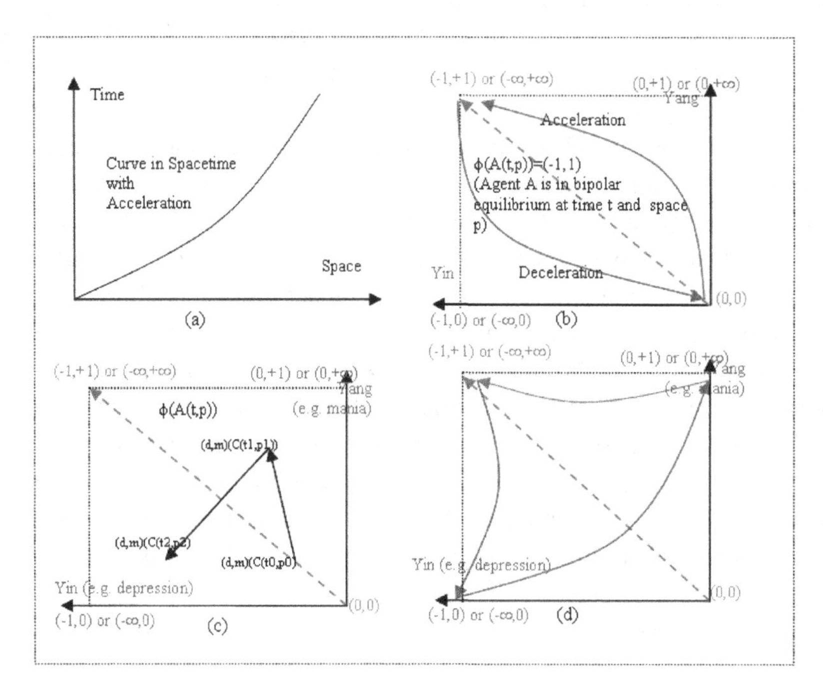

constant speed would become a straight line in parallel with the space dimension. Acceleration turns on the gravitational field and leads to the curved spacetime trajectory.

YinYang bipolar geometry is shown in Figure 1b in which an acceleration curve and a deceleration curve can be both visualized. The YinYang bipolar geometry actually consists of three dimensions, the Yin dimension or pole, Yang dimension or pole, and the bipolar equilibrium dimension which is characterized by the diagonal line.

YinYang bipolar equilibrium transcends the Yin and Yang (the two poles or two energies) and defines the sate of an agent in equilibrium or non-equilibrium such as mental equilibrium or disorder. As discussed in Chapter 1, logically speaking, without YinYang bipolarity, quantum gravity defined in spacetime geometry does not transcend space and time because space and time are not bipolar interactive energies and can't provide bipolar fusion, interaction, quantum entanglement, and definable causality. Therefore, spacetime relativity and quantum mechanics defined in spacetime geometry can't be the unifying logical foundation for gravity, quantum mechanics, socioeconomics, and life sciences. This could well be the fundamental reason that propelled Einstein to conclude (Einstein, 1940) that *"For the time being we have to admit that we do not possess any general theoretical basis for physics which can be regarded as its logical foundation."*

The emergence of space and time in the YinYang bipolar geometry is shown in Figure 1c. Space and time emerging has been difficult to realize in practice in the quest for quantum gravity (Smolin, 2006, p. 240). The difficulty has been associated with an unknown missing fundamental concept. Now, the unknown missing concept has been postulated as YinYang bipolarity (Ch. 1, Postulate 1.1). With bipolar geometry, it is quite amazing to see from Figure 1c that the "difficult emergence" simply follows the arrival of an agent A as a parameter in an YinYang bipolar function *ϕ(A(t,p))* at time *t* and space *p* in a natural and dramatic way.

In the YinYang bipolar geometry, deceleration-acceleration can also be characterized as a non-linear depression-mania function denoted as *(d,m)* where agent (such as a running car) becomes a parameter; space and time emerge following the agent. Thus, *(d,m)*(C(t,p)) denotes the deceleration or acceleration of car C based on its speed at time t and space p. A comparison between Figure. 1(a-c) reveals that equilibrium or non-equilibrium of an agent is to Yin and Yang in YinYang bipolar geometry as gravitation or acceleration is to spacetime in spacetime geometry with fundamentally different semantics and syntax. Since the diagonal line in the YinYang bipolar geometry characterizes strong or weak equilibrium (or constant speed) in different levels, not only can YinYang bipolar geometry account for the strength of deceleration and acceleration, respectively, but also the accumulative speed at time t and space p for visualizing equilibrium or non-equilibrium.

Furthermore, depression and mania in YinYang bipolar geometry are geometrically equivalent to mental deceleration and acceleration. Thus, YinYang bipolar geometry provides a unifying mathematical basis of mental quantum gravity. A comparison of Figures 1b and 1c reveals that acceleration and deceleration in YinYang bipolar geometry can be symmetrical, asymmetrical, linear or non-linear. They can also be in bipolar disorder. These properties can't be visualized in spacetime geometry.

While *the principle of equivalence* in general relativity posits that *the effects of acceleration are indistinguishable from the effects of gravity*, a *principle of equilibrium or non-equilibrium* in bipolar relativity can be posited as follows.

Principle of Equilibrium or Non-Equilibrium: *YinYang bipolar equilibrium transcends the Yin and Yang of nature; local equilibria or non-equilibria can form a global equilibrium or non-equilibrium; any logical, physical, social, mental, or biological order or disorder of a bipolar agent is caused by YinYang bipolar relativity and can be visualized as bipolar equilibrium, quasi-equilibrium, or non-equilibrium in YinYang bipolar geometry;*

While pre-observation cosmological research work (Jack Ng & van Dam, 2001) is difficult to find, a *black hole - big bang – equilibrium - black hole triangle* of the universe is shown in Figure 1d. Based YinYang bipolar relativity, a black hole can be hypothesized as a bipolar agent in depression characterized as (-1, 0); big bang can be hypothesized as a switch from depression to mania characterized as (0,+1). Such a switch is made possible with bipolar dynamic logic. For instance; $(-1,0) \otimes (-1,0) = (0,+1)$. It can be further hypothesized that, after the big bang, the universe can adapt to an (contraction, expansion) equilibrium state $(-1,+1) = (0,+1)\oplus(-1,0)$. Based on bipolar relativity, an agent in an equilibrium state can degenerate to a depressed state. Thus, the universe in (contraction, expansion) equilibrium can go back to a black hole $(-1,+1)\&(-1,0)=(-1,0)$ and start another cycle of the triangle.

The process-based YinYang bipolar cosmology triangle is supported by the theory of Hawking radiation (Hawking, 1974). According to Hawking, a black hole can have particle and/or antiparticle radiation. That is to say, when spacetime ends with a black hole, matter-antimatter bipolarity will miraculously survive and set the stage for another cosmological process. The bipolarity of nature, however, has been largely ignored in logic and mathematics.

Can the universe stay in equilibrium forever? Philosophically and logically, no agent can live forever without staying as a dynamic equilibrium all the time. Wormhole could be such a dynamic equilibrium. It can be imagined, therefore, if the universe can become a wormhole it could stay in a dynamic equilibrium for a long time. In that case the wormhole would become a bipolar equilibrium engine driving and keeping the universe in an orderly equilibrium as long as it can regenerate and reuse the quantum energy available in the universe. If this were the case, the observed expansion of the universe could be due to spacetime curvature at different points of the wormhole.

The wormhole universe assumption might not be a realistic one. Current interpretations of astronomical observations indicate that the age of the Universe is 13.73 ($\pm$ 0.12) billion years, and that the diameter of the observable Universe is at least 93 billion light years, or 8.80×10^{26} meters (Chang, 2008; Lineweaver & Davis, 2005). It may seem paradoxical that two galaxies on opposite sides can be separated by 93 billion light years after only 13 billion years, since special relativity states that matter cannot be accelerated to exceed the speed of light in a localized region of space-time. However, according to general relativity, space can expand with no intrinsic limit on its rate; thus, two galaxies can separate more quickly than the speed of light if the space between them grows. It is uncertain whether the size of the universe is finite or infinite.

According to the prevailing scientific model of the universe, known as the Big Bang theory, the Universe expanded from an extremely hot, dense phase called the Planck epoch, in which all the matter and energy of the observable Universe was concentrated. Since the Planck epoch, the Universe has been expanding to its present form, possibly with a brief period (less than 10^{-32} seconds) of cosmic inflation. Several independent experimental measurements support this theoretical expansion and, more generally, the Big Bang theory. Recent observations indicate that this expansion is accelerating because of dark energy, and that most of the matter and energy in the Universe is not directly observable and fundamentally different from that observed on Earth. The imprecision of current observations has hindered precise predictions in science and the ultimate fate of our universe.

Now, we have the question: *Can the observed expansion of the universe last forever?* A yes answer does not seem to be a logical one due to the ultimate equilibrium state of the 2nd law of thermodynamics. If the expansion would, in deed, ever stop, a logical interpretation would be that the three sides of the black hole - big bang – equilibrium - black hole triangle are actually three transition period:

1. The side from big bang to bipolar equilibrium is the expansion period of the universe after the big bang.
2. The side from equilibrium to black hole is the contraction period. In this period, black holes would slowly suck in everything in the universe into its mouth and bigger black holes would suck in smaller ones to aggregate into even bigger black holes.
3. The side from black hole to big bang is the dormant period. During this period, the last aggregated black hole could be like a deeply depressed person (or a dormant volcano) who one day would become manic (or eruptive) due to equilibrium or non-equilibrium. Alternatively, the last two black holes that try to suck each other in could crash and create another big bang.
4. It can be hypothesized that our current universe is in the mix of expansion and contraction as evidenced by the observable coexistence of space expansion in some region of the universe and black holes in some other regions that try to pull everything together.

It should be remarked that, on the one hand, in clinical psychiatry bipolar equilibrium is considered non-existence where mental disorder is not considered the lost of mental equilibrium but "big bang from nowhere"; on the other hand, cyclic models of the universe have been controversial for many decades. In the 1930s, theoretical physicists, most notably Einstein, considered the possibility of a cyclic model for the universe as an (everlasting) alternative to the Big Bang. However, work by Richard C. Tolman (1934) showed that these early models were impossible because of the entropy problem. That is, in statistical mechanics, entropy only increases because of the 2nd law of thermodynamics (Fermi, 1936). This implies that successive cycles grow longer and larger. Extrapolating back in time, cycles before the present one become shorter and smaller culminating again in a Big Bang and thus not replacing it.

The ***second law of thermodynamics*** is an expression of the universal principle of entropy, stating that the entropy of an isolated system which is not in equilibrium will tend to increase over time, approaching a maximum value at equilibrium, and that the entropy change dS of a system undergoing any infinitesimal reversible process is given by $\delta q/T$, where δq is the heat supplied to the system and T is the absolute temperature of the system. In classical thermodynamics the second law is taken to be a basic postulate. In statistical thermodynamics the second law is a consequence of applying the equal prior probability postulate to the future while empirically accepting that the past was low entropy for reasons not yet well understood.

Thus, the puzzling situation on cyclic models of the universe remained for many decades until the early 21st century when the recently discovered dark energy component provided new hope for a consistent cyclic cosmology (Steinhardt & Turok, 2002). From Figures 1(b-d) it is evident that the YinYang geometry of bipolar relativity presents a coherent logical model for the universe or any agent from a bipolar perspective. Although it is questionable whether YinYang bipolar relativity will lead to *"The Next Superstring Revolution"*, it has definitely come from a topic outside string theory as predicted (Smolin, 2006). Moreover, it does lead to a real world open-ended YinYang bipolar string theory with a deep philosophical root that spans physical, social, mental, and life sciences.

PREDICTIONS

Conjectures

Any significant theory is accompanied with a number of significant falsifiable predictions. While string theory has been criticized for its lack of falsifiable predictions, the theory of YinYang bipolar relativity is rich in predictions which are presented as nine axioms and 21 conjectures as follows.

The first conjecture attempts to answer the questions: Can quantum gravity be applied in life sciences such as neuroscience? Can bipolar relativity play a unifying role for quantum gravity and cognitive informatics? Can bipolar agents and bipolar adaptivity be applied in equilibrium-based computational neuroscience and computational psychology?

Axiom 6.1. Let $\psi = (\psi^-,\psi^+)$ = (self-negation, self-assertion) be a bipolar predicate for the mental equilibrium measures of a patient set P at the neurophysiologic level; let (χ^-,χ^+) be that of the set P at the mood or behavior level; let $\phi = (\phi^-,\phi^+)$ = (negative, positive) be a bipolar predicate for the biochemical capacities of a medicine set M for bipolar disorders; let (φ^-, φ^+) = (un-excite, un-depress) be that for the effects of M at the mental level. $\forall a,b$, $a \in P$ and $b \in M$,

$$[(\psi(a(t_x))\Rightarrow\chi(a(t_y)))]\&[(\phi(b(t_x))\Rightarrow\varphi(b(t_y)))] \Rightarrow \psi(a(t_x))*\phi(b(t_x)))\Rightarrow(\chi(a(t_y))*\varphi(b(t_y)))].$$

Conjecture 6.1. Axiom 6.1 is a fundamental law of equilibrium-based brain and behavior, which can be applied in nanobiomedicine for psychiatric mood regulation on an individual and/or a cohort of mental disorder patients.

The next conjecture is on subatomic particles, nanotechnology, and quantum computing. The B-sub-s meson discovered at the Fermi National Accelerator Laboratory (Fermilab, 2006) was described by a New York Times reporter (Overbye, 2006) as a *"YinYang dance"* or *"A Real Flip-Flopper, at 3 Trillion Times a Second"*. That sets a new frequency record of oscillation. Basically, the subatomic particle changes polarity three trillion times a second, switching into its antimatter opposite and then back again. The measurement of this yin-yang dance was considered a triumph for Fermilab's Tevatron, which smashes together trillion-volt protons and antiprotons to create fireballs of primordial energy, and for the Standard Model that explains all that is known to date about elementary particles and their interactions. Could bipolar relativity be the governing theory for particle physics?

Axiom 6.2. Let $\psi = (\psi^{-},\psi^{+})$ = (negative, positive) be a bipolar predicate and a,b,c,d be any four Yin-Yang bipolar subatomic particles that can change polarity trillions of times a second. $\forall$a,b,c,d, we have:

$$[(\psi(a(t_x,p_1))\Rightarrow\psi(c(t_y,p_3)))\&(\psi(b(t_x,p_2))\Rightarrow\psi(d(t_y,p_4)))]$$

$$\Rightarrow[(\psi(a(t_x,p_1))*\psi\ (b(t_x,p_2)))\Rightarrow(\psi(c(t_y,p_3))*\psi(d(t_y,p_4)))]$$

$$\Rightarrow\ [\psi(a(t_x,p_1)\blacklozenge b(t_x,p_2))\ \Rightarrow\ \psi(c(t_y,p_3)\blacklozenge d(t_y,p_4))].$$

Conjecture 6.2. Axiom 6.2 is a fundamental law of quantum entanglement that can be implemented for quantum computing, communication, and nanobiomedicine. Specifically, the axiom provides a theoretical basis for quantum teleportation without conventional communication (Ch. 7).

The next conjecture is on genomics. The discovery of the ubiquitous genetic regulator protein YinYang1 (YY1) in 1991 (Shi *et al.*, 1991; Park & Atchison, 1991) marks the formal entry of the ancient Chinese YinYang into modern genomics. From that point on, YinYang has reemerged as a unifying philosophical foundation for both Traditional Chinese Medicine (TCM) and bioinformatics in microscopic as well as in macroscopic terms. In less than two decades, many important works have been reported on YY1 such as those in (Ai, Narahari & Roman, 2000; Kim, Faulk & Kim, 2007; Palko *et al.*, 2004; Zhou & Yik, 2006; Santiago *et al.*, 2007; Liu *et al.*, 2007; Wilkinson *et al.*, 2006). Could YinYang bipolar relativity find its position in modern bioinformatics?

Axiom 6.3. Let $\psi=(\psi^{-},\psi^{+})$=(repression, activation) be a bipolar predicate for the abilities of regulator genetic agents (Vasudevan, Tong and Steitz 2007) such as YY1; let $\phi=(\phi^{-},\phi^{+})$=(repressability, activatability) be a predicate for the bipolar capacities of regulated genetic agents; let (χ^{-},χ^{+}) and $(\varphi^{-},\varphi^{+})$ be any bipolar predicates; let a,b,c,d be any agents. We have the laws in Table 1.

Conjecture 6.3. Axiom 6.3 is a fundamental law for equilibrium-based regulation of gene expression, mutation, and molecular interaction in bioinformatics.

Table 1. Equilibrium and non-equilibrium in genomics

$\forall a,b,c,d,$

1) $[\psi(a(t_x,p_1))\Rightarrow\phi(c(t_y,p_3))]\&[\psi(b(t_x,p_2))\Rightarrow\phi(d(t_y,p_4))]$

$\Rightarrow [\psi(a(t_x,p_1))*\psi(b(t_x,p_2))\Rightarrow\phi(c(t_y,p_3))*\phi(d(t_y,p_4))]$

$\Rightarrow [\psi(a(t_x,p_1)\blacklozenge b(t_x,p_2))\Rightarrow\phi(c(t_y,p_3)\blacklozenge d(t_y,p_4))];$

2) $[\psi(a(t_x,p_1))\Rightarrow\psi(c(t_y,p_3))]\&[\phi(b(t_x,p_2))\Rightarrow\phi(d(t_y,p_4))]$

$\Rightarrow [\psi(a(t_x,p_1))*\phi(b(t_x,p_2))\Rightarrow\psi(c(t_y,p_3))*\phi(d(t_y,p_4))]$

The next conjecture is on astrophysics or particle physics. The concept of matter and antimatter is central in astrophysics or particle physics. In particle physics, *antimatter* is the extension of the concept of the antiparticle to matter, where antimatter is composed of antiparticles in the same way that normal matter is composed of particles. There is considerable speculation as to why the observable universe is apparently almost entirely matter, whether there exist other places that are almost entirely antimatter instead, and what might be possible if antimatter could be harnessed, but at this time the apparent asymmetry of matter and antimatter in the visible universe is one of the greatest unsolved problems in physics (Sather, 1999). Could bipolar relativity be the governing theory for exploratory scientific discovery in particle physics?

Axiom 6.4. Let $\psi = (\psi^-,\psi^+)$ = (negative, positive) be a bipolar predicate and a,b,c,d be any four antimatter and/or matter bindings or couplings. $\forall a,b,c,d,$ we have:

$$[(\psi(a(t_x,p_1)) \Rightarrow \psi(c(t_y,p_3))) \,\&\, (\psi(b(t_x,p_2))\Rightarrow\psi(d(t_y,p_4)))]$$

$$\Rightarrow [(\psi(a(t_x,p_1))*\psi\,(b(t_x,p_2))) \Rightarrow (\psi(c(t_y,p_3))*\psi(d(t_y,p_4)))]$$

$$\Rightarrow [(\psi(a(t_x,p_1)\blacklozenge b(t_x,p_2)) \Rightarrow (\psi(c(t_y,p_3)\blacklozenge d(t_y,p_4))].$$

Conjecture 6.4. Axiom 6.4 is an equilibrium-based fundamental law for scientific discovery in astrophysics or particle physics.

The next conjecture is on macroeconomics. It has been a long held idea that the dynamic laws of economics should follow those of physics. Unfortunately, so far there has been little success in borrowing physical laws for the regulation of global economy. Is economics quantum in nature? Can bipolar relativity be borrowed as a governing law for macroeconomics?

Axiom 6.5. Similar to Axiom 6.3, let $\psi = (\psi^-,\psi^+)$ = (inhibition, stimulation) be a predicate for the bipolar abilities of an economic intervention; let $\phi = (\phi^-,\phi^+)$ = (inhibitability, stimulatability) be a predicate for the bipolar capacities of economic agents; let (χ^-,χ^+) and (φ^-,φ^+) be any bipolar predicates; let a,b,c,d be any economic agents. We have the laws in Table 1.

Conjecture 6.5. *Axiom 6.5 is a fundamental law in equilibrium-based regulation and intervention in macroeconomics.*

The next conjecture is on global warming. The great achievement of science and technology has led to significant improvement of the human living standard in many countries. Such achievement, however, is accompanied by the unfortunate deadly side effect – *global warming*. This deadly side effect is so grave that it has threatened the very existence of human civilization including science. Can YinYang bipolar relativity play a role in environmental protection?

Axiom 6.6. Similar to Conjecture 6.3, let $\psi = (\psi^-,\psi^+)$ = (pollution-decrease, pollution-increase) be a predicate for the bipolar abilities of an environment policy; let $\phi = (\phi^-,\phi^+)$ = (protectability, polutability) be a predicate for the bipolar capacities of an environment; let (χ^-,χ^+) and (φ^-,φ^+) be any bipolar predicates; let a,b,c,d be any environmental agents. We have the laws on global warming in Table 1.

Conjecture 6.6. Axiom 6.6 is an equilibrium-based fundamental law in environmental protection and regulation.

The next conjecture is on social dynamics. It has been a long held idea that the laws of social dynamics should follow those of physics. Unfortunately, so far there has been little success in borrowing physical laws for social dynamics. Is social dynamic quantum in nature? Can YinYang bipolar relativity be a unification of both physical and social quantum gravity?

Axiom 6.7. Let $\psi = (\psi^-,\psi^+)$ = (competition, cooperation) be a predicate for the relation between two agents, let A_1, A_2, A_3, A_4, A_5, A_6, A_7, and A_8, be any agents. $\forall$ A_1, A_2, A_3, A_4, A_5, A_6, A_7, A_8, we have:

$$[\psi(A_1,A_2)\Rightarrow\psi(A_3,A_4)]\&[\psi(A_5,A_6)\Rightarrow\psi(A_7,A_8)]$$

$$\Rightarrow \{[\psi(A_1,A_2)*\psi(A_5,A_6)]\Rightarrow[\psi(A_3,A_4)*\psi(A_7,A_8)]\}$$

$$\Rightarrow \{\psi[(A_1, A_2)\blacklozenge(A_5,A_6)] \Rightarrow \psi[(A_3,A_4)\blacklozenge(A_7,A_8)]\}.$$

It is interesting to notice that the physical interaction operator ◆ in this case characterize physical reorganization of the involved agents into new coalition, harmony, or conflict sets (Ch. 3). The universal operator * can only be a logical description of the physical interaction.

Conjecture 6.7. *Axiom 6.7 is a fundamental law for equilibrium-based coordination of multiagent cooperation and competition.*

The next conjecture is on Traditional Chinese Medicine (TCM). After surviving a history of thousands of years, TCM (including but not limited to herbal medicine, qigong, jingluo) has been gradually accepted into the Western society. Since YinYang is the theoretical foundation of TCM, YinYang bipolar relativity should definitely be applicable to TCM. Is YinYang bipolar relativity the common physical law governing both modern bioinformatics and TCM?

Axiom 6.8. Let $\psi = (\psi^-,\psi^+)$ = (yin, yang) be a bipolar predicate in traditional Chinese medicine (TCM); let a,b,c,d be any bipolar agents in TCM. $\forall a,b,c,d,$ we have:

$$[\psi(a(t_x,p_1))\Rightarrow\psi(c(t_y,p_2))]\&[\psi(b(t_x,p_1))\Rightarrow\psi(d(t_y,p_2))]$$

$$\Rightarrow[\psi(a(t_x,p_1))*\psi(b(t_x,p_1))\Rightarrow\psi(c(t_y,p_2))*\psi(d(t_y,p_2))]$$

$$\Rightarrow [\psi(a(t_x,p_1)\blacklozenge b(t_x,p_1))\Rightarrow \psi(c(t_y,p_2)\blacklozenge d(t_y,p_2))].$$

Conjecture 6.8. Axiom 6.8 is an equilibrium-based fundamental law for TCM.

Hilbert's Problem 6 – *"Axiomatize All of Physics"* (Hilbert 1901) has remained unsolved for more than 100 years. It is now a widely considered unsolvable problem. Based on Conjectures 6.1-6.8 regarding physical, social, and mental quantum gravities, we now have the question: *Is YinYang bipolar relativity an axiomatization for all physics?*

The next two conjectures are on the axiomatization of physics in general.

Axiom 6.9. Let $\psi = (\psi^-,\psi^+)$, $\phi = (\phi^-, \phi^+)$, $\chi = (\chi^-,\chi^+)$, and $\varphi = (\varphi^-,\varphi^+)$ be any bipolar predicates; let a,b,c,d be any bipolar agents in physical and social sciences. $\forall a,b,c,d,$ we have:

$$[\psi(a(t_x,p_1))\Rightarrow\chi(c(t_y,p_3))]\&[\phi(b(t_x,p_2))\Rightarrow\varphi(d(t_y,p_4))]$$

$$\Rightarrow\{[\psi(a(t_x,p_1))*\phi(b(t_x,p_2))]\Rightarrow[\chi(c(t_y,p_3))*\varphi(d(t_y,p_4))]\}.$$

Conjecture 6.9. All equilibrium-based bipolar fusion, interaction, oscillation, and quantum entanglement in microscopic and macroscopic worlds satisfy Axiom 6.9.

Conjecture 6.10. The bipolar axiomatization (Ch. 3, Table 6) is the most primitive (with minimal semantics) and most general (domain independent) equilibrium-based axiomatization of agents and agent interaction in physics, life sciences and socioeconomics; any other less primitive axiomatization with added semantics (such as space, time, mass, and energy) must necessarily be less general (or more domain-specific).

The next conjecture is on S5 modality, super symmetry, and equilibrium-based Platonic reality. While classical S5 modality (Garson, 2009) relies on a single static equivalence relation, a bipolar equilibrium relation can be a non-linear bipolar dynamic fusion of many equivalence relations (Ch. 3). The non-linearity makes bipolar S5 modality particularly intriguing and challenging.

Conjecture 6.11. Bipolar S5 modality is a logical, physical, neural, social, and biological reality or modern Platonic bipolar reality of multiagent systems in dynamic equilibria or non-equilibria with or without super symmetry.

Next three conjectures are on bipolar strings. While mainstream string theory is far from reality, YinYang bipolar string theory is a real world string theory based on observable phenomena.

Conjecture 6.12. Bipolar strings are the ***makings of bipolar relativity*** and ***bindings of nature.***

Conjecture 6.13. Gravitational and electromagnetic fields are formed with bipolar strings.

Conjecture 6.14. Bipolar strings are scalable; all observable bipolar dynamic equilibria and non-equilibria in physical, social, and mental worlds are ***large scale bipolar strings***.

The next conjecture is on bipolar twistor space. In (Smolin, 2006, p. 244) the author introduced the English mathematical physicist Roger Penrose's *twistor theory* (Penrose, 2005) and stated that *"No one yet knows what a quantum twistor space looks alike."*

Conjecture 6.15. YinYang bipolar geometrical space is a minimal but most general quantum twistor space; bipolar universal modus ponens (BUMP) is a minimal but most general quantum twistor; bipolar causality and bipolar relativity is a minimal but most general twistor theory.

The next conjecture is on big bang and black holes. Earlier, in 1917, Albert Einstein had found that his newly developed theory of general relativity indicated that the universe must be either expanding or contracting. Unable to believe what his own equations were telling him, Einstein introduced a cosmological constant (a "fudge factor") to the equations to avoid this "problem". Edwin Hubble discovered that the universe is not static but expanding (Sandage, 1989), in particular, Hubble discovered a relationship between redshift and distance, which forms the basis for the modern expansion paradigm. This led Einstein to declare his cosmological model especially the introduction of the cosmological constant, his *"biggest blunder"*. Can bipolar relativity give a general interpretation on the relationship of big bang, black hole, and wormhole?

Conjecture 6.16. Black hole is to a galaxy (or universe) as bipolar depression is to a dysfunctional brain; big bang is to a galaxy (or universe) as bipolar mania is to a dysfunctional brain; wormhole is to a galaxy (or universe) as bipolar mental equilibrium is to a functional brain; if the universe was created by a big bang, a big bang could be proceeded by a transient dormant process from black hole to big bang and followed by an expansion process from big bang to equilibrium, and then an contraction process from equilibrium to black hole (Figure 1d).

The next two conjectures are on bipolar unification with an answer to the question: Can spacetime, agents, singularity, causality, and general relativity be logically (not mathematically yet) unified under bipolar relativity?

Conjecture 6.17. If the universe had been created by a big bang, the big bang must have been caused by bipolar equilibrium or non-equilibrium forces; the curvature of spacetime of the cosmological order must be part of the contour of a multilayer multidimensional global and local bipolar equilibrium or non-equilibrium of nature's basic forces (See Figure 2).

Conjecture 6.18. Spacetime relativity, singularity, gravity, electromagnetism, quantum mechanics, bioinformatics, and socioeconomics are different ***phenomena of bipolar relativity***; microscopic and macroscopic agent interactions in physics, socioeconomics, and life sciences are directly or indirectly caused by bipolar causality and regulated by bipolar relativity.

Figure 2. Curvature of multitilayer multidimensional global and local bipolar equilibrium or non-equilibrium

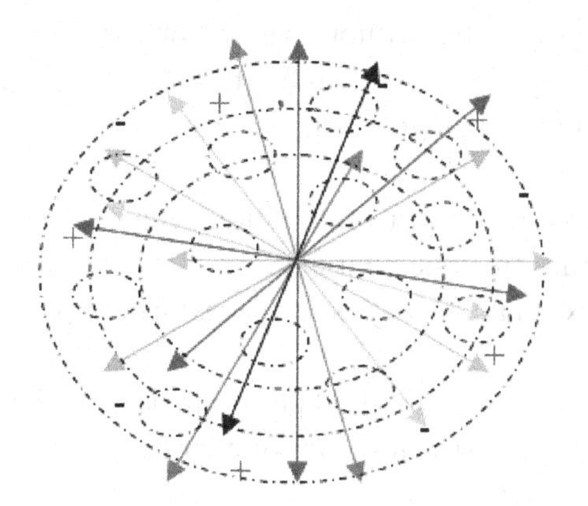

The next conjecture is on an equilibrium-based bipolar cosmological order. David Bohm proposed the concept of implicate-explicate cosmological order (Bohm, 1980) with a causal interpretation of quantum mechanics (Bohm, 1957) based on a wave function but so far there is no logical unification of the two orders. Based on the fact that (1) BDL provides an equilibrium-based bipolar dynamic fusion of linear truth-based logic and non-linear equilibrium-based quantum logic (Ch. 3), (2) BUMP has classical modus ponens (MP) self contained, and (3) YinYang bipolar relativity transcends spacetime, a unifying theory of cosmological order can now be hypothesized.

Conjecture 6.19. YinYang bipolar relativity is the simplest mathematically conceivable cosmological order – a non-linear bipolar dynamic fusion of (1) a unipolar truth-based explicate order and (2) an equilibrium-based implicate order with bipolar quantum entanglement.

Now a key question could be raised: *Could bipolar relativity be a phenomenon of general relativity instead?* The answer to this question is "unlikely" for the following reasons:

1. General relativity leads to singularity that was deemed *"bizarre"* by Einstein himself.
2. Space and time are not symmetrical, can't be quantum entangled, and are less fundamental than quantum bipolarity.
3. Spacetime doesn't provide most fundamental logically definable causality and can only be effect but not cause.

A 2008 quantum physics experiment performed in Geneva, Switzerland reported that the *"speed" of the quantum non-local connection* what Einstein called *"spooky action at a distance"* has a minimum lower bound of 10,000 times the speed of light (Salart et al., 2008). While general relativity provides no prediction about any speed beyond the speed of light, the symmetrical nature of bipolar relativity provides the logical cause-effect relation of quantum entanglement that can pass information with or without passing observable mass or energy (Zhang, 2010a, 2010b) (Ch. 7). The symmetrical property leads to the following two predictions.

Conjecture 6.20. When observable mass or energy is propagated through bipolar relativity or causality (Eq. 6.7) the speed of the propagation is limited by the speed of light (e.g. the propagation of photon or electron); when information is propagated without passing observable mass or energy the speed of the propagation is not limited by the speed of light but by the "speed" of equilibrium-based quantum non-local connection or bipolar quantum entanglement.

Conjecture 6.21. All physical, social, mental, and biological action-reaction forces are fundamentally different forms of bipolar quantum entanglement in large or small scales governed by bipolar relativity. Graviton does not necessarily exist and gravity is not necessarily limited by the speed of light because:

1. Graviton hasn't been experimentally verified but is already predicted a massless quantum agent.
2. Bipolar relativity can be a symmetrical operation that doesn't have to transport observable mass or energy (Eq. 6.7).
3. As a widely accepted theory, all celestial objects are created from a big bang and, therefore, all are entangled with each other in some way from the very beginning.
4. The propagation of observable mass or energy is not a necessity for bipolar quantum entanglement due to bipolar symmetry.
5. Even without observable mass or energy propagation, the collective effect of large-scale quantum entanglement can be observable.

It should be remarked that Conjecture 6.21 is a logical conjecture that coincides with Van Flandern's physical theory regarding the speed of gravity (Van Flandern, 1998). Following Conjecture 6.21, the rotations of planets around a star as well as that of a moon around a planet, ocean high tide and low tide, neurobiological interaction in decision making, physical, economical, social, and mental activities are all fundamentally due to bipolar quantum entanglement. Thus, the ubiquitous effects of quantum entanglement are revealed for the first time.

It should be noted that the symmetrical property of bipolar relativity is a key to enable quantum information being passed with or without passing observable energy or mass (Zhang, 2010a, 2010b). When photon or electron is passed the speed is limited by the speed of light that has been proven in physics. Physicists have so far failed to experimentally verify the existence of graviton and the speed of gravity. If all action-reaction forces are fundamentally equilibrium-based (or non-equilibrium-based) bipolar quantum entanglement in nature, gravity would be unified with quantum mechanics in both micro- and macroscopic worlds under bipolar relativity. *Thus, based on BDL (Ch. 3) and energy symmetry (Ch. 4) bipolar relativity provides a theoretical foundation toward a logically complete unifying quantum theory (see Eq. 6.7 and Eq. 6.8).* The symmetrical aspects of bipolar quantum entanglement are further discussed in Chapter 7.

Notably, the bipolar quantum interpretation of Conjectures 6.20 and 6.21 also coincides with MIT Professor Seth Lloyd's startling thesis that the universe is itself a quantum computer (Lloyd, 2006). According to Lloyd, the universe is all about quantum information processing. Once we understand the laws of physics completely, we will be able to use small-scale quantum computing to understand the universe completely as well. *Could bipolar relativity be such a basic law of physics?*

Falsifiability

Since falsifiability has become a major controversy over string theory in recent years, it is natural to question the falsifiability of bipolar relativity as well. Could it be another "theory of everything" that is not even wrong?

Despite the enormous theoretical magnitude, the falsifiability of bipolar relativity doesn't seem to be a mission impossible. Although some of the 21 conjectures may need many years to be verified or falsified, bipolar relativity in general is verifiable or falsifiable with observation.

First, Conjectures 6.1-6.9 are based on Axioms 6.1-6.9, respectively. The nine axioms are bipolar logical tautologies derivable from BUMP. Therefore, the nine conjectures are logically provable. It is interesting to notice that knowledge discovery from data (KDD) or data mining (DM) was originated from mining association rules from data bases (Han & Kember, 2001). The nine axioms and their corresponding conjectures, on the other hand, provide holistic bipolar association rules or guidelines for scientific knowledge discovery in uncharted scientific territories. Regardless of the final outcome, the guidelines are logically sound.

The testability or falsifiability of some of the conjectures is high in likelihood. In particular, the utility of Conjecture 6.1 in clinical psychiatry is further discussed in Chapter 10 and (Zhang *et al.,* 2010); the utility of Conjecture 6.2 is further discussed in Chapter 7 and (Zhang, 2010) for bipolar quantum computing and quantum teleportation; the falsifiability of Conjecture 6.3 is expected in equilibrium-based gene regulation (Shi *et al.,* 1991; Vasudevan, Tong & Steitz, 2007; Gore & van Oudenaarden, 2009). These conjectures are expected to be either experimentally verified or falsified in the foreseeable future due to broad interests in mental health, quantum computing, and bioinformatics.

Secondly, all 21 conjectures are mostly based on the bipolarity and symmetry of nature's basic forces and antimatter-matter particles, which support the concepts of bipolar agents and bipolar causality. The falsifiability of bipolar relativity is, therefore, closely related to the falsifiability of bipolarity as the most fundamental property of Mother Nature.

Thirdly, the conjectures regarding the triangle of black hole - big bang – equilibrium - black hole provide the first logically coherent process theory of our universe supported by bipolar causality. Although, the process theory of the universe may never be fully verified, bipolar neurobiological disorder, bipolar mental equilibrium, economic depression, and global warming provide observable bipolar equilibrium-based phenomena for the falsifiability of bipolar relativity in the observable world.

Fourthly, the equilibrium-based bipolar quantum entanglement interpretation of gravity in Conjectures 6.20 and 6.21 presents a deviation from general relativity and quantum field theory but stays in line with physical experimentation (Van Flandern, 1998) and in line with the theory of an information processing universe (Lloyd, 2006). As a logically complete theory of quantum gravity, it does not depend on the existence of graviton which is hypothesized as a massless elementary particle but so far undiscovered. Moreover, equilibrium-based bipolar quantum gravity theory can avoid the pitfalls of singularity such as gravitational and quantum infinities (Smolin, 2006) (Ch. 2).

Finally, bipolar relativity is not "*theory of everything*". Bipolar strings as part of bipolar relativity enable particle-wave duality but retain upward and downward scalability. Upward scalability supports macroscopic bipolar binding or interaction; downward scalability supports microscopic scientific discovery. Since no fundamental element of the universe is assumed such as in ether theory or string theory, bipolar relativity define the logical and physical bindings or couplings, respectively, for real world "agents" and "strings" of the universe but retains the open-world non-linear dynamic property of nature tailored for open-ended exploratory scientific discovery especially in quantum computing (Zhang, 2010a, 2010b; Ch. 7).

AXIOMATIZATION OF PHYSICS

Nature is an open-world and science is open-ended; anyone trying to close it will sooner or later find it impossible. The history of searching for the smallest most fundamental elements of the universe has clearly shown that a closed world assumption is bound to fail. Different ancient philosophies once regarded air and water as the most basic elements of nature that have been proven false; aether and string theories are far from real world reality and cannot be experimentally tested. We have to ask ourselves: *If we do not know how large the universe is, how could we possibly know how small the basic elements of nature are? Would it be better to make an open-world open-ended approach to exploratory scientific discovery?*

Without making any assumption on the smallest basic element of nature, bipolar relativity presents an open-world open-ended approach based on well-established observations. On the one hand, it has been shown that strictly logical agents such as a digital computer can't adapt to mental equilibrium like a human agent. On the other hand, it has been shown that the non-linear dynamic property of bipolar relativity doesn't compromise the law of excluded middle: a unique basis for a scalable and observable real-world bipolar string theory.

Einstein's general relativity theory is a generalization of his own special relativity. In the theory of special relativity, Einstein partially unified space and time into a four dimensional entity called *spacetime* that has a geometry different from Euclidian geometry. In Euclidian geometry, two straight lines can never meet if they are initially parallel. With general relativity, however, two light rays can meet due to their bending in a gravitational field toward a star even though they are initially parallel. Under general relativity gravity is indistinguishable from acceleration and, thus, gravitational field is unified.

Since space and time are not physical, biological, neural, mental, or quantum properties, space and time can host gravitation, but cannot form space-time quantum entanglement. Therefore, any relativity defined in spacetime geometry can be referred to as spacetime relativity that cannot be the unifying theory for general relativity, quantum mechanics, socioeconomics, and life sciences. Furthermore, if space is caused by the big bang and space is expanding or contracting, spacetime has to be less fundamental than the cause of the big bang and the expanding or contracting forces.

Fundamentally different from space and time, the Yin and Yang in YinYang bipolar geometry can be any opposite poles (or energies and forces) of nature including but not limited to logical, physical, social, biological, and mental poles. Based on bipolar relativity, the two poles can form equilibrium-based bipolar quantum entanglement (Zhang 2009a, 2009b, 2009c, 2009d, 2010a, 2010b). Thus, YinYang bipolar geometry provides a unifying platform. The unifying nature is manifold:

1. Since bipolarity is a common property of nature's basic forces of action and reaction, YinYang bipolar relativity can be considered a unification of nature's basic forces including gravity, electromagnetism, strong, and weak forces.
2. Since space and time are associated with agents and they can emerge in YinYang bipolar geometry following the arrivals of agents, YinYang bipolar geometry can be considered a unification of spacetime geometry with microscopic and macroscopic agents and nature's basic forces.

3. Since bipolar equilibrium or non-equilibrium is a generic property of nature with definable causality, bipolar relativity can be deemed a logical unification of nature, agents and causality.
4. Since all agent interaction in physics, socioeconomics and life sciences are governed by social or physical dynamics under equilibrium or non-equilibrium, bipolar relativity can be deemed a unification of physics, socioeconomics, and life sciences.

A comparison of Figure 1(a) with Figure 1(b) reveals that

1. the spacetime coordinate is located in the first quadrant with the two dimensions space and time but the YinYang coordinate is background independent and irrelevant to quadrant with three dimensions Yin, Yang, and bipolar Equilibrium where equilibrium is a YinYang bipolar fusion;
2. the acceleration curve in spacetime coordinate is linear and the bipolar disorder curve in the YinYang coordinate is non-linear caused by the breakdown of bipolar fusion ability of self-negation and self-assertion abilities of the brain;
3. spacetime is geometrical and YinYang is physically logical whose four corners form a bipolar quantum lattice (Ch. 4);
4. general relativity doesn't lead to logically definable causality, bipolar relativity leads to logically definable causality (Ch. 3).

The significance of bipolar relativity lies in its ***equilibrium-based open-world open-ended logical unification*** of nature, life sciences, and socioeconomics as well as space, time, gravitation, electromagnetism, quantum mechanics, causality, and agent interaction. Conjectures 6.1-6.9 are "speculative" only in physical terms but not in mathematical terms because they are based on the nine axioms which are bipolar tautologies derived from BUMP. Conjectures 6.10-6.21 as domain-independent predictions span mostly uncharted territories of nature and science. The axioms and conjectures provide guidelines for exploratory scientific discoveries on agent interaction, coordination, and global regulation in microscopic and macroscopic, autonomous and non-autonomous, social and physical worlds. It is hopeful that the predictions are falsifiable because bipolar relativity has been shown an equilibrium-based coherent theory firmly based on the bipolarity of nature's basic forces and matter-antimatter particles.

Evidently, to remain an equilibrium-based logical space the bipolar lattice $B_1=\{-1,0\}\times\{0,1\}$ can no longer be further reduced. In contrast, to remain a truth-based logical space the bivalent lattice $\{0,1\}$ can no longer be further reduced. Therefore, bivalent logic can be deemed the minimal but most general truth-based system; BDL can be considered the minimal but most general equilibrium-based logic. Without bipolarity, however, the truth values 0 and 1 are incapable of carrying any direct physical syntax and semantics. A truth-based model, therefore, cannot avoid the LAFIP or LAFIB paradox (Ch. 1) that could be the reason why there is so far no truth-based axiomatization for physics.

Hilbert stated (Hilbert, 1901): *"If geometry is to serve as a model for the treatment of physical axioms, we shall try first by a small number of axioms to include as large a class as possible of physical phenomena, and then by adjoining new axioms to arrive gradually at the more special theories. ... The mathematician will have also to take account not only of those theories coming near to reality, but also, as in geometry, of all logically possible theories. He must be always alert to obtain a complete survey of all conclusions derivable from the system of axioms assumed."* It is interesting to make the following observations:

1. YinYang coordinate is a generic dimensional model in Hilbert space (Ch. 3, Figures 1-3) and bipolar relativity is a continuation of Hilbert's geometric approach to his Problem 6.
2. The axiomatization (Ch. 3, Table 6) is based on the syntactic approach to proof theory in Hilbert style. With *"a small number of axioms"* it does *"include as large a class as possible of physical phenomena"* such as the 21 conjectures.
3. Bipolar relativity does *"take account not only of those theories coming near to reality, but also, as in geometry, of all logically possible theories"* such as general relativity, string theory, quantum physics, global economy, global warming, genomics, and computational neuroscience that were undeveloped yet at Hilbert's era.

With these observations it can be concluded that YinYang bipolar relativity constitutes an axiomatization of physics – an equilibrium-based partial but most general solution to Hilbert's Problem 6.

RESEARCH TOPICS

YinYang bipolar relativity brings bipolar causality into the agent worlds and opens a wide spectrum of fundamental research topics. A few of them are listed as follows.

Toward a Complete Bipolar Quantum Theory. Bipolar causality and relativity has provided a basis toward a complete quantum theory for quantum computation and communication (Zhang, 2010a, 2010b) (Ch. 7).

Bipolar Quantum Gravity and Quantum Computing. Bipolar relativity provides a unifying theory of agents and causality and, therefore, forms a theoretical basis for quantum gravity and quantum computing in most broad terms which include logical, physical, social, mental, and biological aspects for further investigation.

Beyond Spacetime Geometry. With logically definable bipolar causality, bipolar relativity can now be considered a fundamental cause-effect relation to spacetime. The cause-effect relationship deserves further research efforts for many years to come because classical science has been primarily hosted within spacetime for thousands of years.

Mathematical Development of Bipolar Relativity. So far bipolar relativity is limited to the logical level. As a logical foundation it has prompted further mathematical development with quantization.

SUMMARY

Bipolar agents and bipolar adaptivity have been mathematically defined; logically definable bipolar causality has been presented. Bipolar agents and bipolar causality enable the emergence of space and time. Based on bipolar agents and bipolar causality, the central theme of this book – the theory of bipolar relativity has been presented as a completely background independent theory of quantum gravity. The YinYang geometry of bipolar relativity has been analyzed and compared with that of spacetime relativity. 21 conjectures or predictions have been posted that span quantum computing, cognitive informatics, socioeconomics, and life sciences. Falsifiability of the predictions has been briefly discussed.

A scalable real world bipolar string theory has been introduced. Bipolar strings are not assumed the most fundamental elements of nature and the theory does not contend to be "theory of everything."

Instead, it is defined as a special type of bipolar agent that plays the binding role for all agents in both microscopic and macroscopic worlds. Thus, with bipolar relativity frequency and mass become emerging aspects in bipolar string theory which is distinguished from classical string theory and aether theory.

ACKNOWLEDGMENT

This chapter reuses part of published material in Zhang, W. -R. (2009d). The logic of YinYang and the science of TCM – an eastern road to the unification of nature, agents, and medicine. Interscience: *Int. J. Functional Informatics and Personal Medicine (IJFIPM)*, Vol. 2, No. 3, pp261–291 (2009). Permission to reuse is acknowledged.

REFERENCES

Ai, W., Narahari, J., & Roman, A. (2000). Yin Yang 1 negatively regulates the differentiation-specific E1 promoter of Human Papillomavirus Type 6. *Journal of Virology*, *74*(11), 5198–5205. doi:10.1128/JVI.74.11.5198-5205.2000

Bohm, D. (1957). *Causality and chance in modern physics*. Philadelphia: University of Pennsylvania Press. doi:10.4324/9780203201107

Bohm, D. (1980). *Wholeness and the implicate order*. London: Routledge.

Carey, B. (2007, Aug 1). Man regains speech after brain stimulation. *The New York Times – Health*. Retrieved from http://www.nytimes.com/2007/08/01/health/01cnd-brain.html

Chaib-draa, B. (2002). Causal maps: Theory, implementation, and practical applications in multiagent environments. *IEEE Transactions on KDE*, *14*(6), 1201–1217.

Chang, K. (2008, March 9). Gauging age of universe becomes more precise. *New York Times*. Retrieved from http://www.nytimes.com/2008/03/09/science/space/09cosmos.html

Dirac, P. A. M. (1927). The quantum theory of the emission and absorption of radiation. *Proceedings of the Royal Society of London. Series A*, *114*, 243–265. doi:10.1098/rspa.1927.0039

Dirac, P. A. M. (1928). The quantum theory of the electron. *Proceedings of the Royal Society of London. Series A*, *117*(778), 610–624. doi:10.1098/rspa.1928.0023

Fermi National Accelerator Laboratory. (2006). *Press Release 06-19*. Retrieved on http://www.fnal.gov/pub/presspass/press_releases/CDF_meson.html

Feynman, R. P. (1962). *Quantum electrodynamics*. Addison Wesley.

Fink, M., Montaldo, G., & Tanter, M. (2003). Time-reversal acoustics in biomedical engineering. *Annual Review of Biomedical Engineering*, *5*, 465–497. doi:10.1146/annurev.bioeng.5.040202.121630

Garson, J. (2009). Modal logic. *Stanford Encyclopedia of Philosophy*. Retrieved on http://plato.stanford.edu/ entries/logic-modal/

Gore, J., & van Oudenaarden, A. (2009). Synthetic biology: The yin and yang of nature. *Nature*, *457*, 271–272. doi:10.1038/457271a

Han, J., & Kamber, M. (2001). *Data mining, concepts and techniques*. Morgan Kaufmann.

Hawking, S. (1974). Black-hole evaporation. *Nature*, *248*, 30–31. doi:10.1038/248030a0

Hawking, S., & Penrose, R. (1970). The singularities of gravitational collapse and cosmology. *Proceedings of the Royal Society of London. Series A*, *314*(1519), 529–548. doi:10.1098/rspa.1970.0021

Hilbert, D. (1901). Mathematical problems. *Bulletin of the American Mathematical Society*, *8*, 437–479. doi:10.1090/S0002-9904-1902-00923-3

Huhns, M. N. (Ed.). (1987). *Distributed artificial intelligence*. London: Pitman.

Huhns, M. N. (2000). An agent-based global economy. *IEEE Internet Computing*, *4*(6), 83–84.

Huhns, M. N., & Shehory, O. (Eds.). (2007). *Proceedings of the Sixth International Joint Conference on Autonomous Agents and Multi-Agent Systems (AAMAS 07)*, International Foundation for Autonomous Agents and Multiagent Systems, 2007.

Huhns, M. N., & Singh, M. P. (1997). *Readings in agents*. Morgan Kaufmann.

Jack Ng, Y., & van Dam, H. (2001). A small but nonzero cosmological constant. *International Journal of Modern Physics D*, *10*, 49. doi:10.1142/S0218271801000627

Kim, J. D., Faulk, C., & Kim, J. (2007). Retroposition and evolution of the DNA-binding motifs of YY1, YY2 and REX1. *Nucleic Acids Research*, *35*(10), 3442–3452. doi:10.1093/nar/gkm235

Lineweaver, C., & Davis, T. M. (2005). Misconceptions about the Big Bang. *Scientific American*. Retrieved from http://space.mit.edu/~kcooksey/teaching/AY5/MisconceptionsabouttheBigBang_ScientificAmerican.pdf

Liu, H., Schmidt-Supprian, M., Shi, Y., Hobeika, E., Barteneva, N., & Jumaa, H. (2007). Yin Yang 1 is a critical regulator of B-cell development. *Genes & Development*, *21*, 1179–1189. doi:10.1101/gad.1529307

Lloyd, S. (2006). *Programming the universe*. New York: Alfred A. Knopf, Inc.

Ou, B., Huang, D., Hampsch-Woodill, M., & Flanagan, J. A. (2003). When East meets West: The relationship between Yin-Yang and antioxidation-oxidation. *The FASEB Journal*, *17*, 127–129. doi:10.1096/fj.02-0527hyp

Palko, L., Bass, H. W., Beyrouthy, M. J., & Hurt, M. M. (2004). The Yin Yang-1 (YY1) protein undergoes a DNA-replication-associated switch in localization from the cytoplasm to the nucleus at the onset of S phase. *Journal of Cell Science*, *117*, 465–476. doi:10.1242/jcs.00870

Park, K., & Atchison, M. L. (1991). Isolation of a candidate repressor/activator, NF-E1 (YY-1, delta), that binds to the immunoglobulin kappa 3' enhancer and the immunoglobulin heavy-chain mu E1 site. *Proceedings of the National Academy of Sciences of the United States of America*, *88*(21), 9804–9808. doi:10.1073/pnas.88.21.9804

Pearl, J. (1988). *Probabilistic reasoning in intelligent systems: Networks of plausible inference*. San Mateo, CA: Morgan Kaufmann Publishers.

Pearl, J. (2000). *Causality: Models, reasoning, and inference*. New York, Cambridge, UK: Cambridge University Press.

Penrose, R. (2005). *The road to reality: A complete guide to the laws of the universe*. New York: Alfred A. Knopf.

Sandage, A. (1989). Edwin Hubble 1889-1953. *The Journal of the Royal Astronomical Society of Canada. Royal Astronomical Society of Canada, 83*(6), 351.

Santiago, F. S., Ishii, H., Shafi, S., Khurana, R., Kanellakis, P., & Bhindi, R. (2007). Yin Yang-1 inhibits vascular smooth muscle cell growth and intimal thickening by repressing p21WAF1/Cip1 transcription and p21WAF1/Cip1-Cdk4-Cyclin D1 assembly. *Circulation Research, 101*, 146–155. doi:10.1161/CIRCRESAHA.106.145235

Sather, E. (1999). The mystery of the matter asymmetry. *Beam Line 26*(1), 31. Retrieved from http://www.slac. stanford.edu/pubs/beamline/26/1/26-1-sather.pdf.

Shi, Y., Seto, E., Chang, L.-S., & Shenk, T. (1991). Transcriptional repression by YY1, a human GLI-Kruppel-related protein, and relief of repression by adenovirus E1A protein. *Cell, 67*(2), 377–388. doi:10.1016/0092-8674(91)90189-6

Singh, M. P., & Huhns, M. N. (2005). *Service-oriented computing: Semantics, processes, agents*. West Sussex, UK: John Wiley & Sons, Ltd.

Smolin, L. (2005). *The case for background independence*.

Smolin, L. (2006). *The trouble with physics: The rise of string theory, the fall of a science, and what comes next?*New York: Houghton Mifflin Harcourt.

Steinhardt, P. J., & Turok, N. (2002). A cyclic model of the universe. *Science, 24*(296), 1436–1439. doi:10.1126/science.1070462

The Royal Swedish Academy of Sciences. (2008). The Nobel Prize in Physics 2008. *Press Release*, 7 October. Retrieved from http://nobelprize.org/nobel_prizes/physics/laureates/2008/press.html

Tolman, R. C. (1987). *Relativity, thermodynamics, and cosmology*. New York: Dover.

Van Flandern, T. (1998). The speed of gravity – what do the experiments say? *Physics Letters. [Part A], 250*, 1–11. doi:10.1016/S0375-9601(98)00650-1

Vasudevan, S., Tong, Y., & Steitz, J. A. (2007). Switching from repression to activation: MicroRNAs can up-regulate translation. *Science, 318*(5858), 1931–1934. doi:10.1126/science.1149460

Wellman, M. P. (1994). Inference in cognitive maps. *Mathematics and Computers in Simulation, 36*, 137–148. doi:10.1016/0378-4754(94)90028-0

Wellman, M. P., & Hu, J. (1998). Conjectural equilibrium in multiagent learning. *Machine Learning, 33*, 179–200. doi:10.1023/A:1007514623589

Wilkinson, F. H., Park, K., & Atchison, M. L. (2006). Polycomb recruitment to DNA in vivo by the YY1 REPO domain. *Proceedings of the National Academy of Sciences of the United States of America, 103*, 19296–19301. doi:10.1073/pnas.0603564103

Zadeh, L. A. (2001). Causality is undefinable–toward a theory of hierarchical definability. *Proceedings of FUZZ-IEEE,* (pp. 67-68).

Zhang, W.-R. (1998). Nesting, safety, layering, and autonomy: A reorganizable multiagent cerebellar architecture for intelligent control–with application in legged locomotion and gymnastics. *IEEE Transactions on Systems, Man, and Cybernetics. Part B, 28*(3), 357–375.

Zhang, W.-R. (2003a). Equilibrium relations and bipolar cognitive mapping for online analytical processing with applications in international relations and strategic decision support. *IEEE Transactions on SMC. Part B, 33*(2), 295–307.

Zhang, W.-R. (2003b). Equilibrium energy and stability measures for bipolar decision and global regulation. *International Journal of Fuzzy Systems, 5*(2), 114–122.

Zhang, W.-R. (2005a). YinYang bipolar lattices and L-sets for bipolar knowledge fusion, visualization, and decision making. *International Journal of Information Technology and Decision Making, 4*(4), 621–645. doi:10.1142/S0219622005001763

Zhang, W.-R. (2005b). YinYang bipolar cognition and bipolar cognitive mapping. *International Journal of Computational Cognition, 3*(3), 53–65.

Zhang, W.-R. (2007). YinYang bipolar universal modus ponens (bump)–a fundamental law of non-linear brain dynamics for emotional intelligence and mental health. *Walter J. Freeman Workshop on Nonlinear Brain Dynamics, Proceedings of the 10th Joint Conference of Information Sciences*, (pp. 89-95). Salt Lake City, Utah.

Zhang, W.-R. (2009a). Six conjectures in quantum physics and computational neuroscience. *Proceedings of 3rd Int'l Conf. on Quantum, Nano and Micro Technologies (ICQNM 2009)*, (pp. 67-72). Cancun, Mexico.

Zhang, W.-R. (2009b). YinYang Bipolar Dynamic Logic (BDL) and equilibrium-based computational neuroscience. *Proceedings of International Joint Conference on Neural Networks (IJCNN 2009)*, (pp. 3534-3541). Atlanta, GA.

Zhang, W.-R. (2009c). YinYang bipolar relativity–a unifying theory of nature, agents, and life science. *Proceedings of International Joint Conference on Bioinformatics, Systems Biology and Intelligent Computing (IJCBS)*. (pp. 377-383). Shanghai, China.

Zhang, W.-R. (2009d). The logic of YinYang and the science of TCM – an Eastern road to the unification of nature, agents, and medicine. [IJFIPM]. *International Journal of Functional Informatics and Personal Medicine, 2*(3), 261–291. doi:10.1504/IJFIPM.2009.030827

Zhang, W.-R. (2010a). YinYang bipolar quantum entanglement–toward a logically complete quantum theory. *Proceedings of the 4th Int'l Conference on Quantum, Nano and Micro Technologies (ICQNM 2010)*, (pp. 77-82). St. Maarten, Netherlands Antilles.

Zhang, W.-R. (2010b). *YinYang bipolar quantum computing and bitwise quantum-digital cryptography*. Paper presented at the Quantum Information and Computation VIII, Orlando, Florida, April 5-9, 2010.

Zhang, W.-R., Chen, S., & Bezdek, J. C. (1989). POOL2: A generic system for cognitive map development and decision analysis. *IEEE Transactions on SMC*, *19*(1), 31–39.

Zhang, W.-R., Chen, S., Wang, W., & King, R. (1992). A cognitive map based approach to the coordination of distributed cooperative agents. *IEEE Transactions on SMC*, *22*(1), 103–114.

Zhang, W.-R., & Chen, S. S. (2009). Equilibrium and non-equilibrium modeling of YinYang Wuxing for diagnostic decision analysis in traditional Chinese medicine. *International Journal of Information Technology and Decision Making*, *8*(3), 529–548. doi:10.1142/S0219622009003521

Zhang, W.-R., Pandurangi, A., & Peace, K. (2007). YinYang dynamic neurobiological modeling and diagnostic analysis of major depressive and bipolar disorders. *IEEE Transactions on Bio-Medical Engineering*, *54*(10), 1729–1739. doi:10.1109/TBME.2007.894832

Zhang, W.-R., Pandurangi, K. A., Peace, K. E., Zhang, Y., & Zhao, Z. (in press). MentalSquares–a generic bipolar support vector machine for psychiatric disorder classification, diagnostic analysis and neurobiological data mining. *International Journal on Data Mining and Bioinformatics.*

Zhang, W.-R., Wang, P., Peace, K., Zhan, J., & Zhang, Y. (2008). On truth, uncertainty, equilibrium, and harmony–a taxonomy for YinYang scientific computing. *International Journal of New Mathematics and Natural Computing*, *4*(2), 207–229. doi:10.1142/S1793005708001033

Zhang, W.-R., Wang, W., & King, R. (1994). An agent-oriented open system shell for distributed decision process modeling. *Journal of Organizational Computing*, *4*(2), 127–154. doi:10.1080/10919399409540220

Zhang, W.-R., Zhang, H. J., Shi, Y., & Chen, S. S. (2009). Bipolar linear algebra and YinYang-n-element cellular networks for equilibrium-based biosystem simulation and regulation. *Journal of Biological System*, *17*(4), 547–576. doi:10.1142/S0218339009002958

Zhang, W.-R., & Zhang, L. (2004a). A Multiagent Data Warehousing (MADWH) and Multiagent Data Mining (MADM) approach to brain modeling and NeuroFuzzy control. *Information Sciences*, *167*, 109–127. doi:10.1016/j.ins.2003.05.011

Zhang, W.-R., & Zhang, L. (2004b). YinYang bipolar logic and bipolar fuzzy logic. *Information Sciences*, *165*(3-4), 265–287. doi:10.1016/j.ins.2003.05.010

Zhou, Q., & Yik, J. H. N. (2006). The Yin and Yang of P-TEFb regulation: Implications for Human Immunodeficiency Virus gene expression and global control of cell growth and differentiation. *Microbiology and Molecular Biology Reviews*, *70*(3), 646–659. doi:10.1128/MMBR.00011-06

ADDITIONAL READING

Carlip, S. (2000). Aberration and the speed of gravity. *Physics Letters. [Part A]*, *267*, 81–87. doi:10.1016/S0375-9601(00)00101-8

Dine, M. *(2007). Supersymmetry and String Theory: Beyond the Standard Model.* Cambridge University Press, 2007.

Gefter, A. (2005). Is string theory in trouble? *New Scientist.* Magazine issue 2530. http://www.newscientist.com/article/mg18825305.800-is-string-theory-in-trouble.html?full=true.

Green, M. B., Schwarz, J. H., & Witten, E. (1987). *Superstring Theory.* Cambridge University Press,1987.

Marsch, G. E., & Nissim-Sabat, C. (1999). Comments on "The speed of gravity". *Physics Letters. [Part A], 262*, 103–106.

Polchinski, J. (1998). *String Theory.* Cambridge University Press,1998.

Polyakov, A. M. (1987). *Gauge Fields and Strings.* Harwood Academic, 1987.

Van Flandern, T. (1999). Reply to comments on "The speed of gravity". *Physics Letters. [Part A], 262*, 261–263. doi:10.1016/S0375-9601(99)00676-3

Van Flandern, T., & Vigier, J. P. (2002). Experimental Repeal of the Speed Limit for Gravitational, Electrodynamic, and Quantum Field Interactions. *Foundations of Physics, 32*, 1031–1068. doi:10.1023/A:1016530625645

Woit, P. (2006). *Not Even Wrong: The Failure of String Theory and the Search for Unity in Physical Law.* New York: Basic Book.

KEY TERMS AND DEFINITIONS

Background Independence: *"To say that the laws of physics are background independent means that the geometry of space is not fixed but evolves."* (Smolin 2006, p81-82)

Bipolar Adaptivity: The property of bipolar agents being able to adapt to bipolar equilibrium. (Zhang 2009b,c,d)

Bipolar Agent: Any agent with two poles. (Zhang 2009a,b,c,d)

Bipolar Strings: Bipolar agents with scalability and observability. (Zhang 2009c)

Emergence of Space and Time: *"Space and time emerge from the laws rather than providing an arena in which things happen."* (Smolin, 2006 p82)

Equilibrium-Based Logical Axiomatization of Physics: A minimal but most general axiomatization of physics using equilibrium-based BDL.

YinYang Bipolar Geometry: Equilibrium and non-equilibrium of an agent to YinYang bipolar geometry is as gravitation or acceleration to spacetime geometry. A fundamental difference is that space and time are not quantum in nature and not symmetrical to each other; Yin and Yang as two poles of any bipolar agent can be purely quantum in nature and symmetrical to each other. Thus, bipolar relativity defined in YinYang geometry can be considered a deeper theory of quantum gravity that provides an equilibrium-based unification of nature, agents, and causality (Zhang, 2009a, 2009b, 2009c, 2009d). In YinYang bipolar geometry any acceleration or gravitation in microscopic and macroscopic worlds can be considered a kind of quantum gravity that leads to the subtheories of logical, physical, social, mental, and biological quantum gravities.

Chapter 7
YinYang Bipolar Quantum Entanglement:
Toward a Logically Complete Theory for Quantum Computing and Communication

ABSTRACT

YinYang bipolar relativity leads to an equilibrium-based logically complete quantum theory which is presented and discussed in this chapter. It is shown that bipolar quantum entanglement and bipolar quantum computing bring bipolar relativity deeper into microscopic worlds. The concepts of bipolar qubit and YinYang bipolar complementarity are proposed and compared with Niels Bohr's particle-wave complementarity. Bipolar qubit box is compared with Schrödinger's cat box. Since bipolar quantum entanglement is fundamentally different from classical quantum theory (which is referred to as unipolar quantum theory in this book), the new approach provides bipolar quantum computing with the unique features: (1) it forms a key for equilibrium-based quantum controllability and quantum-digital compatibility; (2) it makes bipolar quantum teleportation theoretically possible for the first time without conventional communication between Alice and Bob; (3) it enables bitwise encryption without a large prime number that points to a different research direction of cryptography aimed at making prime-number-based cryptography and quantum factoring algorithm both obsolete; (4) it shows potential to bring quantum computing and communication closer to deterministic reality; (5) it leads to a unifying Q5 paradigm aimed at revealing the ubiquitous effects of bipolar quantum entanglement with the sub theories of logical, physical, mental, social, and biological quantum gravities and quantum computing.

DOI: 10.4018/978-1-60960-525-4.ch007

INTRODUCTION

After conquering his general theory of relativity regarding space, time, and gravity, there must be few physical phenomena left to freak out Albert Einstein. Yet quantum entanglement caused Einstein to use the word *"spooky."* Take a pair of entangled photons, for instance, it seems that performing an experiment on one of them instantaneously affects another no matter how far apart are the two, whether they are in the same room, or at two opposite ends of the universe. The simultaneous multiple appearances are called *superposition*. Einstein once called this intuitively supernatural behavior *"spooky action at a distance."*

The EPR paradox (Einstein, Podolsky & Rosen, 1935) challenged long-held ideas about the relation between the observed values of physical quantities and the values that can be accounted for by a physical theory. It was their theoretical possibility or impossibility that led Einstein to reject the idea that quantum mechanics might be a fundamental physical law. Instead, Einstein deemed it an incomplete theory. The missing part has been referred to as *"hidden variables"* in the literature.

Three decades later, Bell's theorem (Bell, 1964) extended the argument of the EPR paradox and proved the validity of quantum entanglement from a statistical perspective with probability distributions. Despite its practical significance in quantum computing, Bell's theorem so far hasn't led to a logical unification of general relativity and quantum mechanics. The so-called *"hidden variables"* never really surfaced in simple logically definable terms.

Thus, quantum mechanics at the Planck scale needs to be further reconciled with general relativity. That is the goal of quantum gravity. Until this day, however, the searching for quantum gravity has failed to find a decisive battleground; quantum computing is still years away from reality. Nevertheless, since the 1980s, many successful experimental results have shown the physical existence of quantum entanglement.

Besides the quantum observability problem, another major obstacle in quantum information processing is the difficulty or even impossibility of cloning an unknown quantum state. While classical information can be copied or cloned for storage and retrieval, the quantum *"no cloning"* theorem (Dieks, 1982; Wootters & Zurek, 1982) asserts the impossibility of cloning an unknown quantum state. Without being copied or cloned quantum information cannot be stored for retrieval.

Consequently, despite numerous reported experimental successes in testing quantum entanglement (e.g. Furusawa *et al.,* 1998; Buchanan, 1998; Ghosh *et al.,* 2003; Salart et al., 2008; Jost *et al.,* 2009), the quantum observability controversy and no cloning dilemma remain unresolved. Now many physicists have subscribed to the instrumentalist interpretation of quantum mechanics with the slogan *"Shut up and calculate!"* Some others including several of the best living theoretical physicists feel compelled to question the basic assumptions of relativity and quantum theory. As described by theoretical physicist Lee Smolin they *"learn it, and they can carry out its arguments and calculations as well as anyone. But they don't believe it."* (Smolin, 2006, p. 319).

Nevertheless, the *"spooky"* quantum phenomenon has become a fundamental concept in quantum computing even though, until recently, physicists have only been able to demonstrate quantum entanglement through either highly esoteric examples or under extreme conditions. Without a decisive victory in the quest for quantum gravity, it is fair to say that something fundamental must still be missing from the big picture.

The missing fundamental concept is often traced back to the ultimate unknown cause-effect relationship in quantum entanglement. Without logically definable causality, quantum entanglement could be deemed something beyond the realm of science because, if Aristotle's causality principle is the doctrine of all sciences, there should be no science beyond the doctrine. Again, causality pops up and comes into play. Subsequently, without a resolution to the EPR paradox or the cause-effect relationship in quantum entanglement, research and development in quantum computing can only move forward slowly under theoretical uncertainty. According to an *IEEE Spectrum's* special report – Winners & Losers VII (Guizzo, 2010), practical quantum computing could still be many years away.

Many believe that quantum mechanics as a mathematical description of reality provide predictions and implications that go against the "common sense" of how humans see a set of bodies (a system) behave. This isn't necessarily a failure – it's more of a reflection of how humans understand space and time on larger scales (e.g., centimeters, seconds) rather than much smaller. This view is interesting but stopped short of pointing out two major limitations in existing quantum models:

1. A deeper theory than spacetime relativity is needed because space and time are not symmetrical to each other, can't be entangled, and do not provide the cause-effect relationship of agent interaction.
2. A new logical foundation beyond truth-based reasoning is needed for a complete quantum theory because the particle-wave duality of quantum agents is illogical from a truth-based perspective and any truth-based reasoning on quantum agent interaction would be logical axiomatization for illogical physics (LAFIP) (Zhang, 2009a)(Ch. 1).

As introduced in Chapters 1 and 2, without bipolarity Aristotle's causality principle is logically undefinable by any truth-based system due to the LAFIP paradox. It has been argued in Chapter 1 that the most fundamental property of the universe is not truth, not fuzziness, not space, not time, not spacetime relativity, not matter, not aether, and not strings. Instead, the most fundamental property of nature should be YinYang bipolarity. Then, it is natural to ask the following questions:

1. How can the EPR paradox be resolved with YinYang bipolar relativity?
2. How can the "no cloning" bottleneck be resolved for an unknown quantum state?
3. What is the nature of quantum entanglement cause-effect relationship?
4. Is truth-based quantum entanglement governed by bipolar equilibrium-based quantum entanglement or vice versa?
5. Is *bipolar quantum entanglement* an alternative form or a primary form?
6. What can bipolar quantum entanglement offer to observability, stability, controllability, and measurability in quantum computing?

Based on the logical foundation presented in Chapters 3-5 and YinYang bipolar relativity presented in Chapter 6, we seek answers in this chapter for the above questions from theoretical, technological, and computational perspectives. We focus on the central and basic issues of bipolar quantum entanglement, teleportation, computing and communication. Since computation and communication are inseparable, bipolar *quantum computing* is used for both throughout this work. It is shown that:

1. The conventional concept of qubit can be extended to the concept of *bipolar qubit*;
2. *Bipolar quantum teleportation* and computation can be realized theoretically without the transmission of conventional binary information for the first time.

3. Bipolar quantum computing and communication are compatible with digital computation and communication.
4. Bipolar quantum computation and communication can be simulated with digital technology.
5. Bipolar quantum entanglement makes *bitwise encryption* and *bipolar quantum-digital cryptography* possible.
6. Bipolar quantum entanglement brings quantum computing closer to reality.
7. Bipolar quantum entanglement can be defined as a bipolar symmetry or quasi-symmetry operations that can transport quantum information without passing mass or energy and, therefore, supports the observation that quantum entanglement is not limited to the speed of light (Salart et al., 2008);
8. The speed of gravity could be the speed of quantum entanglement in large scale (Ch. 6, Conjectures 6.20 and 6.21), and bipolar quantum mechanics can be logically unified with gravity under equilibrium-based bipolar quantum entanglement or quantum gravity.

The remaining presentations and discussions in this chapter are organized in the following sections:

- **Review on Quantum Theory**. This section presents a review on classical quantum entanglement and its applications in quantum teleportation, cryptography, quantum information processing, and quantum computing.
- **Toward a Logically Complete Quantum Theory.** This section presents a logically complete bipolar quantum theory with a logical explanation to the missing variable in quantum mechanics.
- **Bipolar Quantum Entanglement and Teleportation**. This section presents the key concepts of bipolar qubits, bipolar complementarity, YinYang bipolar qubit box, bipolar quantum synchronization, entanglement, teleportation and cryptography. Bohr's particle-wave complementarity principle and Schrödinger's cat box are revisited. It is illustrated with simulation for the first time that (1) bipolar teleportation is theoretically possible without conventional communication between Alice and Bob; (2) bipolar bitwise cryptography points to a different direction to cryptography.
- **Taxonomy for Bipolar Quantum Computing.** This section classifies bipolar quantum computing into logical, relational, and algebraic models. In terms of quantum gravity, a *Q5 paradigm* is proposed including the concepts of logical, physical, social, biological, and mental quantum gravities. The potential of bipolar nanotechnology in bipolar quantum computing, biocomputing, and communication is discussed.
- **Research Topics**. This section lists a few research topics.
- **Summary**. This section presents a summary and draws a few conclusions.

REVIEW ON QUANTUM THEORY

The Dilemma of Truth-Based Quantum Logic

Quantum logic is a set of rules for reasoning about propositions which takes the principles of quantum mechanics into account. Since classical quantum logic is truth-based and unipolar in nature, we describe it as *truth-based quantum logic*. On the other hand, YinYang bipolar dynamic logic (BDL) can be termed as *equilibrium-based quantum logic*. Quantum logic research originated from the 1936 paper by Garrett Birkhoff and John von Neumann (1936), who were attempting to reconcile the inconsistency of classical Boolean logic with the facts concerning the measurement of complementary variables in quantum mechanics, such as position and momentum.

Quantum logic is typically formulated as a modified version of propositional logic. But some of its properties clearly distinguish it from classical logic; most notably it does not satisfy the distributive law of propositional logic:

$$p \wedge (q \vee r) = (p \wedge q) \vee (p \wedge r),$$

where the symbols p, q and r are propositional variables. To illustrate consider a particle moving on a line and let

p = "the particle is moving to the left;"
q = "the particle is in the interval [-|x|,|x|];"
r = "the particle is not in the interval [-|x|,|x|];"

then the proposition "$q \vee r$" is true, so

$$p \wedge (q \vee r) = p$$

On the other hand, in quantum mechanics, the *Heisenberg uncertainty principle* (Hilgevoord & Uffink, 2006; Faye, 2008) states that certain pairs of physical properties, like position and momentum, cannot both be known to arbitrary precision. That is, the more precisely one property is known, the less precisely the other can be known. It is impossible to measure simultaneously both position and velocity of a microscopic particle with any degree of accuracy or certainty. This is not only a statement about the limitations of a researcher's ability to measure particular quantities of a system, but once the wave-nature of matter is accepted, the general properties of waves cause the uncertainty principle to be a statement about the nature of the system itself. Therefore, the propositions "$p \wedge q$" and "$p \wedge r$" are both false and the distributive law fails as we have

$$(p \wedge q) \vee (p \wedge r) = \text{false}.$$

From the above, we have two observations: (1) the basic truth-based distributive law in Boolean logic does not apply to quantum mechanics; (2) truth-based quantum logic cannot avoid the LAFIP paradox (Logical Axiomatization of Illogical Physics) (Zhang, 2009a).

Quantum Bit

The concept of quantum bit is central in quantum computing. A *quantum bit* or *qubit* is a unit of quantum information (Schumacher, 1995). It is the quantum counterpart of a digital bit but with fundamental different syntax and semantics. Like a bit, a qubit can have two possible values—normally a 0 or a 1. The difference is that whereas a bit *must* be either 0 or 1, a qubit can be 0, 1, or a superposition of both. It is described by a state vector in a two-level quantum-mechanical system, which is formally equivalent to a two-dimensional vector space over the complex numbers. The *Bloch sphere* (Figure 1) is a geometrical

Figure 1. Bloch sphere

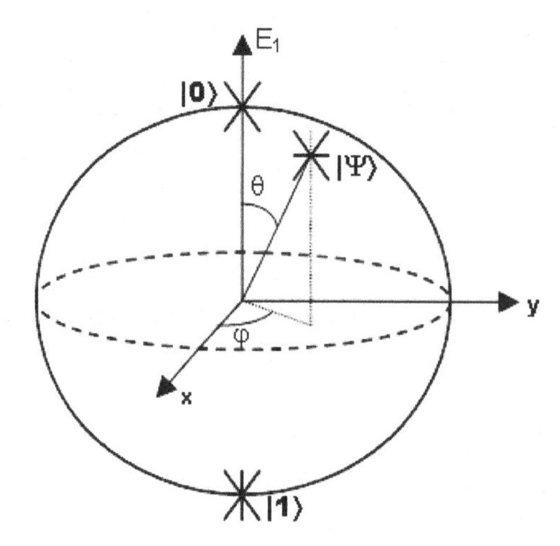

representation of the pure state space of a two-level quantum mechanical system named after the physicist Felix Bloch. Alternatively, it is the pure state space of a one qubit quantum register.

The states in which a qubit may be measured are known as basis states (or vectors). As a convention Dirac notation is used to represent them. In Dirac notation, the two basis states are written as $\left|0\right\rangle$ and $\left|1\right\rangle$ (pronounced "ket 0" and "ket 1"). A pure *qubit state* is a linear *superposition* of those two states that can be represented as a linear combination of $\left|0\right\rangle$ and $\left|1\right\rangle$ such as:

$$\left|\Psi\right\rangle = \alpha\left|0\right\rangle + \beta\left|1\right\rangle,$$

where α and β are probability amplitudes for the two basis states, respectively, and can in general both be complex numbers.

When we measure this qubit in the standard basis, the probability of $\left|0\right\rangle$ is $\left|\alpha\right|^2$ and the probability of $\left|1\right\rangle$ is $\left|\beta\right|^2$. The probability amplitudes α and β must be constrained by the equation

$$\left|\alpha\right|^2 + \left|\beta\right|^2 = 1$$

simply to ensure that you must measure either one state or the other.

The state space of a single qubit register or the Bloch sphere (Figure 1) as a two-dimensional space has an underlying geometry of the surface of a sphere. This essentially means that the single qubit register space has two local degrees of freedom. Represented on such a sphere, a classical bit could lie on only one of the poles.

A distinguishing feature between a qubit and a classical bit is that multiple qubits can exhibit quantum entanglement. Entanglement is a nonlocal property that allows a set of qubits to express higher correlation than is possible in classical systems. Take, for example, two entangled qubits in the Bell state

$\frac{1}{\sqrt{2}}(|00\rangle + |11\rangle)$.

In this state, called an *equal superposition,* there are equal probabilities of measuring either $|00\rangle$ or $|11\rangle$, as $(\frac{1}{\sqrt{2}})^2 = 1/2$.

Imagine that these two entangled qubits are separated, with one each given to Alice and Bob. Alice makes a measurement of her qubit, obtaining – with equal probabilities – either $|0\rangle$ or $|1\rangle$. Because of the qubits' entanglement, Bob must also get the exact same measurement as Alice; i.e., if she measures a $|0\rangle$, Bob must measure the same, as $|00\rangle$ is the only state where Alice's qubit is a $|0\rangle$.

Entanglement also allows multiple states (such as the Bell state mentioned above) to be acted on simultaneously, unlike classical bits that can only have one value at a time. Entanglement is a necessary ingredient of any quantum computation that cannot be done efficiently on a classical computer.

Many of the successes of quantum computation and communication, such as *quantum teleportation* and *superdense coding,* make use of entanglement, suggesting that entanglement is a resource that is unique to quantum computation. A number of entangled qubits taken together is a *qubit register*. Quantum computers perform calculations by manipulating qubits within a register.

Any two-level system can be used as a qubit. Multilevel systems can be used as well, if they possess two states that can be effectively decoupled from the rest (e.g., ground state and first excited state of a nonlinear oscillator).

Quantum Entanglement and the EPR Paradox

It is believed by many that *quantum entanglement,* like energy, is a physical resource associated with the peculiar non-classical correlations that are possible between separated quantum systems. According to Bub (2006), *"Entanglement can be measured, transformed, and purified. A pair of quantum systems in an entangled state can be used as a quantum information channel to perform computational and cryptographic tasks that are impossible for classical systems."*

The term "entanglement" was coined by Schrödinger in his 1935 article to describe this peculiar connection between quantum systems:

When two systems, of which we know the states by their respective representatives, enter into temporary physical interaction due to known forces between them, and when after a time of mutual influence the systems separate again, then they can no longer be described in the same way as before, viz. by endowing each of them with a representative of its own. I would not call that one but rather the characteristic trait of quantum mechanics, the one that enforces its entire departure from classical lines of thought. By the interaction the two representatives (the quantum states) have become entangled. (Schrödinger, 1935, p. 555)

Schrödinger's argument presented an extension to the EPR paradox.

Without a resolution to the quantum mystery, since the 1980s physicists, computer scientists, and cryptographers begins to regard quantum entanglement as *a kind of non-classical resource* that could be exploited, rather than an embarrassment to be explained away (Bub, 2006). This resource view is an interesting one. It is discomforting, however, that the nature of the resource is so far unclear. If it is a physical resource like energy, the cause-effect relationship between entangled quantum systems has to be defined. If it is not similar to other physical resources, the fundamental difference has to be revealed. Without a clear picture on the cause-effect relationship, quantum entanglement will remain mysterious from a logical perspective even though Bell's theorem proved its validity from a probabilistic perspective.

Thus, observations pertaining to entangled states appear to conflict with the property of relativity that information cannot be transferred faster than the speed of light. Although two entangled systems appear to interact across large spatial separations, it can be argued that no useful information can be transmitted in this way as causality cannot be violated through entanglement. Nevertheless, quantum entanglement, as a key concept, has been successfully applied in experimental quantum information processing such as *quantum teleportation*, *quantum cryptography*, and *quantum computing* under special conditions.

Quantum Teleportation

Quantum teleportation is a technique used to transfer information on a quantum level, usually from one particle (or series of particles) to another particle (or series of particles) in another location via quantum entanglement (Bennett *et al.*, 1993). More precisely, it is a quantum protocol by which a qubit a can be transmitted exactly in principle from Alice at one location to Bob at another without transporting any matter or energy. *Quantum teleportation* uses quantum *entanglement* as a key. Let the two entangled qubits be (a,b); let a be with Alice and b be with Bob; let c be another qubit co-located with a on Alice side; assume there is a conventional communication channel capable of transmitting from Alice to Bob two classical bits. The protocol has three steps:

1. Alice enters c into the entanglement (a,b) to form a three-way entanglement $((c,a),b)$ that breaks the original two-way entanglement;
2. Alice measures c and a jointly to yield two classical bits (i.e. one of four entangled quantum states);
3. Alice transmits the two bits to Bob through classical communication (This step is potentially the most time-consuming);
4. Bob uses the two bits to select one of four ways of recovering b and revealing c.

The outcome of this protocol on Bob's side is to change the three-way entanglement $((c,a),b)$ to $((a',b'),c)$ and recover $((c,a),b)$ to a new Bell pair (a',b') and separates c from the three-way entanglement. Thus, Alice has successfully teleported c to Bob without transporting any matter or energy and even without knowing the secret in the "mail". The following is an example (Nielsen & Chuang, 2000, pp. 26-27, 97-98):

Suppose Alice has a qubit C that she wants to teleport to Bob. This qubit can be written generally as: $\left|\Psi\right\rangle = \alpha\left|0\right\rangle + \beta\left|1\right\rangle$. The quantum teleportation scheme requires Alice and Bob to share a maximally entangled state (A,B) beforehand, for instance one of the four Bell basis, Bell states, or EPR pairs, in honor of several of the pioneers who first appreciated the novelty of quantum entanglement (Nielsen & Chuang, 2000, pp. 26-27, 97-98):

$$00:\left|\Phi^{+}\right\rangle = \frac{1}{\sqrt{2}}(\left|0\right\rangle_A \otimes \left|0\right\rangle_B + \left|1\right\rangle_A \otimes \left|1\right\rangle_B) = \frac{1}{\sqrt{2}}(\left|00\right\rangle + \left|11\right\rangle);$$

$$01: \left|\Phi^-\right\rangle = \frac{1}{\sqrt{2}}(\left|0\right\rangle_A \otimes \left|0\right\rangle_B - \left|1\right\rangle_A \otimes \left|1\right\rangle_B) = \frac{1}{\sqrt{2}}(\left|00\right\rangle - \left|11\right\rangle);$$

$$10: \left|\Psi^+\right\rangle = \frac{1}{\sqrt{2}}(\left|0\right\rangle_A \otimes \left|1\right\rangle_B + \left|1\right\rangle_A \otimes \left|0\right\rangle_B) = \frac{1}{\sqrt{2}}(\left|01\right\rangle + \left|10\right\rangle);$$

$$11: \left|\Psi^-\right\rangle = \frac{1}{\sqrt{2}}(\left|0\right\rangle_A \otimes \left|1\right\rangle_B - \left|1\right\rangle_A \otimes \left|0\right\rangle_B) = \frac{1}{\sqrt{2}}(\left|01\right\rangle - \left|10\right\rangle).$$

Alice takes one of the two entangled particles A and Bob takes the other B. We will assume that Alice and Bob share the entangled state $\left|\Phi^+\right\rangle$. So, Alice has two particles: (C, the one she wants to teleport, and A, the one in the entangled pair, and Bob has one particle, B. In the total system, the state of these three particles is given by

$$\left|\Psi\right\rangle \otimes \left|\Phi^+\right\rangle = (\alpha\left|0\right\rangle + \beta\left|1\right\rangle) \otimes \frac{1}{\sqrt{2}}(\left|0\right\rangle \otimes \left|0\right\rangle + \left|1\right\rangle \otimes \left|1\right\rangle)$$

Alice will then make a partial measurement in the Bell basis on the two qubits in her possession. To make the result of her measurement clear, we rewrite the two qubits of Alice in the Bell basis via the following general identities (these can be easily verified):

$$\left|0\right\rangle \otimes \left|0\right\rangle = \frac{1}{\sqrt{2}}(\left|\Phi^+\right\rangle + \left|\Phi^-\right\rangle);$$

$$\left|0\right\rangle \otimes \left|1\right\rangle = \frac{1}{\sqrt{2}}(\left|\Psi^+\right\rangle + \left|\Psi^-\right\rangle);$$

$$\left|1\right\rangle \otimes \left|0\right\rangle = \frac{1}{\sqrt{2}}(\left|\Psi^+\right\rangle - \left|\Psi^-\right\rangle);$$

$$\left|1\right\rangle \otimes \left|1\right\rangle = \frac{1}{\sqrt{2}}(\left|\Phi^+\right\rangle - \left|\Phi^-\right\rangle).$$

The three particle state shown above thus becomes the following four-term superposition:

$$\frac{1}{2}(\left|\Phi^+\right\rangle \otimes (\alpha\left|0\right\rangle + \beta\left|1\right\rangle) + \left|\Phi^-\right\rangle \otimes (\alpha\left|0\right\rangle - \beta\left|1\right\rangle) \\ + \left|\Psi^+\right\rangle \otimes (\beta\left|0\right\rangle + \alpha\left|1\right\rangle) + \left|\Psi^-\right\rangle \otimes (-\beta\left|0\right\rangle + \alpha\left|1\right\rangle).$$

Notice all we have done so far is a change of basis on Alice's part of the system. No operation has been performed and the three particles are still in the same state. The actual teleportation starts when Alice measures her two qubits in the Bell basis. Given the above expression, the result of her local measurement is that the three-particle state would collapse to one of the following four states (with equal probability of obtaining each):

$$\left|\Phi^{+}\right\rangle \otimes (\alpha\left|0\right\rangle + \beta\left|1\right\rangle); \tag{1}$$

$$\left|\Phi^{-}\right\rangle \otimes (\alpha\left|0\right\rangle - \beta\left|1\right\rangle); \tag{2}$$

$$\left|\Psi^{+}\right\rangle \otimes (\beta\left|0\right\rangle + \alpha\left|1\right\rangle); \tag{3}$$

$$\left|\Psi^{-}\right\rangle \otimes (-\beta\left|0\right\rangle + \alpha\left|1\right\rangle). \tag{4}$$

Alice's two particles are now entangled to each other, in one of the four Bell states. The entanglement originally shared between Alice's and Bob's is now broken or has given way to a three-way entanglement. Bob's particle takes on one of the four superposition states shown above. Note how Bob's qubit is now in a state that resembles the state to be teleported. The four possible states for Bob's qubit are unitary images of the state to be teleported.

The crucial step is the local measurement by Alice on the Bell basis. After the measurement is done, Alice has complete knowledge of the state of the three particles; the result of her Bell measurement tells her which of the four states the system is in. She simply has to send her results to Bob through a classical channel. Two classical bits can communicate which of the four results she obtained.

After Bob receives the message from Alice, he will know which of the four states his particle is in. Using this information, he performs a unitary operation on his particle to transform it to the desired state $\alpha\left|0\right\rangle + \beta\left|1\right\rangle$:

- If Alice indicates her result is $\left|\Phi^{+}\right\rangle$, Bob knows his qubit is already in the desired state and does nothing. This amounts to the trivial unitary operation, the identity operator.
- If the message indicates $\left|\Phi^{-}\right\rangle$, Bob would send his qubit through the unitary gate given by the Pauli matrix $\sigma_3 = \begin{bmatrix} 1 & 0 \\ 0 & -1 \end{bmatrix}$ to recover the state.
- If Alice's message corresponds to $\left|\Psi^{+}\right\rangle$, Bob applies the gate $\sigma_1 = \begin{bmatrix} 0 & 1 \\ 1 & 0 \end{bmatrix}$ to his qubit.
- Finally, for the remaining case, the appropriate gate is given by $\sigma_3\sigma_1 = i\sigma_2 = \begin{bmatrix} 0 & 1 \\ -1 & 0 \end{bmatrix}$.

Teleportation is therefore achieved. Experimentally, the projective measurement done by Alice may be achieved via a series of laser pulses directed at the two particles.

Quantum Information

The amount of classical information we gain, on average, when we learn the value of a random variable (or, equivalently, the amount of uncertainty in the value of a random variable) is represented by a quantity called the *Shannon entropy*, measured in bits (Shannon & Weaver, 1949). A random variable is defined by a probability distribution over a set of values. In the case of a binary random variable, the Shannon entropy is 1 bit, representing maximal uncertainty. For the case of maximal knowledge or zero uncertainty in the binary case the Shannon entropy is zero.

Since information is always embodied in the state of a physical system, we can also think of the Shannon entropy as quantifying the physical resources required to store classical information. According to Shannon's source coding theorem or noiseless coding theorem (assuming a noiseless telephone line with no loss of information), the minimal physical resource required to represent a source file to be transferred (effectively, a lower bound on the possibility of compression) is given by the Shannon entropy of the source.

What happens if we use the quantum states of physical systems to store information, rather than classical states? It turns out that quantum information is radically different from classical information. The unit of quantum information is the "qubit", representing the amount of quantum information that can be stored in the state of the simplest quantum system. As we have seen, an arbitrarily large amount of classical information can be encoded in a qubit. This information can be processed and communicated but, because of the peculiarities of quantum measurement, at most one bit can be accessed. The accessible information in a probability distribution over a set of alternative qubit states is limited by the von Neumann entropy, which is equal to the Shannon entropy only when the states are orthogonal in the space of quantum states, and is otherwise less than the Shannon entropy.

While classical information can be copied or cloned, the quantum 'no cloning' theorem (Dieks, 1982; Wootters & Zurek, 1982) asserts the impossibility of cloning an unknown quantum state. This evidently forms a bottleneck in quantum information processing. The article published in the October 23, 2008 issue of the journal *Nature* (Morton *et al.*, 2008), however, reported the first relatively long and coherent transfer of a superposition state in an electron spin "processing" qubit to a nuclear spin "memory" qubit (Morton *et al.*, 2008). This event can be considered the first relatively consistent quantum data storage, a vital step towards the development of quantum computing.

Quantum Cryptography

Linearity prevents the possibility of cloning or measuring an unknown quantum state (Bub, 2006). Similarly, it can be shown that if Alice sends Bob one of two nonorthogonal qubits, Bob can obtain information about which of these qubits was sent only at the expense of disturbing the state. In general, for quantum information there is no information gain without disturbance. The impossibility of copying an unknown quantum state, or a state that is known to belong to a set of nonorthogonal states with a certain probability, and the existence of a trade-off relation between information gain and state disturbance, is the basis of the application of quantum information to cryptography. There are quantum protocols involving the exchange of classical and quantum information that Alice and Bob can exploit to share a secret random key, which they can then use to communicate privately (See Lo's article "Quantum Cryptology" in Lo, Popescu, & Spiller, 1998). Any attempt by an eavesdropper, Eve, to monitor the communication between Alice and Bob will be detectable, in principle, because Eve cannot gain any quantum information without some disturbance to the quantum communication channel. Moreover, the 'no cloning' theorem prohibits Eve from copying the quantum communications and processing them off-line, so to speak, after she monitors the classical communication between Alice and Bob.

While the difference between classical and quantum information can be exploited to achieve successful key distribution, there are other cryptographic protocols that are thwarted by quantum entanglement. Bit commitment is a key cryptographic protocol that can be used as a subroutine in a variety of important cryptographic tasks. In a bit commitment protocol, Alice supplies an encoded bit to Bob. The information available in the encoding should be insufficient for Bob to ascertain the value of the bit, but sufficient, together with further information supplied by Alice at a subsequent stage when she is supposed to reveal the value of the bit, for Bob to be convinced that the protocol does not allow Alice to cheat by encoding the bit in a way that leaves her free to reveal either 0 or 1 at will. It turns out, however, that unconditionally secure two-party bit commitment, based solely on the principles of quantum or classical mechanics (without exploiting special relativistic signaling constraints, or principles of general relativity or thermodynamics) is impossible (Bub, 2006).

Quantum Computing

David Deutsch (1985) showed how to exploit quantum entanglement to perform a computational task that is impossible for a classical computer. Deutsch's example shows how quantum information, and quantum entanglement, can be exploited to compute a global property of a function in one step that would take two steps classically.

After Deutsch, quantum algorithms have been developed to achieve a speed-up over any known classical algorithm, and in some cases the speed-up is shown to be exponential over any classical algorithm. This is again due to the phenomenon of entanglement, where the amount of information required to describe a general entangled state of n qubits grows exponentially with n. Since the state space as Hilbert space has 2^n dimensions, a general entangled state is a superposition of 2^n n-qubit states.

While Deutsch's problem is somewhat trivial, now several quantum algorithms have been developed for non-trivial applications. Most notably, Shor's factorization algorithm for factoring large composite integers in polynomial time has direct application to *"public key"* cryptography, a vital and widely used cryptographic scheme by governmental, military, and commercial institutions of the world (Shor, 1994).

Shor's algorithm is important because with a quantum computer it can be used to break the widely used public-key cryptography scheme known as RSA on digital computers. RSA is based on the assumption that factoring large numbers is computationally infeasible. So far as is known, this assumption is valid for digital (non-quantum) computers because no classical algorithm is known that can factor in polynomial time. However, Shor's algorithm shows that factoring is much more efficient on a quantum computer and an appropriately large quantum computer can break RSA easily.

As a theoretically threatening factor to the security of existing digital communication, Shor's algorithm provides a powerful motivator for the design and construction of quantum computers and for the study of new quantum computer algorithms. Despite its questionable feasibility of physical implementation, Shor's algorithm has triggered the continuing worldwide interest in quantum computers and computing algorithms.

Different Views on Quantum Computing

Different interpretive perspectives on quantum computing are enumerated in Jeffrey Bub's article (Bub, 2006):

1. The so-called "many-worlds" interpretation explains an entangled state as a manifestation of parallel computations in different worlds.
2. An alternative view is the quantum logical approach, which emphasizes the non-Boolean structure of properties of quantum systems.
3. An information-theoretic interpretation of quantum mechanics presented by Andrew Steane (1998, p. 119) makes the radical suggestion: *"Historically, much of fundamental physics has been concerned with discovering the fundamental particles of nature and the equations which describe their motions and interactions. It now appears that a different program may be equally important: to discover the ways that nature allows, and* prevents, *information to be expressed and manipulated, rather than particles to move."* Steane concludes his review with the proposal (1998, p. 171) that *"I would like to propose a more wide-ranging theoretical task: to arrive at a set of principles like energy and momentum conservation, but which apply to information, and from which much of quantum mechanics could be derived. Two tests of such ideas would be whether the EPR-Bell correlations thus became transparent, and whether they rendered obvious the proper use of terms such as 'measurement' and 'knowledge'."*
4. In line with Steane's proposal, Clifton, Bub, and Halvorson (2003) have shown that one can derive the basic kinematic features of a quantum description of physical systems from three fundamental information-theoretic constraints:
    a. the impossibility of superluminal information transfer between two physical systems by performing measurements on one of them;
    b. the impossibility of perfectly broadcasting the information contained in an unknown physical state (which, for pure states, amounts to 'no cloning');
    c. the impossibility of communicating information so as to implement a bit commitment protocol with unconditional security.

In approach (4), the analysis is carried out in an algebraic framework which allows a mathematically abstract characterization of a physical theory that includes, as special cases, all classical mechanical theories of both wave and particle varieties, and all variations on quantum theory, including quantum field theories (plus any hybrids of these theories, such as theories with superselection rules). Within this framework, the three information-theoretic constraints are shown to jointly entail three physical conditions that are taken as definitive of what it means to be a quantum theory in the most general sense, specifically that:

a. The algebras of observables pertaining to distinct physical systems commute (a condition usually called *microcausality* or *kinematic independence*);
b. Any individual system's algebra of observables is *noncommutative*; and
c. The physical world is *nonlocal*, in that space like separated systems can occupy entangled states that persist as the systems separate.

TOWARD A LOGICALLY COMPLETE QUANTUM THEORY

The Missing Fundamental in Quantum Mechanics

From the review in last section it is clear that the uncertainties on quantum mechanics all boil down to the unresolved central issue surrounding the EPR paradox: *find the "hidden variables" and define the "cause-effect" relationship in simple unambiguous logical terms for quantum entanglement.* If this is accomplished, it would lead to stable and controllable quantum mechanics with far reaching impact on nanotechnology and quantum computing both theoretically and practically. Unfortunately, after so many years of worldwide effort, a resolution for this central issue is now widely considered mission impossible.

From the review it is also clear that even though bipolarity, as the most fundamental physical property of nature, is involved in one way or another in all successful experiments in quantum entanglement, none of the reported logical and algebraic systems in quantum mechanics are based on bipolar equilibrium or symmetry; none of them used logically definable causality and bipolar entanglement such as BUMP and bipolar relativity.

As a matter of fact, even though BDL is the only logic that leads to logically definable causality, it has been deemed unscientific because it is equilibrium-based but not truth-based. As we all know that truth-based logic is originated from Aristotle's syllogistic logic that was developed two thousands of years ago when the Earth was believed the center of the universe, electron-positron pair was unimaginable, and negative numbers were not available. A major argument is then, if truth is not a physical property, bipolarity as a fundamental physical property should play a key role in the logic of quantum mechanics.

Bipolar Quantum Symmetry or Quasi-Symmetry

Without a resolution to the EPR paradox, it has become a popular view in quantum mechanics that: *quantum entanglement is a non-classical physical resource, we should just use it instead of explaining it away.* Although it cannot be effectively used without knowing its cause-effect relationship like other resources, the resource view is an interesting and useful one. It provides a starting point for further observation and discussion.

First, if quantum entanglement is a non-classical physical resource, it must be fundamentally different from resources like matter or energy. This conclusion is supported by the observation that successful experimental tests of quantum teleportation with entanglement do not transport matter or energy.

Secondly, if we recall the *symmetry* concept from the standard model of particle physics, we have the useful definition that *a symmetry is an operation that doesn't change how something behaves relative to the outside world.* Then, a quantum teleportation operation on a physical resource would either have to transport matter or energy or be a symmetry. Since no matter or energy is transported, it has to be a symmetry, and quantum entanglement has to be a pure information resource.

With the above observations and analysis, our focus can shift to the symmetry of the cause-effect relationship in quantum entanglement. In this regard we recall the following:

1. The subatomic particle, the B-sub-s meson, discovered at the Fermi National Accelerator Laboratory can switch between matter and antimatter three trillion times a second (Fermilab, 2006).
2. Successful creation of quantum gas of polar molecules has been reported in *Science* (Ni *et al.*, 2008) where each molecule can carry a pair of negative and positive charges.

3. The YinYang 1 (YY1) regulator protein (Shi *et al.,* 1991) as a ubiquitous genetic agent in all cell types of biological species shows bipolar (repressor, activator) behavior, and such bipolar regulating behavior is believed a key in maintaining a dynamic equilibrium of gene transcription (Vasudevan, Tong & Steitz, 2007).
4. A bipolar operation on a bipolar quantum lattice (BQL) B is a symmetry if it doesn't change any property of B; a bipolar operation on a bipolar quantum lattice B is an energy symmetry if it doesn't change the energy total of B even if it does change the energy polarity (Ch. 4). It has been shown that the negation operation "–" is a symmetry on a balanced BQL and an energy symmetry on an unbalanced BQL (Ch. 4).
5. The two basic bipolar strings (-1,1) and (-1,0) provide a bipolar unification of particle-ware duality (Ch. 6) that form a basis for a discrete equilibrium or non-equilibrium quantum field theory where local non-equilibria can form a global equilibrium or non-equilibrium.

Axiom 7.1. Any polarity change of a bipolar agent or bipolar string (e.g. matter-antimatter subatomic particle polarity change) can be considered an energy symmetry or quasi-symmetry that gives out information without passing observable energy as we have

$$(-1,0)^{n} = \begin{cases} (-1,0) \text{ if n is odd;} \\ (0,1) \text{ if n is even.} \end{cases}$$

Axiom 7.2. Any bipolar interaction of a balanced NP (-,+) bipolar agent or string can be considered a bipolar symmetry or quasi-symmetry that gives out no information from a background independent perspective as we have

$$(-1,1)^{n} \equiv (-1,1).$$

So far our discussion assumed *background independence* where NP (-, +) and PN (+,-) bipolarities do not make any difference. If we assume a *background dependent* Hilbert space, we have the following:

Axiom 7.3. In a background dependent space, any polarity change of a bipolar agent or string (e.g. matter-antimatter subatomic particle polarity change) can be considered an NP, PN, or mixed energy symmetry or quasi-symmetry that gives out information without passing observable energy as we have

NP Symmetry:

$$(-1,0)^{n} = \begin{cases} (-1,0); \text{ // if n is odd;} \\ (0,+1). \text{ // if n is even.} \end{cases}$$

PN Symmetry:

$$(0.-1)^{n} = \begin{cases} (0,-1); \text{ // if n is odd;} \\ (+1,0). \text{ // if n is even.} \end{cases}$$

Axiom 7.4. In a background dependent space, any bipolar oscillation of a balanced bipolar agent or bipolar string can be considered an NP, PN, or mixed bipolar symmetry or quasi-symmetry that gives out information without passing observable energy as we have

$$-(-1,+1) = (-1,+1) \otimes (-1,0) = (+1,-1)$$
$$//\ \text{from 2nd quadrant to 4th quadrant.}$$
$$(\text{-}1,\text{+}1) \otimes (\text{-}1,0)^{n} = \begin{cases} (+1,-1);\ //\ \text{if n is odd;} \\ (-1,+1).\ //\text{if n is even.} \end{cases}$$

Hypothesis 7.1: Quantum entanglement consists of one or more operations of bipolar symmetry or quasi-symmetry that gives out information without passing observable energy.

Hypothesis 7.2: Quantum teleportation with entanglement consists of a sequence of operations in bipolar symmetry or quasi-symmetry that gives out information without passing observable energy.

Hypothesis 7.3: The hidden variables of a quantum entanglement form a collective bipolar connection which can be considered *a scalable chain of bipolar strings in symmetry or quasi-symmetry*; the chain is governed by equilibrium-based YinYang bipolar relativity; information can be passed through the chain due to bipolar relativity and transitivity that logically define causality.

Hypotheses 7.1 - 7.3 provide a symmetrical causal interpretation to the EPR paradox. It forms a YinYang bipolar cosmological order theory that not only constitutes a scientific extension to the ancient Chinese YinYang cosmology but also constitutes an equilibrium-based logical foundation for David Bohm's cosmological order theory (Bohm, 1957, 1980).

Formally, Let $\psi = (\psi^-, \psi^+)$ = (negative, positive) be a bipolar predicate and a,b,c,d be any four bipolar strings. $\forall i, i=0,1,..,n$ *and* $\forall a_i, b_i, c_i, d_i$ we have Eq. 7.1a and Eq. 7.1b:

$$[(\psi(a_i(t_x,p_1)) \Rightarrow \psi(c_i(t_y,p_3))) \ \&\ (\psi(b_i(t_x,p_2)) \Rightarrow \psi(d_i(t_y,p_4)))] \Rightarrow [(\psi(a_i(t_x,p_1)) * \psi(b_i(t_x,p_2))) \Rightarrow (\psi(c_i(t_y,p_3)) * \psi(d_i(t_y,p_4)))] \Rightarrow [(\psi(a_i(t_x,p_1) \blacklozenge b_i(t_x,p_2)) \Rightarrow (\psi(c_i(t_y,p_3) \blacklozenge d_i(t_y,p_4))]. \quad (7.1a)$$

$$[\psi(a_i(t_x,p_1)) \Rightarrow \psi(b_i(t_x,p_2)) \Rightarrow \psi(c_i(t_y,p_3))] \Rightarrow [\psi(a_i(t_x,p_1)) \Rightarrow (\psi(c_i(t_y,p_3))]. \quad (7.1b)$$

Eq. 7.1a defines bipolar relativity and 7.1b defines bipolar transitivity. Together the two define a chain of bipolar physical and logical quantum causal-effect operations in symmetry or quasi-symmetry that gives out information without passing observable energy. The physical bipolar interactive operator ♦ and the logical bipolar universal operator * can be unified in the chain of physical operations. Thus, the cause-effect relationship of quantum entanglement is logically defined with bipolarity and symmetry.

It should be remarked that a 2008 quantum physics experiment performed in Geneva, Switzerland has determined that the *"speed" of the quantum non-local connection* what Einstein called *"spooky action at a distance"* has a minimum lower bound of 10,000 times the speed of light (Salart et al., 2008). However, modern quantum physics cannot expect to determine the maximum given that we do not know the sufficient causal condition of the system we are proposing. Bipolar relativity or causality provides a coherent qualitative interpretation for the result of the quantum physics experiment.

Based on bipolar relativity, it is reasonable to believe that the speed of bipolar *quantum non-local connection* between two entangled photons or electrons is much faster than the speed of light because, as a symmetrical operation or quantum synchronization, there is no observable matter or energy transmitted (such as photons in light or electrons in electromagnetic waves) (See Conjectures 6.20-6.21).

In a nutshell, the bipolar quantum theory provides a unique interpretation to the hidden variables in the EPR paradox. *With this interpretation, the hidden variable is bipolarity – the most fundamental property of nature; the "spooky action at a distance" is caused by the bipolar symmetrical property of YinYang bipolar relativity or causality.*

Moreover, if gravitation is interpreted as bipolar quantum entanglement, the speed of gravity could be the speed of quantum entanglement in large scale (Conjectures 6.20 and 6.21). In that case, bipolar quantum mechanics would be logically unified with general relativity under the equilibrium-based bipolar quantum gravity theory. As discussed in Chapter 6, the logically complete theory does not depend on the existence of graviton which has been predicted as a massless elementary particle but so far undiscovered. Thus, the new theory can avoid the pitfalls of gravitational and quantum infinities (Smolin, 2006).

BIPOLAR QUANTUM ENTANGLEMENT AND TELEPORTATION

In this section we present the concept of bipolar quantum bits and discuss the theoretical and technical possibility of bipolar quantum entanglement and teleportation. Simplified minimum cases are used in illustrations. We limit our discussion to the four basis states of a bipolar qubit that form BDL. Without losing generality, the four basis states can always be extended to non-orthogonal states. It is shown that bipolar quantum register enable the compatibility between classical digital computation and bipolar quantum computation. Such compatibility provides a technological basis for quantum-digital computing, quantum-digital communication, and quantum-digital teleportation.

Bipolar Qubit

The theory of bipolar quantum entanglement presented in the last section provides a logical basis for the concept of *bipolar qubit*. It is well-known that it is easy to trap an electron-positron pair but not easy to trap either one of them. On the other hand, the realization of quantum gas of polar molecules has been reported in *Science* (Ni *et al.*, 2008) where each molecule can carry a pair of negative and positive charges. Both findings support the idea of bipolar qubit.

Recalling that a classical qubit is the counterpart of a binary bit, we use binary bit as a common reference. Figure 2 shows a binary view of a quantum variable x with unknown polarity. If we could determine the position of x in terms of left or right (but not both) of the sphere and let one half encode 0 and the other half encode 1, Figure 2 would be a 1-bit binary register. The middle line can be counted as either left or right but not both.

Figure 3 show, respectively, the binary views of a negative quantum (-) and that of a positive quantum (+). If we could determine the position of + or - in terms of left or right (but not both) of the sphere and let one half encode 0 and the other half encode -1 or +1, respectively, Figure 3 would be a 2-bit bipolar register that can register the four bipolar logic values (0,0), (0,+1), (-1,0), and (-1,+1) of BDL. The rationale for the encoding scheme is that:

Figure 2. A binary register

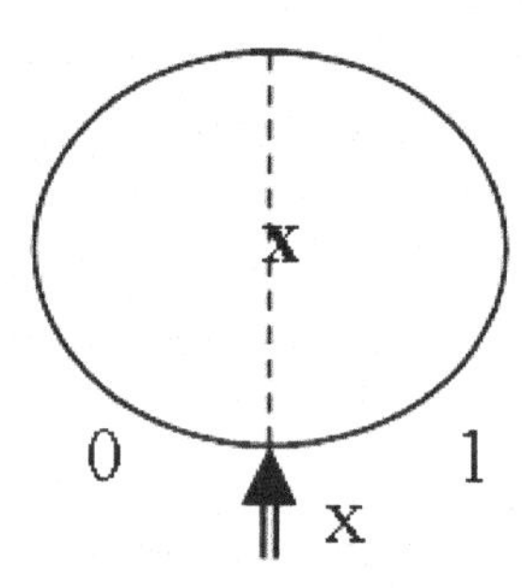

Figure 3. Side views (or binary views) of a bipolar register

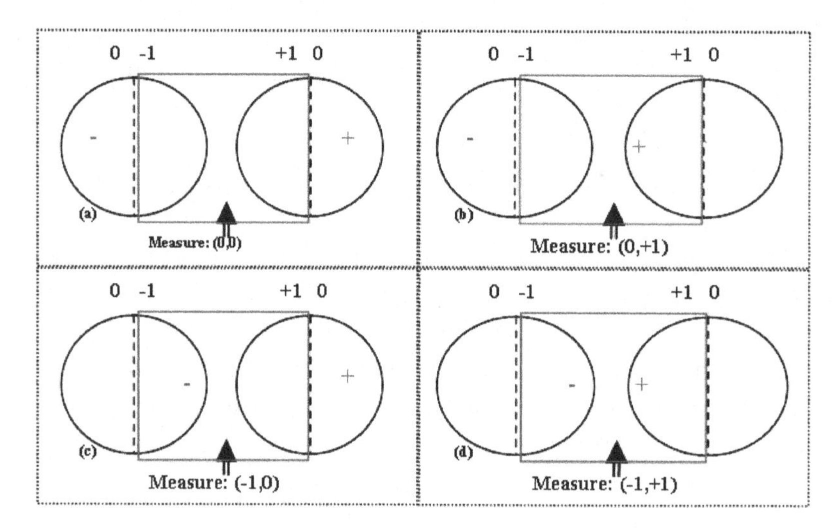

1. The weakest bipolar equilibrium (0,0) is physically implemented with the loose bipolar coupling of the two quanta far away from each other (Figure 3.a);
2. The strongest bipolar equilibrium (-1,+1) is physically implemented with the strong bipolar coupling of the two quanta close from each other (Figure 3d); and
3. The two non-equilibrium states (0, +1) and (-1, 0) are physically implemented with intermediate levels of bipolar coupling as shown in Figure 3b and 3c, respectively.

While a binary view of a bipolar register is a partial view from one side, a quantum view needs two Bloch spheres with two quanta in opposite polarities. This is shown in Figure 4. We call the quantum view a *bipolar qubit register* or *bipolar qubit box*. The quantum view is actually a background dependent 4-dimensional view (3-d + bipolarity). Interestingly, the extra dimension with bipolar symmetry does not add complexity but stability and controllability.

A *bipolar qubit* can be denoted $\left|(\Psi^-, \Psi^+)\right\rangle$. If we only consider the 2-dimensional side view, we would have *four bipolar basis states* $\left|(0,0)\right\rangle, \left|(-1,0)\right\rangle, \left|(0,+1)\right\rangle$, and $\left|(-1,+1)\right\rangle$. Alternatively, the four

Figure 4. A bipolar qubit register with a negative and a positive pole

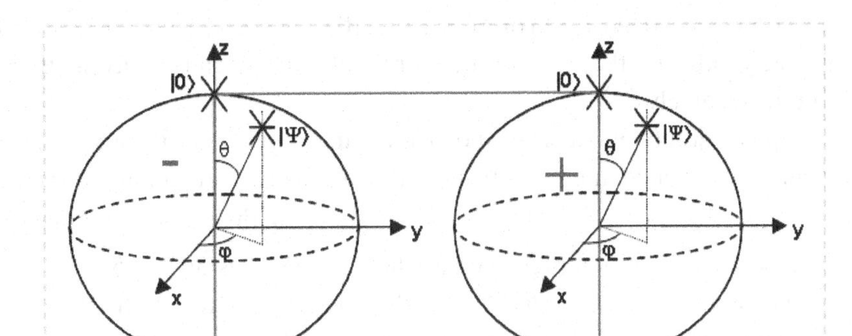

bipolar basis states can be considered approximations of $\left|(\Psi^-, \Psi^+)\right\rangle$. Evidently, the four basic states can find a one-to-one match in the four values of BDL or $\forall(\Psi^-, \Psi^+) \in B_1 = \{-1, 0\} \times \{0, 1\}$.

Let the entropy of a bipolar qubit register be B and that of a classical qubit register be C, it is clear that $B = C^2$. In terms of basis states, a classical qubit has two basis states $\left|0\right\rangle$ and $\left|1\right\rangle$; a bipolar qubit has four basis states $\left|(0,0)\right\rangle, \left|(-1,0)\right\rangle, \left|(0,+1)\right\rangle$, and $\left|(-1,+1)\right\rangle$ where $4 = 2^2$. Thus, a bipolar qubit register has a much larger information capacity. At first glance, it seems to be much more complex than a single qubit register. Actually, it leads to a number of significant simplifications for quantum computing and communication.

Bipolar Synchronization

Without bipolarity, the state of a quantum is highly unstable and uncontrollable. Bipolar relativity brings certain stability and controllability into quantum mechanics and nanotechnology in general. Actually, all successful experiments in quantum mechanics borrow bipolarity one way or another. For instances, polarized photons and polarized ions are typically used in the experiments of quantum entanglement; electromagnetic fields and laser beams are typically used for stabilizing the states of quanta, that are both bipolar in nature; measurement of a quantum state heavily depends on certain stabilization technique that are also bipolar in nature.

We define *bipolar synchronization* as a process of achieving synchronized behavior between a negative and a positive quantum in a bipolar quantum register. Theoretically speaking, the two quanta or the two poles of a bipolar quantum register can be synchronized. By *bipolar synchronization* we mean a bipolar qubit can be steered to any of the four initial basis states: $\left|(0,0)\right\rangle, \left|(0,+1)\right\rangle, \left|(-1,0)\right\rangle$, and $\left|(-1,+1)\right\rangle$. Although physical experiment of bipolar synchronization is beyond the scope of this book, theoretically speaking bipolar synchronization is achievable and assumed a prerequisite for the following discussions. Bipolar synchronization leads to partial bipolar quantum controllability in terms of hardware implementation. On the other hand, software implementation or simulation of bipolar quantum controllability can be realized with BDL. With controllable bipolar qubit registers, bipolar quantum entanglement and teleportation are achievable.

Notice that complementary bipolar qubits could be used for the steering and measurement of a bipolar quantum register as we have ¬(0,0)≡(-1,+1) or ¬(-1,+1) ≡ (0,0) and ¬(-1,0)≡(0,+1) or ¬(0,+1) ≡ (-1,0). Bipolar complementarity indicates that each of the four bipolar qubits can be physically regulated by its bipolar complement, respectively.

If we assume a bipolar qubit can steer a bipolar quantum register to its complementary state, we would be able to achieve certain quantum controllability such as steering a quantum to any of the four bipolar basis states $|(0,0)\rangle, |(-1,0)\rangle, |(0,+1)\rangle$, or $|(-1,+1)\rangle$ using their complementary bipolar qubits. Such a controller can be symbolically represented as shown in Figures 5a and 5b.

Bipolar quantum register has a number of distinguishing features from a traditional quantum register. The most noticeable unique property is that the controller (or observer) and controlled (or observed) can be quantum registers that provides an important theoretical basis for logically definable quantum causality and bipolar quantum teleportation without classical communication channel to be discussed later on.

YinYang Bipolar Complementarity vs. Bohr's Particle-Wave Duality

Niels Bohr was the first to bring YinYang into quantum mechanics for his quantum complementarity principle (Hilgevoord & Uffink, 2006; Faye, 2008). According to Bohr, a causal description of a quantum process cannot be attained and we have to contend ourselves with complementary descriptions. Bohr concluded (Bohr, 1948): *"The viewpoint of complementarity presents itself as a rational generalization of the very ideal of causality."*

Bohr's principle focused on the complementary nature of particle and wave or the Yin and Yang, respectively. Since the two concepts of particle and wave are not (-,+) bipolar opposites, they are complementary but not bipolar complementary. Since equilibrium or non-equilibrium-based YinYang (-,+) bipolarity was not formally defined, Bohr's principle didn't go beyond truth-based logical reasoning. It is essentially an instance of logical axiomatization for illogical physics (LAFIP) from a truth-based perspective. As a result, it stopped short of offering logically definable causality and, inevitably, has to be subject to Heisenberg uncertainty principle in the Copenhagen interpretation of quantum mechanics (Hilgevoord & Uffink, 2006; Faye, 2008).

YinYang bipolar complementarity, on the other hand, focuses on (-,+) bipolarity to foster controllability but embraces uncertainty. The potential of bipolar controllability is shown in Figure 5; uncertainty is

Figure 5. Bipolar complementary qubits as quantum controllers

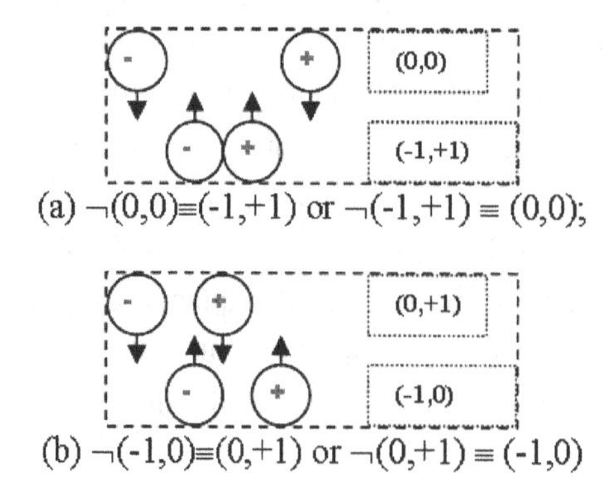

embedded in the universal non-linear bipolar dynamic operator $* \in \{\&, \oplus, \otimes, \oslash, \&^-, \oplus^-, \otimes^-, \oslash^-\}$. While physical experimental validation is beyond the scope of this monograph, BDL has been proven a sound and complete formal system (Ch. 3). Theoretically, with the postulate that bipolarity is the most fundamental property of nature (Ch. 1), bipolar complementarity constitutes a basic principle of quantum gravity.

YinYang Bipolar Qubit Box vs. Schrödinger's Cat Box

Since the rectangle area of the bipolar quantum register (Figure 4) can also be called a *YinYang bipolar qubit box*, we have the intriguing question: *Is the YinYang bipolar qubit box comparable with Schrödinger's cat box?*

Schrödinger's cat is a thought experiment, often described as a paradox, devised by Schrödinger in 1935. It illustrates what he saw as the problem of the Copenhagen interpretation of quantum mechanics (Hilgevoord & Uffink, 2006; Faye, 2008) applied to everyday objects. The thought experiment presents a cat that might be alive or dead, depending on an earlier random event. In the course of developing this experiment, Schrödinger coined the term entanglement (Schrödinger, 1935). In the experiment, he proposed a scenario with a cat in a sealed box, wherein the cat's life or death was dependent on the state of an atomic particle. According to Schrödinger, the Copenhagen interpretation implies *that the cat remains both alive and dead* to the universe outside the box (due to the superposition $|\phi\rangle = (|0\rangle + |1\rangle) / \sqrt{2}$) until the box is opened.

"Is Schrödinger's cat really out of the box?" is often meant to be *"Is the quantum entanglement mystery resolved?"* While it is too early to conclude that the bipolar qubit box provides a resolution to the quantum entanglement mystery, we do have the following observations:

1. Schrödinger's cat box is sealed where a cat cannot get in and out freely.
2. The YinYang bipolar qubit box is a virtual box that doesn't really have sealed boundaries and the two quanta can get in and out freely as long as they stay in their respective Bloch spheres or poles.
3. Schrödinger's cat can be dead and alive at the same time which is paradoxical and inconceivable from everyday truth objects (Note: Schrödinger's cat paradox is similar to Bohr's particle-wave duality and is essentially another instance of logical axiomatization for illogical physics (LAFIP) (Ch. 1)).
4. The negative and positive qubits, or *the Yin and Yang poles,* can coexist in bipolar equilibrium or non-equilibrium that is neither subjected to the LAFIP paradox nor to Schrödinger's cat paradox because an YinYang bipolar logical variable assumes a deterministic value of four options: alive with mental equilibrium (-1,+1), alive with depression (-1,0), alive with mania (0,+1), or dead with eternal equilibrium (0,0).
5. While Schrödinger's cat box is a black box; a bipolar qubit box is a deterministic white box because when a bipolar qubit is out of the box it can't be inside or vice versa.

The above observations and comparisons reveal a probable resolution to Schrödinger's cat paradox. Such a resolution is potentially the key to bring the ubiquitous effects of quantum entanglement and quantum computing closer to logical, physical, mental, social, and biological reality.

Bipolar Quantum Entanglement and Teleportation

With partially controllable bipolar quantum registers, we are now ready to explore bipolar quantum entanglement and teleportation. We limit our discussion to the four orthogonal or basis qubit values and we omit time and space in our discussion. Without losing generality, non-orthogonal qubit states can be used with time and space that can also be approximated to the four orthogonal qubit values.

Definition 7.1. Two bipolar qubits $a = (a^-,a^+)$ and $b = (b^-,b^+)$ are said *two-way bipolar entangled* iff

$$(a^-,a^+) \Leftrightarrow (b^-,b^+); \tag{7.2a}$$

Two bipolar qubits $a = (a^-,a^+)$ and $b = (b^-,b^+)$ are said *one-way bipolar entangled from a to b* if

$$(a^-,a^+) \Rightarrow (b^-,b^+) \text{ but not } (a^-,a^+) \Leftarrow (b^-,b^+). \tag{7.2b}$$

Definition 7.2. Two-way quantum entanglement is *symmetrical quantum entanglement.* One-way quantum entanglement is *quasi-symmetrical or asymmetrical quantum entanglement.*

While symmetrical or asymmetrical quantum entanglement presents the simplest logical definition of quantum entanglement, they also distinguish two different types of quantum entanglements for the first time. With this distinction, subatomic level entanglement between two photons, electrons, or (electron, positron) pairs can be deemed symmetrical small-scale quantum entanglement; gravitational force between a planet and its star can be logically described as asymmetrical large scale quantum entanglement (See Ch. 6, Eq. 6.7, 6.8a, 6.8b).

It is interesting to examine the nature of entanglement at the atomic level. Recall that an atom containing an equal number of protons and electrons is electrically neutral, otherwise it has a positive or negative charge and is an ion. Thus, even though the electromagnetic force between electrons of an atom and its nucleus is unbalanced, bipolar quantum entanglement at the atom level is actually symmetrical that is different from the gravitational force between a planet and its star.

With the notion or symmetrical and asymmetrical bipolar quantum entanglements, gravity, electromagnetism, and quantum mechanics are logically unified under the equilibrium-based approach. Equilibrium-based bipolar quantum gravity is then justified in both microscopic and macroscopic worlds. While large scale asymmetrical quantum entanglement has been discussed in the last chapter, this chapter is focused on symmetrical cases in microscopic quantum worlds. These cases are generally referred to as bipolar quantum entanglement.

Theorem 7.1. Bipolar quantum entanglement is a sufficient and necessary condition for unipolar quantum entanglement, but not vice versa.

Proof. Let $a^- \leftrightarrow b^-$ *and* $a^+ \leftrightarrow b^+$ *denote unipolar quantum entanglement, the sufficiency follows directly from* $[(a^-, a^+) \Leftrightarrow (b^-, b^+)] \Rightarrow [a^- \leftrightarrow b^- \text{ and } a^+ \leftrightarrow b^+]$*. Since the universe is a near perfect matter-antimatter quasi-symmetry with "one extra particle of matter for every ten billion particles of antimatter" (The Royal Swedish Academy of Sciences, 2008), without bipolar quantum entanglement* $(a^-,a^+) \Leftrightarrow (b^-,b^+)$*, there would be no unipolar quantum entanglement* $a^- \leftrightarrow b^-$ *and* $a^+ \leftrightarrow b^+$*. Thus, the necessity follows.*

On the other hand, unipolar quantum entanglement is not a sufficient condition for bipolar quantum entanglement. Formally, without reciprocal bipolar interaction, $[(a^- \leftrightarrow b^-) \wedge (a^+ \leftrightarrow b^+)] \neq [(a^-,a^+) \Leftrightarrow (b^-,b^+)] \Leftrightarrow [(a^-,a^+)^n \Leftrightarrow (b^-,b^+)^n]$. □

We call the four basis states of bipolar qubit *orthogonal* states; we call other states *non-orthogonal.* A bipolar qubit in an orthogonal state is called an *orthogonal bipolar qubit.*

Theorem 7.2. If quantum entanglement is governed by bipolar relativity (Eq. 7.1), given two entangled bipolar qubits $a = (a^-,a^+)$ and $b = (b^-,b^+)$, let $c = (c^-,c^+)$ be another bipolar qubit, let ♦ and * be physical and logical bipolar interaction operators, respectively, we must have

$$[(a^-, a^+) \Leftrightarrow (b^-, b^+)] \Rightarrow [(a^-,a^+)\blacklozenge(c^-,c^+) \Leftrightarrow (b^-,b^+)\blacklozenge(c^-,c^+)]. \quad (7.3a)$$

$$[(a^-, a^+) \Leftrightarrow (b^-, b^+)] \Rightarrow [(a^-,a^+)*(c^-,c^+) \Leftrightarrow (b^-,b^+)*(c^-,c^+)]. \quad (7.3b)$$

Proof. It follows BUMP directly. □

We call Theorem 7.2 the *dual physical-logical laws of bipolar quantum entanglement.* While BUMP is logical in nature, the dual laws of bipolar quantum entanglement define the nature of bipolar quantum entanglement as an information resource. This unique feature makes quantum entanglement explicit both physically and logically. It is shown later that this law makes bipolar teleportation theoretically achievable.

Given two entangled bipolar qubits $a = (a^-,a^+)$ and $b = (b^-,b^+)$ in orthogonal states; let a be with Alice and b be with Bob; let $c = (c^-,c^+)$ be another bipolar qubit co-located with a on Alice side; assume there is a conventional communication channel capable of transmitting from Alice to Bob three binary bits. A *bipolar teleportation protocol* can be carried out with three steps:

1. Alice enters c into the entanglement (a,b) to form a three-way entanglement $((c,a),b)$ that breaks the original two-way entanglement. Unlike in classical quantum teleportation, in the bipolar case, $((c,a),b)$ can be formally denoted using the BDL operation $(a^-,a^+)*(c^-,c^+) = (x^-,x^+) \in B_l$.
2. Alice measures $((c,a),b)$ to yield the bipolar value (x^-,x^+) and with $(a^-,a^+)*(c^-,c^+) = (x^-,x^+)$ Alice will be able to determine the operator $* \in \{\&, \oplus, \otimes, \varnothing, \&^-, \oplus^-, \otimes^-, \varnothing^-\}$, or one of eight possible operators.
3. Alice transmits the one of eight operators with three binary bits to Bob through the conventional communication channel (As in classical teleportation, this is potentially the most time-consuming step).
4. Due to the bipolar quantum entanglement $(a^-,a^+) \Leftrightarrow (b^-,b^+)$, Bob's side now measures (x^-,x^+), and he can use the three bits to select one of eight ways to reveal $c=(c^-, c^+)$ using a simple binary decoder without quantum computation.
5. The original bipolar quantum entanglement $(a^-,a^+) \Leftrightarrow (b^-,b^+)$ is now $[a = (x^-,x^+)] \Leftrightarrow [b= (x^-,x^+)]$, that does not need any reset if the quality of entanglement is good and can be immediately used for teleporting the next bipolar qubit.

BIPOLAR QUANTUM COMPUTATION AND COMMUNICATION

In this section we discuss bipolar quantum information processing and show with simulation for the first time that (1) bipolar teleportation is theoretically possible without conventional communication channel between Alice and Bob; (2) bipolar bitwise cryptography points to a different direction of cryptography.

Bipolar Quantum Information Processing

As discussed earlier, without bipolarity, linearity prevents the possibility of cloning or measuring an unknown quantum state and, if Alice sends Bob one of two nonorthogonal qubits, Bob can obtain information about which of these qubits was sent only at the expense of disturbing the state. In general, for quantum information there is no information gain without disturbance. This problem can be remedied with bipolar synchronization, entanglements, and teleportation.

The outcome of the bipolar quantum teleportation protocol on Bob's side is to determine (c^-,c^+) with a conventional decoder from the equation $(b^-,b^+)*(c^-,c^+) = (x^-,x^+)$ where (b^-, b^+), (x^-, x^+), and * are known. Thus, Alice has successfully teleported $c = (c^-,c^+)$ to Bob without transporting any matter or energy. But she has to know the secret of (c^-,c^+) in the "mail" to determine *. We call this unfortunate drawback the *bipolar confidentiality anomaly*. This anomaly can be resolved assuming bipolar quantum controllability as discussed in the follows.

Bipolar quantum entanglement is actually one stone targeting two birds. It enables both teleportation and quantum cryptography at the same time. Since teleportation is a quantum protocol, any attempt by an eavesdropper, Eve, to monitor the communication between Alice and Bob will be detectable in principle because Eve cannot gain any quantum information without some disturbance to the quantum communication channel. Moreover, Eve also needs the three bits information transmitted from Alice to Bob to crack the teleported data. If the three bits are ciphered, he has to crack the ciphered code as well. Thus, bipolar entanglement leads to efficient bipolar teleportation and effective *bipolar quantum cryptography*.

Bipolar quantum cryptography makes the following differences from conventional approaches:

- While conventional cryptography uses a large prime number for encryption that is expected to grow exponentially if and when quantum computing becomes a reality, such a big prime number would make quantum-digital conversion a big headache. Bipolar quantum cryptography uses bitwise encryption which points to a new direction of research for both binary and quantum communication.
- In conventional quantum computing, if Alice sends Bob one of two nonorthogonal qubits, Bob can obtain information about which of these qubits was sent only at the expense of disturbing the state. In general, for quantum information there is no information gain without disturbance. The impossibility of copying an unknown quantum state, or a state that is known to belong to a set of nonorthogonal states with a certain probability, and the existence of a trade-off relation between information gain and state disturbance, is the basis of the application of quantum information to cryptography. Bipolarity makes it much easier to achieve certain stability and controllability. With bipolar approximation as described in Figure 5, a bipolar qubit could be much more sustainable to disturbance and *bipolar quantum controllability* can be achievable. Bipolar controllability, in turn, can lead to achievable bipolar cloning.

- While conventional quantum communication is not directly compatible with binary code, quantum-digital compatibility is theoretically achievable with bipolar quantum controllability because a bipolar qubit can encode two binary bits or one binary bit following the bipolar-unipolar recovery theorem (Ch. 3, Theorem 3.7).

Assuming certain bipolar quantum controllability and cloning achieved through bipolar complementarity (Figure 5), the bipolar confidentiality anomaly can be resolved. The resolution is shown in Figure 6. In the figure Alice has the entangled bipolar qubits (a^-,a^+) and (b^-,b^+) with data (c^-,c^+) to be teleported to Bob and the three-way entanglement (x^-,x^+). First, without controllability and cloning, the sender Alice has to know the data secret (c^-,c^+) together with (a^-, a^+) and (x^-,x^+) to figure out the exact bipolar operator * so that she can send a three bit code of the operator to the receiver Bob. Bob then use (b^-,b^+), (x^-,x^+) and the three bits operator code to reveal (c^-, c^+). Now, with bipolar controllability the operator * can be selected by the system and transmitted to Bob without the knowledge of Alice. With cloning, Bob can store an earlier entangled qubit for later decoding purposes.

Eliminating Conventional Communication with Bipolar Quantum Teleportation

The transmission of some binary bits to Bob through the conventional communication channel has been the only non-quantum time-consuming step in quantum teleportation. Unconditional teleportation is reported on continuous quantum state where conventional communication channel is still required (Furusawa et al., 1998). Bipolar complementarity and cloning (if achieved) (see Figures 5 & 6) can resolve this barrier. If the operator * can be represented as a two bit binary code, it can also serve as the entangled bipolar qubit (a^-,a^+) as well. We would have the much simplified bipolar quantum entanglement and teleportation as depicted in Figure 7. With this protocol we could actually achieve pure quantum teleportation and cryptography without conventional communication channel.

Protocol for Bipolar Quantum Teleportation without Classical Communication:

- *Step 1. Either encode * as a bipolar qubit in Channel-1 such that (0,0) for* $\oplus$, *(-1,1) for &, (-1,0) for* $\otimes$, *(0,1) for* $\varnothing$, *or encode * as a bipolar qubit in Channel-2 such that (0,0) for* $\oplus^-$; *(-1,1) for* $\&^-$; *(-1,0) for* $\otimes^-$; *(0,1) for* $\varnothing^-$.
- *Step 2. Entangle * as two bipolar qubits a and b; Alice has a and Bob has b.*

Figure 6. Bipolar Quantum Teleportation and Cryptography (Adapted from Zhang, 2010a)

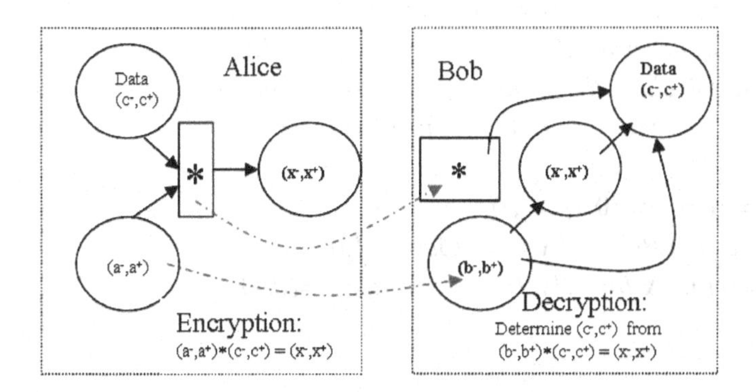

Figure 7. Bipolar Quantum Teleportation without Conventional Communication (Adapted from Zhang, 2010a)

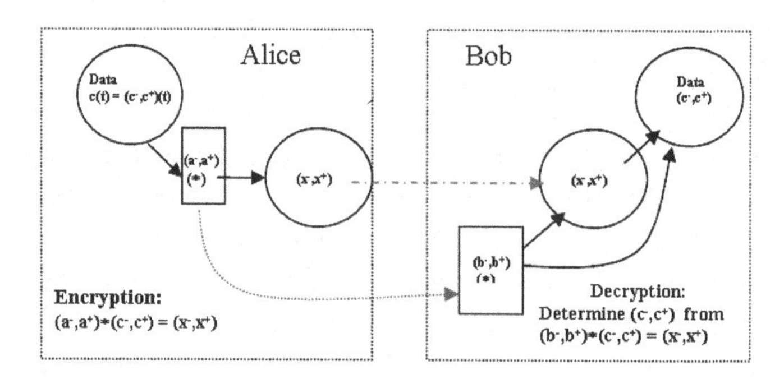

- *Step 3. Alice signals bob to start teleportation through either Channel 1 or 2.*
- *Step 4. Alice teleport c to Bob.*
- *Step 5. Bob recovers c using qubit b, new qubit x, and *.*
- *Step 6. Both Alice and Bob check the quality of the current entanglement of a = x and b = x; purify the entanglement whenever necessary but no need to reset.*
- *Step 7. Stop or go to Step 1, 2, 3, or 4.*

Simulation of Bipolar Quantum Teleportation

While conventional quantum computing and communication is difficult (if not impossible) to simulate with a digital computer, bipolar synchronization and entanglement makes it easy to simulate quantum computing and communication with digital technology. The key is that BDL provides a unification of classical logic and bipolar quantum entanglement.

Based on Figure 7 a simulated example of the protocol for bipolar quantum teleportation of a bit sequence without classical communication is shown as follows (Zhang, 2010a).

Simulation (∀t, t = 0,1, 2, 3, …):
Iteration-0 (t=0):
*Step 1. Encode * = (-1,0) for ⊗ as a bipolar qubit in Channel-1.*
*Step 2. Let * be two entangled bipolar qubits a and b; Alice takes a and Bob takes b.*
Step 3. Alice signals bob to start teleportation through Channel-1.
Step 4. Alice teleport c(t) to Bob and x = c(t)⊗a = (0,1).
Step 5. From c(t)⊗b=(0,1) & b=(-1,0), Bob recovers c(t)=(-1,0).
Step 6. Both Alice and Bob check the quality of the current entanglement of a = b = x = (0,1); purify the entanglement whenever necessary but no need to reset.
Step 7. Go to Step 4.
Iteration-1 (t=1: a = b = * = (0,1) = ∅)
Step 4. Alice teleport c(t) to Bob and x = c(t)∅a = (-1,0).
Step 5. From c(t)∅b=(-1,0) & b=(0,1), Bob recovers c(t)=(0,1).

Step 6. Both Alice and Bob check the quality of the current entanglement of a = b = x = (-1,0); purify the entanglement whenever necessary but no need to reset.

Step 7. Go to Step 4.

Iteration-2 (t=2: a = b = * = (-1,0) = ⊗)

Step 4. Alice teleport c(t) to Bob and x = c(t)⊗a = (0,0).

Step 5. From c(t)⊗b=(0,0) & b=(-1,0), Bob recovers c(t)=(0,0).

Step 6. Both Alice and Bob check the quality of the current entanglement of a = b = x = (0,0) for ⊕; purify the entanglement whenever necessary but no need to reset.

Step 7. Go to Step 4.

Iteration-3 (t=3: a = b = * = (0,0) = ⊕)

Step 4. Alice teleport c(t) to Bob and x = c(t)⊕a = (-1,1).

Step 5. From c(t)⊕ b=(-1,1) & b=(0,0), Bob recovers c(t)=(-1,1).

Step 6. Both Alice and Bob check the quality of the current entanglement of a = b = x = (-1,1) for &; purify the entanglement whenever necessary but no need to reset.

Step 7. Stop.

Remarks:

1. The above simulation is logically consistent. However, because YinYang BDL is platform independent and any physical implementation has to be platform dependent, logical-physical consistency should be enforced. For instance, c(t) can be physically coded as ¬c(t) in Iterations 2 and 3 to enforce bipolar complementarity. That is not further discussed in this book.
2. In the above simulation, only one bit binary information is needed from Alice to Bob for each session of teleportation, 0 for Channel-1 and 1 for Channel-2. If entanglement purification is needed, Step 7 would be *"Go to Step 2."* If channel change is needed within the same session, Step 7 would be *"Go to Step 3."*

After the above four iterations, Alice would have successfully teleported the four bipolar qubits c(0)c(1)c(2)c(3) = (-1,0)(0,1)(0,0)(-1,1) to Bob without knowing what has been transported and without conventional binary communication. Furthermore, the transported bipolar qubits can encode the eight binary bits "10010011" or the complement "01101100", four binary bits "1101" or "0010" based on the recovery theorem (Theorem 3.8) to digital logic. Evidently, with bipolarity, bitwise quantum cryptography is logically achievable and quantum-digital information compatibility can be established. To the authors' knowledge, this is the first logical simulation of quantum teleportation of its kind without conventional communication.

In this chapter we focus on the four orthogonal bipolar qubits and left non-orthogonal states for future research efforts. While the two basis states $|0\rangle$ and $|1\rangle$ in unipolar quantum computing could not be bipolar entangled due to their lack of bipolarity. The four bipolar orthogonal states $|(0,0)\rangle, |(-1,0)\rangle, |(0,+1)\rangle,$ and $|(-1,+1)\rangle$ enable bipolar quantum entanglement which form a key for quantum-digital compatibility, bitwise encryption, and bipolar teleportation without conventional communication.

Bipolar Coding and Bitwise Encryption

Intuitively, we have the following bipolar coding mechanisms:

1. Each bipolar qubit encodes the information of two binary bits such as: (0,0) for 00; (0,+1) for 01; (-1, 0) for 10; (-1,+1) for 11 or, formally, (n,p) encodes the decimal number $|n|\times 2^1 + |p|\times 2^0$ that can always be converted to a binary number $|n|$ concatenated with $|p|$.
2. Each bipolar qubit encodes the information of one binary bits such as: (0,0) for 0; (0,+1) for 1; (-1,0) for 1; (-1,+1) for 1 or, formally, (n,p) encode the binary bit $|n|\vee|p|$.
3. Each bipolar qubit encodes the information of one binary bits such as: (0,0) for 0; (0,+1) for 0; (-1,0) for 0; (-1,+1) for 1 or, formally, (n,p) encode the binary bit $|n|\wedge|p|$.

If the above coding schemes are used as encryption protocols, they are evidently rather simple. It won't take much time to break the code. However, a closer examination of the coding schemes reveals a fundamentally different ***bitwise encryption*** scheme (Zhang, 2010a, 2010b) that encrypts each binary bit – a radical move away from prime-number-based encryption. Could bitwise encryption be practically useful?

Let us examine the example bipolar qubit pattern: (0,0) (0,+1) (-1,0) (-1,+1). Is it for 00011011, 11100100, 0111, or 0001? Can any quantum computing algorithm like Shor's fast factoring algorithm help in breaking the code? Evidently, no matter how powerful a quantum computer is, it won't help if the encryption is not based on prime numbers.

Certainly, a simple algorithm can be designed to break the above short code if it is semantically sensitive. It would become quite difficult to break it, however, if all four different encoding schemes are used in combination with the bipolar operators defined in BDL (CH. 3). For instance,

- $(0,0) = \neg(-1,+1)$ that can encode the binary bit 0 or 1 or the two binary bits 00 or 11;
- $(0,+1) = \neg(-1,0)$ that can encode the binary bit 0 or 1 or the two binary bits 01 or 10;
- $(-1,0) = \neg(0,+1)$ that can encode the binary bit 0 or 1 or the two binary bits 10 or 01;
- $(-1,+1) = \neg(0,0)$ that can encode the binary bit 0 or 1 or the two binary bits 11 or 00.

Evidently, if we use the four simple coding schemes combined with $\neg$ we have

1. One bipolar qubit (n,p) can be converted to $2^2 = 4$ different binary bit patterns.
2. Two bipolar qubits (n,p)(u,v) can be converted to $2^2\times 2^2 = 2^4 = 16$ different binary bit patterns.
3. N bipolar qubits (n,p)(u,v)..(x,y) can be converted to $2^2\times 2^2\times..\times 2^2 = 2^{2N}$ different binary bit patterns.
4. One K (1024) bipolar qubits can be converted to $2^{2\times 1024}$ different binary bit patterns.
5. One M (mega) (1024^2) bipolar qubits can be converted to $2^{2\times 1024\times 1024}$ different binary bit patterns.

So far we have only discussed one layer encryption. The result of an encryption can always be encrypted again. For instance, given the data encrypted in bipolar qubit patterns, the following can be done:

1. Perform bipolar-binary reverse coding;
2. Perform binary-bipolar coding again.

While digital commutation has been based on logical axiomatization for illogical physics (LAFIP), the "illogical" bipolar operators would make it extremely difficult for a future quantum computer to break bitwise encryption without knowing the key.

Bitwise Quantum-Digital Cryptography

Bipolar encryption and decryption leads to the term quantum-digital cryptography. A message coded in 1M bipolar qubits using the above bitwise bipolar encryption and teleported using Eq. 7.3 from Alice to Bob based on bipolar quantum entanglement (Eq. 7.2) would be very difficult to break even by a powerful quantum computer unless the eavesdropper knows the key for decryption. Surprisingly such a level of security is achievable without a large prime number. It is too early to say whether it is possible to develop a *bitwise quantum-digital public key* bipolar quantum cryptography protocol to replace RSA. If so, prime numbers and quantum factoring algorithms would both be made obsolete and the threatening factor of quantum factorization could be removed.

Interestingly, bipolar quantum-digital cryptography is applicable in pure software implementation that does not have to have quantum hardware implementation. This is because that BDL as a sound and complete formal logical system enables bipolar quantum entanglement but is also a non-linear bipolar dynamic generalization of Boolean logic and recoverable to Boolean logic. While this book is focused on YinYang bipolar relativity, quantum computing and bitwise quantum-digital cryptography is further discussed in another volume.

A TAXONOMY FOR BIPOLAR QUANTUM COMPUTING

Bipolar Logical Quantum Computing

Bipolar logical quantum computing can be considered the counterpart of conventional quantum computing and communication which spans bipolar qubits, bipolar quantum information processing, bipolar teleportation, and bipolar cryptography based on BDL, BQL, and BDFL in different applications. This chapter has so far been focused on bipolar logical quantum computing and communication.

While bipolar qubit does not exclude conventional approaches to quantum computation, it has been demonstrated that bipolar synchronization and entanglement provide a potential logical basis for controllable and stable bipolar quantum computation as well as communication. On the one hand, bipolar approximation and synchronization make it possible for quantum-digital compatibility. On the other hand, since each pole of a bipolar quantum register is a usual Bloch sphere, bipolar quantum computing is linked to the existing quantum technological basis. This chapter has been focused on the logical basis for bipolar quantum computing and communication. Software and physical implementation can be further discussed in future reports.

Bipolar Relational Quantum Computing

It is asserted in (Smolin, 2005) that the correct quantum theory of gravity must be background independent and relational rather than absolute. Relationism denies that space and time are basic entities like matter and radiation. In contrast, absolutism regards space and time as fundamental and logically anterior to matter and radiation. While Isaac Newton was an absolutist, Albert Einstein was a relationist by philosophical disposition. According relationism, every physical or conceptual object in natural and social sciences should be defined relative to other objects.

Bipolar agents, bipolar causality, and bipolar relativity constitute a completely background independent quantum theory that spans both microscopic and macroscopic worlds. However, conventional quantum computing is based on qubit register or Bloch sphere which is background dependent. In the earlier section, we used "left", "right, NP and PN concepts as a few exceptions of background independence, which are necessary to specify a bipolar qubit in Hilbert space for the benefit of (1) compatibility between bipolar quantum computing and conventional quantum computing and (2) compatibility between bipolar quantum computing and classical digital computing. Theoretically speaking, a valid background independent theory should reduce to a background dependent theory with certain limitation but not vice versa.

Since bipolar relations and equilibrium relations are completely background independent, they belong to the structures of relationism. It has been shown in Chapter 5 that bipolar fuzzy relations and equilibrium relations can be used in socioeconomics for multiagent coordination and global regulation. In the remaining chapters the relational aspects are further developed to algebraic models and applied in computational bioeconomics and cognitive mapping, respectively. Together with bipolar relativity, these relational models provide a mathematical basis for relational quantum computing in physical, social, cognitive, and life sciences. They are quantum in nature because bipolarity is a fundamental quantum property and bipolar quantum entanglement is involved in relational computing with the non-linear bipolar interactive operator $\otimes$. Thus, bipolar relations bring quantum computing into social, mental, and life sciences as well as physical and computing sciences.

Bipolar Algebraic Quantum Computing

Bipolar mathematical abstraction has led to bipolar linear algebra and bipolar cellular networks (Zhang & Chen, 2009; Zhang *et al.,* 2009). The algebraic quantum aspects are further developed to a bipolar quantum automata theory in Chapters 8 and applied in bioeconomics, biosystem simulation and regulation in Chapter 9.

Logical, Physical, Social, Mental, and Biological Quantum Gravities (Q5)

As argued in Chapter 1, since acceleration is equivalent to gravitation under general relativity, any physical, socioeconomic, mental, and biological acceleration, growth, or aging are qualified to be a kind of quantum gravity. It can be further argued that as a most fundamental scientific unification not only can quantum gravity be applied in physical science, but also in computing science, social science, brain science, and life sciences as well. Indeed, it would be hard to imagine that quantum gravity as the grand unification would not be the governing theory for all sciences. This argument leads us to five sub-theories of quantum gravity: *physical quantum gravity, logical quantum gravity, social quantum gravity, biological quantum gravity, and mental quantum gravity (Q5 paradigm).*

In the Q5 paradigm, the theory of physical quantum gravity should be concerned with quantum physics; the theory of logical quantum gravity should be focused on quantum computing; the theory of social quantum gravity should span social sciences and economics; the theory of biological quantum gravity should be focused on life sciences; the theory of mental quantum gravity should be focused on the interplay of quantum mechanics and brain dynamics.

The Q5 paradigm may sound like a mission impossible. It actually follows a single undisputable observation and a single condition: (1) bipolar equilibrium or non-equilibrium is a generic ubiquitous concept from which nothing can escape; (2) equilibrium-based bipolar quantum entanglement is definable with bipolar universal modus ponens (BUMP).

Following Schrödinger's cat paradox, English mathematical physicist Roger Penrose described two mysteries of quantum entanglement (Penrose, 2005, p. 591). The first mystery is the phenomenon itself; the second one is: *"Why do these ubiquitous effects of entanglement not confront us at every turn?"* Penrose remarked: "I do not believe that this second mystery has received nearly the attention that it deserves." Evidently, YinYang bipolar quantum entanglement provides a resolution to the first mystery and the Q5 paradigm should provide a resolution to the second one. While the earlier chapters and this chapter have been focused on logical and physical quantum computing, later chapters are focused on social, mental, and biological quantum computing.

RESEARCH TOPICS

Bipolar quantum entanglement has opened a number of important research topics including but not limited to the following:

Experimental Testing of Bipolar Quantum Entanglement. The experimental testing of bipolar quantum entanglement is the first and far most open research topic. If it is physically achievable quantum computing would be brought much closer to reality.

Bipolar Quantum Controllability and Cloning. YinYang bipolar complementarity brings theoretical possibility for quantum controllability and cloning that should be further investigated.

Bipolar Quantum Teleportation. The possibility of eliminating conventional communication with bipolar quantum teleportation is an important topic that could revolutionize communication if physically achieved.

Bipolar Quantum-Digital Compatibility. Theoretical and practical research on bipolar quantum-digital compatibility deserves further investigation.

Bipolar Quantum Information. Bipolar quantum information gain and entropy are interesting topics for further investigation.

Bipolar Quantum Computing. This chapter has been focused on the four basis bipolar qubit values, non-orthogonal bipolar qubit values need a complex number based BDL which is a interesting topic.

Bipolar Cryptography. Bipolar bitwise encryption has opened a new direction for quantum cryptography. The possibility of replacing prime number based encryption with bipolar bitwise encryption and make factorial algorithms obsolete is enormously significant theoretically, practically, and commercially.

Logical, Physical, Social, Biological, and Mental Quantum Gravities. Generalize quantum computing from computer technology to all logical, physical, social, biological, and mental quantum gravities will need long term research efforts for many years to come.

SUMMARY

A logically complete quantum theory has been presented for bipolar quantum computation and communication. Based on bipolar strings and bipolar symmetry, the bipolar quantum theory provides a unique logical interpretation of the EPR paradox and the missing variable in conventional quantum mechanics. It has been suggested that the missing fundamental concept in conventional quantum mechanics is bipolarity; the missing principle is bipolar complementarity which has been compared with Niels Bohr's particle-wave duality and Schrödinger's cat paradox.

With bipolar complementarity, it has been shown that stability and controllability can be introduced into quantum computing. Technically, with the concept of bipolar synchronization, conventional qubit and quantum entanglement have been extended to bipolar qubit and bipolar quantum entanglement. It has been shown that bipolar quantum entanglement leads to bipolar quantum teleportation, bipolar cryptography, and bipolar quantum computing. It can be concluded with the following suggestions:

1. The missing variable in conventional quantum mechanics is bipolarity and missing fundamental principle is bipolar complementarity.
2. Bipolar quantum entanglement can be defined as a bipolar symmetry or quasi-symmetry operation that gives out information without passing observable energy.
3. Gravity and quantum mechanics can be unified under bipolar quantum entanglement of different types, in different speeds and different scales (See Conjectures 6.20, 6.21).
4. Conventional qubit register can be extended to bipolar qubit register.
5. Bipolar quantum teleportation and computation can be realized theoretically.
6. Quantum-digital compatibility can be achieved with bipolar quantum computing and communication.
7. Bipolar quantum computation and communication can be simulated with digital technology.
8. Bipolar teleportation is theoretically possible without conventional communication between Alice and Bob.
9. Bipolar bitwise encryption points to a different direction of cryptography that has the potential to make prime number based cryptography and quantum factorization algorithms both obsolete.

Advanced topics of bipolar quantum algorithms, computational complexity, and bipolar cryptography are left for further studies. It should be pointed out that the feasibility of bipolar quantum computing is due to the unifications achieved in earlier chapters. The enabling unifications include

1. the unification of classical logic and quantum entanglement with BDL and BUMP (Chapters 3-5),
2. the logical unification of nature's basic forces and matter-antimatter pairs under bipolarity,
3. the logical unification of gravity with quantum entanglement under bipolar relativity, and
4. the logical unification of particle-wave duality with equilibrium, symmetry, and quasi-symmetry.

It is worthwhile to point out that nanotechnology is a general term including but not limited to quantum technology and nanobiotechnology. A rapidly developing area of nanocomputing is DNA computing (molecular computing or biocomputing (Adleman, 1994). In the meantime, biocomputing, quantum computing, brain and mind researchers have joined forces in tackling the difficult topic of consciousness where quantum gravity is considered a key (Hagan, Hameroff & Tuszyński, 2002; Hameroff *et al.*, 2002; Schwartz, Stapp & Beauregard, 2005). This chapter has been focused on the central and basic issues of bipolar quantum entanglement as part of physical quantum gravity. Basic concepts can be applied in molecular or biocomputing as well due to the generality of bipolarity, equilibrium, and symmetry.

ACKNOWLEDGMENT

This chapter has been partially presented in Zhang, W. -R. (2010a). YinYang Bipolar Quantum Entanglement – Toward a Logically Complete Quantum Theory. *IEEE CPS: Proc. 4th Int'l Conf. on Quantum, Nano and Micro Technologies (ICQNM 2010)*, 10–15 February, St. Maarten, Netherlands Antilles pp.77-82. Permission to reuse is acknowledged.

REFERENCES

Bell, J. S. (1964). On the Einstein Podolsky Rosen paradox. *Physics*, *1*(3), 195–200.

Bennett, C. H., Brassard, G., Crepeau, C., Jozsa, R., Peres, A., & Wootters, W. K. (1993). Teleporting an unknown quantum state via dual classical and EPR channels. *Physical Review Letters*, *70*, 1895–1899. doi:10.1103/PhysRevLett.70.1895

Birkhoff, G., & von Neumann, J. (1936). The logic of quantum mechanics. *The Annals of Mathematics*, *37*(4), 823–843. doi:10.2307/1968621

Bohm, D. (1957). *Causality and chance in modern physics*. Philadelphia: University of Pennsylvania Press. doi:10.4324/9780203201107

Bohm, D. (1980). *Wholeness and the implicate order*. London: Routledge.

Bohr, N. (1948). On the notions of causality and complementarity. *Dialectica*, *2*(3-4), 312–319. doi:10.1111/j.1746-8361.1948.tb00703.x

Bub, J. (2006). Quantum entanglement and information. *Stanford Encyclopedia of Philosophy.* Retrieved from http://plato.stanford.edu/entries/qt-entangle/

Buchanan, M. (1998). Beyond reality–watching information at play in the quantum world is throwing physicists into a flat spin. *New Scientist*, 2125.

Clifton, R., Bub, J., & Halvorson, H. (2003). Characterizing quantum theory in terms of information-theoretic constraints. *Foundations of Physics*, *33*, 1561–1591. doi:10.1023/A:1026056716397

Deutsch, D. (1985). Quantum theory, the Church-Turing principle and the universal quantum computer. *Proceedings of the Royal Society of London. Series A*, *400*, 97–117. doi:10.1098/rspa.1985.0070

Dieks, D. (1982). Communication by EPR devices. *Physics Letters. [Part A]*, *92*, 271–272. doi:10.1016/0375-9601(82)90084-6

Einstein, A., Podolsky, B., & Rosen, N. (1935). Can quantum-mechanical description of physical reality be considered complete? *Physical Review*, *47*(10), 777–780. doi:10.1103/PhysRev.47.777

Faye, J. (2008). Copenhagen interpretation of quantum mechanics. *Stanford Encyclopedia of Philosophy.* Retrieved from http://plato.stanford.edu/entries/qm-copenhagen/

Fermi National Accelerator Laboratory. (2006). *Press release.* Retrieved from http://www.fnal.gov/pub/presspass/press_releases/CDF_meson.html.

Furusawa, A., Sørensen, J. L., Braunstein, S. L., Fuchs, C. A., Kimble, H. J., & Polzik, E. S. (1998). Unconditional quantum teleportation. *Science, 23*, 706–709. doi:10.1126/science.282.5389.706

Ghosh, S., Rosenbaum, T. F., Aeppli, G., & Coppersmith, S. N. (2003). Entangled quantum state of magnetic dipoles. *Nature, 425*(6953), 48. doi:10.1038/nature01888

Guizzo, E. (2010). *IEEE spectrum's special report: Winners & losers VII.*

Hagan, S., Hameroff, S. R., & Tuszyński, J. A. (2002). Quantum computation in brain microtubules: Decoherence and biological feasibility. *Physical Review E: Statistical, Nonlinear, and Soft Matter Physics, 65*(6). doi:10.1103/PhysRevE.65.061901

Hameroff, S., Nip, A., Porter, M., & Tuszynski, J. (2002). Conduction pathways in microtubules, biological quantum computation, and consciousness. *Bio Systems, 64*(1-3), 149–168. doi:10.1016/S0303-2647(01)00183-6

Hilgevoord, J., & Uffink, J. (2006). The uncertainty principle. *Stanford Encyclopedia of Philosophy.* Retrieved from http://plato.stanford.edu/entries/qt-uncertainty/

Jost, J. D., Home, J. P., Amini, J. M., Hanneke, D., Ozeri, R., & Langer, C. (2009). Entangled mechanical oscillators. *Nature, 459*(7247), 653. doi:10.1038/nature08006

Lo, H.-K., Popescu, S., & Spiller, T. (1998). *Introduction to quantum computation and information.* Singapore: World Scientific. doi:10.1142/9789812385253

Morton, J. J. L., Tyryshkin, A. M., Brown, R. M., Shankar, S., Lovett, B. W., & Ardavan, A. (2008). Solid state quantum memory using the 31P nuclear spin. *Nature, 455*, 1085–1088. doi:10.1038/nature07295

Ni, K.-K., Ospelkaus, S., de Miranda, M. H. G., Pe'er, A., Neyenhuis, B., & Zirbel, J. J. (2008). A high phase-space-density gas of polar molecules. *Science, 322*(5899), 231–235. doi:10.1126/science.1163861

Penrose, R. (2005). *The road to reality: A complete guide to the laws of the universe.* New York: Alfred A. Knopf.

Salart, D., Baas, A., Branciard, C., Gisin, N., & Zbinden, H. (2008). Testing the speed of spooky action at a distance. *Nature, 454*, 861–864. doi:10.1038/nature07121

Schrödinger, E. (1935). Discussion of probability relations between separated systems. *Proceedings of the Cambridge Philosophical Society, 31*, 555–563. doi:10.1017/S0305004100013554

Schumacher, B. (1995). Quantum coding. *Physical Review A., 51*, 2738–2747. doi:10.1103/PhysRevA.51.2738

Schwartz, J. M., Stapp, H. P., & Beauregard, M. (2005). Quantum physics in neuroscience and psychology: A neurophysical model of mind–brain interaction. [References and further reading may be available for this article. To view references and further reading you must purchase this article.]. *Philosophical Transactions of the Royal Society of London. Series B, Biological Sciences, 360*, 1309–1327. doi:10.1098/rstb.2004.1598

Shannon, C. E., & Weaver, W. (1949). *The mathematical theory of communication.* Urbana, IL: University of Illinois Press.

Shi, Y., Seto, E., Chang, L.-S., & Shenk, T. (1991). Transcriptional repression by YY1, a human GLI-Kruppel-related protein, and relief of repression by adenovirus E1A protein. *Cell, 67*(2), 377–388. doi:10.1016/0092-8674(91)90189-6

Shor, P. (1994). Algorithms for quantum computation: Discrete logarithms and factoring. *Proceedings of the 35th Annual Symposium on Foundations of Computer Science*, 124-134.

Smolin, L. (2006). *The trouble with physics: The rise of string theory, the fall of a science, and what comes next?*New York: Houghton Mifflin Harcourt.

Steane, A. M. (1998). Quantum computing. *Reports on Progress in Physics, 61*, 117–173. doi:10.1088/0034-4885/61/2/002

Vasudevan, S., Tong, Y., & Steitz, J. A. (2007). Switching from repression to activation: MicroRNAs can up-regulate translation. *Science, 318*(5858), 1931–1934. doi:10.1126/science.1149460

Wootters, W. K., & Zurek, W. H. (1982). A single quantum cannot be cloned. *Nature, 299*, 802–803. doi:10.1038/299802a0

Zhang, W.-R. (2005a). YinYang bipolar lattices and L-sets for bipolar knowledge fusion, visualization, and decision making. *International Journal of Information Technology and Decision Making, 4*(4), 621–645. doi:10.1142/S0219622005001763

Zhang, W.-R. (2007). YinYang bipolar universal modus ponens (bump) – a fundamental law of non-linear brain dynamics for emotional intelligence and mental health. *Walter J. Freeman Workshop on Nonlinear Brain Dynamics, Proceedings of the 10th Joint Conf. of Information Sciences*, (pp. 89-95). Salt Lake City, Utah.

Zhang, W.-R. (2009a). Six conjectures in quantum physics and computational neuroscience. *Proceedings of 3rd International Conference on Quantum, Nano and Micro Technologies (ICQNM 2009)*, (pp. 67-72). Cancun, Mexico.

Zhang, W.-R. (2009b). YinYang Bipolar Dynamic Logic (BDL) and equilibrium-based computational neuroscience. *Proceedings of International Joint Conference on Neural Networks (IJCNN 2009)*, (pp. 3534-3541). Atlanta, GA.

Zhang, W.-R. (2009c). YinYang bipolar relativity–a unifying theory of nature, agents, and life science. *Proceedings of International Joint Conference on Bioinformatics, Systems Biology and Intelligent Computing (IJCBS)*. (pp. 377-383). Shanghai, China.

Zhang, W.-R. (2009d). The logic of YinYang and the science of TCM – an Eastern road to the unification of nature, agents, and medicine. *International Journal of Functional Informatics and Personal Medicine, 2*(3), 261–291. doi:10.1504/IJFIPM.2009.030827

Zhang, W.-R. (2010a). YinYang bipolar quantum entanglement–toward a logically complete quantum theory. *Proceedings of the 4th Int'l Conf. on Quantum, Nano and Micro Technologies (ICQNM 2010)*, (pp. 77-82). St. Maarten, Netherlands Antilles.

Zhang, W.-R. (2010b). *YinYang bipolar quantum computing and bitwise quantum-digital cryptography.* Paper presented at Quantum Information and Computation VIII, Orlando, Florida, April 5-9, 2010.

Zhang, W.-R., & Chen, S. S. (2009). Equilibrium and non-equilibrium modeling of YinYang Wuxing for diagnostic decision analysis in traditional Chinese medicine. *International Journal of Information Technology and Decision Making*, *8*(3), 529–548. doi:10.1142/S0219622009003521

Zhang, W.-R., Pandurangi, A., & Peace, K. (2007). YinYang dynamic neurobiological modeling and diagnostic analysis of major depressive and bipolar disorders. *IEEE Transactions on Bio-Medical Engineering*, *54*(10), 1729–1739. doi:10.1109/TBME.2007.894832

Zhang, W.-R., Zhang, H. J., Shi, Y., & Chen, S. S. (2009). Bipolar linear algebra and YinYang-N-Element cellular networks for equilibrium-based biosystem simulation and regulation. *Journal of Biological System*, *17*(4), 547–576. doi:10.1142/S0218339009002958

ADDITIONAL READING

Adleman, L. M. (1994). Molecular Computation Of Solutions To Combinatorial Problems. *Science*, *266*(11), 1021–1024..doi:10.1126/science.7973651

Lloyd, S. (2006). *Programming the Universe*. New York: Alfred A. Knopf, Inc.

Woit, P. (2006). *Not Even Wrong: The Failure of String Theory and the Search for Unity in Physical Law*. New York: Basic Book.

KEY TERMS AND DEFINITIONS

Bipolar Quantum Computing: Any computing with bipolar quantum entanglement. It is shown in this book that bipolar quantum computing makes it possible to bring the ubiquitous effects of quantum entanglement out of mystery.

Bipolar Quantum Entanglement: Two spatially separated bipolar quantum agents A and B are bipolar symmetrically entangled if A⇔B or bipolar asymmetrically entangled if A⇒B. Bipolar quantum entanglement is the first known logically definable quantum entanglement. It provides a uniform or unifying logical representation for both gravitation and quantum mechanics. Furthermore, it brings the ubiquitous effects of quantum entanglement to the real world logically, physically, mentally, and biologically in a Q5 paradigm (Chs. 6-11).

Bipolar Quantum Teleportation: Teleportation based on bipolar quantum entanglement. (Zhang, 2010a).

Bipolar Qubit Register (or Bipolar Qubit Box): While a classical qubit register can be modeled with one Bloch sphere, two Bloch spheres are involved in a bipolar qubit register: one for the negative pole and another for the positive pole with synchronized harmonic or coherent bipolar quantum behavior.

Bipolar Qubit: A qubit with two poles – negative pole and positive pole (Zhang, 2010a).

Bitwise Encryption: Encrypt each bit of a binary bit pattern using bipolar logic and bipolar-binary conversion. (Zhang, 2010a, 2010b)

Bloch Sphere: A geometrical representation of the pure state space of a two-level quantum mechanical system named after the physicist Felix Bloch (Nielsen & Chuang, 2000).

Q5 Paradigm: A quantum gravity theory consisting of five sub-theories: *physical quantum gravity, logical quantum gravity, social quantum gravity, biological quantum gravity, and mental quantum gravity*. The theory of physical quantum gravity should be concerned with quantum physics; the theory of logical quantum gravity should be focused on quantum computing; the theory of social quantum gravity should span social sciences and economics; the theory of biological quantum gravity should be focused on life sciences; the theory of mental quantum gravity should be focused on the interplay of quantum mechanics and brain dynamics.

Quantum-Digital Cryptography: Cryptography use bitwise encryption/decryption.

Shannon Entropy: The amount of information we gain, on average, when we learn the value of a random variable (or, equivalently, the amount of uncertainty in the value of a random variable). (Shannon & Weaver, 1949)

Chapter 8
Bipolar Quantum Linear Algebra (BQLA) and Bipolar Quantum Cellular Automata (BQCA)

ABSTRACT

This chapter brings bipolar relativity from the logical and relational levels to the algebraic level. Following a brief review on traditional cellular automata and linear algebra, bipolar quantum linear algebra (BQLA) and bipolar quantum cellular automata (BQCA) are presented. Three families of YinYang-N-Element bipolar cellular networks (BCNs) are developed, compared, and analyzed; YinYang bipolar dynamic equations are derived for YinYang-N-Element BQCA. Global (system level) and local (element level) energy equilibrium and non-equilibrium conditions are established and axiomatically proved for all three families of cellular structures that lead to the concept of collective bipolar equilibrium-based adaptivity. The unifying nature of bipolar relativity in the context of BQCA is illustrated. The background independence nature of YinYang bipolar geometry is demonstrated with BQLA and BQCA. Under the unifying theory, it is shown that the bipolar dimensional view, cellular view, and bipolar interactive view are logically consistent. The algebraic trajectories of bipolar agents in YinYang bipolar geometry are illustrated with simulations. Bipolar cellular processes in cosmology, brain, and life sciences are hypothesized and discussed.

INTRODUCTION

Bipolar relativity now consists of the theories of bipolar sets, bipolar dynamic logic (BDL), bipolar quantum lattice (BQL), bipolar dynamic fuzzy logic (BDFL), bipolar agents, bipolar causality, bipolar strings, and bipolar quantum theory. Following the arrival of bipolar agents, the emergence of space and time has become a natural process. A completely background independent quantum theory of bipolar relativity has been achieved. However, the theory so far is a logical theory. One unanswered question is whether the logical foundation is leading to bipolar mathematical extension. This chapter is to make up this gap by introducing *bipolar quantum linear algebra* (*BQLA*) and YinYang-N-Element *bipolar quantum cellular automata* (*BQCA*) characterized with bipolar dynamic equations. With limited mathematical depth BQLA and BQCA are intended to be a starting point for further quantization and mathematical development.

YinYang is about equilibrium, harmony, symmetry, and stability which are all fundamental concepts in physical, socioeconomic, and bioeconomics systems. The current computation and modeling tools for physical, socio- and biosystem processes such as growing, aging, degenerating, equilibrium, and non-equilibrium processes in molecular and gene regulation networks, however, are based on classical truth-based mathematical abstraction where the Yin (such as genomic repression ability) and Yang (such as genomic activation ability) as well as their bipolar interaction, oscillation, local and global equilibria cannot be explicitly captured for holistic visualization and analysis. For instance, the YinYang1 (YY1) ubiquitous regulator protein (Shi *et al.,* 1991) exhibits both repressor and activator behaviors in gene expression regulation, but classical mathematical abstraction do not support holistic YinYang bipolarity. A YinYang mathematical model is needed for simulating the repression and activation abilities with explicit bipolar interaction, oscillation, quantum entanglement, equilibrium, and non-equilibrium.

As a universal computational architecture, *cellular automata* find applications in all scientific fields. There are many possible generalizations and extensions of cellular automata. Among all the different varieties, holistic nature is a common property of cellular automata. The holistic nature makes cellular automation an excellent candidate for computing with equilibrium or non-equilibrium-based bipolar relativity. While mainstream cellular automation has been rooted in classical truth-based mathematical abstraction, YinYang bipolar mathematical abstraction provides a theoretical foundation for a bipolar quantum linear algebra (BQLA) and a theory of YinYang-N-Element bipolar quantum cellular automation (BQCA).

In this chapter we present BQLA and BQCA. It is shown that BQLA makes it possible to describe YinYang bipolar relativity as a unifying mathematical physics theory in the context of YinYang-N-Element BQCA. It is shown that the dimensional view, bipolar logical view, and cellular view are semantically consistent under the unifying theory. Therefore, bipolar set theory, bipolar dynamic logic, bipolar quantum linear algebra, bipolar agents, bipolar causality, and BQCA are all unified under bipolar relativity. The algebraic trajectory of bipolar agents in YinYang bipolar geometry is illustrated with simulated data. The background independent nature of bipolar relativity is demonstrated with bipolar algebraic YinYang bipolar geometry.

The remaining presentations and discussions of this chapter are organized in the following sections:

- **Background Review.** This section presents a brief review on related concepts in cellular automation and linear algebra.
- **Bipolar Quantum Linear Algebra (BQLA) and Bipolar Quantum Cellular Automata (BQCA).** This section presents: (1) the definitions of BQCA; (2) Bipolar Quantum Linear Algebra (BQLA); (3) *YinYang-N-Element Bipolar Cellular Networks* (*BCNs*); (4) YinYang-N-Element Dynamic Equations; (5) Non-Classical vs. Classical YinYang-N-Element Cellular Networks; (6) *YinYang-N-Element Cellular Combinatorics*. In general, three families of BCNs are developed, compared, and analyzed: one family has predefined nourishing and regulating cycles following the classical YinYang-5-Element protocol in traditional Chinese medicine (TCM); another family has

random connectivity and link weights; a third family has predefined number of fan-in and fan-out but with random connections, that leads to YinYang-N-Element bipolar combinatorics.

- **Equilibrium, Non-Equilibrium, and Oscillatory Cellular Automata Conditions.** This section presents: (1) the laws of cellular symmetry and broken symmetry; (2) discrete and continuous BQCA. YinYang bipolar dynamic equations are derived for YinYang-N-Element BQCAs; global (system level) and local (elementary level) energy equilibrium and non-equilibrium conditions are established and axiomatically proved for all three families of BCNs.
- **A Unifying Paradigm of Bipolar Relativity.** This section introduces (1) logical equivalence of three different views of YinYang bipolar relativity; (2) trajectories of bipolar agents; (3) conditions for collective bipolar adaptivity to equilibria; (4) background independence of YinYang bipolar geometry; (5) mind as bipolar cellular processes; (6) universe as bipolar cellular processes; and (7) life as bipolar cellular processes.
- **Research Topics**. This section lists a few research topics.
- **Summary**. This section summarizes the key points of this chapter and draws a few conclusions.

BACKGROUND REVIEW

Cellular Automata

Cellular automata were studied in the early 1950s as a possible model for biological systems (Wolfram, 2002, p. 48). Von Neumann was one of the first to consider such a model, and incorporated a cellular model into his "universal constructor" (Von Neumann, 1966). Comprehensive studies of cellular automata have been performed by Wolfram starting in the 1980s, and Wolfram's fundamental research in the field is documented in his influential book *A New Kind of Science* (Wolfram, 2002) with a number of groundbreaking new discoveries.

A *cellular automaton* (plural: *cellular automata*) consists of a collection of cells, each in one of a finite number of *states* or *colors* such as *On* and *Off*. The cells are organized into a grid in any finite number of dimensions. Each cell has a set of neighborhood cells (usually including the cell itself). The neighborhood of a cell can be specified with a distance such as distance *1* or *2* from the cell. The initial state of the network can be specified by assigning a state for each cell at time $t = 0$. A new generation is created by advancing t by *1* according to a set of rules or mathematical functions governing the state change in a neighborhood. For example, the rule might be that the cell is *On* in the next generation if exactly two of the cells in the neighborhood are *On* in the current generation; otherwise the cell is *Off* in the next generation. Typically, the rule for updating the state of cells is the same for each cell and does not change over time, and is applied to the whole grid simultaneously. Exceptions are not common but can be specified.

Cellular automata come in a variety of shapes and varieties. One of the most fundamental properties of a cellular automaton is the type of *grid* on which it is computed. The simplest such grid is a one-dimensional line. In two dimensions, square, triangular, and hexagonal grids may be considered. The number of colors (or distinct states) that a cellular automaton may assume must also be specified. This number is typically an integer, with a binary number being the simplest choice. For a binary automaton, color *0* is commonly called *White*" or *Off*, and color *1* is commonly called *Black* or *On*. However, cellular automata having a continuous range of possible values may also be considered (Wolfram, 2002).

In addition to the grid on which a cellular automaton lives and the states or colors its cells may assume, the neighborhood over which cells affect one another must also be specified. The simplest choice is *nearest neighbors* or *distance 1* neighbors in which only cells directly adjacent to a given cell may be affected at each time step. Two common neighborhoods in the case of a two-dimensional cellular automaton on a square grid are a square neighborhood and a diamond-shaped neighborhood.

The theory of cellular automata has rich semantics and many applications. It is possible to produce a great variety of unexpected behaviors with simple rules and structures. Wolfram (2002) defined four classes called *Wolfram Classes* into which cellular automata and several other simple computational models can be divided depending on their behavior. While earlier studies in cellular automata tried to identify types of patterns for specific rules, Wolfram's classification was the first attempt to classify the rules themselves. Based on their complexity, the classes are:

- **Class 1:** Nearly all initial patterns evolve quickly into a stable, homogeneous state. Any randomness in the initial pattern disappears.
- **Class 2:** Nearly all initial patterns evolve quickly into stable or oscillating structures. Some of the randomness in the initial pattern may be filtered out. Local changes to the initial pattern tend to remain local.
- **Class 3:** Nearly all initial patterns evolve in a pseudo-random or chaotic manner. Any stable structures that appear are quickly destroyed by the surrounding noise. Local changes to the initial pattern tend to spread indefinitely.
- **Class 4:** Nearly all initial patterns evolve into structures that interact in complex and interesting ways.

Stable or oscillating structures of Class 2 may be the eventual outcome, but the number of steps required to reach this state may be very large, even when the initial pattern is relatively simple. Local changes to the initial pattern may spread indefinitely. Wolfram has conjectured that many, if not all class 4 cellular automata are capable of universal computation. This has been proved for Rule 110 and Conway's game of life (Wolfram, 2002).

Generalization, Extensions, and Applications of Cellular Automata

There are many possible generalizations, extensions, and applications of cellular automata. Among them are genetically evolving cellular automata, cryptographic cellular automata, *quantum cellular automata*, discrete cellular automata, continuous cellular automata, and others.

Evolving cellular automata using genetic algorithms have been studied in recent years (Mitchell, Crutchfield & Hraber, 1994; Crutchfeld, Mitchell & Das, 2002). These cellular automata are motivated by building decentralized systems. Using evolutionary computation it is shown that cellular arrays can be programmed (Crutchfeld, Mitchell & Das, 2002). Computation in decentralized systems is very different from classical systems, where the information is processed at some central location depending on the system's state. In a decentralized system, the information processing occurs in the form of global and local pattern dynamics. The inspiration for this approach comes from complex natural systems like insect colonies, nervous system and economic systems. The results produced from this approach are interesting, in that a very simple array of cellular automata produces results showing coordination over global scale, fitting the idea of emergent computation. Future work in the area may include more sophisticated models using cellular automata of higher dimensions, which can be used to model complex natural systems.

Cellular automata have been proposed for cryptography (Wolfram, 1986) where a one way function is the evolution of finite cellular automata whose inverse is believed to be hard to find. Given the rule, anyone can easily calculate future states, but it appears to be very difficult to calculate previous states. However, the designer of the rule can create it in such a way as to be able to easily invert it. Therefore, it is apparently a trapdoor function, and can be used as a public-key cryptosystem. The security of such systems is an ongoing research topic.

Rules can be probabilistic rather than deterministic. A probabilistic rule gives, for each pattern at time t, the probabilities that the central cell will transition to each possible state at time $t+1$. Sometimes a simpler rule is used; for example: "The rule is the Game of Life, but on each time step there is a 0.001% probability that each cell will transition to the opposite color."

Continuous automata (Wolfram, 2002) are like totalistic CA, but instead of the rule and states being discrete (e.g. a table, using states {0,1,2}), continuous functions are used, and the states become continuous (usually values in [0,1]). The state of a location is a finite number of real numbers. Certain cellular automata can yield diffusion in liquid patterns in this way.

It is found that some living things use naturally occurring cellular automata in their functioning. Plants regulate their intake and loss of gases via a cellular automation mechanism (Peak & Messinger, 2004). Neural networks can be used as cellular automata, too. The Belousov-Zhabotinsky reaction is a spatio-temporal chemical oscillator which can be simulated by means of a cellular automaton.

Cellular automation processors are a physical, not software only, implementation of cellular automata concepts, which can process information computationally. One such cellular automata processor array configuration is the systolic array. Cell interaction can be via electric charge, magnetism, vibration (phonons at quantum scales), or any other physically useful means. This can be done in several ways so no wires are needed between any elements. This is very unlike processors used in most computers today based on von Neumann designs, which are divided into sections with elements that can communicate with distant elements, over wires.

Quantum Cellular Automata (QCA) (Arrighi & Fargetton, 2007; Pérez-Delgado & Cheung, 2007; Shepherd, Franz & Werner, 2006; Watrous, 1995) refers to any one of several models of quantum computation, which have been devised in analogy to conventional models of cellular automata introduced by von Neumann. QCA may also refer to quantum dot cellular automata, which is a proposed physical implementation of "classical" cellular automata by exploiting quantum mechanical phenomena.

In the context of models of computation or of physical systems, *quantum cellular automaton* is a term currently without a single agreed-upon meaning. However, models of quantum cellular automata tend to attempt to merge or borrow elements of both (1) the study of cellular automata in conventional computer science and (2) the study of quantum information processing. In particular, the following are common features of models of quantum cellular automata:

- The computation is considered to come about by parallel operation of multiple computing devices, or cells. The cells are usually taken to be identical, finite-dimensional quantum systems (e.g. each cell is a *qubit*);
- Each cell has a neighborhood of other cells. Altogether these form a network of cells, which is usually taken to be regular (e.g. the cells are arranged as a lattice with or without periodic boundary conditions);

- The evolution of all of the cells has a number of physics-like symmetries. Locality is one: the next state of a cell depends only on its current state and that of its neighbors. Homogeneity is another: the evolution acts the same everywhere, and is independent of time;
- The state space of the cells, and the operations performed on them, should be motivated by principles of quantum mechanics.

One feature that is often considered important for a model of quantum cellular automata is that it should be universal for quantum computation or it can efficiently simulate quantum Turing machines (Watrous, 1995; Pérez-Delgado & Cheung, 2007), some arbitrary quantum circuit (Shepherd, Franz & Werner, 2006) or simply all other quantum cellular automata (Arrighi & Fargetton, 2007).

Bipolar Cellular Networks

Traditionally cellular automata are based on classical unipolar mathematical abstraction which does not take bipolar equilibrium and non-equilibrium states into consideration. It is shown in (Zhang, 1996; Zhang & Chen, 2009; Zhang *et al.*, 2009) that classical linear algebra and cellular automata can be extended to YinYang-N-Elements bipolar linear algebra and bipolar cellular networks (BCNs) based bipolar sets and bipolar lattices (Zhang, 2005a). In this chapter we extend bipolar linear algebra and BCNs to bipolar quantum linear algebra (BQLA) and bipolar quantum cellular automata (BQCA).

While linear algebra is not necessarily a major mathematical tool in classical cellular automata, BQLA is found quite suitable for BQCA. BQLA is a bipolar extension of classical linear algebra. While classical linear algebra is based on unipolar mathematical abstraction, BQLA is based on bipolar mathematical abstraction with bipolar quantum entanglement and symmetry (Ch. 3; Ch. 4; Ch.5). Some basic concepts of classical linear algebra are introduced in the next section to provide a basis for BQLA.

Linear Algebra

Linear algebra is a branch of mathematics concerned with the study of vectors, vector spaces (also called *linear spaces*), linear maps (also called *linear transformations*), and systems of linear equations. It has extensive applications in both natural and social sciences because nonlinear models can often be approximated by linear ones.

Linear algebra had its beginnings in the study of vectors in 2-dimensional and 3-dimensional Cartesian spaces. A vector, here, is a directed line segment, characterized by both its magnitude (also called length or norm) and its direction. The zero vector is an exception; it has zero magnitude and no direction. Vectors can be used to represent physical entities such as forces, and they can be added to each other and multiplied by scalars, thus forming the first example of a real vector space, where a distinction is made between "scalars", in this case real numbers, and "vectors".

Modern linear algebra has been extended to consider spaces of arbitrary or infinite dimension. A vector space of dimension n is called an n-space. Most of the useful results from 2- and 3-space can be extended to these higher dimensional spaces. Although people cannot easily visualize vectors in n-space, such vectors are useful in representing data. Since vectors, as n-tuples, consist of n ordered components, data can be efficiently summarized and manipulated in this framework.

For example, in bioeconomics, one can create and use, say, 5-dimensional vectors or 5-tuples to represent the energy levels of five biological agents such as the five elements in TCM: kidney (v_1), lung

(v_2), heart (v_3), spleen (v_4), and liver (v_5). One can decide to display the energy levels of the agents for a particular time t, where the agents' order is specified, for example, by using a vector (v_1, v_2, v_3, v_4, v_5) where each agent is in its respective position.

A vector space (or linear space) as a purely abstract concept about which theorems are proved, is part of abstract algebra, and is well integrated into this discipline. Some striking examples of this are the group of invertible linear maps or matrices, and the ring of linear maps of a vector space. Linear algebra also plays an important part in analysis, notably, in the description of higher order derivatives in vector analysis and the study of tensor products and alternating maps.

In this abstract setting, the scalars with which an element of a vector space can be multiplied need not be numbers. The only requirement is that the scalars form a mathematical structure, called a field. In applications, this field is usually the field of real numbers or the field of complex numbers. Linear maps take elements from a linear space to another (or to itself), in a manner that is compatible with the addition and scalar multiplication given on the vector space(s). The set of all such transformations is itself a vector space. If a basis for a vector space is fixed, every linear transform can be represented by a table of numbers called a matrix. The detailed study of the properties of and algorithms acting on matrices, including determinants and eigenvectors, is considered to be part of linear algebra.

The following are some useful theorems:

- Every vector space has a basis. In linear algebra, a **basis** is a set of vectors that, in a linear combination, can represent every vector in a given vector space or free module, and such that no element of the set can be represented as a linear combination of the others. In other words, a basis is a linearly independent spanning set.
- Any two bases of the same vector space have the same cardinality; equivalently, the dimension of a vector space is well-defined.
- A matrix is invertible if and only if the linear map represented by the matrix is an isomorphism.
- If a square matrix has a left inverse or a right inverse then it is invertible (see invertible matrix for other equivalent statements).
- An $n \times n$ matrix is diagonalizable (i.e. there exists an invertible matrix P and a diagonal matrix D such that $A = PDP^{-1}$) if and only if it has n linearly independent eigenvectors.
- The spectral theorem states that a matrix is orthogonally diagonalizable if and only if it is symmetric.

The following is an example that shows how to use a matrix to formulate a linear system for solution:

Given

$a_1x + b_1y - c_1z = d_1$

$a_2x + b_2y - c_2z = d_2$

$a_3x + b_3y - c_3z = d_3$

we have

$$\begin{bmatrix} a_1 & b_1 & c_1 \\ a_2 & b_2 & c_2 \\ a_3 & b_3 & c_3 \end{bmatrix} \begin{bmatrix} x \\ y \\ z \end{bmatrix} = \begin{bmatrix} d_1 \\ d_2 \\ d_3 \end{bmatrix};$$

$$\text{or } [x, y, z] \times \begin{bmatrix} a_1 & b_1 & c_1 \\ a_2 & b_2 & c_2 \\ a_3 & b_3 & c_3 \end{bmatrix}^T = [d_1, d_2, d_3],$$

$$\text{or } [x, y, z] \times \begin{bmatrix} a_1 & a_2 & a_3 \\ b_1 & b_2 & b_3 \\ c_1 & c_2 & c_3 \end{bmatrix} = [d_1, d_2, d_3].$$

In case the square matrix is a connectivity matrix, we say [x, y, z] is an input vector and $[d_1, d_2, d_3]$ is an output vector. It is well-known that, if all the elements in the square connectivity matrix are normalized to the range of [0,1] and every row and column adds up to 1, we must have $d_1 + d_2 + d_3 \equiv x + y + z$. This can be generalized to any square matrix.

In general, given any $n \times n$ square connectivity matrix M of n elements, if the row elements and column elements of M are all normalized to the range [0,1], the summation of every row and every column equals 1, and the summation of the output vector must be equal to that of the input vector.

BIPOLAR QUANTUM LINEAR ALGEBRA AND BIPOLAR QUANTUM CELLULAR AUTOMATA

Definitions

Bipolar set theory provides the basis for bipolar mathematical abstraction and bipolar relativity provides a unifying theory for nature, agents, and causality. The theory so far is limited to the logical level. Bipolar cellular automation is to extend bipolar relativity to a mathematical physics or biophysics theory with a geometrical and algebraic basis.

Definition 8.1a. A *discrete BQCA* consists of a collection of N bipolar cells, each in one of a finite number of bipolar *equilibrium or non-equilibrium states.* Formally, let c be a bipolar cell, B be a crisp bipolar quantum lattice (Ch. 4), and ϕ be a bipolar function or mapping defined on B, we have, $\forall c$, $\phi(c): c \Rightarrow B$. The bipolar cells are organized into a grid in any finite number of dimensions. Each bipolar cell has a set of neighborhood bipolar cells including the cell itself. The neighborhood of a cell can be specified with a distance such as distance 1,2,..,N-1 from the cell. The initial state of the network can be specified by assigning a state for each cell at time $t = 0$. A new generation is created by advancing t by 1 according to a set of bipolar quantum logic or algebraic functions and bipolar interconnections governing the bipolar state change in a neighborhood.

Definition 8.1b. A *continuous BQCA* is an extension of discrete BQCA from a discrete bipolar quantum lattice (Ch. 4) to any continuous bipolar quantum lattice in $B = B_\infty = [-\infty, 0] \times [0, \infty]$ *or* $B = B_F = [-1,0] \times [0, 1]$. We say a continuous BQCA is *normalized* when $B = B_F$ (Ch. 4) (Zhang 2005a)

Except the lattice difference, all other concepts of continuous BQCA are basically the same as discrete BQCA. For instance, if $B = B_l = \{-1,0\} \times \{0,1\}$, we have the four different equilibrium or non-equilibrium states $(-1,0)$, $(0,1)$, $(0,0)$, and $(-1,1)$ and a finite number of bipolar logical operators. If $B = B_F = [-1,0] \times [0,1]$, we have an infinite number of bipolar fuzzy states for full equilibrium, quasi-equilibrium, and non-equilibrium with a finite number of bipolar logical operators or bipolar dynamic triangular norms (Ch. 4). If $B = B_\infty = [-\infty,0] \times [0, \infty]$, we have an infinite number of bipolar states for full equilibrium, quasi-equilibrium, and non-equilibrium with a finite number of bipolar algebraic operators such as +, -, ×, and /. Thus, BQCA is an extension or application of bipolar relativity that encompasses bipolar sets, bipolar logic, BQLA, bipolar agents or cells, and BCNs or bipolar grids.

Definition 8.1c. A BQCA is an *oscillating* BQCA if all its cells change their states from (x,y) to $-(x,y)$ periodically.

Definition 8.1d. A BQCA is in *energy equilibrium or energy symmetry* if its cellular function does not change the absolute value of its total energy even if it changes the polarity. Otherwise, we say it is in a *broken energy equilibrium or broken energy symmetry*.

It is noteworthy to recall that, even though traditional cellular automata theory has claimed universality, with classical mathematical abstraction, it is so far are inadequate for modeling local and global equilibrium of a dynamic system or a biological agent. The major barrier is that, the dynamic system in whole is an agent and its components are also agents. For instance, a person is an agent that consists of smaller agents such as neurons, genes, molecules, organs, etc. The global and local equilibria are interdependent. Without bipolarity, each cell in classical cellular automata cannot carry any direct bipolar equilibrium-based syntax and semantics and cannot be a bipolar agent at the global level and form a global agent collectively. Furthermore, without bipolarity, bipolar causality is undefinable, bipolar relativity and quantum entanglement cannot be realized, and the LAFIP paradox cannot be avoided.

Bipolar Quantum Linear Algebra (BQLA) (Adapted from Zhang & Chen 2009)

The bipolar lattice $B_l = \{-1\ 0\} \times \{0\ 1\}$ (Ch. 3) and bipolar fuzzy lattice $B_F = [-1\ 0] \times [0\ 1]$ (Ch. 4) can be naturally extended to the infinite bipolar lattice $B_\infty = [-\infty\ 0] \times [0\ +\infty]$. While $B_l = \{-1\ 0\} \times \{0\ 1\}$ and $B_F = [-1\ 0] \times [0\ 1]$ are bounded complemented unit square crisp/fuzzy lattices, respectively, B_∞ is unbounded. $\forall (x,y),(u,v) \in B_\infty$, a few major operations can be defined as shown in Equation (8.1a,b) and (8.2a,b) (Zhang & Chen 2009).

Entangled Bipolar Multiplication: $(x,y) \times (u,v) \equiv (xv+yu,\ xu+yv);$ (8.1a)

Scalar Bipolar Division: $(x,y)/\,a \equiv (x/a,\ y/a),\ a>0;$ (8.1b)

Bipolar Addition: $(x,y) + (u,v) \equiv (x+u,\ y+v);$ (8.2a)

Bipolar Subtraction: $(x,y) - (u,v) \equiv (x-u, y-v)$, $|x| \geq |u|$ and $|y| \geq |v|$. (8.2b)

In Equation 8.1a, $\times$ is a bipolar cross-pole tensor multiplication operator with the infused non-linear bipolar quantum entanglement semantics of --=+, -+=+-=1, and ++=+; in Equation 8.2a+ is a linear bipolar addition or fusion operator. In Equation 8.1b the scalar value a has to be a positive number; in Equation 8.2b bipolar subtraction has to satisfy the bipolar conditions $|x| \geq |u|$ and $|y| \geq |v|$. With these basic operations, classical linear algebra is naturally extended to BQLA with bipolar fusion, diffusion, interaction, separation, oscillation, and quantum entanglement properties. These properties enable biological agents to interact through bipolar bioelectromagnetic fields such as heart-heart, heart-brain, brain-brain, organ-organ, molecule-molecule, and genome-genome bio-electromagnetic fields as well as biochemical pathways. Thus, the bipolar properties are suitable for equilibrium/non-equilibrium based dynamic modeling with quantum aspects. The bipolar property is depicted in Chapter 3, Figure 1.

Given an input bipolar row vector matrix $E = [e_i] = [(e_i^-, e_i^+)]$, $i=1,2,..,k$, and a bipolar connectivity matrix $M = [m_{ij}] = [(m_{ij}^-, m_{ij}^+)]$, $i=1,2,..,k$ and $j = 1,2,..,n$, *we have* $V = E \times M = [V_j] = [(v_j^-, v_j^+)]$. While E is the input vector to a dynamic system characterized with the connectivity matrix M, V is the result row vector matrix with n bipolar elements following Equation 8.3.

$$V_j = \sum_{i=1}^{k} (e_j \times m_{ij}). \quad (8.3)$$

Equation 8.3 has the same form as in classical linear algebra except for: (i) e_j and m_{ij} are bipolar elements; (ii) the multiplication operator is defined in Equation (8.1) on bipolar variables; and (iii) the $\sum$ operator is based on the addition operation defined on bipolar variables in Equation (8.2).

BQLA provides a new mathematical tool for modeling bipolar elements (or agents) with explicit YinYang representation and equilibrium, quasi- or non-equilibrium states for vital energy (or Qi in TCM) equilibrium and stability analysis. The concepts of equilibrium energy and stability of a bipolar relation (Ch. 3; Ch. 5) can be naturally extended to that of bipolar connectivity and vector matrices. In this case, energy in a row matrix can be considered as biological energy of biological elements or agents such as energy for repression and activation of regulator proteins; energy embedded in a connectivity matrix can be considered organizational energy (McKelvey, 2004) of the biological agents from a bioeconomics perspective such as the bipolar capacities of biological pathways.

Definition 8.2. YinYang Bipolar Elementary Energy (Zhang & Chen, 2009). Given a bipolar element $e=(e^-, e^+)$,

i. $\varepsilon^-(e) = e^-$ is *the Yin or negative energy of e;*
ii. $\varepsilon^+(e) = e^+$ is *the Yang or positive energy of e*;
iii. $\varepsilon(e) = (\varepsilon^-(e), \varepsilon^+(e)) = (e^-, e^+)$ is the *YinYang bipolar energy measure of e;*
iv. The absolute total $|\varepsilon|(e) = |\varepsilon^-|(e) + |\varepsilon^+|(e)$ is the *total energy* of e;
v. $\varepsilon_{imb}(e) = |\varepsilon^+|(e) - |\varepsilon^-|(e)$ is *the YinYang imbalance* of e;
vi. $EnergyBalance(e) = (|\varepsilon|(e) - |\varepsilon_{imb}(e)|)/2.0 = \min(|e^-|, e^+)$;
vii. $Harmony(e) = Balance(e) = (|\varepsilon|(e) - |\varepsilon_{imb}(e)|)/|\varepsilon|(e)$.

YinYang Bipolar System Energy (Zhang & Chen, 2009). Given an $k\times n$ bipolar matrix $M = [m_{ij}] = (M^-, M^+) = ([m_{ij}^-], [m_{ij}^+])$, where M^- is the *Yin half* with all the negative elements and M^+ is the *Yang half* with all the positive elements,

i. $\varepsilon^-\left(M\right) = \sum_{i=1}^{k}\sum_{j=1}^{n}\varepsilon_{ij}^- = \sum_{i=1}^{k}\sum_{j=1}^{n}m_{ij}^-$ is the *negative or Yin energy* of *M*;
ii. $\varepsilon^+\left(M\right) = \sum_{i=1}^{k}\sum_{j=1}^{n}\varepsilon_{ij}^+ = \sum_{i=1}^{k}\sum_{j=1}^{n}m_{ij}^+$ is the *positive or Yang energy* of *M*;
iii. the polarized total, denoted $\varepsilon(M) = (\varepsilon^-(M), \varepsilon^+(M))$ is the *YinYang bipolar energy of M of M;*
iv. the absolute total, denoted $|\varepsilon|(M) = |\varepsilon^-|(M) + |\varepsilon^+|(M)$, is the *total energy* of *M*;
v. the energy subtotal for row i of M is denoted $|\varepsilon|\left(M_{i*}\right) = \left|\sum_{j=0}^{n}\varepsilon_{ij}\right|$;
vi. the energy subtotal for column j of M is denoted $|\varepsilon|\left(M_{*j}\right) = \left|\sum_{i=0}^{k}\varepsilon_{ij}\right|$;
vii. the summation $\varepsilon_{\text{imb}}\left(M\right) = \sum_{i=1}^{k}\sum_{j=1}^{n}\varepsilon_{imp}(m_{ij}) = \sum_{i=1}^{k}\sum_{j=1}^{n}(m_{ij}^+ - |m_{ij}^-|)$ is the *YinYang imbalance* of *M*;
viii. *balance or harmony or stability of M* is defined as *Harmony*(M) = *Balance*(*M*) = *Stability*(*M*) = $(|\varepsilon|(M) - |\varepsilon_{\text{imb}}(M)|)/|\varepsilon|(M)$;
ix. the *average* energy *of M* is measured as $h = (\varepsilon^-(M)/(kn), \varepsilon^+(M)/(kn))$ where $kn=k\times n$ is the total number of elements in M.

Based on the concept of energy and energy equilibrium or energy symmetry introduced in Chapter 4 we have:

Theorem 8.1. Elementary Energy Equilibrium or Symmetry Law (Zhang & Chen 2009). $\forall (x,y)\in B_\infty = [-\infty, 0]\times[0, +\infty]$ and $\forall (u,v)\in B_F = [-1,0]\times[0,1]$, we have

a. energy symmetry $[\ |\varepsilon|(u,v) \equiv 1.0] \Rightarrow [\ |\varepsilon|((x,y)\times (u,v)) \equiv |\varepsilon|(x,y)\ ]$;
b. energy non-symmetry $[\ |\varepsilon|(u,v)<1.0] \Rightarrow [\ |\varepsilon|((x,y)\times (u,v)) < |\varepsilon|(x,y)\ ]$;
c. energy non-symmetry $[\ |\varepsilon|(u,v)>1.0] \Rightarrow [\ |\varepsilon|((x,y)\times (u,v)) > |\varepsilon|(x,y)\ ]$.

Proof. Following Equation (8.1) and the definition of ***YinYang Bipolar Elementary Energy*** *we have* $|\varepsilon|((x,y)\times (u,v)) = |\varepsilon|(xv+yu, xu+yv) = |xv|+|yu| + xu+yv = (|u|+v)|x| + (|u|+v)\, y$. *Then we have*

a. *If* $(|u|+v) = 1$, *we have* $(|u|+v)\,|x| + (|u|+v)\,y = |\varepsilon|(x,y)$.
b. *If* $(|u|+v) < 1$, *we have* $(|u|+v)\,|x| + (|u|+v)\,y < |\varepsilon|(x,y)$.
c. *If* $(|u|+v) > 1$, *we have* $(|u|+v)\,|x| + (|u|+v)\,y > |\varepsilon|(x,y)$. □

The significance of Theorem 8.1 lies in the facts that:

i. It enables explicit bipolar representation of the Yin and Yang in fusion for YinYang bipolar analysis regarding balance, equilibrium, and harmony (Maciocia, 1989).
ii. It enables (unipolar) energy equilibrium and non-equilibrium analysis in classical scientific terms as defined in thermodynamics.
iii. It enables the characterization of a metabolic nourishing/enhancing relation with a balanced or harmonic bipolar coefficient or link weight. Such a nourishing relation can be realized through biological pathways.
iv. It enables the characterization of a biological restricting or regulating relation with an unbalanced positive coefficient or link weight. Such a restricting relation can be realized through bioelectromagnetic fields as well as biological pathways.
v. It enables the characterization of a reflexive or inertial relation with an unbalanced positive coefficient or link weight.

Observations: Let (x,y) be the YinYang states of biological agent A; let (u,v) be the link weight from A to another biological agent B; let the input from A to B be $(x,y)\times(u,v)$. We have

a. Regardless of the balance or imbalance of (x,y), $(x,y)\times(u,v) = (xv+yu, xu+yv)$ is always balanced as long as (u,v) is balanced, e.g. $(-7\ 2) \times (-0.1\ 0.1) = (-0.9\ -0.9)$. If (u,v) is balanced with total energy $|\varepsilon|(u,v) = 1.0$, we have the energy equilibrium $|\varepsilon|((x,y)\times(u,v)) = |\varepsilon|(x,y)$. e.g. $|\varepsilon|\ ((-7\ 2)\times(-0.5\ 0.5)) = |\varepsilon|(-4.5, 4.5)$. (Note: In the following we may omit the comma in (x, y) with numerical bipolar values.)
b. Following (i), if (u,v) is not balanced but has total energy 1.0, $|u|>0$ and $v>0$, $\varepsilon((x,y)\times(u,v)) = \varepsilon(x,y)$ shows an energy equilibrium with a balancing factor min(|u|, v);
c. Following (i), if $u = 0$ and $v = 1.0$, we have $(x,y) \times (u,v) = (xv+yu, xu+yv) = (x,y)$; if $u = 0$ and $0 < v < 1$, $(u\ v)$ defines a restricting or regulating factor on $(x,y)\times(u,v) = (xv+yu, xu+yv)$. If (u,v) is the reflexive link from A to itself, (u,v) defines an inertial factor.
d. Following (i), if $|\varepsilon|(u,v) < 1.0$, $|\varepsilon|((x,y)\times(u,v)) < |\varepsilon|(x,y)$ leads to an energy loss and a non-equilibrium process.
e. Following (i), if $|\varepsilon|(u,v) > 1.0$, $|\varepsilon|((x,y)\times(u,v)) > |\varepsilon|(x,y)$ leads to an energy increase and a non-equilibrium process.

Definition 8.3. Equilibrium/Non-Equilibrium Systems. A bipolar dynamic system S is said to be an ***equilibrium system*** if the system's total energy $|\varepsilon|S$ remains in an equilibrium state with the derivative $d(|\varepsilon|S)/dt = 0$ without external disturbance. Otherwise it is a ***non-equilibrium system.*** A non-equilibrium system is a ***strengthening system*** if $d(|\varepsilon|S)/dt>0$; it is a ***weakening system*** if $d(|\varepsilon|S)/dt<0$.

Theorem 8.2. Energy Transfer Equilibrium Law (Zhang & Chen 2009). Given an n×n input bipolar matrix $E = [e_{ik}] = [(e_{ik}^-, e_{ik}^+)]$, $0<i,k\leq n$, an n×n bipolar connectivity matrix $M = [m_{kj}] = [(m_{kj}^-, m_{kj}^+)]$, $0<k,j\leq n$, *and* $V = E \times M = [V_{ij}] = [(v_{ij}^-, v_{ij}^+)]$, ∀k,j, let $|\varepsilon|(M_{k*})$ be the k-th row energy subtotal and let $|\varepsilon|(M_{*j})$ be the j-th column energy subtotal, we have, ∀k,j,

a. $[|\varepsilon|(M_{k*}) \equiv |\varepsilon|(M_{*j}) \equiv 1.0\ ] \Rightarrow [|\varepsilon|(V) \equiv |\varepsilon|(E)]$;

b. $[|\varepsilon|(M_{k*}) \equiv |\varepsilon|(M_{*j}) < 1.0\,] \Rightarrow [|\varepsilon|(V) < |\varepsilon|(E)];$
c. $[|\varepsilon|(M_{k*}) \equiv |\varepsilon|(M_{*j}) > 1.0\,] \Rightarrow [|\varepsilon|(V) > |\varepsilon|(E)].$

Proof. Following Theorem 8.1 and Equation 8.3, the three conditions enable the energy of each element of E be transferred and distributed to the elements of V, respectively, in (a) 100%, (b) less than 100%, or (c) more than 100%. □

From the above definitions and laws it is clear that without YinYang bipolarity, classical linear algebra cannot deal with the coexistence of the Yin and the Yang of bipolar elements and their interactions and entanglement such as genetic repression and activation.. It is clear that information gain has doubled with bipolar explicit representation. For instance, a classical energy measure E = 200 in absolute total could mean a balanced state of repression and activation abilities such as (-100 100) or an imbalance (0 200) or (-200, 0), etc. Such distinctions are crucial for YinYang diagnostic analysis in TCM (Maciocia, 1989) and YY1 gene regulation network simulation. Thus, BQLA with bipolar energy analysis brings ancient YinYang and modern equilibrium/non-equilibrium thermodynamics and bioinformatics concepts into a unified framework.

The energy transfer equilibrium law introduced above states that if the energy of each row and each column of a bipolar connectivity matrix M is normalized to 1.0, $V = E \times M$ will remain in a global bipolar energy equilibrium regardless of elementary level equilibrium, non-equilibrium, or oscillation. On the other hand, if the energy of each row and column is less than 1.0 or greater than 1.0 we will observe energy decrease or increase, respectively. Now we have the advanced topics:

1. How can BQLA be generalized from the YinYang-5-Element model in (Zhang & Chen, 2009) to YinYang-N-Element settings?
2. How can M be initialized with randomization and normalization?
3. What are the conditions for achieving local (element level) energy equilibrium under global (system level) equilibrium?

These questions are answered in the next section.

Figure 1. YinYang-5-elements (Zhang & Chen, 2009)

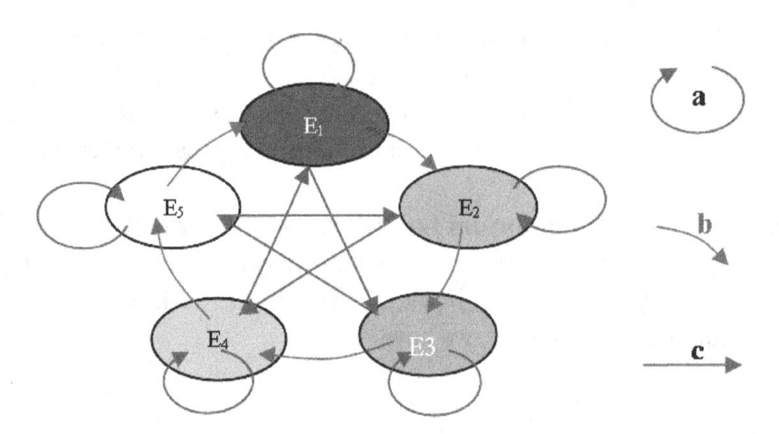

YinYang-N-Element Bipolar Cellular Networks (BCNs) (Adapted from (Zhang & Chen, 2009)

For thousands of years, the *YinYang Wuxing* (YinYang-5-Element or YinYang-5-Agents) system of human body has been a major theoretical foundation for TCM including herbal medicine, (acupuncture) channels, qigong, and acupuncture (Li, 1987). Regardless of its scientific basis, the YinYang-5-Element system provides a fundamental cellular structure or graph for modern scientific research in computational biology. A diagram is shown in Figure 1 (Zhang & Chen, 2009), where the set of five elements or agents (Kidney (E1), Lung (E2), Heart (E3), Spleen (E4), and Liver (E5)) subsystems are linked with five nourishing relations (green arrows) that form the external cycle, five regulating relations (red arrows) that form the internal cycle, and five reflexive arrows as inertial factors.

Similar to the YinYang-5-Element cellular structure, we may have YinYang-N-Element BCNs. Figure 2 shows a 6-element BCN and Figure 3 shows a 7-element structure. When N is an even number the internal regulating relations are partitioned into two sets. For instance, {E1,E3,E5} and {E2,E4,E6} form two regulatory partitions in the YinYang-6-Element cellular structure (Figure 3). Surprisingly, YinYang-

Figure 2. YinYang-6-Elements (Zhang et al., 2009)

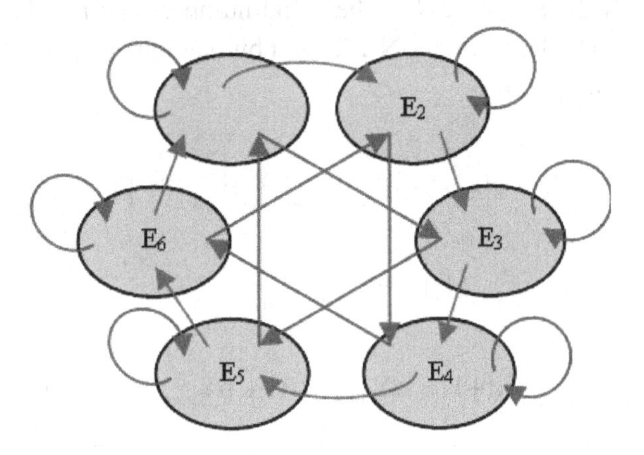

Figure 3. YinYang-7-Elements (Zhang et al., 2009)

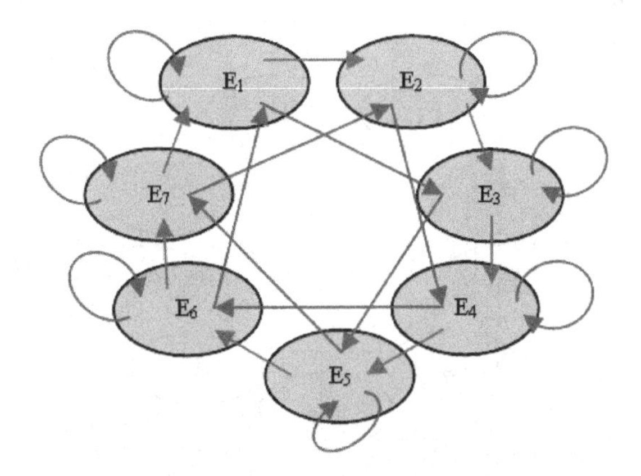

9-Element structure (Figure 4) does not partition to three regulatory groups as we have the complete cycle 1-3-5-7-9-2-4-6-8-1. This finding indicates that N does not have to be a prime number to prevent regulatory partitioning. However, when N is a prime number, such partitioning is most unlikely. It can be shown that when N<5 the cyclic regulating relations cannot be realized. Thus, YinYang-5-Element is the minimal structure of its kind.

YinYang-N-Element Dynamic Equations (Adapted from Zhang *et al.*, 2009)

Let $\{Y_n\} = \{Y_1, Y_2, Y_3, ..., Y_n\}$ be the set of YinYang-N-Elements, respectively. Let Y_n be a YinYang bipolar variable or element with a negative side and a positive side. Let $t = 0, 1, 2, 3, ..., i, ...$ and $Y_n(t)$ be the states of Yin and Yang of Y_n at time t; let $n = N, N+1, N+2 ....$ *An* ordinary differential equation for the energy level or Qi (vital energy in TCM and qigong that regulates the functions of the body as a whole system) of each element in a YinYang-N-Element system can be defined in Equation (8.4a).

Similar to that in (Li, 1987), the energy- or Qi-based dynamic equation Equation 8.4a suffers from its lack of explicit YinYang bipolar representation for the Yin and the Yang of each subsystem. If we add YinYang bipolarity to each subsystem Y and each coefficient a, b, or c such as in Equation 4b, it would be difficult if not impossible to solve due to non-linear dynamic YinYang bipolar interaction between the five elements. This dilemma can be circumvented with discrete time and linear system assumption.

Based on Figures 1 – 4, let $n = 0,1,2,3, ...$ be serial numbers and $t = 0,1,2,3, ...$ be serial discrete time points, we convert 8.4a to the YinYang-N-Element bipolar dynamic equation as shown in Equation 8.4c with linear system assumption.

$$\frac{d|\varepsilon|Y_{(n\%N+1)}}{dt} = |e|a'|e|Y_{(n\%N+1)}(t) + |e|b'|e|Y_{((n-1)\%N+1)}(t) + |e|c'|e|Y_{((n-2)\%N+1)}(t). \quad (8.4a)$$

$$\frac{dY_{(n\%N+1)}}{dt} = aY_{(n\%N+1)}(t) + bY_{((n-1)\%N+1))}(t) + cY_{((n-2)\%N+1)}(t). \quad (8.4b)$$

$$Y_{(n\%N+1)}(t+1) = (Y^-_{(n\%N+1)}, Y^+_{(n\%N+1)})(t+1) = aY_{(n\%N+1)}(t) + bY_{((n-1)\%N+1))}(t) + cY_{((n-2)\%N+1)}(t) =$$

$$a(Y^-_{(n\%N+1)}, Y^+_{(n\%N+1)})(t) + b(Y^-_{((n-1)\%N+1)}, Y^+_{((n-1)\%N+1)})(t) + c(Y^-_{((n-2)\%N+1)}, Y^+_{((n-2)\%N+1)})(t). \quad (8.4c)$$

Figure 4. YinYang-9-Elements

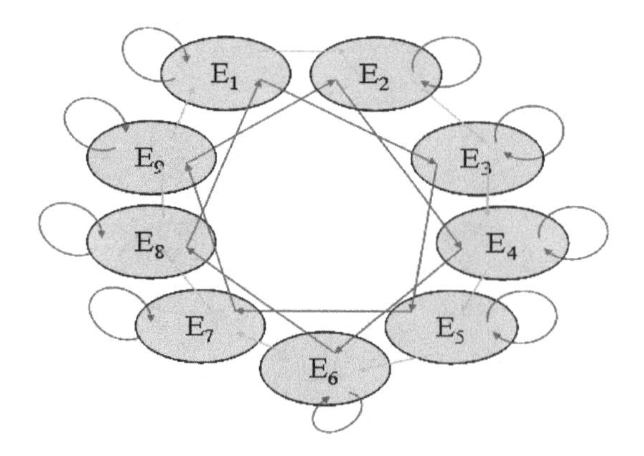

In Equation 8.4, % is a modulo operator (same as in computer languages) that enables the five elements to have their indices bounded within *1* to *N*. The parameters *a, b,* and *c* are inertial, nourishing, and regulating coefficients, respectively. These coefficients are different from those in ordinary differential equations. They are YinYang bipolar variables or constants from a bipolar lattice. The inertial coefficient a tends to keep the Yin and Yang of an element at time point $t+1$ the same as that at time t. It can also be considered a reflexive link weight as shown in Figure 1. The nourishing coefficient b enables enhancing and balancing effect from one organ to the next in the outer cycle. It is the weight of a green link in Figure 1. The restricting coefficient c is a link weight of the star in Figure 1. The coefficients should satisfy the following properties:

1. Since a characterizes an *inertial* reflexive relation it should be positive such that $a \times Y(t)$ is kept the same as $Y(t)$ to a certain extent.
2. Since b characterizes a metabolic *nourishing relation* it should be YinYang balanced or partially balanced such that $b \times Y(t)$ is a balanced input to the next element regardless of the balance/imbalance of $Y(t)$.
3. Since c characterizes a *restricting relation* it should be positive such that $c \times Y(t)$ is kept the same as $Y(t)$ to a certain extent.

Equation 8.4 is actually a set of N dynamic equations each of which characterizes the Yin and Yang states in a unified representation. When N=5, the five equations for YinYang-5-Element cellular network can be explicitly described in Equation 5a-e (Zhang & Chen, 2009), respectively, for $n=0,1,2,3,4$.

$$Y_1(t+1)= (Y^-_1, Y^+_1)(t+1) = aY_1(t)+bY_5(t)+cY_4(t)+cY_3(t) = a(Y^-_1,Y^+_1)(t)+b(Y^-_5,Y^+_5)(t)+c(Y^-_4,Y^+_4)(t). \quad (8.5a)$$

$$Y_2(t+1)= (Y^-_2, Y^+_2)(t+1) = aY_2(t) + bY_1(t) + cY_5(t) = a(Y^-_1, Y^+_1)(t) + b(Y^-_2, Y^+_2)(t)+ c(Y^-_5, Y^+_5)(t). \quad (8.5b)$$

$$Y_3(t+1)= (Y^-_3, Y^+_3)(t+1) = aY_3(t) + bY_2(t + cY_1(t) = a(Y^-_3, Y^+_3)(t) + b(Y^-_2, Y^+_2)(t) + c(Y^-_1, Y^+_1)(t). \quad (8.5c)$$

$$Y_4(t+1)= (Y^-_4, Y^+_4)(t+1) = aY_4(t) + bY_3(t) + cY_2(t) = a(Y^-_4, Y^+_4)(t) + b(Y^-_3, Y^+_3)(t) + c(Y^-_2, Y^+_2)(t). \quad (8.5d)$$

$$Y_5(t+1)= (Y^-_5, Y^+_5)(t+1) = aY_5(t) + bY_4(t) + cY_3(t) = a(Y^-_5, Y^+_5)(t) + b(Y^-_4, Y^+_4)(t) + c(Y^-_3, Y^+_3)(t). \quad (8.5e)$$

Non-Classical vs. Classical YinYang-N-Element Cellular Networks (Zhang *et al.*, 2009)

It is clear that the YinYang-5-Element BCN structure can be naturally extended to YinYang-N-Elements, $\forall N, 1 \leq N < \infty$. If the extension strictly follows the parent-grandparent or nourishing-regulating relationships we have ***a family of classical YinYang-N-Element cellular structure*** with predefined nourishing-regulating cycles. Limitations of this family include:

1. The elements are sparsely connected and full connectivity is not observed.
2. Random connectivity with random weights is not considered for simulation and analysis.
3. Without randomization BQLA would not be able to support statistical or quantum bioinformatics.

The three limitations can be addressed with a family of ***non-classical YinYang-N-Element*** BCNs with random nourishing-regulating cycles. Random nourishing-regulation cycles are of particular interest in statistical quantum mechanics and gene regulation network simulation. In such a network bipolar link weights of a connectivity matrix M are randomly assigned to achieve random connectivity without predefined nourishing and regulation cycles. Figure 5 shows such a fully connected network where each bipolar link weight can be randomly assigned to (u,v), $\forall(u,v)\in B_F=[-1, 0] \times [0, 1]$.

The randomly assigned bipolar weights of a connectivity matrix can be normalized to meet the equilibrium or non-equilibrium conditions. For instance, Algorithm A (see Table 1) is for meeting the energy equilibrium condition $[|\varepsilon|(M_{k*}) \equiv |\varepsilon|(M_{*j}) \equiv 1.0\ ]$. Using vector forms, the N bipolar elements in the network can be represented as a bipolar row matrix $Y(t) = [Y_1(t), Y_2(t),.., Y_N(t)]$. The connectivity matrix M can be represented in an N×N square matrix $M(t)$. Then we have $Y(t+1) = Y(t) \times M(t)$. In case we assume M is constant or not time-variant, M(t) = M. An algorithm (Algorithm B in Table 1) has been implemented using BQLA in MS Visual C++ to test the property of $Y(t+1) = Y(t) \times M(t)$.

YinYang-N-Element Cellular Combinatorics

As discussed earlier, there are three types of bipolar link weights. A harmonic link weight is a stabilizing factor that makes the link a metabolic pathway leading to bipolar equilibrium of a target cell (or agent); a positive link weight is a regulating pathway through which the regulating cell can impose its bipolar state to the regulated agent; a negative link weight causes the target cell to oscillate. Although chaotic oscillation will definitely cause physical or mental disorder, orderly oscillation is not necessarily a bad thing. Without orderly neural and mental oscillation in different frequencies at different levels we would have no memory scanning and no mental equilibrium.

Besides the classical and random YinYang-N-Element BCNs, there is another type of non-classical BCNs, namely the YinYang-N-Element combinatorial BCNs, in which each element is connected to any other k elements (k-fan-in and k-fan-out connections, $k \leq N$) but in any possible configuration. An instance of this type is named *mixed BCNs* in (Jaeger, Chen & Zhang, 2009) which is a combination of classical YinYang-N-Element BCNs following YinYang-5-Elelemts with fixed $k = 3$.

Figure 5. YinYang-N-Elements with Random Weight Full Connectivity (Zhang et al., 2009)

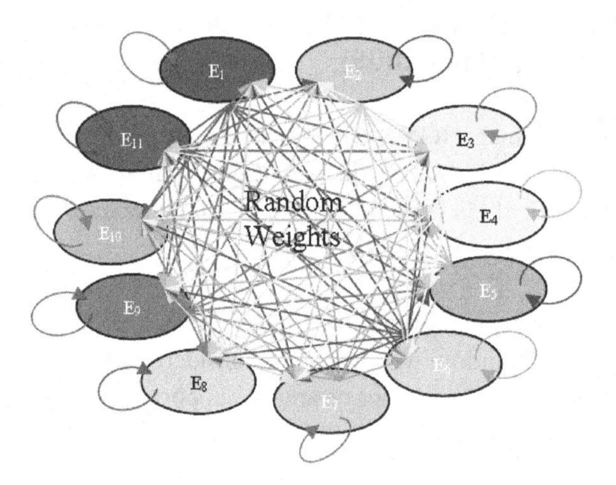

While the idea of mixed BCN is to retain the nourishing-regulating relationships between elements as in the classical 5-Element theory and at the same time allowing limited randomness in connectivity, the idea of combinatorial BCN is to allow maximum random reconfiguration. Note that each element in the traditional Wuxing (5-elelemt) star diagram has exactly 3 incoming and 3 outgoing connections when counting reflexive links (Zhang & Chen, 2009). Mixed YinYang-5-Element networks are also limited to 3-fan-in and 3-fan-out connections (Jaeger, Chen & Zhang, 2009). Combinatorial BCNs extend the scope of classical YinYang-N-Element theory to areas requiring reconfiguration, randomization, such as statistical bioinformatics.

To estimate the number of combinatorial YinYang-N-Element networks, let us assume that it is possible to represent the connectivity of such a network by an N x N binary adjacency matrix that has exactly k 1's in each row and k 1's in each column. The number of possible combinatorial networks is then equivalent to the number of possible adjacency matrices satisfying the required row and column sums. We are now going to estimate this number, providing a lower bound for the minimum number of networks we can expect for a given N. In fact, this approach is reminiscent of an idea by Shannon, who used a similar idea to estimate the number of possible crosswords in different dimensions, and which he already briefly mentioned in his seminal 1948 paper (Shannon, 1948).

Let C be the number of valid entries for a single row of the adjacency matrix, which is the number of possible rows with exactly k 1's. Obviously, this number is the same as the number of valid entries for a single column, and is equal to the following binomial coefficient:

$$C = \binom{N}{k}$$

Table 1. Two algorithms for testing energy equilibrium (Zhang et al., 2009)

```
//Algorithm A: Normalize to /|ε|(M_k*) ≡ |ε|(M_*j)≡1.0 ]
//---------------------------------------------------------------------------
YinYangMatrix M(N,N);          // create an N×N bipolar matrix
M.randomize();                 // assign random weights
M.normalizeRows();             // normalize each row |ε|(M_k*) = 1
M.normalizeCols().             // normalize each column |ε|(M_*j) = 1
//---------------------------------------------------------------------------
Algorithm B: Test Y(t+1) = Y(t) × M(t)
//---------------------------------------------------------------------------
YinYangMatrix M(N,N);          // create an N×N bipolar connectivity matrix
YinYangMatrix Yt0(1,N);
YinYangMatrix Yt1(1,N);
M.randomize();                 // assign random link weights to M
M.normalizeRows();             // normalize each row |ε|(M_k*) = 1
M.normalizeCols().             // normalize each column |ε|(M_*j) = 1
file1 >> Yt0;                  // input row matrix from file1
int times;
cin >> times;                  // enter times to iterate
for (int i = 0; i<times; i++){
Yt1 = Yt0*M;                   // row matrix multiply M
file2 << Yt1;                  // output result row matrix to file2
file2 << Yt1.totalEnergy() << "\n";   // output energy to file2
Yt0 = Yt1;                     // reassign for next iteration
}
```

Furthermore, let p be the probability that a randomly generated row is valid; i.e. it contains exactly k 1's. Clearly, this probability is the number of valid rows divided by the number of all possible rows, which is

$$p = \frac{C}{2^N} = \frac{\begin{pmatrix} N \\ k \end{pmatrix}}{2^N} \tag{8.6}$$

For a combinatorial YinYang-N-Element network, all rows need to be valid. According to Equation 8.6, the probability of an adjacency matrix with only valid rows is p^N. Since we are assuming independent rows and columns here, we need to multiply this probability with the probability that all columns are valid, which leads to the following probability p_A of a valid adjacency matrix A:

$$p_A = p^{2N} = \frac{\begin{pmatrix} N \\ k \end{pmatrix}^{2N}}{2^{2N^2}} \tag{8.7}$$

Given this probability of a valid combinatorial network, it is now possible to compute a lower boundary for the number of valid networks by multiplying the probability p_A in Equation 8.7 with the number of possible adjacency matrices. This leads to the following result for the number $E$of combinatorial YinYang-N-Element networks that we can expect for a given number of elements N:

$$E = p_A \cdot 2^{N^2} = \frac{\begin{pmatrix} N \\ k \end{pmatrix}^{2N}}{2^{N^2}} \tag{8.8}$$

It can be shown that, for any given N≥5, E is maximal when k = N/2. Thus, N=5 and k=3 is a optimal combination that leads to the YinYang-5-Elelemt configuration with both regulating and harmonic link weights. This can be a vindication of the YinYang-5-Elelemt theory as the foundation of traditional Chinese medicine (TCM). It can also be a vindication of the five organs of the human body of the most intelligent and complex agent known in the universe. Based on Equation 8.8 each of the networks in Figures 1 – 4 is just one out of E different combinatorial networks when k=3. While this chapter is focused on the mathematical properties of YinYang-N-Element BQCA, the application of optimal k=N/2 is discussed in the next chapter.

EQUILIBRIUM, NON-EQUILIBRIUM, AND OSCILLATORY BQCAS ADAPTED FROM (ZHANG *ET AL.*, 2009)

Laws of Symmetry and Broken Symmetry

BQLA is actually non-linear on each pole due to bipolar interaction and quantum entanglement; however, if the energy total of a BQLA matrix is considered an equilibrium or non-equilibrium, BQLA becomes a linear algebra. That is why it is named *bipolar quantum linear algebra.* Evidently, without linear approximation, the dynamic equations described in Equation 8.4a and 8.4b would be very difficult to solve. BQLA provides a mathematical tool for linear approximation of YinYang-N-Element BQCA dynamic equations.

For instance, using matrix forms, the YinYang-5-Element cellular network can be represented as a 5-dimensional bipolar row matrix $Y(t) = [Y_1(t), Y_2(t), Y_3(t), Y_4(t), Y_5(t)]$. The connectivity of Equation 8.5a-e can be represented in a 5×5 square matrix $M(t)$ (Zhang & Chen, 2009). Then we have $Y(t+1) = Y(t) \times M(t)$. This is shown in Equation 8.9. In case we assume M is constant or not time-variant, M(t) = M.

$$\mathrm{M} = \begin{bmatrix} a & b & c & 0 & 0 \\ 0 & a & b & c & 0 \\ 0 & 0 & a & b & c \\ c & 0 & 0 & a & b \\ b & c & 0 & 0 & a \end{bmatrix} = \begin{bmatrix} (a^-,a^+) & (b^-,b^+) & (c^-,c^+) & (0,0) & (0,0) \\ (0,0) & (a^-,a^+) & (b^-,b^+) & (c^-,c^+) & (0,0) \\ (0,0) & (0,0) & (a,a) & (b^-,b^+) & (c^-,c^+) \\ (c^-,c^+) & (0,0) & (0,0) & (a,a) & (b^-,b^+) \\ (b,b) & (c^-,c^+) & (0,0) & (0,0) & (a,a) \end{bmatrix};$$

$$Y(t+1) = Y(t)\times M(t) = [Y_1(t), Y_2(t), Y_3(t), Y_4(t), Y_5(t)] \times \mathrm{M}(t) = [(Y_1^-(t), Y_1^+(t)), (Y_2^-(t), Y_2^+(t)), (Y_3^-(t), Y_3^+(t)), (Y_4^-(t), Y_4^+(t)), (Y_5^-(t), Y_5^+(t)] \times \mathrm{M}(t). \quad (8.9)$$

For any finite integer N, classical, random, combinatorial YinYang-N-Element BCN structures satisfy the following four laws. The four laws establish energy equilibrium, non-equilibrium, and oscillation conditions at both the local and global levels.

Theorem 8.3. ***YinYang-N-Element Energy Equilibrium Law (Law of Energy Symmetry)*** (Zhang *et al.*, 2009). *Let* $t=0,1,2,\ldots,$ $Y(t+1)=Y(t)\times M(t)$, $|\varepsilon|Y(t)$ be the total energy of an YinYang-N-Element vector $Y(t)$, $|\varepsilon|M(t)$ be the total energy of the connectivity matrix $M(t)$, $|\varepsilon|M_{i*}(t)$ be the energy subtotal of row i of $M(t)$, $|\varepsilon|M_{*j}(t)$ be the energy subtotal of column j of $M(t)$.

1. Regardless of the local YinYang balance/imbalance of the elements at any time point *t*, the system will remain a global energy equilibrium if, $\forall t,\ d(|\varepsilon|Y(t))/dt \equiv 0$, *or* (a) $\forall i,j,\ [|\varepsilon|(M_{i*}) \equiv |\varepsilon|(M_{*j}) \equiv 1.0\ ]$ and (b) no external disturbance or input to the system after the initial vector $Y(0)$ is given.
2. Under the same conditions of *(1)*, if, $\forall t$, $|\varepsilon^-(M_{*j})| > 0$ and $|\varepsilon^+(M_{*j}))| > 0$, all bipolar elements connected by M will eventually reach a local YinYang balance $(-|\varepsilon|Y(t)/(2N), |\varepsilon|Y(t)/(2N))$ at certain time point t.

Proof.

1. *It follows directly from Theorem 8.2.*
2. *It follows from the observations:*
   a. *Regardless of the balance or imbalance of (x,y), (x,y)×(u,v) = (xv+yu, xu+yv) is always balanced as long as (u,v) is balanced (e.g. (-7 2)×(-0.1 0.1) =(-0.9 -0.9));*
   b. *if (u,v) is not balanced but has total energy 1.0, |u|>0 and |v|>0, we have ε((x,y)×(u,v))=ε(x,y) that must be more balanced with a balancing factor min(|u|, v).*
   c. *the condition,* ∀*t,* $|\varepsilon^{-}(M_{*j})| > 0$ *and* $|\varepsilon^{+}(M_{*j}))| > 0$ *ensures every element at time t+1 is more balanced than at t.* □

Following the definition of symmetry as an operation that doesn't change how something behaves relative to the outside world, Theorem 8.3 can be called the energy equilibrium law or the law of energy symmetry. Under the conditions of this law, any operation is an energy symmetry.

Theorem 8.4. ***YinYang-N-Element System Non-Equilibrium Strengthening Law (Law1 of Broken Symmetry)*** (Zhang *et al.*, 2009). For the same system as for Theorem 8.3, if, $\forall i,j$, $|\varepsilon|(M_{i*}) \equiv |\varepsilon|(M_{*j}) > 1.0$, regardless of the local YinYang balance/imbalance of the elements at any time point t, the system energy will increase and eventually reach a bipolar infinite (-∞,∞) state without external disturbance or we have, $\forall t$, $d(|\varepsilon|Y(t))/dt > 0$.

Proof. It follows Theorem 8.2 directly. □

Theorem 8.5. ***YinYang-N-Element System Non-Equilibrium Weakening Law (Law2 of Broken Symmetry)*** (Zhang *et al.*, 2009). For the same system as for Theorem 8.3, if, $\forall i,j$, $|\varepsilon|(M_{i*}) \equiv |\varepsilon|(M_{*j}) < 1.0$, regardless of the local YinYang balance/imbalance of the elements at any time point t, the system energy will decrease and eventually reach a (0,0) state without external disturbance or we have, $\forall t$, $d(|\varepsilon|Y(t))/dt < 0$, until $|\varepsilon|Y(t) = 0$.

Proof. It follows Theorem 8.2 directly. □

Note that the energy equilibrium law can be used to simulate the energy symmetry of closed dynamic systems that tend to be oscillatory or to reach internal balance but the total absolute energy remains the same all the time; the non-equilibrium laws can be used to simulate the broken energy symmetry of open dynamic systems with degenerating or growing properties, respectively.

Theorem 8.6. ***YinYang-N-Element Energy Unbalance Law (Law of Oscillating Symmetry).*** *Let* $t=0,1,2,\ldots$, $Y(t+1)=Y(t)\times M(t)$, $|\varepsilon|Y(t)$ be the total energy of an YinYang-N-Element vector $Y(t)$, $|\varepsilon|M(t)$ be the total energy of the connectivity matrix $M(t)$, $|\varepsilon|M_{i*}(t)$ be the energy subtotal of row i of $M(t)$, $|\varepsilon|M_{*j}(t)$ be the energy subtotal of column j of $M(t)$.

1. Regardless of the local YinYang balance/imbalance of the elements at any time point t, the system will remain a global energy equilibrium if, $\forall t$, $d(|\varepsilon|Y(t))/dt \equiv 0$, *or* (a) $\forall i,j$, $[|\varepsilon|(M_{i*}) \equiv |\varepsilon|(M_{*j}) \equiv 1.0]$ and (b) no external disturbance or input to the system after the initial vector $Y(0)$ is given.

2. Under the same conditions for (1), if, $\forall t$, $|\varepsilon^-(M_{*j})| = 0$ but $|\varepsilon^+(M_{*j}))| = 1$, and every bipolar cell starts with an unbalanced state with unbalanced total energy *Y(0)*, they will have no chance to reach local bipolar balance.
3. Under the same conditions for (2), if M is a discrete matrix and, $\forall t$, $|\varepsilon^-(M_{*j})| = 0$ but $|\varepsilon^+(M_{*j}))| = 1$, and every bipolar cell starts with an unbalanced state, the system will remain in a state passing mode from a regulator cell at time t to its regulated cell at time t+1.
4. Under the same conditions for (3), if the unbalanced state (x,y) is assigned to at least one cell and –(x,y) is assigned to the remaining cells, the system will be an oscillating BQCA.

Proof.

1. *follows directly from Theorem 8.3.*
2. *follows from (1) and the observations: (a) Given any unbalanced cell (x,y), (x,y)* $\times$ *(0,v) cannot be balanced; (b) If* $\varepsilon Y(t)$ *is unbalanced, there is no chance for* $\varepsilon Y(t+1)$ *to be balanced.*
3. *With a discrete M,* $\forall i,j$, $[|\varepsilon|(M_{i*}) \equiv |\varepsilon|(M_{*j}) \equiv 1.0\,]$ *and* $|\varepsilon^-(M_{*j})| = 0$ *but* $|\varepsilon^+(M_{*j}))| = 1$ *will enforce the rule that the state of a cell at time t+1 will be exactly the same as the sate of its regulator at time t.*
4. *Following (3) with only two cell states (x,y) and –(x,y), every cell of the BQCA will sooner or later change from (x,y) to –(x,y) or vice versa..* □

Discrete vs. Continuous BQCA

Definition 8.4a. If the operations or functions in a discrete BQCA are defined in BDL, it is called a *BDL-based discrete BQCA*. If a discrete BQCA are defined in BQLA, it is called a *BQLA-based discrete BQCA*.

Definition 8.4b. If the operations or functions in a continuous BQCA are defined in BDFL, it is called a *BDFL-based continuous BQCA*. If the operations or functions in a continuous BQCA are defined in BQLA, it is called a *BQLA-based continuous BQCA*.

Recall that Theorem 8.3 provides the conditions for all cells to reach local equilibria; if the 2nd condition of Theorem 8.3 is broken as in Theorem 8.6, oscillation can be achieved. This is simulated in the next chapter in the context of genetic regulation.

A UNIFYING PARADIGM OF BIPOLAR RELATIVITY

Logical Equivalence of Three Different Views

In Chapter 6, bipolar relativity is described as a logical theory based on BDL, BDFL, bipolar agents, and bipolar causality. BQLA makes it possible to describe bipolar relativity as a mathematical physics or biophysics theory in the context of YinYang-N-Element BQCA.

From Chapter 1, Figure 1, it is clear that, an N-dimensional agent A can be defined as a bipolar agent vector $(a_1,a_2, .., a_N)$ in a multidimensional equilibrium or non-equilibrium bipolar agent vector space. The vector can be deconstructed into a number of component bipolar agents that form the set $\{a_1,a_2, .., a_N\}$ of bipolar equilibria, quasi-equilibria, or non-equilibria. Here the concept of bipolar agent and bipolar equilibrium is unified and we can say *"a bipolar agent is a bipolar equilibrium or non-equilibrium."* Let the number of bipolar equilibria be N, the multidimensional agent A can be defined as a YinYang-N-Element BQCA such as the one in Figures 1 – 5. Let N = 10, the YinYang-10-Element random bipolar BCN would be a cellular representation of the 10-dimensional equilibrium of the agent A. Moreover, the different dimensions are bipolar interactive as depicted in Chapter 3, Figure 1 through the random connectivity matrix of Figure 5. This is depicted in Figure 6 by bringing Figure 5 together with Chapter 1, Figure 1 and Chapter 3, Figure 1.

From Figure 6 it is evident that the dimensional view, bipolar interactive view, and YinYang-N-Element BQCA view are logically consistent. Therefore, bipolar set theory, bipolar dynamic logic, equilibrium relations, BQLA, bipolar agents, bipolar causality, and BQCA are all unified under bipolar relativity theory. On the other hand, since bipolar relativity also defines bipolar quantum entanglement, the ubiquitous effect of quantum entanglement is revealed in set-theoretic, logical, relational, algebraic, agent, causal, dimensional, and cellular forms, respectively, for the first time.

Trajectories of Bipolar Agents

In Chapter 6 the geometry of YinYang bipolar relativity is discussed. Now with BQLA and BQCA we can further detail agent trajectories in YinYang bipolar geometry. Based on simulated data for the YinYang-5-Element BQCA (Figure 1 and Equation 8.5), Figure 7a shows the simulation of a kidney attack (an event or a cause) with Yin deficiency that disturbed kidney Yang and then the kidney Yin deficiency caused heart Yin deficiency in accordance with TCM theory (Maciocia, 1989, pp. 95-96,

Figure 6. Three views of bipolar relativity or bipolar quantum entanglement: (a) Dimensional view; (b) Cellular view; (c) Bipolar interactive view

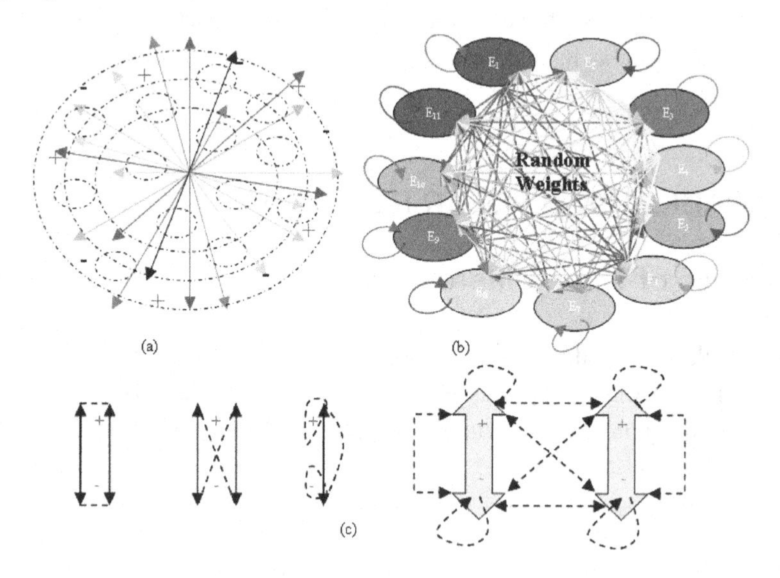

202-207). Figure 7b shows the same effect in equilibrium-based view. Figure 8 shows the equilibrium energy degeneration of two elements E_0 and E_2. These will be further discussed in the context of bio-economics in the next chapter.

Collective Bipolar Adaptivity

While a BQCA can be considered a multidimensional agent, each cell in a BQCA can be deemed a component bipolar agent. The agents in a BQCA can collectively achieve bipolar adaptivity. Thus, the bipolar adaptivity of individual bipolar agents as defined in Chapter 6, Definition 6.4a and 6.4b can be extended to collective bipolar adaptivity.

Figure 7. (a) Agent trajectories: kidney attack caused heart disturbance – YinYang magnitude view; (b) Agent trajectories: kidney attack caused heart disturbance – YinYang equilibrium view (Adapted from Zhang et al., 2009)

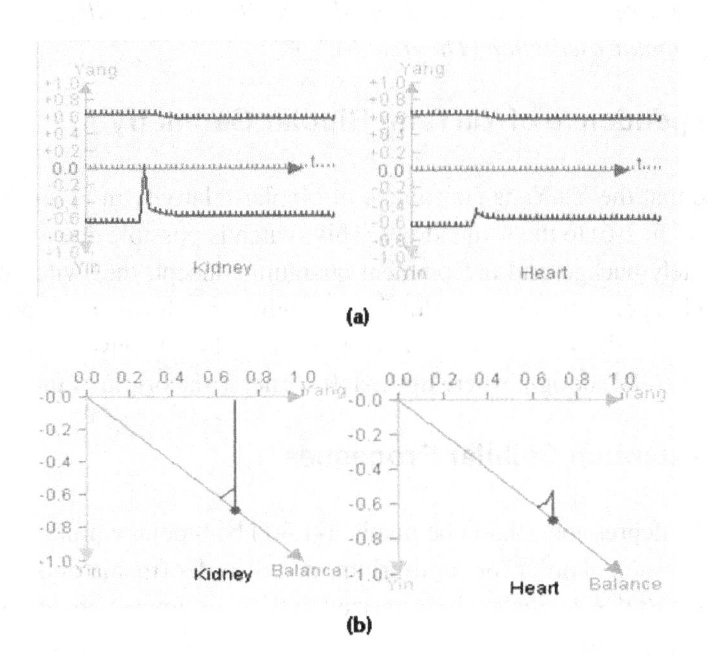

Figure 8. Agent trajectories of cause-effect in equilibrium energy decreases (E0 – Cause, E2 – Effect)

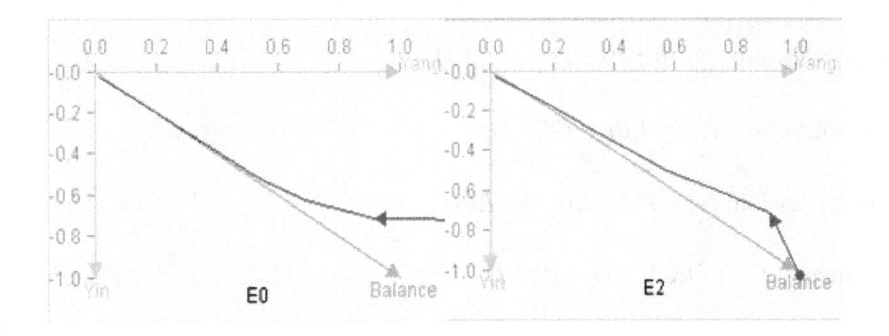

Definition 8.5. If all cells of a BQCA can collectively adapt from a quasi- or non-equilibrium to bipolar equilibrium we say the BQCA has *collective bipolar adaptivity.*

Theorem 8.7a. The two conditions of Theorem 8.3 are sufficient for collective bipolar adaptivity of a BQCA.

Proof. It follows from Theorem 8.3 directly. □

Theorem 8.7b. The two conditions of Theorem 8.3 are necessary for collective bipolar adaptivity of a BQCA.

Proof. If condition (1) is not met, we must have broken energy symmetry (Theorem 8.4 or 8.5), where the total energy is constantly decreasing or increasing globally and locally, collective adaptivity cannot be achieved. If condition (2) is not met, there would be no balancing factor in the connectivity matrix, even if we have global energy equilibrium as defined in condition (1), collective adaptivity cannot be achieved due to local bipolar oscillation (Theorem 6). □

Background Independence of YinYang Bipolar Geometry

It should be remarked that the YinYang coordinate of bipolar relativity in Figures 7b and 8 is flapped from the 2nd quadrant (Ch. 1-6) to the 4th quadrant. This switch is possible due to two reasons: (1) (-,+) bipolarity is a completely background independent quantum concept, the switch makes no difference unless the background dependent concepts "left" and "right" or "above" and "below" are introduced into the geometry; (2) using the 4th quadrant is more compatible to the convention followed in computer graphical user interface (GUI) design, where upper-left corner is the origin of the 2-D computer screen.

Mind as Bipolar Quantum Cellular Processes

With BDL, let (-1,0) be depression, (0,+1) be mania, (-1,+1) be bipolar equilibrium, and (0,0) be eternal equilibrium. Bipolar mental order (or equilibrium) and disorder (mania or depression) can be both modeled with a discrete BQCA logically characterized with the following equations:

Depression to mania: $(-1,0) \otimes (-1,0) = (0,+1);$ (8.10a)

Mania to depression: $(0,+1) \otimes (-1,0) = (-1,0);$ (8.10b)

Depression to equilibrium: $(-1,0) \oplus (0,+1) = (-1,+1);$ (8.10c)

Mania to equilibrium: $(0,+1) \oplus (-1,0) = (-1,+1);$ (8.10d)

Equilibrium to depression: $(-1,+1) \& (-1,0) = (-1,0);$ (8.10e)

Equilibrium to mania: $(-1,+1) \& (0,+1) = (0,+1);$ (8.10f)

Depression to eternal equilibrium: $(-1,0)\&(0,+1) = (0,0)$; (8.10g)

Mania to eternal equilibrium: $(0,+1)\&(-1,0) = (0,0)$. (8.10h)

In Equations 8.10a-h we did not use eternal equilibrium $(0,0)$ as an active force in bipolar interaction with the assumption that eternal equilibrium can't be revived. Otherwise, $(-1,+1)$ & $(0,0)$ would characterize a sudden death; $(0,0) \oplus (-1,+1) = (-1,+1)$ would characterize a revival from eternal equilibrium to full equilibrium. Moreover, we use the three basic binary operators $\oplus$, &, and $\otimes$ only; all the eight operators &, $\oplus$, $\&^-$, $\oplus^-$, $\otimes$, $\varnothing$, $\otimes^-$, and $\varnothing^-$ of BDL are derivable from the three basic operations (Ch. 3). The three operators $\oplus$, &, and $\otimes$ support the semantics of bipolar fusion, separation, and oscillation.

If Equations 8.10a-h are used as governing rules for the transition of a person P's mental state starting at time $t = 0$ to $t = 1$, the mind of a bipolar patient $P(t+1)$ can be modeled as the single cell BDL-based discrete BQCA in Figure 9 where (x,y) can be considered an external or internal environmental or biological trigger for the transition of the mental state from $P(t)$ to $P(t+1)$.

Universe as Bipolar Quantum Cellular Processes

Based on the *"big bang – equilibrium – black hole – big bang"* triangle (Ch. 6, Figure 1d), it can be hypothesized that

1. The side of the triangle from big bang to bipolar equilibrium is the (contraction, expansion) = (x,y) phase of the universe after the big bang where $y>|x|$.
2. The side of the triangle from equilibrium to black hole is the (contraction, expansion) = (-1,0) phase where $|x|>y$. In this phase, a black hole slowly sucks in all matter in a galaxy or the whole universe into its mouth. What is left outside becomes a mystery.
3. The side of the triangle from black hole to big bang is the dormant period for creating the next big bang. Alternatively, if the last two black holes, a matter-based one and an antimatter-based one, that try to suck in each other, the collision of the two could create a big bang fuelled by particle-antiparticle annihilation.
4. The observable accelerated expansion and black holes in our universe suggest that our universe is currently in the mix of expansion and contraction. The matter-antimatter annihilation in some galactic center or stars suggest a certain (contraction, expansion) equilibrium.

Figure 9. BDL-Based Discrete BQCA of the mind

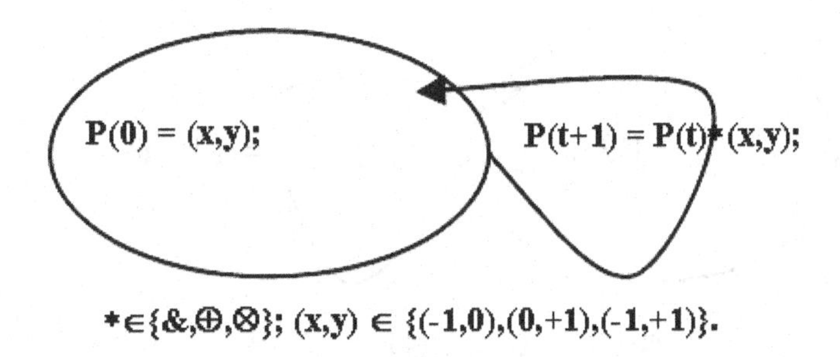

5. It is difficult to expect the observable expansion-contraction quasi-equilibrium to last forever unless "big bang" is observed in some galaxies while black holes and stars are observed elsewhere at the same time. In that case, "big bang" would become "small bangs" that create different galaxies at different time points and time would have multiple starting and ending points

With bipolar representation, let (-1,0) be depression characterizing a black hole, (0,+1) be mania characterizing a big bang, (-1,+1) be (contraction, expansion) bipolar equilibrium. The universe can be modeled as an equilibrium or non-equilibrium based cellular process logically characterized with the following three equations – a subset of Equations 8.10a-h of a mental BQCA:

Black hole to big bang: $(-1,0)\otimes(-1,0) = (0,+1);$ (8.11a)

Big bang to equilibrium: $(0,+1)\oplus(-1,0) = (-1,+1);$ (8.11b)

Equilibrium to black hole: $(-1,+1)\&(-1,0) = (-1,0).$ (8.11c)

If Eqs. 8.11a-c are used as governing rules for the 3-phase transition of our universe U, U(t) would be like the single cell BDL-based discrete BQCA in Figure 10.

With bipolarity many previously undefinable phenomena are now definable. The bipolar operator sequence "$\oplus,\&,\otimes,\oplus,\&,\otimes,..$" reveals a pattern of three basic functions: $\oplus$ is for bipolar fusion; & is for bipolar separation; $\otimes$ is for bipolar oscillation. Since all the eight operators $\&$, $\oplus$, $\&^-$, $\oplus^-$, $\otimes$, $\varnothing$, $\otimes^-$, and $\varnothing^-$ of BDL (Ch. 3) can be derived from the three basic operations. This leads us to the question: *If the three basic operations are sufficient to define a minimal BQCA for the mind and the universe at the macroscopic level, could they also be the most fundamental operations of the universe at the microscopic level?*

According to "The Nobel Prize in Physics 2008":

We are all the children of broken symmetry. It must have occurred immediately after the Big Bang some 14 billion years ago when as much antimatter as matter was created. The meeting between the two is fatal for both; they annihilate each other and all that is left is radiation. Evidently, however, matter won against antimatter, otherwise we would not be here. But we are here, and just a tiny deviation from perfect symmetry seems to have been enough – one extra particle of matter for every ten billion particles

Figure 10. BDL-based discrete BQCA of the universe

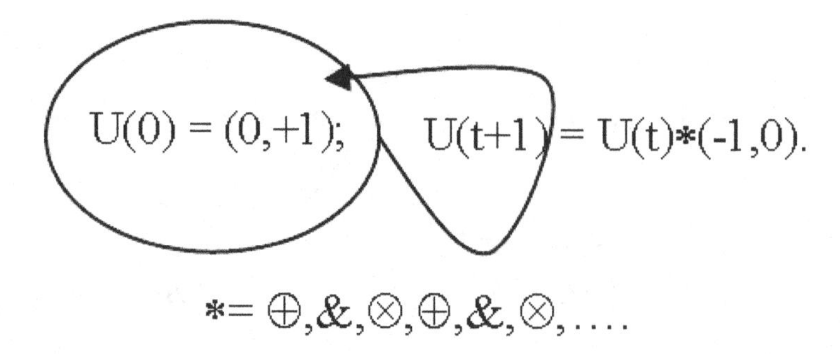

of antimatter was enough to make our world survive. This excess of matter was the seed of our whole universe, which filled with galaxies, stars and planets – and eventually life. But what lies behind this symmetry violation in the cosmos is still a major mystery and an active field of research. (The Royal Swedish Academy of Sciences, 2008)

The above findings in particle physics tells us that our universe could be a near perfect quasi-equilibrium of antimatter-matter particle pairs $(-p,+p)$ immediately after the big bang until matter won the majority. With unipolar truth-based tradition, the bipolar symmetry has no logical representation. With bipolarity, however, the symmetry can be numerically represented as a bipolar predicate or mapping defined in $B = [-\infty,0]\times[0,+\infty]$. Then, we have

φ *(antimatter, matter)* $= (-10^{13}, 10^{13}+1)$. (8.12)

Now, our question is: *What lies behind the symmetry violation? What is universe? What is life? Can bipolar relativity offer a logically, mathematically, and physically consistent and coherent unifying theory? What is the relationship between antimatter-matter quasi-symmetry and contraction-expansion quasi-symmetry?*

Since Equation 8.12 defines a broken symmetry to matter's advantage on the positive side, the annihilation of particle and antiparticle could have resulted the matter-dominated universe. The resulting universe can then be abstracted with the bipolar logical value (0,+1) characterizing an expanding matter-dominated universe.

On the other hand, the observed black hole region must be in contraction. If the mouth of the black hole sucks in mostly matter, what is left outside would be mostly antimatter. Thus, the near perfect antimatter-matter symmetry could be tipped to antimatter advantage outside. The surviving inside-outside matter-antimatter bipolar equilibrium or non-equilibrium could become the next trigger to another big bang.

Alternatively, if multiple black holes can aggregate into the last two, it is possible that one is a matter black hole and another is an antimatter black hole. The collision of the two could cause another big bang fueled by matter and antimatter particle annihilation.

Conjecture 8.1. Antimatter-matter bipolar symmetry or broken symmetry is bipolar equivalent to contraction-expansion bipolar symmetry or broken symmetry. Formally, let ϕ and φ be two bipolar predicates, we have

ϕ(antimatter, matter) $\Leftrightarrow$ φ(contraction, expansion).

Conjecture 8.2. There exists a region somewhere in the universe that shows antimatter-matter broken symmetry but to antimatter's advantage.

Conjecture 8.3. Bipolar fusion, coupling, or binding ($\oplus$), bipolar separation (&), and bipolar oscillation ($\otimes$) are three fundamental operations of nature at both the macroscopic and microscopic levels.

Conjecture 8.4. The universe is a process of bipolar cellular automation (as characterized in Figure 10); the mystery behinds the matter-antimatter symmetry violation is bipolar relativity. Formally, let $\psi = (\psi^-,$

ψ^+) = (antimatter, matter) and $\phi = (\phi^-, \phi^+)$ = (contraction, expansion) be bipolar symmetry predicates, t be time, and p be space or galaxy, $\forall t,p,$ we have:

$$[(\psi(t_x,p_1) \Rightarrow \phi\,(t_y,p_3))\ \&\ (\psi(t_x,p_2) \Rightarrow \phi\,(t_y,p_4))]$$

$$\Rightarrow [(\psi(t_x,p_1)^*\ \psi\,(t_x,p_2)) \Rightarrow (\phi\,(t_y,p_3))^*\ \phi\,(t_y,p_4))]$$

$$\Rightarrow [(\psi((t_x,p_1)\blacklozenge(t_x,p_2)) \Rightarrow (\phi\,((t_y,p_3)\blacklozenge(t_y,p_4))]. \qquad (8.13)$$

Equation 8.13 states that if the antimatter-matter symmetry at (t_x,p_1) implies the contraction-expansion symmetry at (t_y,p_3) and the antimatter-matter symmetry at (t_x,p_2) implies the contraction-expansion symmetry at (t_y,p_4), the antimatter-matter interaction $(t_x,p_1)\blacklozenge(t_x,p_2)$ is equivalent to the contraction-expansion interaction $(t_y,p_3)\blacklozenge(t_y,p_4)$.

Life as Bipolar Quantum Cellular Processes

Based on Theorems 8.4 and 8.5, an equilibrium vital energy increasing or growing curve is sketched in Figure 11a; an equilibrium energy decreasing or degenerating curve is sketched in Figure 11b. Such growing or degenerating processes can be fully illustrated with bipolar quantum linear algebra and YinYang-N-Element BQCA for biological system simulation and regulation in the next chapter (e.g. Figure 8).

Conjecture 8.5. Life is a process of bipolar cellular automation consisting of (1) a growing phase from a weaker vital energy equilibrium to a stronger one, (2) an equilibrium phase with vital energy quasi-equilibrium, and (3) a degenerating phase from a stronger vital energy equilibrium to eternal equilibrium. (Figure 11).

RESEARCH TOPICS

BQLA and BQCA have opened a wide spectrum of research topics including but not limited to the following:

Figure 11. Growing and aging curves (Zhang, 2009d)

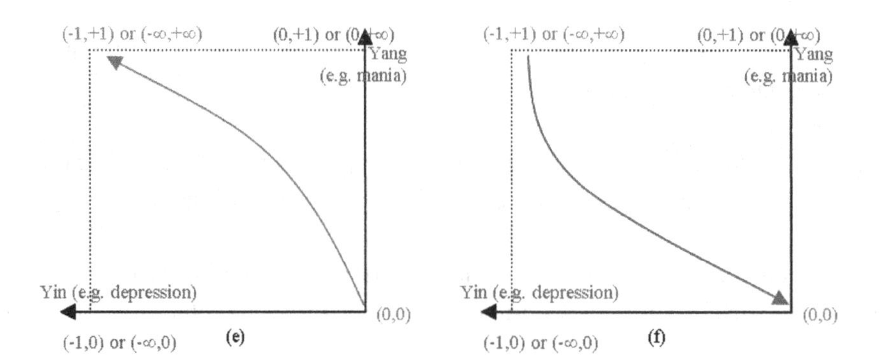

- Develop bipolar quantum tensor calculus (BQTC) to unify Einstein's equations of general relativity under bipolar equilibrium/non-equilibrium and bipolar quantum entanglement.
- Develop complexity-number-based BQLA and BQCA.
- Apply BQLA and BQCA in bipolar quantum computing.
- Apply BQLA and BQCA in psychopharmacology.
- Apply BQLA and BQCA in genomics.
- Apply BQLA and BQCA in molecular interaction.
- Apply BQLA and BQCA in macroeconomics.
- Apply BQLA and BQCA in cognitive informatics.
- Apply BQLA and BQCA in social dynamics.
- Apply BQLA and BQCA in bioeconomics (Ch. 9).

SUMMARY

Following a brief review on traditional cellular automata and linear algebra, bipolar quantum linear algebra (BQLA) and BQCA have been presented. Three families of YinYang-N-Element bipolar cellular networks (BCNs) have been developed, compared, and analyzed: one family has predefined nourishing and regulation cycles following the classical 5-Element protocol in traditional Chinese medicine (TCM); another family has random connectivity and link weights; a third family has a predefined number of fan-in and fan-out but with random connections, that leads to YinYang-N-Element bipolar combinatorics. YinYang bipolar dynamic equations have been derived for YinYang-N-Element BQCA. Global (system level) and local (element level) energy equilibrium and non-equilibrium conditions have been established and axiomatically proved for all three families of cellular structures that leads to the concept of collective bipolar adaptivity in BQCA. The unifying nature of bipolar relativity in the context of BQCA has been illustrated. Background Independence nature of YinYang bipolar geometry has been demonstrated with bipolar algebraic YinYang bipolar geometry.

It is shown that BQLA makes it possible to describe bipolar relativity as a unifying mathematical physics or biophysics theory in the context of YinYang-N-Element BQCA. Furthermore, it has been shown that under the unifying theory, the dimensional view, bipolar logical view, and YinYang-N-Element BQCA view are logically consistent. Therefore, bipolar set theory, bipolar dynamic logic, BQLA, bipolar agents, bipolar causality, and BQCA are all unified under bipolar relativity theory. The algebraic trajectory of bipolar agents in YinYang bipolar geometry is illustrated with simulated data.

The significance of this work is 3-fold: (1) BQLA provides a unique and unifying mathematical foundation for BQCA in physical, socio- and biosystems; (2) YinYang-N-Element cellular automation provides a unique and unifying holistic architecture for equilibrium and non-equilibrium simulation and regulation at the system, particle, molecular, quantum, or genetic levels; (3) BQLA and BQCA enable bipolar relativity be further generalized from a logical theory to a mathematical physics or biophysics theory that leads to the conjectures about universe and life. Bioeconomics simulation and global regulation are further discussed in the next chapter.

ACKNOWLEDGMENT

This chapter reuses part of published material in the two articles: (1) Zhang, W. –R. & Chen, S. S. (2009). Equilibrium and Non-Equilibrium Modeling of YinYang Wuxing for Diagnostic Decision Analysis in Traditional Chinese Medicine. *World Scientific Publishing (WSP): Int'l J. of Infor. Tech. and Decision Making (IJITDM), Vol. 8, No. 3 (2009)* pp529-548; (2) Zhang, W. –R., H. J. Zhang, Y. Shi & S. S. Chen (2009). Bipolar Linear Algebra and YinYang-N-Element Cellular Networks for Equilibrium-Based Biosystem Simulation and Regulation. *World Scientific Publishing (WSP): Journal of Biological Systems (JBS), Volume: 17, Issue: 4 (2009)* pp. 547-576. Permission to ruse is acknowledged.

REFERENCES

Arrighi, P. & Fargetton, R. (2007). *Intrinsically universal one-dimensional quantum cellular automata.* DCM'07.

Crutchfeld, J. P. Mitchell, M. & Das, R. (2002). The evolutionary design of collective computation in cellular automata. In J.P. Crutcheld & P.K. Schuster (Eds.), *Evolutionary dynamics exploring the interplay of selection, neutrality, accident, and function.* New York: Oxford University Press.

Jaeger, S., Chen, S.-S., & Zhang, W.-R. (2009). TCM in innate immunity. *Proceedings of International Joint Conference on Bioinformatics, Systems Biology and Intelligent Computing (IJCBS).* (pp. 397-401). Shanghai, China.

Li, F. (1987). Functional structure mode of human body and YinYang -Wuxing equations. *Physica Scripta, 36*, 966–969. doi:10.1088/0031-8949/36/6/015

Maciocia, G. (1989). *The foundations of Chinese medicine: A comprehensive text for acupuncturists and herbalists.* Churchill LivingStone.

McKelvey, B. (2004). Toward a 0th law of thermodynamics: Order-creation complexity dynamics from Physics and Biology to Bioeconomics. *Journal of Bioeconomics, 6*(1), 65–96. doi:10.1023/B:JBIO.0000017280.86382.a0

Mitchell, M., Crutchfield, J. P., & Hraber, P. T. (1994). Evolving cellular automata to perform computations: Mechanisms and impediments. *Physica D. Nonlinear Phenomena, 75*, 361–391. doi:10.1016/0167-2789(94)90293-3

Peak, W., & Messinger, M. (2004). Evidence for complex, collective dynamics and emergent, distributed computation in plants. *Proceedings of the National Academy of Sciences of the United States of America, 101*(4), 918–922. doi:10.1073/pnas.0307811100

Pérez-Delgado, C. & Cheung, D. (2007). Local unitary quantum cellular automata. *Physics Review of Letters A, 76.*

Shannon, C.E. (1948). A mathematical theory of communication. *Bell System Technology Journal,* 27(623-656), 379–423.

Shepherd, D. J., Franz, T., & Werner, R. F. (2006). Universally programmable quantum cellular automaton. *Physical Review Letters*, 97.

Shi, Y., Seto, E., Chang, L.-S., & Shenk, T. (1991). Transcriptional repression by YY1, a human GLI-Kruppel-related protein, and relief of repression by adenovirus E1A protein. *Cell*, *67*(2), 377–388. doi:10.1016/0092-8674(91)90189-6

The Royal Swedish Academy of Sciences. (2008). The Nobel Prize in Physics 2008. *Press Release*, 7 October. Retrieved from http://nobelprize.org/nobel_prizes/physics/laureates/2008/press.html

Von Neumann, J. (1966). *The theory of self-reproducing automata*. Urbana, IL: University of Illinois Press.

Watrous, J. (1995). On one-dimensional quantum cellular automata. *Proceedings of 36th FOCS*, 528–537.

Wolfram, S. (1986). Cryptography with cellular automata. In *Crypto '85, Proceedings,* (LNCS 218), (pp. 429–432). Springer.

Wolfram, S. (2002). *A new kind of science*. Champaign, IL: Wolfram Media.

Zhang, W.-R. (1996). NPN fuzzy sets and NPN qualitative algebra: A computational framework for bipolar cognitive modeling and multiagent decision analysis. *IEEE Transactions on SMC*, *16*, 561–574.

Zhang, W.-R. (2005a). YinYang bipolar lattices and L-sets for bipolar knowledge fusion, visualization, and decision making. *International Journal of Information Technology and Decision Making*, *4*(4), 621–645. doi:10.1142/S0219622005001763

Zhang, W.-R. (2009d). The logic of YinYang and the science of TCM – an Eastern road to the unification of nature, agents, and medicine. [IJFIPM]. *International Journal Functional Informatics and Personal Medicine*, *2*(3), 261–291. doi:10.1504/IJFIPM.2009.030827

Zhang, W.-R., & Chen, S. S. (2009). Equilibrium and non-equilibrium modeling of YinYang Wuxing for diagnostic decision analysis in traditional Chinese medicine. *International Journal of Information Technology and Decision Making*, *8*(3), 529–548. doi:10.1142/S0219622009003521

Zhang, W.-R., Zhang, H. J., Shi, Y., & Chen, S. S. (2009). Bipolar linear algebra and YinYang-N-Element cellular networks for equilibrium-based biosystem simulation and regulation. *Journal of Biological System*, *17*(4), 547–576. doi:10.1142/S0218339009002958

ADDITIONAL READING

Anthony, C. K., & Moog, H. (2002). *I Ching: The Oracle of the Cosmic Way*. Stow, MA: Anthony Publishing Company, Inc.

Balkin, J. M. (2002). *The Laws of Change: I Ching and the Philosophy of Life*. New York: Schocken Books.

Chopard, B., & Droz, M. (1998). *Cellular Automata Modeling of Physical Systems*, Cambridge University Press, 1998.

Deutsch, A. & Dormann, S. (2005). *Cellular Automaton Modeling of Biological Pattern Formation*. Birkhäuser Boston, 2005.

Gutowitz, H. (1991). Cellular Automata*: theory and experiment.* MIT Press, 1991

Karcher, S. (2002). *I Ching: The Classic Chinese Oracle of Change: The First Complete Translation with Concordance*. London: Vega Books.

Moran, E., & Yu, J. (2001). *The Complete Idiot's Guide to the I Ching*. New York: Alpha Books.

Schiff, J. L. (2007). *Cellular Automata: A Discrete View of the World.* Wiley & Sons, Inc., 2007.

KEY TERMS AND DEFINITIONS

BCN: Bipolar cellular network: an instance of BQCA.

BQCA: Bipolar quantum cellular automata where each cell is a bipolar agent (Ch. 6).

BQLA: Bipolar quantum linear algebra where each element in a matrix is a bipolar element (Zhang, 2005a).

Cellular: Automata: A cellular automaton (plural: *cellular automata*) consists of a collection of cells, each in one of a finite number of states. The cells are organized into a grid in any finite number of dimensions. Cellular automata were studied in the early 1950s as a possible model for biological systems (Wolfram, 2002, p. 48). Von Neumann was one of the first to consider such a model, and incorporated a cellular model into his "universal constructor" (Von Neumann, 1966). Comprehensive studies of cellular automata have been performed by S. Wolfram starting in the 1980s (Wolfram, 2002).

Quantum Cellular Automata: Quantum Cellular Automata (QCA) refers to any one of several models of quantum computation, which have been devised in analogy to conventional models of cellular automata introduced by von Neumann. It may also refer to quantum dot cellular automata, which is a proposed physical implementation of "classical" cellular automata by exploiting quantum mechanical phenomena (Arrighi & Fargetton, 2007; Pérez-Delgado & Cheung, 2007; Shepherd, Franz & Werner, 2006; Watrous, 1995).

YinYang-N-Element BCN: An extension of YinYang-5-element BCN to N element BCN which still follow the classical connectivity protocol in YinYang-5-element structure (Zhang *et al.*, 2009).

YinYang-N-Element Cellular Combinatorics: The study of different connectivity of YinYang-N-element cellular structures.

YinYang-WuXing: YinYang-WuXing are five movements, five phases, and five steps/stages which are widely referred to as YinYang-5-Elements or YinYang-5-Agents in the literature. The five elements metal, wood, water, fire, and earth were considered the five basic elements of nature in ancient Chinese cosmology; the five elements liver, lung, kidney, heart, and spleen are the matching elements in TCM based on the theory of nature and human body unity. The system of five phases was used for describing interactions and relationships between phenomena. The system is still used as a reference in some forms of TCM and qigong. For thousands of years, there was no mathematical model to characterize the explicit Yin and Yang bipolarity of each element and their interactions. Such a model is first proposed in (Zhang & Chen, 2009).

Part 4
Applications

Chapter 9
YinYang Bipolar Quantum Bioeconomics for Equilibrium-Based Biosystem Simulation and Regulation

ABSTRACT

As a continuation of Chapter 8, this chapter presents a theory of bipolar quantum bioeconomics (BQBE) with a focus on computer simulation and visualization of equilibrium, non-equilibrium, and oscillatory properties of YinYang-N-Element cellular network models for growing and degenerating biological processes. From a modern bioinformatics perspective, it provides a scientific basis for simulation and regulation in genomics, bioeconomics, metabolism, computational biology, aging, artificial intelligence, and biomedical engineering. It is also expected to serve as a mathematical basis for biosystem inspired socioeconomics, market analysis, business decision support, multiagent coordination and global regulation. From a holistic natural medicine perspective, diagnostic decision support in TCM is illustrated with the YinYang-5-Element bipolar cellular network; the potential of YinYang-N-Element BQCA in qigong, Chinese meridian system, and innate immunology is briefly discussed.

INTRODUCTION

Discovery of the ubiquitous genetic regulator protein YinYang1 (YY1) at Harvard medical school in 1991 (Shi *et al.*, 1991) marks the formal entry of the ancient Chinese YinYang into modern genomics – a core area of life sciences. From that point on, YinYang has reemerged as a unifying philosophical foundation for both *Traditional Chinese Medicine* (*TCM*) and modern bioinformatics in microscopic as well as in macroscopic terms. In less than two decades, many important works have been reported on YinYang related regulation of gene expression such as those in (Park & Atchison, 1991; Ai, Narahari & Roman, 2000; Kim, Faulk & Kim, 2007; Palko *et al.*, 2004; Zhou & Yik, 2006; Santiago *et al.*, 2007; Liu *et al.*, 2007; Santiago *et al.*, 2007; Wilkinson *et al.*, 2006; Gore & van Oudenaarden, 2009).

Despite the great achievement of genome research in the last decade, the governing rules of molecular interaction and human health are still largely unknown. As reported by a New York Times report: the genome has yielded to biologists one insightful surprise after another in last decade; the primary goal of the Human Genome Project — to ferret out the genetic roots of common diseases like cancer and Alzheimer's and then generate treatments — has been largely elusive (Wade, 2010).

On the other front, in the last few decades, TCM in general and acupuncture in particular, has been gradually accepted by the world as a viable medical practice complementary to Western medicine. As pointed out by a California licensed acupuncturist Matthew Bauer: *"At first roundly rejected by modern medical authorities, a great increase in scientific research has convinced many authorities that acupuncture in particular seems to have legitimate clinical value"* (Bauer, 2006).

Along with the gradual acceptance is the evermore closer scientific scrutiny. Such scrutiny has created a new crisis for TCM – the crisis of lacking scientific theoretical foundation. The same California licensed acupuncturist wrote in his paper *"The Final Days of Traditional Beliefs? – Part One"*: *"While this turnaround has been greeted by many in the Chinese medicine community as the long sought after validation they had been working for, there is real reason to wonder if this acceptance may in fact be the beginning of the end of the central role traditional theories have played in Chinese medicine for at least 2,000 years"* (Bauer, 2006).

It is evident that, on the one hand, TCM has been proven clinically effective; on the other hand, the scientific classification of TCM is troubling to many. As discussed in the introduction (Ch. 1), the problem did not originate from Western science. Truth-based unipolar cognition as a Western tradition has triumphed and made great achievements in all fields of science and technology. A key for the success has been the formal truth-based mathematical abstraction. Even though YinYang has survived more than five thousand years of recorded human history and equilibrium-based TCM has been practiced in China for at least two thousand years, the Eastern philosophy has failed to provide a systematic formal logical and mathematical foundation for its physical, social, and biological claims. Consequently, while truth-based cognition stopped short of offering logically definable causality equilibrium-based YinYang has not stepped up with a complementary solution (Zhang, 2009d).

Now BDL has led to the theory of YinYang bipolar relativity. While bipolar relativity constitutes a holistic approach to science in general, YinYang-N-Element cellular automata provide a unique mathematical physics or biophysics theory for *bioeconomics* (*BE*). In this chapter we introduce the theory of YinYang *bipolar quantum bioeconomics* (*BQBE*) and we provide *equilibrium-based bipolar simulation and regulation* techniques for BQBE to illustrate applications in system biology and TCM. Simulation results are discussed that can be extended from system levels to molecular, genetic, socioeconomic, ecological, and environmental levels. This chapter is, therefore, a natural continuation of last chapter. While the last chapter focused on the mathematical formulation of BQLA, BQCA, and their unification under bipolar relativity, this chapter is focuses on bioeconomics simulation and regulation in modern bioinformatics and TCM using BQLA and BQCA.

The remaining presentations and discussions of this chapter are organized in the following sections:

- **Review on Bioeconomics.** This section presents a brief review on bioeconomics (BE).
- **Bipolar Quantum Bioeconomics (BQBE).** This section presents the theory of YinYang Bipolar Quantum Bioeconomics (BQBE) with equilibrium-based unifying properties.
- **BQBE in TCM Diagnostic Decision Support.** This section presents computer simulations of YinYang-5-Element cellular networks for diagnostic decision support in Traditional Chinese Medicine (TCM) with the subsections: (1) YinYang WuXing (YYWX) and TCM; (2) system architecture of TCM diagnostic decision support; (3) equilibrium process modeling of YYWX; (4) quasi- or non-equilibrium process of YYWX; (5) graphical user interface.
- **Biosystem Simulation and Regulation with YinYang-N-Element BQCA.** This Section presents computer simulations of arbitrary YinYang-N-Element cellular networks for the regulation of repression and activation activities such as those in molecular or genetic networks with the subsections: (1) assumptions and conditions; (2) simulation and regulation of classical YinYang-N-Element BQCA; (3) simulation and regulation of random YinYang-N-Element BQCA; (4) simulation and regulation of oscillatory discrete BQCA.
- **Analysis and Applications.** This section is a discussion on the potential applications of YinYang-N-Element cellular automata in bioeconomics with discussions on: (1) simulation data analysis; (2) BQCA and BQBE in Immunology; (3) Other Applications.
- **Research Topics**. This section lists a few research topics.
- **Summary.** This section summarizes the key points of the chapter and draws a few conclusions.

REVIEW ON BIOECONOMICS

Bioeconomics as a promising research area has attracted substantial academic interest in recent years. While there have been different definitions for bioeconomics, all definitions agree on the interplay of biology in economics and vice versa. Bioeconomics is, on the one hand, the study of the dynamics of living resources using economic models and, on the other hand, the science determining the socioeconomic activity threshold for which a biological system can be effectively and efficiently utilized without destroying the conditions for its regeneration and therefore its sustainability (Clark, 1987).

Historically, bioeconomics is closely related to the early development of theories in fisheries economics (Gordon, 1954; Schaefer, 1957). Contemporarily, it has been related to global warming that has threatened the very existence of humanity and all biological life on Earth (Mohammadian, 2009).

Among bioeconomics literature, Mansour Mohammadian's website gives a concise, up to date, and systematic outline of the area by answering the following questions (Mohammadian, 2009):

- What is the area of interest of Bioeconomics?
- Why develop biological economics?
- What is new about Bioeconomics?
- Nature of Bioeconomics
- Characteristics
- Challenges
- Bioeconomic Values
- Bioeconomics as Economics of the Third Way

We highlight Mohammadian's answers for some key questions and identify the relationships of bioeconomics with YinYang bipolar relativity.

First, what is the area of interest of bioeconomics? In answering this question, Mohammadian gives a modern view:

The main area of interest is the investigation and clarification of the interactions that occur at the interface of the socioeconomic and biological systems. They occur when the socioeconomic system in its daily activity impacts the biological system and gives rise to such phenomena as global warming, depletion of the ozone layer and others. It goes without saying that these interactive phenomena that are moreover complex and are of uncertain nature have serious repercussions for the human enterprise. Economics has not really dealt with the concepts of uncertainty and irreversibility as economists cannot make up their minds that humanity is definitely and irreversibly destroying the foundation of life: killing the goose that lays the golden egg. (Mohammadian, 2009)

Although this modern view may not be shared by all, it does point directly to the most concerned and most debated issue in the world that bioeconomics should address.

Secondly, *why develop biological economics?* To answer this question, Mohammadian enumerated ten reasons which reiterate the central point by Geoffrey Hodgson: *"The reform of economics is not a question of adding new dimensions to the neoclassical economic theory. A theoretical revolution is required at the core of economics itself"* (Hodgson, 1992).

Thirdly, *what is new about bioeconomics?* In answering this question, Mohammadian outlined the new concepts of bioeconomics in five aspects: scientific, economical, social, cultural, and academic. Scientifically, it is pointed out that bioeconomics is holistic (vs. reductionism), synthetic (vs. analytic), interdisciplinary (vs. unidisciplinary), interactional (vs. relational), emergent, and postmodern. Among the new concepts, on the top is the holistic nature of bioeconomics. Naturally, we have the question: *Could the holistic nature of YinYang bipolar relativity play a role in bioeconomics?*

Fourthly, the epistemology of bioeconomics is identified. In Mohammadian's view, the epistemology of the holistic interdisciplinary nature of bioeconomics is based on the principles of (1) systemic (dynamic, open, non-linear and far from equilibrium), (2) holism (the whole is more than sum of its parts), (3) dialectic logic vs. deductive logic, and (4) complementarity.

About dialectic logic vs. deductive logic, Mohammadian wrote: *"The logic of bioeconomics is the logic of the included third. This is to say that it is neither only the logic of biology nor is it only the logic of economics but it is the logic that results from the conciliation of the two which is the third logic; the logic of bioeconomics. In this sense we have to evaluate the socio-economic activity not as a process of opposite tensions; for example to use/not to use resources but to conciliate their use according to the biological principles of conservation and regeneration; that is sustainable (bioeconomic) utilization."* Naturally, we have the question: Could YinYang bipolar logic, bipolar agents, bipolar algebra, bipolar cellular automata, and, in general, bipolar relativity serve such a purpose in bioeconomics?

About complementarity Mohammadian states: *"This principle is very important in the process of socioeconomic activity because if this activity is carried out according to the bioeconomic principles such concepts as competition and cooperation, quantity and quality, egoism and altruism, anthropocentrism and biocentrism among others are complementary although at first sight they may not appear so."* Naturally, we have the question: Could YinYang bipolar equilibrium or non-equilibrium play a role here?

Fifthly, regarding the nature of bioeconomics, Mohammadian outlined four aspects: (1) the demographic nature of bioeconomics is highlighted as *"Lots of people competing for Little biological resources;"* (2) the Biological nature is highlighted as *"Lots of people competing for Lots of biological resources;"* (3) the Economic nature is summarized as *"Lots of people competing for Lots of manufactured resources"* or *"Few people competing for Few manufactured resources;"* (4) the bioeconomic nature is highlighted as *"Few people enjoying for Sustainable resources."* It is concluded that competition is a key in bioeconomics.

Now, we have the question: What role can bipolar relativity play in bioeconomics?

BIPOLAR QUANTUM BIOECONOMICS (BQBE)

The Quantum Nature of Bioeconomics

As argued in Chapter 1, since acceleration is equivalent to gravitation under general relativity, any natural, social, economical, and mental acceleration/deceleration or growth/aging is qualified to be a kind of quantum gravity. At its centennial celebration, however, Einstein's general theory of relativity – the greatest theory of all time has not been further generalized beyond spacetime geometry. It has failed to go deeper into biological and economical worlds. A key reason preventing spacetime relativity from entering bioeconomics is that space and time in general relativity are not quantum in nature and not symmetrical to each other. On the other hand, quantum mechanics has so far failed to play a major role in bioeconomics because the ubiquitous effect of quantum entanglement has been logically undefinable. YinYang bipolar relativity and bipolar quantum entanglement has closed this gap.

With bipolar relativity quantum gravity has led us to the *Q5 paradigm.* As discussed in the earlier chapter, in the Q5 paradigm, the theory of physical quantum gravity should be concerned with quantum physics; the theory of logical quantum gravity should be focused on quantum computing; the theory of social quantum gravity should span social sciences and economics; the theory of biological quantum gravity should be focused on life sciences; the theory of mental quantum gravity should be focused on the interplay of quantum mechanics and brain dynamics. Thus, socioeconomics cannot escape from social quantum gravity and biological systems cannot escape from biological quantum gravity. Bioeconomics as an interdisciplinary area can naturally be fit into the Q5 paradigm.

The quantum nature of economics and biological systems are seldom examined in the vast literature. This difficulty can be traced to quantum mechanics. First quantum mechanics is an incomplete theory where the "hidden variables" prevent unipolar quantum entanglement from being used in real world human societies. With bipolar dynamic logic (BDL) and bipolar quantum theory, bipolar universal modus ponens (BUMP) enables logically definable symmetrical causality and bipolar quantum entanglement is applicable in any equilibrium-based world including bioeconomics.

Equilibrium-Based Nature of Bioeconomics

As discussed in the earlier chapter, the Q5 paradigm follows a single undisputable observation and a single condition: (1) bipolar equilibrium or non-equilibrium is a ubiquitous or pervasive concept from which nothing can escape; (2) equilibrium-based bipolar quantum entanglement is definable with BUMP in BDL. Thus, equilibrium-based reasoning as a key for the Q5 paradigm plays a unifying role for both truth-based logical thinking and bipolar quantum logical thinking in symmetry and harmony.

YinYang is about equilibrium, harmony, symmetry, and stability which are all fundamental concepts in biological systems and economics in micro- and macroscopic levels. The existing theories and simulation tools for biological processes such as growing, aging, degenerating, regulating, equilibrium, and non-equilibrium processes in molecular and gene regulation networks and biochemistry (Ou *et al.,* 2003), however, are based on classical truth-based mathematical abstraction where the Yin (such as repression or regulation ability) and Yang (such as activation or stimulation ability) as well as their bipolar interaction, oscillation, quantum entanglement, local and global equilibria or non-equilibria cannot be explicitly captured for holistic visualization and analysis. For instance, the Yin Yang 1 protein exhibits both repressor and activator behaviors in gene network regulation (Shi *et al.,* 1991), but classical mathematical tools do not support holistic YinYang bipolarity. Another example involves bipolar equilibrium and bipolar disorder. Even though without bipolar mental equilibrium bipolar neurobiological disorder would be "big bang" or "black hole" from nowhere and caused by nothing, in the books of Western psychiatric medicine, however, mental equilibrium is deemed unobservable non-existence. Conversely, even though without economic equilibrium economic overheat or recession would be undefinable in modern macroeconomics and even though Nash equilibrium (Nash, 1950) was recognized with a Nobel Prize (The Nobel Foundation, 1994), YinYang bipolar equilibrium and bipolar relativity is thus far undefined in the books of economics.

Toward a Unifying Theory of Bioeconomics

It has been shown that BQLA has made bipolar relativity a unifying mathematical physics or biophysics theory for modeling global (system level) and local (element level) energy equilibrium, non-equilibrium, and oscillatory conditions that leads to the concept of collective bipolar adaptivity. Under the unifying theory, the dimensional view, bipolar logical view, and YinYang-N-Element cellular automata view are coherent and consistent. Therefore, bipolar set theory, bipolar dynamic logic, BQLA, bipolar agents, bipolar causality, and BQCA are all unified under the theory of YinYang bipolar relativity. With bipolar relativity, the theory of BQBE is within reach.

First, what are the major differences of BQBE from BE? In a nutshell, BQBE is equilibrium-based, quantum in nature, and fundamentally holistic where bipolar relativity presents a unification of agent interaction and causality in both macroscopic and microscopic worlds as well as social, physical, and biological worlds. Although BE is also intended to be holistic, it can't be fundamentally holistic with truth-based reasoning that is subjected to the LAFIP paradox and is incapable of modeling equilibrium, non-equilibrium, bipolar quantum entanglement, oscillation, and interaction.

Secondly, what is the area of interest of BQBE? This question can be answered in a few words: *same as for BE.* In addition, due to bipolar unification, physical, cognitive, neurobiological, economic, and decision-centric systems can all be unified under BQBE. Therefore, it enlarges the scope of BE and is more inclined toward and suitable for decision support in biological and socioeconomic analysis.

Thirdly, why develop BQBE? As pointed out by Geoffrey Hodgson (1992): *"The reform of economics is not a question of adding new dimensions to the neoclassical economic theory. A theoretical revolution is required at the core of economics itself."* The equilibrium-based quantum property of BQBE provides a cognitively and philosophically different way of thinking for bioeconomics simulation and regulation. This different way of thinking could be pivotal in decision making on complex and difficult issues such as economic development and global warming as well as the most fundamental issues such as agent interaction and causality.

Fourthly, what is new in BQBE? Scientifically, it brings bipolar mathematical abstraction, BDL, BDFL, bipolar agents, bipolar causality, bipolar linear algebra, quantum theory, and collective adaptivity all together into the unifying theory of bipolar relativity for bioeconomics. The new theory points to a kind of "new science" that is truly holistic, synthetic, interdisciplinary, interactive, emergent, and postmodern. After all, the "new science" is actually "the oldest science" of the world as YinYang has survived for thousands of years. Recently YinYang has entered almost all fields of social and physical sciences (Ch. 1).

Lastly, about the epistemology of BQBE, it is (1) systemic (dynamic, open, non-linear, equilibrium-, or non-equilibrium based), (2) holistic where the whole is more than the sum of its parts as evidenced by the transient nature of the YinYang geometry of bipolar relativity (Ch. 6), (3) logically dialectic as YinYang BDL subsumes deductive Boolean logic and quantum logic (Ch. 3), and (4) complementary where equilibrium and non-equilibrium are holistic complementary concepts to positivist truth and falsity.

BQBE IN TCM DIAGNOSTIC DECISION SUPPORT

YinYang WuXing and TCM

Through thousands of years, the theory of YinYang WuXing (YinYang-5-Elements or YinYang-5-Agents – five subsystems of human body) has been a major analytical foundation in TCM including acupuncture, herbal therapeutics, and Qigong. It is noted by historian Professor Ebrey (1993, pp. 77-79) that *"The concepts of Yin and Yang and the Five Agents provided the intellectual framework of much of Chinese scientific thinking especially in fields like biology and medicine. The organs of the body were seen to be interrelated in the same sorts of ways as other natural phenomena, and best understood by looking for correlations and correspondences. Illness was seen as a disturbance in the balance of Yin and Yang or the Five Agents caused by emotions, heat or cold, or other influences. Therapy thus depended on accurate diagnosis of the source of the imbalance."* Scientifically speaking, however, the theory of YinYang WuXing has been primarily empirical in nature without a formal mathematical or scientific foundation for thousands of years.

In recent decades, YinYang related scientific research has made significant progress; however, there is no commonly accepted mathematical model for YinYang WuXing until today. Since human body is one of the most complex systems in nature, a better understanding of YinYang WuXing is of interest for bioeconomics, bioinformatics, metabolism, computational biology, aging, biomedical engineering, artificial intelligence, and especially for decision support in TCM. Furthermore, better understanding of YinYang WuXing is expected to advance research and development in equilibrium-based computation and modeling in socioeconomics, decision support, multiagent coordination and global regulation (Zhang, 2005a, 2006).

In (Li, 1987) an interesting and significant Qi-based approach is presented in an attempt to establish a mathematical model of human body from the perspective of TCM (Note: ***Qi*** in TCM is the vital energy that regulates the functions of the body as a whole system; ***Qigong*** refers to a wide variety of traditional cultivation practices that involve methods of accumulating, circulating, and working with Qi or energy within the body.). While the basic idea in (Li, 1987) is to bridge the gap between modern dynamics theory and ancient YinYang WuXing theory, the mathematical tools used are classical ordinary dynamic differential equations, where the Yin and the Yang can't be explicitly represented as a bipolar fusion due the lack of bipolar syntax and semantics in classical mathematical abstraction.

YinYang bipolar relativity including YinYang bipolar sets, lattices, logic, bipolar agents, bipolar causality, BQLA, and BQCA provides a systematic formal mathematical physics or biophysics theory for TCM. The formal theory enables the explicit modeling of the dynamic processes of the Yin, the Yang, and its bipolar fusion or separation into equilibrium or non-equilibrium.

According to YinYang every matter has two sides or two poles. Yin is the feminine, negative, or cold side and Yang is the masculine, positive, or warm side. The fusion, coupling, or binding of the two sides in equilibrium or harmony is considered a key for the mental or physical health of a neurological, biological, or any dynamic system. This principle has played a key role in TCM, where symptoms are often diagnosed as the loss of balance of the Yin and the Yang. For instance, kidney Yin deficiency or Yang deficiency may cause heart Yin or Yang deficiency, respectively (Maciocia, 1989, pp. 95-96, 202-207).

While YinYang can be used to characterize the state of any biological agent in TCM, WuXing theory presents a natural interactive process model of human body. According to their functions, ***enhancing*** and ***regulating*** relations (WuXing relations) exist among the five interacting subsystems. Based on YinYang (-,+) bipolar relativity we attempt to model the WuXing human body system as a BQCA of bioeconomics.

We follow the conventions adopted in (Li, 1987): (a) a functional WuXing element refers not only to one organ but a subsystem (For instance, "kidney" refers not only to the kidney itself but also to the genital, urinary and endocrinal system); (b) other parts of human body are weakly coupled with the five functional subsystems according to TCM. Different from (Li, 1987), we consider Qi (the vital energy in TCM that flows within natural as well as biological systems and regulates the functions of nature and the body) as a YinYang fusion, coupling, or binding with bipolar interaction, equilibrium, non-equilibrium, harmony, and/or balance aspects.

In this section, we simulate YinYang WuXing equations (Ch. 8) with linear approximation using BQLA. We discuss decision support in diagnostic analysis in TCM with computer simulations and graphical visualization for certain symptoms including self-healing processes, *rebalancing*, *growing and aging processes*. It is shown that the new model can be used for TCM diagnostic decision support as well as for better understanding of the interactive nature of YinYang-5-Elements.

System Architecture of TCM Diagnostic Decision Support

Based on bipolar linear algebra and BQCA, a decision support system YYWX has been prototyped and implemented in MS Visual C++ and Java. The top-level system architecture is depicted in Figure 1. The system is designed to be iterative in nature. After a TCM doctor diagnoses a patient, his intermediate diagnosis together with the patient symptoms can be entered into the YYWX system. The decision support provided by the system becomes feedback to the TCM doctor for further diagnostic analysis. This process iterates until a final diagnostic decision is reached.

Diagnostic processes have been simulated with the YYWX decision support system. Our simulation is open-world and can be either equilibrium or non-equilibrium based in nature. It is said open-world because we allow external disturbance to be represented in a vector matrix of the WuXing elements; it can be equilibrium-based because the model is able to rebalance itself to a new equilibrium state under certain conditions; it can be non-equilibrium based because the system is able to model energy increase (strengthening) or decrease (weakening) processes under certain conditions.

Figure 1. Diagnostic decision support system architecture for TCM (Adapted from Zhang & Chen, 2009)

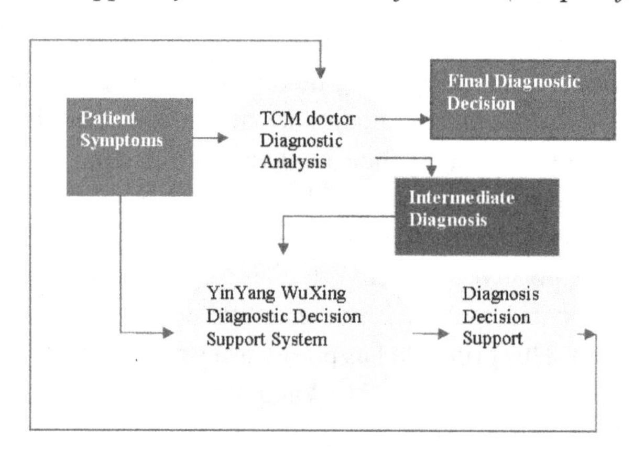

Equilibrium Process Modeling of YYWX (Adapted from Zhang & Chen, 2009)

BQLA and BQCA provide a new mathematical and physical theory for modeling the WuXing elements with explicit YinYang representation and equilibrium, quasi- or non-equilibrium states for vital energy (or Qi in TCM) and stability analysis. A key is the concept of equilibrium energy and stability of a bipolar connectivity matrix or a vector matrix. In this case, energy of the row matrix can be considered as the biological energy of the biological elements or agents; energy of the connectivity matrix can be considered the organizational energy of the biological cells or agents from a bioeconomics perspective (McKelvey, 2004).

WuXing with Balanced Nourishing. Assume $\forall Y_n \in B_\infty$ and $\forall a,b,c \in B_F$, let $a = (0\ 0.35)$, $b=(-0.15\ 0.15)$, and $c = (0\ 0.35)$ be constants, we have the following:

$Y(t) = [Y_1(t), Y_2(t), Y_3(t), Y_4(t), Y_5(t)];$

$Y(t+1) = Y(t) \times M;$

$$M = \begin{bmatrix} (0.00, 0.35) & (-0.15\ ,0.15) & (0.00, 0.35) & (0,0) & (0,0) \\ (0,0) & (0.00, 0.35) & (-0.15\ ,0.15) & (0.00, 0.35) & (0,0) \\ (0,0) & (0,0) & (0.000, 0.35) & (-0.15\ ,0.15) & (0.00, 0.35) \\ (0.00, 0.35) & (0,0) & (0,0) & (0.00, 0.35) & (-0.15\ ,0.15) \\ (-0.15\ ,0.15) & (0, 0.35) & (0,0) & (0,0) & (0.000, 0.35) \end{bmatrix}.$$

Note that M satisfies both the global and local energy equilibrium conditions of Theorems 8.3, where the energy total of each row and column of M equals $0.35+0.35+|-0.15| + 0.15 = 1$.

Case1. Assume the normal Yin and Yang for a patient's subsystem in the WuXing diagram (Ch. 8, Fig. 1) is (-100 100), we have the normal system energy total $|\varepsilon|(Y)= 1000$. Given

Table 1. Rebalancing Sequence after temporary Kidney Yin Deficiency without Excessive Yang (Zhang & Chen, 2009)

T	$Y_1(t)$(Kidney) $Y_2(t)$(Lung) $Y_3(t)$(Heart) $Y_4(t)$(Spleen) $Y_5(t)$ (liver)	Energy Equilibrium	YinYang Balance/ Harmony
0	(**0.000 100.000**)(-100.000 100.000)(-100.000 100.000)(-100.000 100.000)(-100.000 100.000)	900.000	0.889
1	(**-65.000 100.000**)(-85.000 85.000)(-65.000 100.000)(-100.000 100.000)(-100.000 100.000)	900.000	0.922
2	(**-87.750 100.000**)(-89.500 89.500)(-71.000 95.500)(-89.500 89.500)(-87.750 100.000)	900.000	0.946
3	(-90.200 94.488)(-90.200 94.488)(-82.412 95.275)(-87.625 87.625)(-82.412 95.275)	900.000	0.962
4	(-88.892 90.393)(-88.118 94.120)(-88.118 94.120)(-88.892 90.393)(-83.976 92.980)	900.000	0.973
5	(-88.768 89.818)(-87.125 92.378)(-89.289 91.915)(-89.289 91.915)(-87.125 92.378)	900.000	0.981
..	…	….	…
26	(-89.999 90.001)(-89.999 90.001)(-89.999 90.001)(-89.999 90.001)(-89.999 90.001)	900.000	1.000
27	(-89.999 90.001)(-89.999 90.001)(-89.999 90.001)(-89.999 90.001)(-89.999 90.001)	900.000	1.000
28	(-90.000 90.000)(-90.000 90.000)(-90.000 90.000)(-90.000 90.000)(-90.000 90.000)	900.000	1.000
29	(-90.000 90.000)(-90.000 90.000)(-90.000 90.000)(-90.000 90.000)(-90.000 90.000)	900.000	1.000

Y(0) = [**(0.00 100.00)** (-100.00 100.00) (-100.00 100.00) (-100.00 100.00) (-100.00 100.00)],

where $Y_1(0)=(0.00\ 100.00)$ indicates kidney Yin deficiency without Yang excess due to a health disturbance, we may have the rebalancing sequence of the YinYang-5-Element system after the disturbance as shown in Table 1.

Case2. Given *Y(0) = [**(-50.00 150.00)**(-100.00 100.00) (-100.00 100.00) (-100.00 100.00) (-100.00 100.00)]* where $Y_1(0)=(50.00\ 150.00)$ indicates kidney Yin deficiency and Yang excess due to internal loss of balance, we have the rebalancing sequence as shown in Table 2.

Case3. Given *Y(0) = [**(-150.00 50.00)**(-100.00 100.00) (-100.00 100.00) (-100.00 100.00) (-100.00 100.00)]* where $Y_1(0)=(-150.00\ 50.00)$ indicates kidney Yin excess and Yang deficiency due to internal loss of balance, we may have the rebalancing sequence as shown in Table 3.

An analysis of the rebalancing sequences in Figures 2-4 leads to the following findings:

1. Kidney Yin deficiency may cause heart Yin deficiency (see highlighted data area in Tables 1 and 2), which is consistent with traditional Chinese medicine teachings (Maciocia, 1989, p. 206).

Table 2. Rebalancing Sequence after Kidney Yin Deficiency with Yang Excess (Zhang & Chen, 2009)

T	$Y_1(t)$(Kidney) $Y_2(t)$(Lung) $Y_3(t)$(Heart) $Y_4(t)$(Spleen) $Y_5(t)$ (liver)	Energy Equilibrium	YinYang Balance/ Harmony
0	(**-50.000 150.000**)(-100.000 100.000)(-100.000 100.000)(-100.000 100.000)(-100.000 100.000)	1000.000	0.900
1	(**-82.500 117.500**)(-100.000 100.000)(-82.500 117.500)(-100.000 100.000)(-100.000 100.000)	1000.000	0.930
2	(**-93.875 106.125**)(-100.000 100.000)(-87.750 112.250)(-100.000 100.000)(-93.875 106.125)	1000.000	0.951
3	(**-97.856 102.144**)(-97.856 102.144)(-93.569 106.431)(-100.000 100.000)(-93.569 106.431)	1000.000	0.966
4	(-99.250 100.750)(-96.999 103.001)(-96.999 103.001)(-99.250 100.750)(-95.498 104.502)	1000.000	0.976
5	(-99.475 100.525)(-97.374 102.626)(-98.687 101.313)(-98.687 101.313)(-97.374 102.626)	1000.000	0.983
...	…	…	..
26	(-99.999 100.001)(-99.999 100.001)(-99.999 100.001)(-99.999 100.001)(-99.999 100.001)	1000.000	1.000
27	(-99.999 100.001)(-99.999 100.001)(-99.999 100.001)(-99.999 100.001)(-99.999 100.001)	1000.000	1.000
28	(-100.000 100.000)(-100.000 100.000)(-100.000 100.000)(-100.000 100.000)(-100.000 100.000)	1000.000	1.000
29	(-100.000 100.000)(-100.000 100.000)(-100.000 100.000)(-100.000 100.000)(-100.000 100.000)	1000.000	1.000

Table 3. Rebalancing Sequence with Kidney Yin Excess and Yang Deficiency (Zhang & Chen, 2009)

T	***$Y_1(t)$(Kidney)*** *$Y_2(t)$(Lung)* <u>*$Y_3(t)$(Heart)*</u> *$Y_4(t)$(Spleen)* *$Y_5(t)$ (liver)*	**Energy Equilibrium**	**YinYang Balance/ Harmony**
0	(**-150.000 50.000**)(-100.000 100.000)(<u>-100.000 100.000</u>)(-100.000 100.000)(-100.000 100.000)	1000.000	0.900
1	(**-117.500 82.500**)(-100.000 100.000)(<u>-117.500 82.500</u>)(-100.000 100.000)(-100.000 100.000)	1000.000	0.930
2	(**-106.125 93.875**)(-100.000 100.000)(<u>-112.250 87.750</u>)(-100.000 100.000)(-106.125 93.875)	1000.000	0.951
3	(**-102.144 97.856**)(-102.144 97.856)(<u>-106.431 93.569</u>)(-100.000 100.000)(-106.431 93.569)	1000.000	0.966
4	(-100.750 99.250)(-103.001 96.999)(-103.001 96.999)(-100.750 99.250)(-104.502 95.498)	1000.000	0.976
5	(-100.525 99.475)(-102.626 97.374)(-101.313 98.687)(-101.313 98.687)(-102.626 97.374)	1000.000	0.983
...	...	...	...
26	(-100.001 99.999)(-100.001 99.999)(-100.001 99.999)(-100.001 99.999)(-100.001 99.999)	1000.000	1.000
27	(-100.001 99.999)(-100.001 99.999)(-100.001 99.999)(-100.001 99.999)(-100.001 99.999)	1000.000	1.000
28	(-100.000 100.000)(-100.000 100.000)(-100.000 100.000)(-100.000 100.000)(-100.000 100.000)	1000.000	1.000
29	(-100.000 100.000)(-100.000 100.000)(-100.000 100.000)(-100.000 100.000)(-100.000 100.000)	1000.000	1.000

Figure 2. A graphical user interface (Zhang & Chen, 2009)

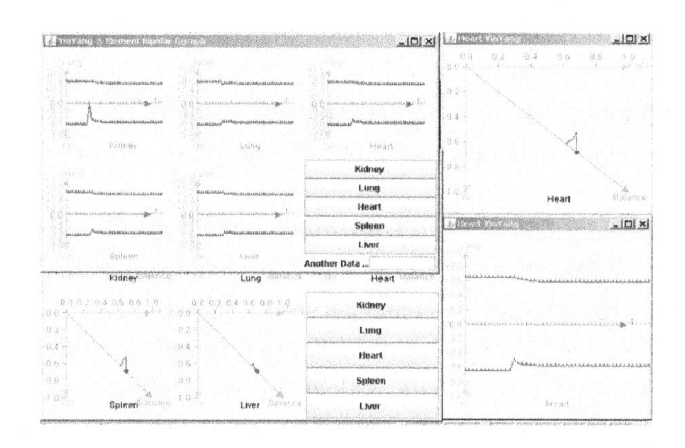

2. Kidney Yang deficiency may cause heart Yang deficiency (see highlighted data area in Table 3), that is consistent with traditional Chinese medicine teachings (Maciocia, 1989, p. 203).
3. The energy for each $Y(t) = [Y_1(t), Y_2(t), Y_3(t), Y_4(t), Y_5(t)]$, $t=0$ *to* 29, remained a global equilibrium in all three local rebalancing processes to local equilibria. The global equilibrium is due to the selection of constant coefficients *a, b,* and *c* in the connection matrix *M* to meet the equilibrium conditions as stated in Theorems 8.3.

(4) In Table 1 the global energy equilibrium remained at 900.00 or 90% of 1000, that indicates the loss of energy caused by a permanent loss of energy $Y_1(0) = (0\ 100)$. In Tables 2 and 3 it is remained at 1000.00 or 100% of 1000, which indicates no global loss of energy because the initial vector only shows internal unbalance.

(5) The harmony level of $Y(t) = [Y_1(t), Y_2(t), Y_3(t), Y_4(t), Y_5(t)]$, $t=0$ *to* 29, showed remarkable self-recovering (or self-healing) capability of the WuXing system.

WuXing with Unbalanced Nourishing. Despite energy levels in the previous examples, the harmony levels always recover to normal with balanced nourishing. Now we have the question: What will happen if the nourishing relation *b* is not balanced? Remarkably, after changing the nourishing relation

Figure 3. YinYang wave forms for the bipolar elements of Table 8 (Zhang et al., 2009)

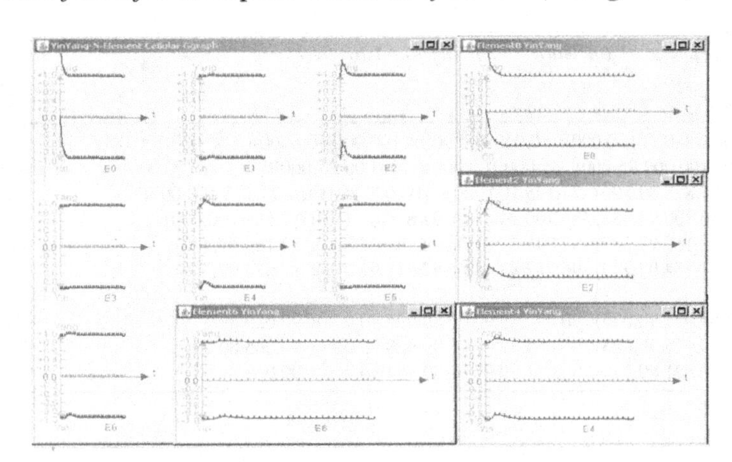

Figure 4. YinYang energy increase curve for the bipolar elements of Table 3 (Zhang et al., 2009)

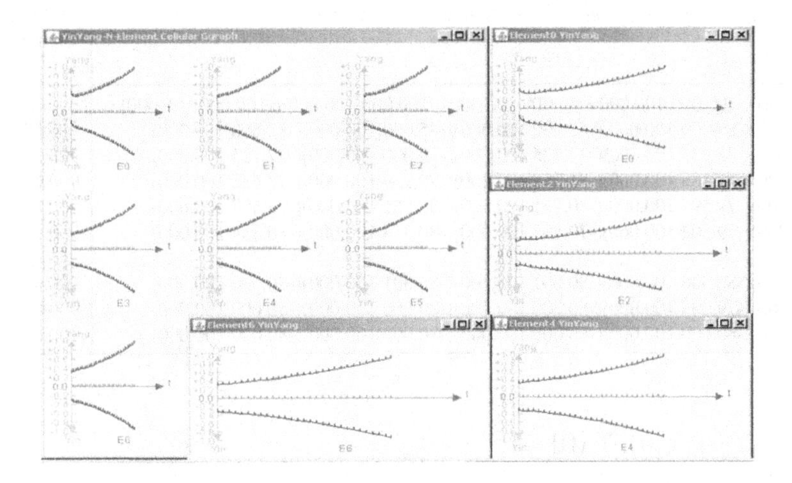

b from *(-0.15 0.15)* to *(-0.3 0.0)*, the WuXing system can still rebalance (after a longer time sequence) to full harmony (see Table 4) as long as the equilibrium conditions in Theorems 8.3 are met. This is consistent with "Yin creates Yang" in YinYang theory. Mathematically, we have $- - = +$. If we change b from *(-0.15 0.15) to (0 0.3)*, however, the balanced nourishing relations in Figure 1 will all become regulating relations. Without satisfying the condition of Theorem 8.4, the WuXing system can no longer rebalance to local equilibrium (see Table 5) even though the global equilibrium conditions in Theorem 8.3 are met. This is because according to YinYang theory "Yang does not create Yin". Mathematically, we have $+ + = +$.

Quasi- or Non-Equilibrium Process of YYWX

Energy Decrease. Assume $\forall Y_n \in B_\infty$ and $\forall a,b,c \in B_F$, let $a = (0\ 0.35)$, $b = (-0.05\ 0.05)$, and $c = (0\ 0.35)$ be constants; We select $Y(0)$ with a perfectly normal initial energy. We have the following:

Table 4. Permanent Loss of Energy and Rebalancing to Harmony (Zhang & Chen, 2009)

t	$Y_1(t)$*(Kidney)* $Y_2(t)$*(Lung)* $Y_3(t)$*(Heart)* $Y_4(t)$*(Spleen)* $Y_5(t)$ *(liver)*	**Energy Equilibrium**	**YinYang Balance/ Harmony**
0	(-50.000 100.000)(-100.000 100.000)(-50.000 100.000)(-100.000 100.000)(-100.000 100.000)	900.000	0.889
1	(-82.500 100.000)(-100.000 85.000)(-65.000 100.000)(-100.000 85.000)(-82.500 100.000)	900.000	0.889
2	(-93.875 89.500)(-93.875 89.500)(-77.125 100.000)(-100.000 79.000)(-77.125 100.000)	900.000	0.916
3	(-97.856 82.113)(-86.700 94.488)(-86.700 94.488)(-97.856 82.113)(-77.688 100.000)	900.000	0.923
4	(-98.499 80.785)(-82.169 97.427)(-92.941 87.820)(-92.941 87.820)(-82.169 97.428)	900.000	0.935
5	(-96.232 83.663)(-81.754 97.749)(-96.232 83.663)(-87.635 92.719)(-87.635 92.719)	900.000	0.943
...	...	...	...
36	(-90.029 89.971)(-89.925 90.075)(-90.092 89.908)(-89.925 90.075)(-90.029 89.971)	900.000	0.999
37	(-89.975 90.025)(-89.975 90.025)(-90.065 89.935)(-89.920 90.080)(-90.065 89.935)	900.000	0.999
38	(-89.944 90.056)(-90.021 89.979)(-90.021 89.979)(-89.944 90.056)(-90.069 89.931)	900.000	1.000

Table 5. A Failed Equilibrium Rebalancing Process with Loss of Energy and Harmony (Zhang & Chen, 2009)

t	$Y_1(t)$*(Kidney)* $Y_2(t)$*(Lung)* $Y_3(t)$*(Heart)* $Y_4(t)$*(Spleen)* $Y_5(t)$ *(liver)*	**Energy Equilibrium**	**YinYang Balance/ Harmony**
0	(-50.000 100.000)(-100.000 100.000)(-50.000 100.000)(-100.000 100.000)(-100.000 100.000)	900.000	0.889
1	(-82.500 100.000)(-85.000 100.000)(-65.000 100.000)(-85.000 100.000)(-82.500 100.000)	900.000	0.889
2	(-83.375 100.000)(-83.375 100.000)(-77.125 100.000)(-79.000 100.000)(-77.125 100.000)	900.000	0.889
3	(-79.969 100.000)(-81.188 100.000)(-81.188 100.000)(-79.969 100.000)(-77.688 100.000)	900.000	0.889
4	(-79.284 100.000)(-79.597 100.000)(-80.761 100.000)(-80.761 100.000)(-79.597 100.000)	900.000	0.889
5	(-79.895 100.000)(-79.503 100.000)(-79.895 100.000)(-80.354 100.000)(-80.354 100.000)	900.000	0.889
...	...	...	...
14	(-79.999 100.000)(-80.000 100.000)(-80.001 100.000)(-80.001 100.000)(-80.000 100.000)	900.000	0.889
15	(-80.000 100.000)(-79.999 100.000)(-80.000 100.000)(-80.000 100.000)(-80.000 100.000)	900.000	0.889
16	(-80.000 100.000)(-80.000 100.000)(-80.000 100.000)(-80.000 100.000)(-80.000 100.000)	900.000	0.889

$$Y(t) = [Y_1(t), Y_2(t), Y_3(t), Y_4(t), Y_5(t)]$$

$$Y(t+1) = Y(t) \times M.$$

$$Y(0) = [(-100.00\ 100.00)(-100.00\ 100.00)(-100.00\ 100.00)(-100.00\ 100.00)(-100.00\ 100.00)]$$

$$M = \begin{bmatrix} (0.00, 0.35) & (-0.05\ ,0.05) & (0.00, 0.35) & (0,0) & (0,0) \\ (0,0) & (0.00, 0.35) & (-0.05\ ,0.05) & (0.00, 0.35) & (0,0) \\ (0,0) & (0,0) & (0.000, 0.35) & (-0.05\ ,0.05) & (0.00, 0.35) \\ (0.00, 0.35) & (0,0) & (0,0) & (0.00, 0.35) & (-0.05\ ,0.05) \\ (-0.05\ ,0.05) & (0, 0.35) & (0,0) & (0,0) & (0.000, 0.35) \end{bmatrix}$$

From M it is clear that the total energy of each row and each column is $0.35 + 0.35 + |-0.5| + 0.5 = 0.80 < 1$. The energy of vector *Y(t)* should decrease to zero gradually regardless of the harmony or disharmony of Y(t) at any time point t. The gradual decrease of energy might be used to simulate natural aging or deteriorating processes in life sciences. Simulated data is shown in Table 6.

Table 6. Rebalancing Sequence with Decreasing Energy (Zhang & Chen, 2009)

t	$Y_1(t)$(Kidney) $Y_2(t)$(Lung) $Y_3(t)$(Heart) $Y_4(t)$(Spleen) $Y_5(t)$ (liver)	Energy Non-Equilibrium	YinYang Balance/ Harmony
0	(-150.000 50.000)(-100.000 100.000)(-100.000 100.000)(-100.000 100.000)(-100.000 100.000)	1000.000	0.900
1	(-97.500 62.500)(-80.000 80.000)(-97.500 62.500)(-80.000 80.000)(-80.000 80.000)	800.000	0.913
2	(-70.125 57.875)(-64.000 64.000)(-76.250 51.750)(-64.000 64.000)(-70.125 57.875)	640.000	0.923
3	(-53.344 49.056)(-53.344 49.056)(-57.631 44.769)(-51.200 51.200)(-57.631 44.769)	512.000	0.933
4	(-41.710 40.210)(-43.961 37.959)(-43.961 37.959)(-41.710 40.210)(-45.462 36.458)	409.600	0.941
5	(-33.293 32.243)(-35.394 30.142)(-34.081 31.455)(-34.081 31.455)(-35.394 30.142)	327.680	0.949
6	(-26.858 25.571)(-28.053 24.376)(-26.858 25.571)(-27.593 24.836)(-27.593 24.836)	262.144	0.955
...	...	...	...
23	(-0.593 0.588)(-0.593 0.588)(-0.593 0.588)(-0.593 0.588)(-0.593 0.588)	5.903	0.995
24	(-0.474 0.470)(-0.474 0.470)(-0.474 0.470)(-0.474 0.470)(-0.474 0.470)	4.722	0.996
25	(-0.379 0.376)(-0.379 0.376)(-0.379 0.376)(-0.379 0.376)(-0.379 0.376)	3.778	0.996
...	...	...	...
64	(-0.000 0.000)(-0.000 0.000)(-0.000 0.000)(-0.000 0.000)(-0.000 0.000)	0.001	1.000
65	(-0.000 0.000)(-0.000 0.000)(-0.000 0.000)(-0.000 0.000)(-0.000 0.000)	0.001	1.000
66	(-0.000 0.000)(-0.000 0.000)(-0.000 0.000)(-0.000 0.000)(-0.000 0.000)	0.000	1.000

Energy Increase. Assume $\forall Y_n \in B_\infty$ and $\forall a,b,c \in B_F$, let $a = (0\ 0.35)$, $b = (-0.17\ 0.17)$, and $c = (0\ 0.35)$ be constants. We select $Y(0)$ with much lower initial energy far below normal. We have the following:

$$Y(t) = [Y_1(t), Y_2(t), Y_3(t), Y_4(t), Y_5(t)]$$

$$Y(t+1) = Y(t) \times M.$$

$$Y(0) = [(-50.0\ 30.0)(-50.0\ 30.0)(-30.0\ 50.0)(-50.0\ 30.0)(-20.0\ 20.0)]$$

$$M = \begin{bmatrix} (0.00, 0.35) & (-0.17\ ,0.17) & (0.00, 0.35) & (0,0) & (0,0) \\ (0,0) & (0.00, 0.35) & (-0.17\ ,0.17) & (0.00, 0.35) & (0,0) \\ (0,0) & (0,0) & (0.000, 0.35) & (-0.17\ ,0.17) & (0.00, 0.35) \\ (0.00, 0.35) & (0,0) & (0,0) & (0.00, 0.35) & (-0.17, 0.17) \\ (-0.17\ ,0.17) & (0, 0.35) & (0,0) & (0,0) & (0.000, 0.35) \end{bmatrix}$$

From M it is clear that the total energy of each row and each column is $0.35 + 0.35 + |-0.17| + 0.17 = 1.04 > 1$. It is expected that the energy of vector $Y(t)$ should increase gradually regardless of the harmony or disharmony of Y(t) at any time point t. The gradual increase of energy might be used to simulate *growing processes* in life sciences. It may hold the potential for modeling Qi in TCM and Qigong masters' internal energy accumulation and circulation. Simulated data is shown in Table 7.

Graphical User Interface

A graphical user interface (GUI) has been prototyped in Java language under the MS Windows environment. The equilibrium and non-equilibrium states of YinYang-5-Elements can be displayed in graphical forms. This is illustrated in Figure 2. The GUI interface can be used for TCM education as well as TCM diagnostic decision analysis.

Table 7. Rebalancing Sequence with Increasing Energy (Zhang & Chen, 2009)

T	$Y_1(t)$(Kidney) $Y_2(t)$(Lung) $Y_3(t)$(Heart) $Y_4(t)$(Spleen) $Y_5(t)$ (liver)	Energy Non-Equilibrium	YinYang Balance/ Harmony
0	(-150.000 50.000)(-100.000 100.000)(-100.000 100.000)(-100.000 100.000)(-100.000 100.000)	1000.000	0.900
1	(-121.500 86.500)(-104.000 104.000)(-121.500 86.500)(-104.000 104.000)(-104.000 104.000)	1040.000	0.907
2	(-114.285 102.035)(-108.160 108.160)(-120.410 95.910)(-108.160 108.160)(-114.285 102.035)	1081.600	0.920
3	(-114.630 110.343)(-114.630 110.343)(-118.918 106.055)(-112.486 112.486)(-118.918 106.055)	1124.864	0.948
4	(-117.736 116.236)(-119.987 113.985)(-119.987 113.985)(-117.736 116.236)(-121.488 112.484)	1169.858	0.963
5	(-122.190 121.140)(-124.291 119.039)(-122.978 120.352)(-122.978 120.352)(-124.291 119.039)	1216.653	0.972
...	...	...	...
22	(-236.996 236.988)(-236.996 236.988)(-236.996 236.988)(-236.996 236.988)(-236.996 236.988)	2369.918	0.999
23	(-246.474 246.469)(-246.474 246.469)(-246.474 246.469)(-246.474 246.469)(-246.474 246.469)	2464.715	1.000
24	(-299.871 299.870)(-299.871 299.870)(-299.871 299.870)(-299.871 299.870)(-299.871 299.870)	2563.303	1.000

BIOSYSTEM SIMULATION AND REGULATION WITH YINYANG-N-ELEMENT BQCA (ADAPTED FROM ZHANG *ET AL.*, 2009)

According to Gore and van Oudenaarden's viewpoint published in *Nature* (Gore & van Oudenaarden, 2009), *"the Yin and the Yang of nature"* form the basis for synthetic biology. They wrote: *"Oscillations in gene expression regulate various cellular processes and so must be robust and tunable. Interactions between both negative and positive feedback loops seem to ensure these features."* These views reinforce the idea of YinYang bipolar equilibrium-based biosystem simulation and regulation which is further presented and discussed in this section.

Assumptions and Conditions

The focus of the last section was simulation and regulation of YinYang-5-Elelemt BQBE for diagnostic decision support in TCM from a bioeconomics perspective. A number of questions remain unanswered. One is whether simulation and regulation with arbitrary YinYang-N-Element BQCA can be applied to *biosystem simulation and regulation* in general such as genetic and molecular network simulation and regulation? Another is whether simulation and regulation with arbitrary YinYang-N-Element BQCA ports to the simulation of *local oscillation under global equilibrium* conditions as specified in Theorem 8.6?

In the last chapter, YinYang-N-Element BQCA was extended to three families of YinYang-N-Element cellular networks: the classical family, the non-classical or randomly connected family, and the combinatorial family. While the YinYang-N-Element Energy Equilibrium Law (Theorem 8.3), the YinYang-N-Element System Energy (Non-Equilibrium) Decreasing Law (Theorem 8.4), the YinYang-N-Element System Energy (Non-Equilibrium) Increasing Law (Theorem 8.5), and the YinYang-N-Element System Energy Oscillation Law (Theorem 8.6) were proved axiomatically in the last section for all three families, we present simulation results based on the conditions of the four laws with the assumptions as follows.

Cellular Assumption. Without losing generality, it is assumed that a bipolar regulation network forms a classical, random, or combinatorial cellular structure. That is, a regulator agent at the system, molecular, or genetic level regulates itself directly or indirectly. It is shown in the following that the bipolar entanglement of BQLA provides a unique computational basis for modeling bipolar fusion, diffusion, interaction, and oscillation from a computational bioeconomics perspective. The simulation distinguishes nourishing relations from regulating relations as discussed in early section. The simulation is limited in the scope of repression/activation properties. Thus, each cell or element E_i represents one type of bio-agent such as a regulator protein.

Bipolar Fusion and diffusion Assumptions. It is shown in (Emran et al., 2006; Kim, Faulk & Kim, 2007; Zhou & Yik, 2006) that the genetic binding of the ubiquitous regulator proteins YY1 and E1A exhibits mediated repression and activation abilities to other agents. Thus, we use the pair (repression, activation) as a general bipolar predicate that maps a bipolar agent's regulatory abilities to $B_\infty = [-\infty\ 0] \times [0\ +\infty]$. On the other hand, we assume the (repression, activation) abilities are bipolar transitive through bipolar interaction.

Bipolar Equilibrium/Non-Equilibrium Assumption. It is assumed that bio-agent regulation can form either a global YinYang bipolar equilibrium (Vasudevan, Tong & Steitz, 2007) or a quasi- or non-equilibrium regardless of local oscillation and instability.

Closed- or Open-World Assumption. Without losing generality, we support either a closed world or an open world assumption or both with a transition. That is, disturbance can be exerted to a YinYang-N-Element Cellular network.

Simulation and Regulation of Classical YinYang-N-Element BQCA

The classical family of BCNs has three income and three outgoing links for each element. Based on energy and stability measures, the stabilities of the bipolar transitive closures of YinYang-N-Element BCNs are computed for N ranging from 5 to 11, respectively, with comparable nourishing and regulating link weights. It is found that YinYang-5-Element cellular network has the highest stability. This finding coincides with the mathematical analysis in the last chapter. That is, k=N/2 is an optimal configuration and for N=5 we have k = 5/2 that round up to 3. Since when N < 5 it is impossible to form an internal regulatory cycle, YinYang-5-elements form the minimum and optimal structure of this class.

Why N = 6 and k = N/2 = 3 does not form an optimal structure in terms of stability. Actually it is optimal in mathematical terms. However, the number 6 is even that leads to two regulating partitions (see Chapter 8, Figure 2). Generally speaking, when N is an odd number a YinYang-N-Element structure is more stable.It is observed that a larger N leads to a lower stability in this family of BCNs, however, when N > 10 the stability of this class of cellular structures tends to be indistinguishable. This could be caused by the partitioning effect of the regulating links. This finding shows that YinYang-5-Elements form a basic cellular architecture for bioeconomics. It is also a vindication of YinYang-5-Elements – the theoretical foundation of TCM regarding the five major organs of the human body as a natural system.

Equilibrium-Based YinYang Rebalancing behavior. This simulation follows the conditions of Theorem 8.3. Table 8 shows an equilibrium-based energy rebalancing process of the classical YinYang-7-Element cellular network (Figure 4), where the energy of each row and column of the connectivity matrix M is normalized to 1.0. The bipolar energy levels of the elements can be visualized from the YinYang energy wave forms shown in Figure 3. The rebalancing is based on the equation $Y(t+1) = Y(t) \times M(t)$. From Table 8 it is clear that at t = 0, the Yin and Yang of element E1=(0 200) and $Y(t=0)$ is unbalanced locally. After 27 iterations, all elements in $Y(t=27)$ are rebalanced to (-100 100). Evidently, the rebalancing condition in Theorem 8.3 is satisfied. Namely we have, $\forall t$, $|\varepsilon^-(M_{*j})| > 0$ and $|\varepsilon^+(M_{*j}))| > 0$. During the rebalancing process, the system remained in global energy equilibrium with energy total 1400. However, the initial system stability 0.86 is improved to 1.0 at t = 27. The rebalancing process illustrates the commonsense physical law that any closed system tends to reach an equilibrium state. In

Table 8. Equilibrium Energy Rebalancing Process (Zhang et al., 2009)

(t)	**E_1**	**E_2**	**E_3**	**E_4**	**E_5**	**E_6**	**E_7**	**Energy |ε|**	**Stability**
(0)	(0.000 200.000)	(-100.000 100.000)	(-100.000 100.000)	(-100.000 100.000)	(-100.000 100.000)	(-100.000 100.000)	(-100.000 100.000)	1400	0.86
(1)	(-65.000 135.000)	(-100.000 100.000)	(-65.000 135.000)	(-100.000 100.000)	(-100.000 100.000)	(-100.000 100.000)	(-100.000 100.000)	..	..
(2)	(-87.750 112.250)	(-100.000 100.000)	(-75.500 124.500)	(-100.000 100.000)	(-87.750 112.250)	(-100.000 100.000)	(-100.000 100.000)	..	..
(3)	(-95.713 104.287)	(-100.000 100.000)	(-87.137 112.863)	(-100.000 100.000)	(-87.138 112.863)	(-100.000 100.000)	(-95.713 104.287)	..	..
..	..	..	..	..	..	..	..	..	..
(26)	**(-99.999 100.000)**	**(-100.000 100.000)**	**(-99.999 100.000)**	**(-100.000 100.000)**	**(-100.000 100.000)**	**(-100.000 100.000)**	**(-100.000 100.000)**	..	..
(27)	**(-100.000 100.000)**	**(-100.000 100.000)**	**(-100.000 100.000)**	**(-100.000 100.000)**	**(-100.000 100.000)**	**(-100.000 100.000)**	**(-100.00 100.00)**	1400	1.0

a biological system such rebalancing behavior can be realized through molecular or genetic interaction, fusion, and diffusion.

Non-Equilibrium Strengthening Behavior. This simulation follows the conditions of Theorem 8.4. Table 9 shows a non-equilibrium rebalancing process with increasing energy of the same YinYang-7-Element cellular network (Ch. 8, Figure 4), where the energy of each row and column of the connectivity matrix M is initialized to be greater than 1.0. The bipolar energy increase of the elements can be visualized from the YinYang energy wave forms shown in Figure 4. The increase of energy can be used to model a growing process due to the condition of the bio-organization. Namely the row energy and column energy of the connectivity matrix M(t) are set to be greater than 1.0. Again the process follows the equation $Y(t+1) = Y(t) \times M(t)$. It is interesting to notice that $Y(t=0)$ is unbalanced because E1 = (-70 130). After 28 iterations, $Y(t=28)$ is rebalanced; total energy of Y increased from 1400 to 4198.18; system stability improved from 0.9 to 1.0. Evidently, the local rebalancing conditions in Theorem 8.4 are satisfied. Namely we have, $\forall t$, $|\varepsilon^-(M_{*j})| > 0$ and $|\varepsilon^+(M_{*j}))| > 0$, or both negative and positive energies of each column are greater than 0.

Non-Equilibrium Weakening Behavior. This simulation follows the conditions of Theorem 8.5. Table 10 shows a non-equilibrium rebalancing process with decreasing energy of the YinYang-7-Element cellular network (Ch. 8, Figure 4), where the energy of each row and column of the connectivity matrix M is initialized to be less than 1.0. The bipolar energy decrease of the elements can be visualized from the YinYang energy wave forms shown in Figure 5. Interestingly, the bipolar energy can also be visualized from an equilibrium-based perspective as shown in Figure 6. In Figure 6 there are three dimensions: Yin dimension, Yang dimension, and YinYang equilibrium or balance dimension. In this case, both

Table 9. Non-Equilibrium Energy Increasing Process (Zhang et al., 2009)

(t)	**E_1**	**E_2**	**E_3**	**E_4**	**E_5**	**E_6**	**E_7**	**Energy |ε|**	**Stability**
(0)	(-70.000 130.000)	(-100.000 100.000)	(-100.000 100.000)	(-100.000 100.000)	(-100.000 100.000)	(-100.000 100.000)	(-100.000 100.000)	1400	0.9
(1)	(-93.500 114.500)	(-104.000 104.000)	(-93.500 114.500)	(-104.000 104.000)	(-104.000 104.000)	(-104.000 104.000)	(-104.000 104.000)	..	..
(2)	(-104.485 111.835)	(-108.160 108.160)	(-100.810 115.510)	(-108.160 108.160)	(-104.485 111.835)	(-108.160 108.160)	(-108.160 108.160)	..	..
..	..	..	..	..	..	..	..	..	..
(27)	**(-288.336 288.337)**	**(-288.337 288.337)**	**(-288.337 288.337)**	**(-288.337 288.337)**	**(-288.337 288.337)**	**(-288.336 288.337)**	**(-288.337 288.337)**	..	..
(28)	**(-299.870 299.870)**	**(-299.870 299.870)**	**(-299.870 299.870)**	**(-299.870 299.870)**	**(-299.870 299.870)**	**(-299.870 299.870)**	**(-299.870 299.870)**	4198.18	1.0

Table 10. Non-Equilibrium Energy Decreasing Process (Zhang et al., 2009)

(t)	E_1	E_2	E_3	E_4	E_5	E_6	E_7	Energy |ε|	Stability
(0)	(-70.000 130.000)	(-100.000 100.000)	(-100.000 100.000)	(-100.000 100.000)	(-100.000 100.000)	(-100.000 100.000)	(-100.000 100.000)	1400	0.9
(1)	(-69.500 90.500)	(-80.000 80.000)	(-69.500 90.500)	(-80.000 80.000)	(-80.000 80.000)	(-80.000 80.000)	(-80.000 80.000)	..	..
(2)	(-60.325 67.675)	(-64.000 64.000)	(-56.650 71.350)	(-64.000 64.000)	(-60.325 67.675)	(-64.000 64.000)	(-64.000 64.000)	..	..
(3)	(-49.914 52.486)	(-51.200 51.200)	(-47.341 55.059)	(-51.200 51.200)	(-47.341 55.059)	(-51.200 51.200)	(-49.914 52.486)	..	..
..	..	..	..	..	..	..	..	..	..
(54)	**(-0.001 0.001)**	**(-0.001 0.001)**	**(-0.001 0.001)**	**(-0.001 0.001)**	**(-0.001 0.001)**	**(-0.001 0.001)**	**(-0.001 0.001)**	..	..
(55)	**(-0.000 0.000)**	**(-0.000 0.000)**	**(-0.000 0.000)**	**(-0.000 0.000)**	**(-0.000 0.000)**	**(-0.000 0.000)**	**(-0.000 0.000)**	0.000	1.0

Figure 5. YinYang energy decrease curve for the bipolar elements of Table 10 (Zhang et al., 2009)

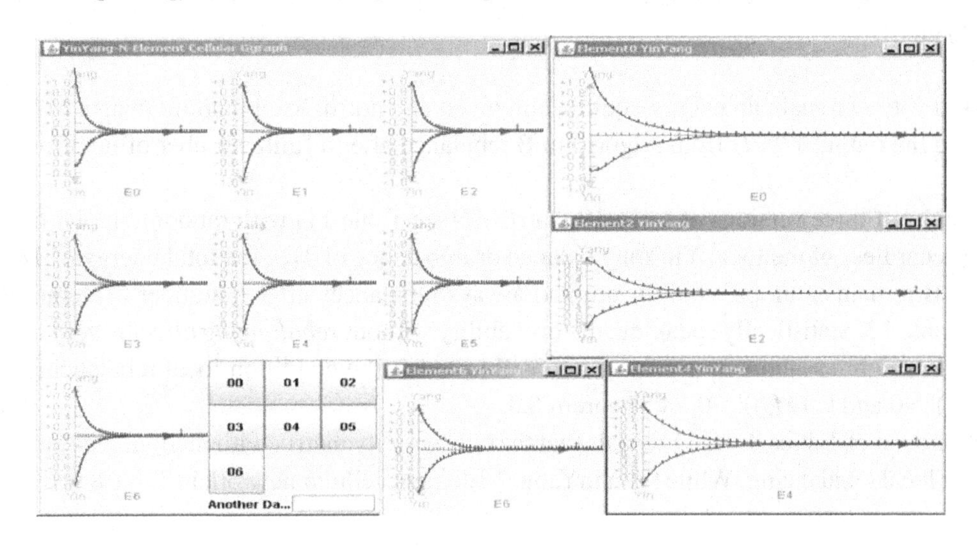

Figure 6. Equilibrium-based dimensional View of Figure 5 (Zhang et al., 2009)

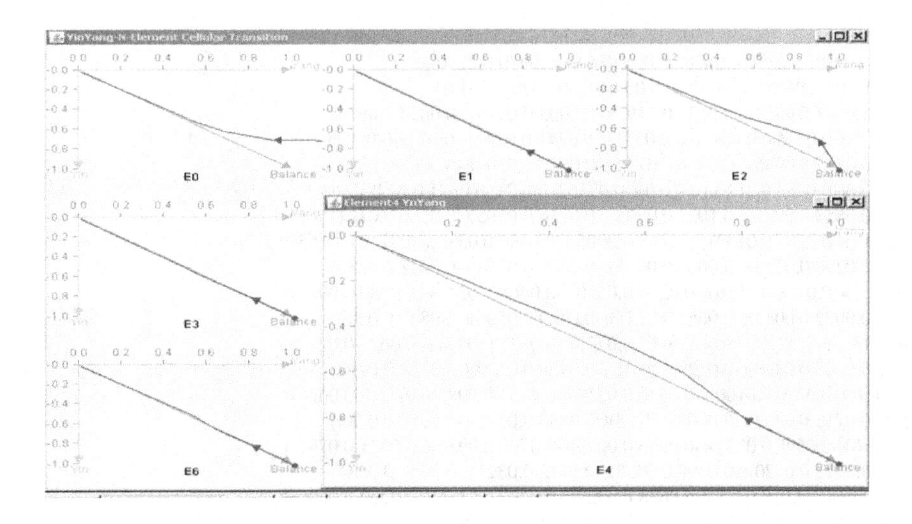

system energy and local elementary energy are decreased to 0 after 55 iterations following $Y(t+1) = Y(t) \times M(t)$. It can be observed that the local energy at each element is almost rebalanced at t = 3 with certain oscillating behavior long before total energy reaches zero. Evidently, the rebalancing condition of Theorem 8.3 is satisfied. Namely we have, $\forall t$, $|\varepsilon^-(M_{*j})| > 0$ and $|\varepsilon^+(M_{*j})| > 0$. The decrease of energy can be used to model *degenerating (or aging) processes*. It can also be used to model molecular or genetic agent population control.

Simulation and Regulation of Random YinYang-N-Element BQCA

With Algorithm A and B (Ch. 8, Table 1) any N×N bipolar connectivity matrices with random weights can be created. Table 11 shows a 10×10 bipolar matrix which meets the equilibrium condition $[|\varepsilon|(M_{k*}) \equiv |\varepsilon|(M_{*j}) \equiv 1.0]$. Now two interesting questions on the output of $Y(t+1) = Y(t) \times M$ (Algorithm B in Ch. 8, Table 1) are:

(1) Would $Y(t+1)$ remain an energy equilibrium given any normalized random matrix M?
(2) Would the output $Y(t+1)$ from Algorithm B rebalance after a finite number of iterations?

Test results of three versions of a 10×10 matrix M (see Table 11) with random bipolar link weights show that regardless of the local YinYang balance or imbalance of $Y(t)$, the total energy of $Y(t+1)$ from Algorithm B remains a global equilibrium and always rebalances after a number of iteration. This is shown in Table 12. Statistically speaking, the probability without *rebalancing* effect is very low because random link weight assignment is almost impossible to fail the condition local rebalancing condition $\forall t$, $|\varepsilon^-(M_{*j})| > 0$ and $|\varepsilon^+(M_{*j})| > 0$, of Theorem 8.3.

A comparison of Tables 8 and 11 reveals that the connectivity matrix with random link weights shows much faster local rebalancing. While the YinYang-7-Element cellular network in Table 8 with predefined

Table 11. A normalized random 10 × 10 connectivity matrix $\forall i,j$, $[|\varepsilon|(M_{i*}) \equiv |\varepsilon|(M_{*j}) \equiv 1.0]$ (Zhang et al., 2009)

[(-0.1217 0.0702)(0.0 0.0)(-0.0350 0.0)(-0.0648 0.0864)(-0.0080 0.2290)
(0.0 0.0742)(0.0 0.0848)(0.0 0.2088)(0.0 0.0047)(-0.0079 0.0044)]
[(-0.0437 0.0366)(-0.0504 0.0564)(-0.0538 0.1193)(-0.0003 0.0919)(-0.0664 0.0616)
(-0.0720 0.0105)(-0.0846 0.0348)(0.0 0.0531)(-0.0241 0.0142)(-0.0278 0.0987)]
[(-0.0067 0.0373)(-0.0817 0.0946)(-0.0916 0.0134)(-0.0033 0.0504)(-0.0708 0.0313)
(-0.0056 0.0060)(-0.0238 0.2120)(0.0 0.0)(-0.0756 0.0892)(-0.0931 0.0138)]
[(-0.0065 0.0910)(-0.0650 0.0891)(-0.1032 0.0875)(-0.0756 0.0505)(-0.0266 0.0121)
(-0.0051 0.1110)(-0.0308 0.0133)(-0.0248 0.0905)(-0.0486 0.0383)(-0.0140 0.0165)]
[(-0.0503 0.0172)(-0.0290 0.0423)(-0.0007 0.0825)(-0.0569 0.0274)(-0.0418 0.0230)
(-0.1032 0.1063)(-0.0178 0.0506)(-0.0236 0.0630)(-0.0468 0.0242)(-0.0980 0.0955)]
[(-0.0934 0.0763)(-0.0022 0.0043)(-0.0651 0.0135)(-0.0420 0.0506)(-0.0832 0.0503)
(-0.0270 0.0113)(-0.0351 0.0780)(-0.0817 0.0702)(-0.0828 0.0401)(-0.0677 0.0252)]
[(-0.0256 0.0709)(-0.0938 0.0490)(-0.0329 0.0901)(-0.0458 0.0139)(-0.0142 0.0290)
(-0.0676 0.1168)(-0.0270 0.0100)(-0.0830 0.0213)(-0.1161 0.0085)(-0.0218 0.0626)]
[(-0.0660 0.0137)(-0.0713 0.0263)(-0.0308 0.0118)(-0.0880 0.0538)(-0.0056 0.0784)
(-0.0447 0.0466)(-0.0598 0.0237)(-0.0966 0.0079)(-0.0700 0.0936)(-0.0321 0.0793)]
[(-0.0041 0.0643)(-0.0823 0.0720)(-0.0490 0.0058)(-0.0612 0.0321)(-0.0080 0.0684)
(-0.0695 0.0565)(-0.0838 0.0328)(-0.0638 0.0429)(-0.0497 0.0136)(-0.0799 0.0600)]
[(-0.0696 0.0350)(-0.0013 0.0888)(-0.0350 0.0792)(-0.0754 0.0298)(-0.0711 0.0210)
(-0.0585 0.0076)(-0.0820 0.0153)(-0.0284 0.0404)(-0.0771 0.0829)(-0.0179 0.0838)]

Table 12. Energy rebalancing with the 10×10 connectivity matrix in Table 11 (Zhang et al., 2009)

t	E1 E2 E3 E4 E5 E6 E7 E8 E9 E10	Energy equilibrium
0	[(0.0 200.0)(-100.0 100.0)(-100.0 100.0)(-100.0 100.0)(-100.0 100.0) (-100.0 100.0)(-100.0 100.0)(-100.0 100.0)(-100.0 100.0)(-100.0 100.0)]	2000.0
1	[(-105.1484 94.8516)(-100.0 100.0)(-103.4954 96.5046)(-97.8443 102.1557)(-77.9012 122.0988) (-92.5839 107.4161)(-91.5168 108.4832)(-79.1157 120.8843)(-99.5279 100.4721)(-100.3496 99.6504)]	2000.0
2	[(-101.1853 98.8147)(-101.0405 98.9595)(-98.1035 101.8965)(-101.8977 98.1023)(-100.0018 99.9982) (-99.7331 100.2669)(-100.9868 99.0132)(-102.5404 97.4596)(-101.3483 98.6517)(-98.7705 101.2294)]	2000.0
3	[(-100.0742 99.9258)(-99.7469 100.2531)(-100.0545 99.9455)(-99.8802 100.1198)(-100.6555 99.3445) (-100.3268 99.6732)(-99.5522 100.4478)(-100.1012 99.8988)(-99.8593 100.1407)(-100.2884 99.7116)]	2000.0
4	[(-99.9182 100.0818)(-100.0479 99.9521)(-100.0065 99.9935)(-99.9693 100.0307)(-99.9719 100.0281) (-99.9706 100.0294)(-100.0587 99.9413)(-100.0413 99.9587)(-100.0334 99.9666)(-99.9699 100.0300)]	2000.0
5	[(-100.0064 99.9935)(-99.9917 100.0083)(-100.0050 99.9950)(-100.0014 99.9986)(-99.9908 100.0092) (-99.9922 100.0078)(-99.9881 100.0119)(-99.9743 100.0257)(-99.9947 100.0053)(-100.0061 99.9939)]	2000.0
7	[(-100.0007 99.9993)(-100.0022 99.9978)(-99.9988 100.0012)(-100.0009 99.9991)(-99.9990 100.0010) (-100.0003 99.9996)(-100.0022 99.9978)(-100.0039 99.9961)(-100.0016 99.9984)(-99.9981 100.0018)]	2000.0
8	[(-100.0001 99.9999)(-99.9995 100.0004)(-100.0 100.0)(-100.0 100.0)(-100.0007 99.9993) (-100.0002 99.9998)(-99.9995 100.0004)(-99.9997 100.0002)(-99.9997 100.0002)(-100.0003 99.9996)]	2000.0
9	[(-99.9999 100.0)(-100.0001 99.9999)(-100.0 100.0)(-99.9999 100.0)(-99.9999 100.0) (-100.0 100.0)(-100.0001 99.9999)(-100.0001 99.9999)(-100.0 99.9999)(-99.9999 100.0)]	2000.0
10	[(-100.0 100.0)(-100.0 100.0)(-100.0 100.0)(-100.0 100.0)(-100.0 100.0) (-100.0 100.0)(-100.0 100.0)(-100.0 100.0)(-100.0 100.0)(-100.0 100.0)]	2000.0

classical nourishing and regulating cycles needs 27 iterations to rebalance, the YinYang-10-Element connectivity matrix with random link weights in Table 11 only needs 10 iterations to rebalance. Similarly it can be shown that a random connectivity matrix also leads to faster degenerating, faster growing, and less oscillating effects with non-equilibrium conditions. These observations suggest that the random family of BCNs is more responsive to disturbance than the predefined classical family. These findings provide a basis for further investigation. The bipolar energy changes of the elements can be visualized from the YinYang energy wave forms shown in Figures 7 and 8. From the two figures it is clear that element zero (E0) is the source of irregularity or cause of unbalance that has ripple effects on the other elements. Among the affected elements, E4 and E7 were more disturbed than the other elements. This

Figure 7. YinYang energy wave forms for the bipolar elements of Table 12 (Zhang et al., 2009)

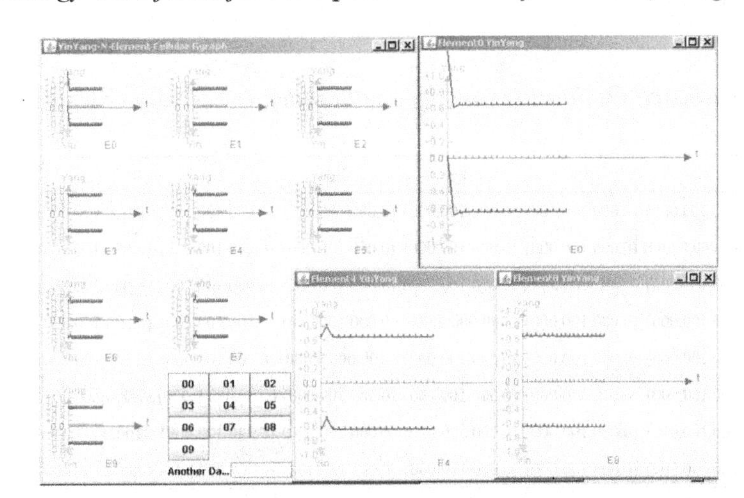

Figure 8. Equilibrium-based dimensional View of Figure 7 (Zhang et al., 2009)

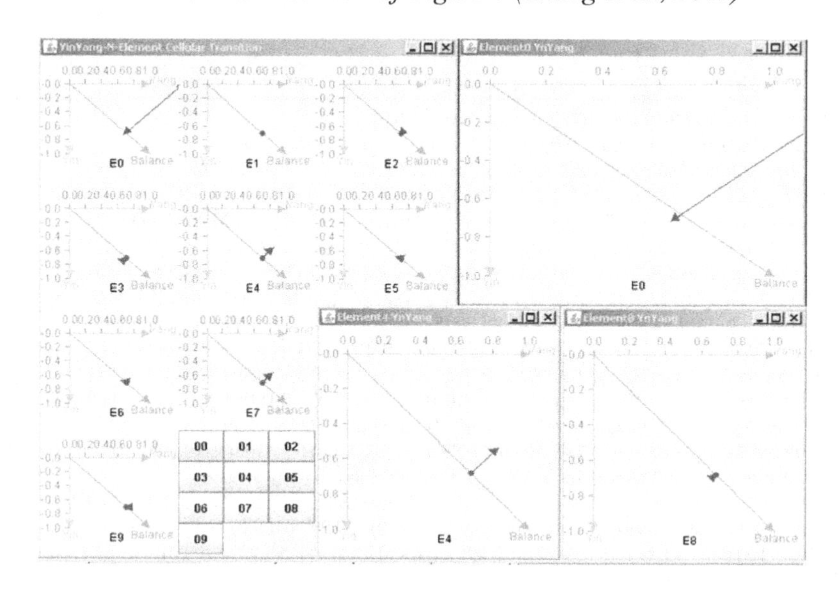

can be more clearly visualized with the dimensional view in Figure 8. It is clear that all elements tend to reach a local equilibrium after the disturbance based on the condition $\forall t$, $|\varepsilon^-(M_{*j})| > 0$ and $|\varepsilon^+(M_{*j}))| > 0$, of Theorem 8.3.

Simulation and Regulation of Oscillatory Discrete BQCA

It is pointed out in (Vasudevan, Tong & Steitz, 2007) that genetic repression and activation abilities can oscillate and form either a global YinYang equilibrium or a quasi- or non-equilibrium. BQLA provides a unique computational basis for bipolar cellular interaction and oscillation in gene regulation networks. Table 13 and Figure 9 show an equilibrium-based one by one YinYang oscillation sequence with global energy equilibrium. In this case, the energy total of the system remained in an equilibrium (700) but local elements remained imbalanced and system stability remained 0. This is realizable by violating the

Table 13. Equilibrium-Based Oscillation (one by one in order 1->2->3..) (Zhang et al., 2009)

(t)	E_1	E_2	E_3	E_4	E_5	E_6	E_7	Energy $\|\varepsilon\|$	Stability
(0)	**(-100.000 0.000)**	(0.000 100.000)	(0.000 100.000)	(0.000 100.000)	(0.000 100.000)	(0.000 100.000)	(0.000 100.000)	700	0.0
(1)	(0.000 100.000)	**(-100.000 0.000)**	(0.000 100.000)	(0.000 100.000)	(0.000 100.000)	(0.000 100.000)	(0.000 100.000)	..	..
(2)	(0.000 100.000)	(0.000 100.000)	**(-100.000 0.000)**	(0.000 100.000)	(0.000 100.000)	(0.000 100.000)	(0.000 100.000)	..	..
(3)	(0.000 100.000)	(0.000 100.000)	(0.000 100.000)	**(-100.000 0.000)**	(0.000 100.000)	(0.000 100.000)	(0.000 100.000)	..	..
(4)	(0.000 100.000)	(0.000 100.000)	(0.000 100.000)	(0.000 100.000)	**(-100.000 0.000)**	(0.000 100.000)	(0.000 100.000)	..	..
(5)	(0.000 100.000)	(0.000 100.000)	(0.000 100.000)	(0.000 100.000)	(0.000 100.000)	**(-100.000 0.000)**	(0.000 100.000)	..	..
(6)	(0.000 100.000)	(0.000 100.000)	(0.000 100.000)	(0.000 100.000)	(0.000 100.000)	(0.000 100.000)	**(-100.000 0.000)**	..	..
(7)	**(-100.000 0.000)**	(0.000 100.000)	(0.000 100.000)	(0.000 100.000)	(0.000 100.000)	(0.000 100.000)	(0.000 100.000)	700	0.0

Figure 9. Oscillation wave forms for the bipolar elements of Table 13 (Zhang et al., 2009)

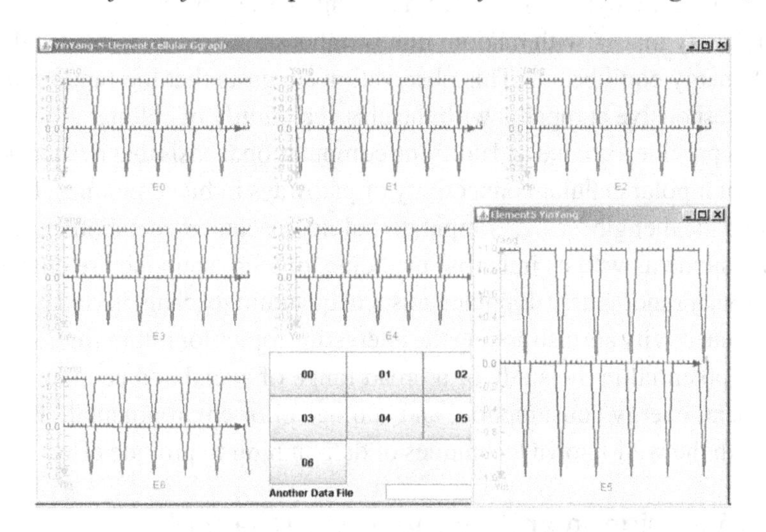

condition for rebalancing in Theorem 8.3. Specifically, the link weights in the connectivity matrix M in this case do not satisfy the condition, $\forall t$, $|\varepsilon^{-}(M_{*j})| > 0$ and $|\varepsilon^{+}(M_{*j}))| > 0$. Such oscillation can be used for simulating molecular or genetic repression and activation (Vasudevan, Tong & Steitz, 2007). Moreover, an equilibrium process can be carried by oscillating wave forms through superposition for simulating more complex biological processes.

ANALYSIS AND APPLICATIONS

Simulation Data Analysis

Application of YinYang-5-Elelemt BQCA and BQBE in TCM has been discussed and YinYang-N-Element BQCA simulation and regulation have been presented in the early sections. The results show that the properties of bipolar fusion, diffusion, interaction, oscillation, and quantum entanglement properties of BQCA enable unified cellular modeling of biological interactions at the system, molecular, and genetic levels. Metabolic nourishing and regulating relations, global and local equilibrium and non-equilibrium conditions, and discrete oscillating conditions established and axiomatically proved in last chapter have all been simulated successfully. Computer simulations of equilibrium and non-equilibrium processes have shown that the new approach provides a basic YinYang cellular network architecture for research on growing and degenerating processes as well as equilibrium and non-equilibrium processes such as molecular or genetic regulation and neural dynamics (Zhang, 2009a, 2009b). The explicit YinYang property enables the simulation of repression and activation behaviors with bipolar interactive and oscillatory properties while maintaining global energy equilibrium. Such bipolar properties cannot be explicitly captured with classical (unipolar) linear algebra and unipolar cellular automata.

A comparison of seven of the classical networks (N ranging from 5 to 11) with predefined nourishing-regulating cycles reveals that YinYang-5-Elements (the theoretical foundation of Traditional Chinese Medicine (TCM)) form the minimum structure of its kind and exhibit highest stability among the classical family of YinYang-N-Element cellular networks.

A comparison between randomly connected YinYang-N-Element BQCA with the classical family reveals that a connectivity matrix with random link weights shows much faster local rebalancing while maintaining global energy equilibrium. This observation suggests that the random family of BQCA is less stable but more responsive compared with the classical family of cellular structures.

BQLA and BQCA provide a unique and unifying computational basis that merits further investigation into different forms of bipolar cellular connectivity or pathways in bioeconomics, bioecology, quantum bioinformatics, biomedical engineering, computational neuroscience, metabolism, growing, degeneration, and nanobiomedicine as well as in herbal medicine, qigong, acupuncture channels, and meridian systems in TCM. Bipolar randomness, bipolar statistical quantum mechanics, and 3-dimensional bipolar molecular cellular connectivity are also among the interesting topics for future studies. In addition, BQLA and BQCA have the potential in the study of a broad range of social and economic problems, such as economic development, energy consumption, and ecological or environmental protection. BQLA can also be combined with the well-known techniques of data mining to provide effective decision support.

BQBE and BQCA in Qigong and Acupuncture Channels

The *Chinese meridian* or *jingluo* system (WHO, 1993) is a concept central to TCM such as acupuncture, taiji, and qigong. According to the theory of jingluo, there are channels along which the vital energy or *qi* of the psychophysical system is believed to flow. Acupuncture achieves its effects by regulating and, ideally, balancing the energy running through a network of complex bodily patterns. Some Western literature claims that, *"there is no physically verifiable anatomical or histological basis for the existence of acupuncture points or meridians"* (Wikipedia – Meridian).

The above is a troubling observation to Western science. First,

The clinical efficacy of acupuncture has been reviewed elsewhere. The scientific merits and physiological phenomena achieved through acupuncture have been demonstrated through the use of various types of advanced technologies, including myoneural electrophysiology studies, radioactive tracer survey and imaging, and single-photon emission computed tomography. Other researches, including thermographic studies and kinesiology phenomena, have demonstrated clinical and physiological responses to acupuncture, although the findings remain controversial. (Mok, 2000)

Secondly, *"The growing acceptance of acupuncture is undeniable as seemingly greater numbers of physicians begin to take interest. The National Institutes of Health currently endorses the application of medical acupuncture in many areas. Since February 1996, the Food and Drug Administration has classified acupuncture devices as general medical, rather than investigational, devices"* (Mok, 2000). Thirdly, if acupuncture assisted surgery has been performed live under TV camera, the claim *"there is no physically verifiable anatomical or histological basis for the existence of acupuncture points or meridians"* has to be an error.

After all, with BQCA and BQBE, it can be argued that the words *"anatomical"* and *"histological"* in Western medical science clearly belong to unipolar truth-based bottom-up approach to medicine with a focus on local and partial evidence, the science of TCM follows an equilibrium-based top-down approach to bioeconomics with a focus on the global holistic evidence. The effectiveness of acupuncture should be admissible as global holistic evidence. For instance, with the truth-based bottom-up approach, it would be difficult to diagnose the effect of global warming or economics reception to human health including mental health; with the equilibrium-based top-down approach such effects are logically definable because (cooling, warming), (recession, expansion) and (depression, mania) are typical bipolar predicates.

The simulation results presented in the earlier sections clearly show that equilibrium and non-equilibrium as well as growing and degenerating conditions of a BQCA are largely based on the connectivity matrix. The biological connectivity can be considered as biological pathways with biological fusion, diffusion, interaction, oscillation, nourishing, regulation, quantum entanglement, and balancing properties governed by organizational efficiency, effectiveness, and harmony of the human body. Acupuncture may well improve the global organizational rebalancing and harmony without any local anatomical or histological evidence. After all if the organizational balance is improved through the bio-electromagnetic or bio-quantum level, they may not have manifest at the local level in anatomical or histological forms for some health problems. Furthermore, the bio-electromagnetic or bio-quantum effects at the molecular, genetic, neurodynamic, and system levels are so far unclear for many health problems in Western medical science itself.

Not only are Western scientists troubled, TCM practitioners are also troubled by the lack of scientific basis in TCM. One tendency of TCM practitioners is to fit TCM entirely into Western medical science. It is argued (Corso & Araujo, 2007; Yung, 2008) that, for TCM to be more acceptable to Western scientific thinking, it is critical to stress the importance of experimental verification of any theory. The author of this monograph agrees with the emphasis on experimental verification of any TCM theory and admires the wonderful work of the birdcage model (Yung, 2004a, 2004b, 2005a, 2005b, 2005c). It is believed, however, that such emphasis should be open-minded because TCM is natural medicine and nature is an open system with many uncharted territories in mathematical, social, physical, and life sciences. For instance, nano-medicine was unimaginable a few decades ago but now a reality. Bipolar mathematical abstraction has provided BDL as logic of YinYang; BQLA and BQCA have brought a holistic mathematical physics or biophysics theory into the world. The Bipolar models are expected to play a unifying role for the integration of bioinformatics, Western medicine, and TCM.

Indeed, many have wondered whether TCM is science. Since without equilibrium at the system, genetic, and molecular levels any mental or physical disorder would be "big bang" from nowhere, TCM can be considered part of equilibrium-based holistic natural medicine. With the holistic science of BQBE, YinYang equilibrium and harmony can remain a central theme in TCM that does not have to fit entirely into Western medicine and lose its philosophical complementary identity.

It is the author's view that

1. Equilibrium-based YinYang bipolar philosophy is scientific because equilibrium is the only physical concept that can form a chicken-egg paradox with the mighty universe.
2. YinYang equilibrium should be central in TCM theory and the science of TCM should be complementary to Western medical science instead of just fitting into the Western scientific thinking.
3. The science of TCM should develop itself through the interplay of equilibrium-based theoretical hypotheses, experimental tests, and modern life sciences.

Based on BQLA and BQCA, a multitier YinYang-N-Element cellular network architecture for the Chinese meridian system is proposed as in Figure 10. On the top tier we have the vital spots for acupuncture, acupressure, qigong, and electromagnetic therapeutics. Through these spots, typically acupoints, the meridian system can be regulated. The middle layer is the connectivity matrix of YinYang-N-Elements under consideration. The connectivity matrix can be considered the Jingluo system or the meridian system

Figure 10. A multitier YinYang-N-Element cellular network architecture for the Chinese meridian system (Zhang, Chen & Zang, 2009)

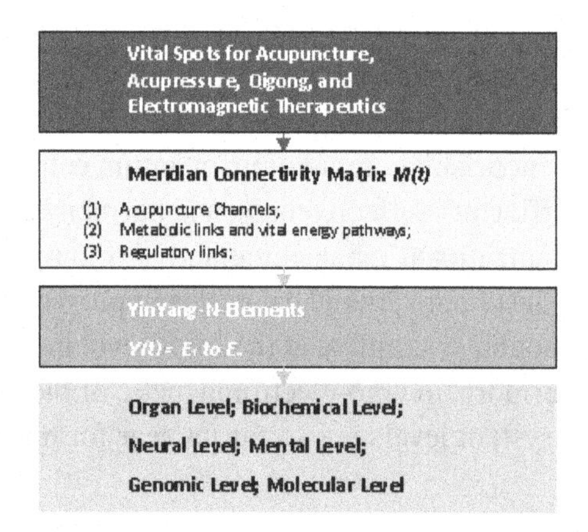

which consists of metabolic links and regulating links. Such links serve as channels of vital energy or qi for the YinYang-N-Elements under consideration. Using modern terminology the connectivity matrix provides organizational bioenergy which can make the system stronger, weaker, balanced, unbalanced, harmonic, or malfunctioning.

With the multitier architecture, regulation could be through acupuncture, acupressure, qi going, electromagnetic therapeutics, or other means. The regulation effects could be in the forms of (1) rebalancing the connectivity matrix *M(t)*, (2) strengthening some link of it, or (3) weakening some link of it. Rebalancing is equivalent to TIAO (adjust) in acupuncture and herbal medicine, strengthening is equivalent to BU (energize) and weakening is equivalent to XIE (release or relax). The effects on the connectivity matrix in turn lead to effects on the YinYang-N-Elements at different levels such as the five elements at the organ level, antidepressant stimulation at the biochemical level, epigenetic effect at the genomic level, or metabolic effect at the molecular level. All effects are for bringing the system into equilibrium and harmony.

To the author's knowledge the multitier BQCA architecture is mathematically and physically unique. It is the first bipolar TCM model based on the strengths and equilibrium of the Yin and the Yang vital energies (or the Yin and Yang of Qi). Since equilibrium is a well-known holistic scientific concept and the only physical concept that can form a philosophical chicken-egg paradox with the mighty universe (no one knows exactly which one created the other), the equilibrium-based architecture is intended to subsume all types of bipolar equilibria such as negative-positive electromagnetic fields, action-reaction forces, negative-positive poles of capacitors or batteries. As an open-world and open-ended model, it does not rule out any future possible mathematical or physical development such as the incorporation of quantum biophysics theories, YinYang genomic theories (Shi et al., 1991), and other theories at the molecular and systems levels for equilibrium-based global regulation.

While basic ideas have been discussed, further testing and development of the new model are left for future research with open questions in TCM such as:

1. How to use BQLA and YinYang-N-Element cellular networks as an analytical model for the electromagnetic birdcage theory of the Chinese meridian system (Yung, 2004a, 2004b, 2005a, 2005b, 2005c)?

2. How to use BQLA and the multitier cellular architecture to model acupuncture and qigong processes and effects?
3. How to model the meridian system with a connectivity matrix?
4. How to characterize the effect of a holistic treatment with the meridian system?
5. How to connect the different layers and systematically model the meridian system?
6. How to incorporate the meridian system with genomic, molecular, metabolic, and bioeconomics systems (Zhang, 2009d, 2010)?
7. How to incorporate the meridian system with cognitive mapping and computational neuroscience (Zhang et al., 1989, 1992, 2007; Zhang, 2007, 2009a, 2009b, 2009c)?
8. How to apply YinYang bipolar relativity theory (Zhang, 2009c) in the meridian system?
9. How to quantify the Yin and Yang of the meridian system and integrate it into the general picture of scientific unification?

BQCA and BQBE in Immunology

YinYang-N-Element bipolar combinatorics of BQCA with organizational and equilibrium energies provides an attractive model for bioeconomics in immunology. In particular, innate immunity, in contrast to adaptive immunity, is the first line of defense for organisms to overcome invasive pathogens (Jaeger, Chen & Zhang, 2009). It is capable of launching immediate attacks against pathogens and is responsible for staging later adaptive responses from the host. The innate immune system employs a limited number of germline-encoded receptors, called pathogen recognition receptors (PRRs), to detect a multitude of pathogen-associated molecular patterns (PAMPs) in microbes ranging from bacteria to viruses, and to distinguish these foreign agents from self-patterns (Akira, Uematsu & Takeuchi, 2006). Despite major research efforts, the principles that govern the perceptive and cognitive mechanisms of PRRs remain largely elusive.

The major receptor family responsible for recognizing PAMPs is the Toll-like receptor (TLR) family. An interesting observation relates to the innate immune system: It has been estimated that most mammalian species have between ten and fifteen types of Toll-like receptors. Thirteen TLRs (named simply TLR1 to TLR13) have been identified in humans and mice, and equivalent forms of many of these have been found in other mammalian species (Du *et al.,* 2000; Chuang & Ulevitch, 2000; Tabeta, Georgel & Janssen, 2004). Each TLR can engage different sets of ligands and induce differential and overlapping responses. Many immune cells have expressions of TLRs, including macrophages, dendritic cells, and mast cells.

Toll-like receptors are an evolutionarily stable defense concept that many species have adopted. The types of Toll-like receptors used, and their number in particular, varies only slightly among species. This observation led to the YinYang-N-Element approach to innate immunology (Jaeger, Chen & Zhang, 2009), the YinYang approach assumes that the most effective immune response happens in the state of YinYang bipolar equilibrium.

Based on this hypothesis, the ten to fifteen types of TLRs of most mammalian species can be simulated as YinYang-N-Element BQCA networks where N ranges from 10 to 15 (Jaeger, Chen & Zhang, 2009). Typically, Figure 11 shows a classical YinYang-13-Element BQCA with 6-fan-in and 6-fan-out connections (including the reflexive link). With YinYang-N-Element combinatorics as discussed in last section, a classical YinYang-N-Element BQCA such as the one in Figure 11 would become a particular case in its *k*-group. The number *k* is the number of fan-in or fan-out connections. When $k = N/2$, its *k*-

Figure 11. Maximum case 6-in and 6-out classical BCN

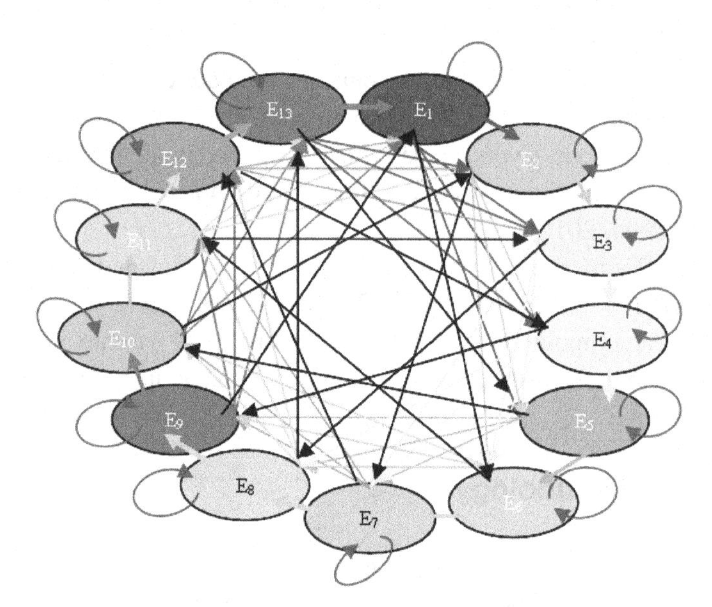

group enjoys the maximum number of possible different configurations. In this case, the number *k* can be hypothesized an optimal number for the BQCA to maximize the flexibility of its immune response in recognizing and launching immediate attacks against different pathogens. It is easy to verify that, when N = 10, k=5; when N=11, k=5 or 6; when N = 12, k=6; when N=13, k=6 or 7; when N= 14, k =7; and when N=15, k=7 or 8.

Application and Research Areas

BQBE is applicable in any application where bipolar equilibrium-based regulation is involved. Typical application areas include molecular and genetic regulation, macroeconomics regulation, and computational neuroscience.

The following is a list of open research topics in uncharted territories:

1. Quantify bipolar equilibrium energy based on (antimatter, matter) particle pairs.
2. Apply YinYang-N-Element BQCA in macroeconomics for global regulation.
3. Apply YinYang-N-Element BQCA for management, business decision support, and multiagent coordination.
4. Apply YinYang-N-Element BQCA in computational neuroscience.
5. Apply YinYang-N-Element BQCA in genomic and molecular network regulation.
6. Apply YinYang-N-Element BQCA in socioeconomic regulation for a harmonic society.
7. Clarify how a time-variant bio-connectivity matrix M(t) can enable Qi accumulation, circulation, and release in large amount within a short time period in Qi Gong exercises.
8. Clarify how YinYang-N-Element BQCA can be used to combine acupuncture channels (Jingluo) with the WuXing system.

RESEARCH TOPICS

This chapter has been focused on YinYang-N-Element bioeconomics simulation and regulation at the system level in general. Domain dependent research topics include but are not limited to the following list:

- Apply BQLA and BQCA in system biology.
- Apply BQLA and BQCA in neurobiological simulation and regulation.
- Apply BQLA and BQCA in genomics simulation and regulation.
- Apply BQLA and BQCA in molecular interaction simulation and regulation.
- Further develop BQLA and BQCA for cancer and other disease simulation and regulation.
- Apply BQLA and BQCA in macroeconomics simulation and regulation.
- Apply BQLA and BQCA in cognitive informatics simulation and regulation.
- Apply BQLA and BQCA in social dynamics simulation and regulation.

SUMMARY

A theory of bipolar quantum bioeconomics has been presented based on BQLA and BQCA. Computer simulation and visualization of equilibrium, non-equilibrium, and oscillation properties as well as growing and degenerating processes of YinYang-N-Element BQCA have been presented. Diagnostic decision support in TCM has been illustrated. Since YinYang-WuXing is the fundamental theory of TCM, the simulation results provide a unique scientific basis for future research and education in herbal medicine, Qi (vital energy), QiGong, JingLuo (acupuncture channels) and acupuncture. From a modern bioinformatics perspective, it provides a scientific basis for simulation and regulation in genomics, bioeconomics, metabolism, computational biology, aging, artificial intelligence, and biomedical engineering. It is also expected to serve as a mathematical basis for biosystem inspired socioeconomics, market analysis, business decision support, multiagent coordination and global regulation.

ACKNOWLEDGMENT

This chapter reuses part of published material in the two articles: (1) Zhang, W. –R. & Chen, S. S. (2009). Equilibrium and Non-Equilibrium Modeling of YinYang Wuxing for Diagnostic Decision Analysis in Traditional Chinese Medicine. *World Scientific Publishing (WSP): Int'l J. of Infor. Tech. and Decision Making (IJITDM), Vol. 8, No. 3 (2009)* pp529-548; (2) Zhang, W. –R., H. J. Zhang, Y. Shi & S. S. Chen (2009). Bipolar Linear Algebra and YinYang-N-Element Cellular Networks for Equilibrium-Based Biosystem Simulation and Regulation. *World Scientific Publishing (WSP): Journal of Biological Systems (JBS), Volume: 17, Issue: 4 (2009)* pp. 547-576. Permission to reuse is acknowledged.

REFERENCES

Akira, S., Uematsu, S., & Takeuchi, O. (2006). Pathogen recognition and innate immunity. *Cell, 124*(4), 783–801. doi:10.1016/j.cell.2006.02.015

Bauer, M. (2006). The final days of traditional beliefs? Part one. *Chinese Medicine Times, 1*(4). Retrieved from http://www.chinesemedicinetimes.com/section.php?xSec=122

Chuang, T. H., & Ulevitch, R. J. (2000). Cloning and characterization of a sub-family of human toll-like receptors: hTLR7, hTLR8, and hTLR9. *European Cytokine Network, 11*(3), 372–378.

Clark, C. W. (1987). Bioeconomics. *The new Palgrave: A dictionary of economics*, vol. 1. (pp. 245-46).

Corso, G., & Araujo, A. I. (2007). Comment on birdcage model for the Chinese meridian system. *The American Journal of Chinese Medicine*, *35*(2), 365–367. doi:10.1142/S0192415X07004886

Du, X., Poltorak, A., Wei, Y., & Beutler, B. (2000). Three novel mammalian toll-like receptors: Gene structure, expression, and evolution. *European Cytokine Network, 11*(3), 362–371.

Ebrey, P. (1993). *Chinese civilization: A sourcebook* (2nd ed.). New York: Free Press.

Einstein, A. (1934). *On the method of theoretical physics. The Herbert Spencer lecture, delivered at Oxford, June 10, 1933*. Amsterdam: Querido Verlag.

Emran, F., Florens, L., Ma, B., Swanson, S. K., Washburn, M. P., & Hernandez, N. (2006). A role for Yin Yang-1 (YY1) in the assembly of snRNA transcription complexes. *Gene*, *1*(377), 96–108. doi:10.1016/j.gene.2006.03.012

Gordon, H. S. (1954). The economic theory of a common-property resource: The fishery. *The Journal of Political Economy*, *62*(2), 124–142. doi:10.1086/257497

Gore, J., & van Oudenaarden, A. (2009). Synthetic biology: The yin and yang of nature. *Nature*, *457*(7227), 271–272. doi:10.1038/457271a

Hodgson, G. (1992). The reconstruction of economics: Is there still a place for neoclassical theory? *Journal of Economic Issues*, *26*(3), 749–767.

Jaeger, S., Chen, S.-S., & Zhang, W.-R. (2009). TCM in innate immunity. *Proceedings of International Joint Conference on Bioinformatics, Systems Biology and Intelligent Computing (IJCBS)*. (pp. 397-401). Shanghai, China.

Kim, J. D., Faulk, C., & Kim, J. (2007). Retroposition and evolution of the DNA-binding motifs of YY1, YY2 and REX1. *Nucleic Acids Research*, *35*(10), 3442–3452. doi:10.1093/nar/gkm235

Li, F. (1987). Functional structure mode of human body and YinYang -Wuxing Equations. *Physica Scripta*, *36*, 966–969. doi:10.1088/0031-8949/36/6/015

Maciocia, G. (1989). *The foundations of Chinese medicine: A comprehensive text for acupuncturists and herbalists*. Churchill LivingStone.

McKelvey, B. (2004). Toward a 0th law of thermodynamics: Order-creation complexity dynamics from Physics and Biology to Bioeconomics. *Journal of Bioeconomics*, *6*(1), 65–96. doi:10.1023/B:JBIO.0000017280.86382.a0

Mohammadian, M. (2009). *Biography*. Retrieved from http://www.scienceofbioeconomics.com/author.asp

Mok, Y.-P. (2000). Acupuncture-assisted anesthesia. *Medical Acupuncture – A Journal For Physicians By Physicians, 12*(1). Retrieved from http://www.medicalacupuncture.com/aama_marf/journal/vol12_1/anesthesia.html

Nash, J. (1950). Equilibrium points in n-person games. *Proceedings of the National Academy of Sciences of the United States of America, 36*(1), 48–49. doi:10.1073/pnas.36.1.48

Ou, B., Huang, D., Hampsch-Woodill, M., & Flanagan, J. A. (2003). When East meets West: The relationship between Yin-Yang and antioxidation-oxidation. *The FASEB Journal, 17*, 127–129. doi:10.1096/fj.02-0527hyp

Shi, Y., Seto, E., Chang, L.-S., & Shenk, T. (1991). Transcriptional repression by YY1, a human GLI-Kruppel-related protein, and relief of repression by adenovirus E1A protein. *Cell, 67*(2), 377–388. doi:10.1016/0092-8674(91)90189-6

Tabeta, K., Georgel, P., & Janssen, E. (2004). Toll-like receptors 9 and 3 as essential components of innate immune defense against mouse cytomegalovirus infection. *Proceedings of the National Academy of Sciences of the United States of America, 101*(10), 3516–3521. doi:10.1073/pnas.0400525101

The Nobel Foundation. (1994). Laureates of 1994. Retrieved from http://nobelprize.org/nobel_prizes/economics/laureates/1994

Wade, N. (2010, June 12). A decade later, human gene map yields few new cures. *New York Times*. Retrieved from http://www.nytimes.com/2010/06/13/health/research/13genome.html?partner=rss&emc=rss

WHO (World Health Organization). (1993). *Standard acupuncture nomenclature* (2nd ed.).

Wikipedia. (2010). *Chinese medicine.* Retrieved from http://en.wikipedia.org/wiki/Meridian_(Chinese_medicine)

Yung, K. T. (2004a). A birdcage model for the Chinese Meridian system part I: A channel as a transmission line. *The American Journal of Chinese Medicine, 32*(5), 815–828. doi:10.1142/S0192415X04002417

Yung, K. T. (2004b). A birdcage model for the Chinese Meridian system part II: The Meridian system as a birdcage resonator. *The American Journal of Chinese Medicine, 33*(6), 985–997. doi:10.1142/S0192415X04002582

Yung, K. T. (2005a). A birdcage model for the Chinese Meridian system part III: Possible mechanism of magnetic therapy. *The American Journal of Chinese Medicine, 33*(4), 589–597. doi:10.1142/S0192415X05003193

Yung, K. T. (2005b). A birdcage model for the Chinese Meridian system part IV: Meridians as the primary regulatory system. *The American Journal of Chinese Medicine, 33*(5), 759–766. doi:10.1142/S0192415X05003302

Yung, K. T. (2005c). A birdcage model for the Chinese Meridian system part V: Applications to animals and plants. *The American Journal of Chinese Medicine, 33*(6), 903–912. doi:10.1142/S0192415X05003491

Yung, K. T. (2008). Reply to comment on birdcage model for the Chinese Meridian system. *The American Journal of Chinese Medicine, 36*(6), 1219–1221. doi:10.1142/S0192415X08006582

Zhang, W.-R. (2009a). Six conjectures in quantum physics and computational neuroscience. *Proceedings of 3rd International Conference on Quantum, Nano and Micro Technologies (ICQNM 2009)*, (pp. 67-72). Cancun, Mexico.

Zhang, W.-R. (2009b). YinYang Bipolar Dynamic Logic (BDL) and equilibrium-based computational neuroscience. *Proceedings of International Joint Conference on Neural Networks (IJCNN 2009)*, (pp. 3534-3541). Atlanta, GA.

Zhang, W.-R. (2009c). YinYang bipolar relativity–a unifying theory of nature, agents, and life science. *Proceedings of International Conference on International Joint Conference on Bioinformatics, Systems Biology and Intelligent Computing (IJCBS)*. (pp. 377-383). Shanghai, China.

Zhang, W.-R. (2009d). The logic of YinYang and the science of TCM–an Eastern road to the unification of nature, agents, and medicine. [IJFIPM]. *International Journal of Functional Informatics and Personal Medicine*, *2*(3), 261–291. doi:10.1504/IJFIPM.2009.030827

Zhang, W.-R., Chen, S., & Zang, H. (2009). A multitier YinYang-N-element cellular architecture for the Chinese meridian system. [IJFIPM]. *International Journal of Functional Informatics and Personal Medicine*, *2*(3), 292–302. doi:10.1504/IJFIPM.2009.030828

Zhang, W.-R., & Chen, S. S. (2009). Equilibrium and non-equilibrium modeling of YinYang Wuxing for diagnostic decision analysis in traditional Chinese medicine. *International Journal of Information Technology and Decision Making*, *8*(3), 529–548. doi:10.1142/S0219622009003521

Zhang, W.-R., Wang, W., & King, R. (1994). An agent-oriented open system shell for distributed decision process modeling. *Journal of Organizational Computing*, *4*(2), 127–154. doi:10.1080/10919399409540220

Zhang, W.-R., Zhang, H. J., Shi, Y., & Chen, S. S. (2009). Bipolar linear algebra and YinYang-N-Element cellular networks for equilibrium-based biosystem simulation and regulation. *Journal of Biological System*, *17*(4), 547–576. doi:10.1142/S0218339009002958

Zhou, Q., & Yik, J. H. N. (2006). The Yin and Yang of P-TEFb regulation: Implications for Human Immunodeficiency Virus gene expression and global control of cell growth and differentiation. *Microbiology and Molecular Biology Reviews*, *70*(3), 646–659. doi:10.1128/MMBR.00011-06

ADDITIONAL READING

Anthony, C. K., & Moog, H. (2002). *I Ching: The Oracle of the Cosmic Way*. Stow, MA: Anthony Publishing Company.

Balkin, J. M. (2002). *The Laws of Change: I Ching and the Philosophy of Life*. New York: Schocken Books.

Karcher, S. (2002). *I Ching: The Classic Chinese Oracle of Change: The First Complete Translation with Concordance*. London: Vega Books.

Moran, E., & Yu, J. (2001). *The Complete Idiot's Guide to the I Ching*. New York: Alpha Books.

Zhang, W.-R. (1996). NPN Fuzzy Sets and NPN Qualitative Algebra: A Computational Framework for Bipolar Cognitive Modeling and Multiagent Decision Analysis. *IEEE Trans. on SMC.*, *16*, 561–574.

Zhang, W.-R. (2005a). YinYang Bipolar Lattices and L-Sets for Bipolar Knowledge Fusion, Visualization, and Decision. *Int'l J. of Inf. Technology and Decision Making*, *4*(4), 621–645. doi:10.1142/S0219622005001763

Zhang, W.-R. (2006a). YinYang Bipolar Fuzzy Sets and Fuzzy Equilibrium Relations for Bipolar Clustering, Optimization, and Global Regulation. *Int'l J. of Inf. Technology and Decision Making*, *5*(1), 19–46. doi:10.1142/S0219622006001885

Zhang, W.-R. (2007). YinYang bipolar universal modus ponens (bump) – a fundamental law of non-linear brain dynamics for emotional intelligence and mental health. *Walter J. Freeman Workshop on Nonlinear Brain Dynamics, Proc. of the 10th Joint Conf. of Information Sciences*, Salt Lake City, Utah, USA, July 2007, 89-95.

Zhang, W.-R., Pandurangi, A., & Peace, K. (2007). YinYang Dynamic Neurobiological Modeling and Diagnostic Analysis of Major Depressive and Bipolar Disorders. *IEEE Transactions on Bio-Medical Engineering*, *54*(10), 1729–1739. doi:10.1109/TBME.2007.894832

KEY TERMS AND DEFINITIONS

Bioeconomics (BE): On the one hand, BE is the study of the dynamics of living resources using economic models; on the other hand, BE is the science determining the socioeconomic activity threshold for which a biological system can be effectively and efficiently utilized without destroying the conditions for its regeneration and therefore its sustainability (Clark, 1987).

Biosystem Simulation and Regulation: Simulation and regulation of a biological system at any level such as organ, genomic, molecular, or bio-electromagnetic levels.

BQBE: Bipolar quantum bioeconomics – an equilibrium-based quantum extension of BE based on YinYang bipolar relativity and bipolar quantum entanglement.

Chinese Meridian (or Jingluo): According to the theory of jingluo, there are channels along which the vital energy or *qi* of the psychophysical system is believed to flow. Acupuncture achieves its effects by regulating and, ideally, balancing the energy running through a network of complex bodily patterns.

Degenerating (or Aging) Process: The process of a system whose energy total decreases.

Equilibrium-Based Bipolar Simulation and Regulation: Simulation and regulation of Equilibrium and non-equilibrium behaviors of cellular structures with explicit Yin and Yang poles using bipolar logic, bipolar relativity, YinYang-N-Element BQLA, and/or BQCA.

Growing Process: The process of a system whose energy total increases.

Local Oscillation Under Global Equilibrium: Local bipolar oscillation at the component level of a system whose energy total remains unchanged.

Rebalancing: The process of a system to reach a new bipolar equilibrium after disturbance.

TCM: Traditional Chinese Medicine.

Chapter 10
MentalSquares:
An Equilibrium-Based Bipolar Support Vector Machine for Computational Psychiatry and Neurobiological Data Mining

ABSTRACT

While earlier chapters have focused on the logical, physical, and biological aspects of the Q5 paradigm, this chapter shifts focus to the mental aspect. MentalSquares (MSQs) – an equilibrium-based dimensional approach is presented for pattern classification and diagnostic analysis of bipolar disorders. While a support vector machine is defined in Hilbert space, MSQs can be considered a generic dimensional approach to support vector machinery for modeling mental balance and imbalance of two opposite but bipolar interactive poles. A MSQ is dimensional because its two opposite poles form a 2-dimensional background independent YinYang bipolar geometry from which a third dimension – equilibrium or non-equilibrium – is transcendental with mental fusion or mental separation measures. It is generic because any multidimensional mental equilibrium or non-equilibrium can be deconstructed into one or more bipolar equilibria which can then be represented as a mental square. Different MSQs are illustrated for bipolar disorder (BPD) classification and diagnostic analysis based on the concept of mental fusion and separation. It is shown that MSQs extend the traditional categorical standard classification of BPDs to a non-linear dynamic logical model while preserving all the properties of the standard; it supports both classification and visualization with qualitative and quantitative features; it serves as a scalable generic dimensional model in computational neuroscience for broader scientific discoveries, and it has the cognitive simplicity for clinical and computer operability. From a broader perspective, the agent-oriented nature of MSQs provides a basis for multiagent data mining (Zhang & Zhang, 2004) and cognitive informatics of brain and behaviors (Wang, 2004).

INTRODUCTION

Mind reading is a stunning achievement in brain research (Singer, 2008). Now scientists can accurately predict which of thousands of pictures a person is looking at by analyzing brain activity using functional magnetic resonance imaging (fMRI). Mind reading should shed light on how the brain processes visual information, and it might one day be used to reconstruct dreams. On the other hand, event-related potential/electroencephalography (EPR/EEG) technology is being intensively studied for event-related (versus image-related) brain research for furthering the understanding of human intelligence.

Despite numerous great achievements in brain research, some mental health problems such as *bipolar disorder* (*BPD*) remain at the top of the list of major diseases that have so far found no cure (Kessler *et al.,* 2006). The difficulty of finding a cure for BPD is even recorded in the Bible. It is described as being caused by some ghostly spirit; sometimes one patient could get several such spirits; and only Jesus could cure the disease by ordering the ghostly spirits to go away.

Modern medicine shows that BPD is psychiatric disease caused by neurobiological or genetic reasons. Since millions suffer from major depressive and bipolar disorders, the modeling, characterization, classification, and diagnostic analysis of such mental disorders bear great significance in medical and pharmaceutical research. Scientifically speaking, *bipolar neurobiological data mining* from uncharted territories bring unprecedented challenges to data and knowledge engineering.

Following the theory of bipolar agents and YinYang bipolar relativity, bipolarity as the most fundamental property of nature is not necessarily a bad concept because YinYang bipolar relativity claims that we are all bipolar, either in bipolar equilibrium or in bipolar disorder. Thus, BPD is the loss of bipolar order or equilibrium. It is neither caused by a ghostly spirit nor caused by a "mental big bang" or "black hole" from nowhere. The purpose of a therapeutic intervention is then to recover the disorder to a healthy energetic bipolar mental equilibrium such as the reciprocal mental balancing abilities denoted (self-negation, self-assertion) of a person. Psychiatric education, therefore, should stress the importance of understanding the bipolar nature of agents and promote YinYang bipolar mental equilibrium for mental health in a bipolar world.

For instance, YinYang bipolar thinking can help people maintain mental equilibrium by exercising positive thinking when depressed and exercising negative thinking when excited. Such YinYang bipolar thinking is actually common public knowledge of emotional intelligence practiced everyday by millions of people in the world. Otherwise, more people would suffer from depression or bipolar disorder.

The theory of bipolar agents and bipolar relativity opened the door for applying *mental quantum gravity* in mental health research, which can be deemed an extension of brain research with the addition of bipolar quantum entanglement and equilibrium. Theoretically, the extension brings brain dynamics and quantum gravity into the equilibrium-based logical unification of bipolar relativity. Practically, the extension provides brain research with a real world context of bipolar agents, bipolar functionality, and bipolar behaviors, respectively, such as negative and positive particles, matter and antimatter, action and reaction, depression and mania, self-negation and self-assertion abilities, etc.

Naturally, we need to ask the question: *If our brain is considered a neurobiological universe, how does the universe interact with its bipolar equilibrium and non-equilibrium context?* Thus, the bipolar context, with logically definable causality and relativity, provides a test bed for actual or potential applications in clinical psychopharmacology, computational neuroscience, psychiatry, and nano-biomedicine for mental health. Theoretically speaking, the context provides an environment for the falsifiability of quantum mind theory as well as bipolar causality and relativity.

Building upon the previous chapters, this chapter presents the theory of bipolar brain dynamics for equilibrium-based computational neuroscience and psychiatry. *MentalSquares* (MSQ) – an equilibrium-based generic dimensional approach is presented. It is shown that MSQ constitutes a *bipolar support vector machine* (*BSVM*) for mental disorder classification, diagnostic analysis, and neurobiological data mining with a number of advantages over *support vector machine* (*SVM*) defined in Hilbert space.

The remaining presentations and discussions of this chapter are organized in the following sections:

- **Bipolar disorder classification.** This section introduces the current standard classification of bipolar disorders and discusses its limitations.
- **Equilibrium-based computational neuroscience and psychiatry.** This section discusses: (1) observability of bipolar equilibrium, (2) logical transformation of brain and behavior, (3) mental fusion, mental separation, and mental squares, and (3) equilibrant, balancer, oscillator, and life extinguisher neurobiological functionalities.
- **Equilibrium-based BPD classification and analysis.** This section presents: (1) three types of major bipolar disorders, (2) scenarios in bipolar diagnostic analysis, (3) fuzzified bipolar diagnostic analysis, (4) bipolar inference for clinical psychopharmacology, and (5) unified measures.
- **Hypothesis-driven exploratory knowledge discovery about brain and behavior.** This section discusses: (1) a neurobiological engine of bipolar relativity, (2) internal and external unification, and (3) patient-drug neurobiological reaction classification and data mining.
- **Stability analysis in bipolar brain dynamics**. The focus of this section is stability analysis of bipolar brain dynamics.
- **Mixed BPD classification.** This section is focused on the classification of *mixed episodes* (or *mixed states*) of BPDs.
- **Scalability to Schizophrenia Classification**. This section discusses the scalability of *MSQs* to schizophrenia classification.
- **Other potential applications.** This section introduces the neural, genetic, and nano implications of *MSQs*.
- **Computer operability.** This section illustrates the operability of *MSQs* with simulations.
- **Research Topics**. This section lists a few research topics.
- **Summary.** This section summarizes the major points of this chapter and draws a few conclusions.

BIPOLAR DISORDER CLASSIFICATION

According to a report by Kessler *et al.* (2006), BPD and major depression are the leading causes for the loss of business productivity in the US. Enhanced models for classification and analysis of BPDs are in critical need for understanding how people respond to medications and how to treat BPDs more effectively. Scientific research efforts on bipolar disorder have been made in recent decades for enhanced models such as Pettigrew and Miller (1998) and Krug *et al.* (2008). Among clinical approaches, a dimensional view has gained favor in a rapidly growing literature emphasizing shared abnormalities (Regier, 2007; Lopez *et al.,* 2007; Kraemer, 2007; Hudziak *et al.,* 2007; Andrews *et al.,* 2007; Krueger *et al.,* 2007; Czobor *et al.,* 2007; Vieta *et al.,* 2007; Jaeger, Berns & Czobor, 2003; Bearden, Hoffman & Cannon, 2001; McDonald *et al,.* 2004) that cut across the current diagnostic divide. On the other hand, quantification and statistical analysis of BPDs with approximate entropy has shown valuable insights into non-linear dynamic bipolar symptoms in time series (Pincus, 2006; Rao *et al.,* 2006; Glenn *et al.,* 2006).

Despite the above tremendous efforts, bipolar disorder classification and characterization have so far proven extremely difficult if not impossible. This difficulty can be attributed to multiple factors including: (1) psychiatric disorders are complex neurobiological problems with both logical and "illogical" but nevertheless physical symptoms; (2) so far there is no commonly accepted formal logical model in computational neuroscience for equilibrium-based holistic bipolar pattern analysis that can link neurobiological reactions to brain imaging, genetic research, and therapeutic interventions; (3) quantification and statistical analysis of BPDs are so far based on the traditional standard categorical spectrum model specified in the *Diagnostic and Statistical Manual* of Mental Disorders (*DSM IV*) (American Psychiatric Association, 2000). The standard doesn't take into account mental fusion and separation of different dimensions (see Figure 1); (4) although the dimensional approach has shown promising results, it is not clear how to derive an unifying generic model from different dimensions with both clinical operability and scalability from the spectrum model to multiple dimensions; (5) although the support vector machine (SVM) approach in Hilbert space has become a powerful computational approach for data classification, without bipolarity, however, support vector machines can't deal with non-linear bipolar dynamic mental fusion, interaction, oscillation, equilibrium, separation, and quantum entanglement properties (Zhang & Zhang, 2004a; Zhang, 2005a, 2005b, 2006a, 2007, 2009a, 2009b, 2009c, 2009d, 2010; Zhang & Peace, 2007; Zhang, Pandurangi & Peace, 2007).

As discussed in earlier chapters, without bipolar fusion, binding, interaction, and oscillation there would be no quantum field, no neural dynamics, no bioinformatics, no equilibrium, no universe, and no brain and mind. With classical mathematical abstraction and (unipolar) truth-based logic, however, we have the 3-fold dilemma: (1) if we consider each pole of a bipolar equilibrium or non-equilibrium as a self-evident element, we will lose holistic bipolar fusion or binding (see Figure 1). (2) If we consider a bipolar equilibrium or non-equilibrium as a self-evident element, its membership in a set can only be true or false where polarity can't be represented. (3) If we preserve the independence rule between a set of equilibria and its elements we will not be able to link the global equilibrium/non-equilibrium to local ones. The dilemma makes it impossible for classical set theory and logic to define operations for equilibrium-based bipolar fusion, interaction, oscillation, and quantum entanglement (see Ch. 3, Figure. 1). As stated in Chapter 1, this can be further illustrated with some intuitive examples.

Figure 1. DSM IV Standard of bipolar disorder classification

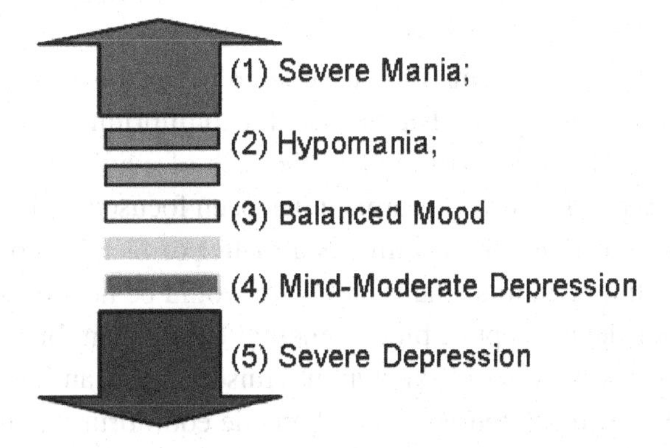

Example. How can depression, mania, mental equilibrium, and eternal equilibrium (or brain death) be directly characterized with logical values? How can the negative effect, positive effect, balancing effect, and deadly effect of a bipolar disorder medicine be characterized with logical expressions? How can the neurobiological reactions of bipolar disorder patients be characterized with logical expressions for personalized biomedicine?

Evidently, there is no way to define such seemingly "illogical" but nevertheless natural non-linear dynamic logical values and operators with any truth-based system. This is the original observation that led to the LAFIP paradox on Hilbert's Problem 6: *"Logical Axiomatization for Illogical Physics"* (LAFIP or LAFIB) (Zhang, 2009a) (Ch. 1). The paradox manifests the inconvenient truth: (a) without bipolarity any truth-based logic is "too logical" for reasoning on "illogical" non-linear neural dynamics; (b) bipolarity is indispensable in computational neuroscience and psychiatry.

From the above, it is clear that the lack of an equilibrium-based computational model hinders our understanding of mental disorders and hinders the power of pattern classification of support vector machines. Thus, three important issues facing computational neuroscience are: (1) How to unify non-linear neural dynamics with equilibrium-based bipolar quantum fields? (2) How to unify neural dynamics with mental consciousness? (3) How to unify internal biological and genomic worlds of the brain with external environment?

EQUILIBRIUM-BASED COMPUTATIONAL NEUROSCIENCE AND PSYCHIATRY

Observability of Bipolar Equilibrium

Equilibrium-based computational neuroscience is holistic in nature which brings top-down inductive reasoning (using BUMP to determine *) and bottom-up deductive reasoning (using BUMP for deduction) into a unifying framework. In this framework a normal person is considered in mental equilibrium of both self-negation and self-assertion abilities; a bipolar disorder patient, on the other hand, lacks bipolar equilibrium.

Is YinYang bipolar equilibrium observable/provable holistic truth? The answer is definitely "Yes". If it were not, how could we have "any closed system tends to reach equilibrium" as a basic law in thermodynamics? How could the words "YinYang" and "equilibrium" appear in genomics, physics, computer science, logic, mathematics, decision science, and other areas? How could they appear in top scientific journals including *Science*, *Nature*, and *Cell*? Evidently, YinYang bipolar equilibrium as holistic truth is observable, empirically verifiable, and/or scientifically provable. The problem is it is a less-developed scientific concept with less observable properties.

In mental health, bipolar disorder symptoms have been defined as observables in the DSM standard for clinical psychiatry for many years. Bipolar mental equilibrium, however, has been a neglected observable due to unipolar truth-based tradition and the lack of equilibrium-based logical foundation and definition. As a result, bipolar disorder treatment has been focused on how to eliminate symptoms, not how to bring a patient to mental equilibrium. As a matter of fact, without bipolar equilibrium the universe or human brain would be in total chaos and there would be neither observables nor observers.

Simply speaking without the concept of bipolar mental equilibrium bipolar disorder as the loss of mental balance would be baseless. It is well-known that misdiagnosis and overdose of bipolar disorder patients can result in serious medical consequence. A viable equilibrium scale may enforce diagnostic

Table 1. An example mental equilibrium scale

Self-Sufficiency						
1	Personal hygiene and appearance	1	2	3	4	5
2	Ability of independent travel	1	2	3	4	5
3	Ability of obtain nourishment	1	2	3	4	5
Working Ethics						
4	Dependability	1	2	3	4	5
5	Punctuality	1	2	3	4	5
6	Effectiveness	1	2	3	4	5
Family Relations						
7	Communicate with family	1	2	3	4	5
8	Participating family activities	1	2	3	4	5
9	Maintaining normal relation s	1	2	3	4	5
Social Relations						
10	Has network of friends	1	2	3	4	5
11	Participates in social activities	1	2	3	4	5
12	Behave properly	1	2	3	4	5
Self-Awareness						
13	Aware of own problem	1	2	3	4	5
14	Good character and conduct	1	2	3	4	5
15	Shows patience	1	2	3	4	5
Learning Ability						
16	Can learn new vocabulary	1	2	3	4	5
17	Can do logical reasoning	1	2	3	4	5
18	Can do numerical calculation	1	2	3	4	5
19	Can be artistic	1	2	3	4	5
20	Can learn music	1	2	3	4	5

consistency and prevent misdiagnoses and overdose. For illustration purpose of illustration, a mental equilibrium scale is proposed in Table 1, which can serve as a complement to mania and depression scales in clinical psychiatry.

Now we have the theory of equilibrium-based computational neuroscience with the following aspects: (1) equilibria/non-equilibria are ubiquitous natural phenomena and bipolar equilibria/non-equilibria are generic forms of multidimensional equilibria; (2) without bipolar fusion, interaction, and oscillation there would be no bipolar equilibrium; (3) without bipolar equilibrium bipolar disorder would be "big bang" that came from nowhere and was caused by nothing; (4) without equilibrium a world would be in total chaos and there would be neither observables no observers; (5) bipolar mental equilibrium can be defined as neurobiological or mental fusion of self-negation and self-assertion abilities; (6) bipolar disorder can be defined as the loss of bipolar fusion, balance, interaction, and oscillation of the two abilities; (7) the focus of equilibrium-based neuroscience should be on bringing disorders to equilibrium instead of just treating symptoms; (8) BDL and BDFL provide a logical basis for the theory; (9) BQLA and BQCA provide a mathematical physics or biophysics basis for the theory; and (10) the theory is governed by YinYang bipolar relativity.

Logical Transformation of Brain and Behavior

While BDL and BDFL are generally applicable in an open-world of dynamic equilibria, it is particularly suitable for computational neuroscience. Figure 2 is a 2-D transformation of the standard to a MSQ. It is clear that the four-corners of the MSQ lead to $BDL = (B_P, \equiv, \oplus, \&, \oslash, \otimes, \oplus^-, \&^-, \oslash^-, \otimes^-, -, \neg, \Rightarrow)$ (Ch. 1).

The four values in B_1 can be used for characterizing (1) balanced mental state or medical intervention (-1, +1), (2) mental depression or negative medication (-1,0), (3) mania or positive medication (0,1), and (4) zero energy (brain death) or deadly medication (0,0). Thus, a bipolar disorder patient set {p} and a psychiatric drug set {m} can be both bipolar sets and the two sets can be bipolar interactive. The seemingly illogical bipolar disorder phenomenon becomes a logical structure. Without the equilibrium-based logical structure, mental balance used to be characterized with a neutral or zero value (see Figure 2(a)) that is misleading.

The 4-valued logical transformation does not provide sufficient gray levels. The real-valued or fuzzy logical transformation of B_F is shown in Figure 3 with a 180 degree flip following GUI convention (Ch. 5). With the fuzzy version, the standard DSM classification can be characterized with C-type bipolar α-level sets (Zhang, 2006a; Zhang, Pandurangi & Peace, 2007):

1. Severe Mania Set (Region (1)): $M_2 = \{(x,y) | \forall (x,y) \in B_F, (x+y) \geq 0.6\}$;
2. Mild to Moderate Mania Set (Region (2)): $M_1 = \{(x,y) | \forall (x,y) \in B_F, 0.2 \leq (x+y) < 0.6\}$;
3. Normal Balanced Mood Set (Region (3)): $N = \{(x,y) | \forall (x,y) \in B_F, -0.2 \leq (x+y) < 0.2\}$;
4. Mild to Moderate Depression Set (Region (4)): $D_1 = \{(x,y) | \forall (x,y) \in B_F, -0.6 < (x+y) \leq -0.2\}$; and
5. Severe Depression Set (Region (5)): $D_2 = \{(x,y) | \forall (x,y) \in B_F, (x+y) \geq -0.6\}$.

The C-type bipolar α-level sets can be considered higher order fuzzy sets with the granular bipolar linguistic terms *Severe Mania, Mild to Moderate Mania, Normal Balanced Mood, Mild to Moderate Depression, and Severe Depression* defined in the 2-D lattice B_F as shown in Figure 3.

Figure 2. 2-D transformation of the DSM standard

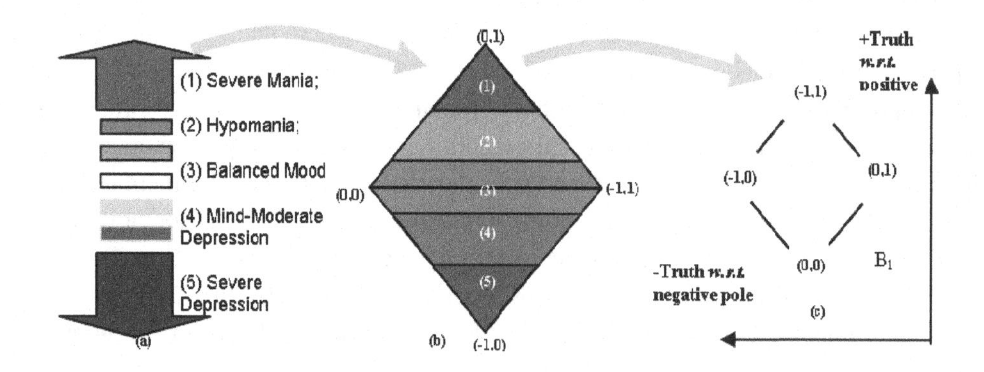

Figure 3 should be interpreted as the geometry of bipolar relativity consisting of the negative (or Yin) dimension, positive (or Yang) dimension, and the bipolar fusion or equilibrium dimension represented by the diagonal from (0,0) to (-1,+1). A point (n,p) characterizes the fusion of both negative and positive membership degrees of mental equilibrium that are different from a point in a usual two dimensional X-Y coordinate. For instance, given (n,p) = (-0.5, +0.5) we have n + p = 0 which characterizes a mental balance; given (n,p) = (-0.8, +0.5) we have n+p = -0.3 which characterizes a mild depression; given (n,p) = (-0.5, 0.8) we have n + p = +0.3 which characterizes a hypomania; given (n, p) = (-1.0,0.2) or (-0.2,1.0) we would have n + p = -0.8 or +0.8 which would characterize a severe depression or mania, respectively.

Interestingly, the scale in Table 1 can be used to enforce consistency between the equilibrium and disorder diagnoses. For instance, if the mental equilibrium diagnosis from Table 1 is normalized to (-0.5,+0.5) but the depression diagnosis of the same patient is (-0.9,0.0) we would have an inconsistency because (-0.5,0.5) + (-0.9,0) = (-1.4, 0.5) where -1.4 is out of the boundary. The inconsistency has to be resolved to prevent misdiagnosis and overdose.

Thus, each point in Figure 3 can be mapped to the DSM standard (Figure 1) with the addition of a mental equilibrium dimension. A weak equilibrium (e.g. (-0.2,0.2)) is easily disturbed. The goal of a psychiatric medication is then to bring a patient's mental state toward (-1,+1) which characterizes a strong equilibrium. Thus, BDFL enables bipolar classification and computation in psychopharmacology.

The bipolar level set classification can also be used to classify psychiatric drugs into different Positive, Negative, and balancer drugs. With the bipolar classifications for both bipolar syndromes and drugs, bipolar dynamic cause-effect reasoning is made logically possible to support diagnostic analysis and computational psychopharmacology (Zhang, Pandurangi, & Peace, 2007). Figure 4 is a sketch of the YinYang trajectory of healing or restoration to balance cause-effect sequence.

Mental Fusion, Mental Separation, and MentalSquares

From the above discussion, it is clear that a MSQ can be defined mathematically as a YinYang bipolar geometry with bipolar relativity. In the YinYang geometry we have a negative pole and a positive pole from which a third dimension – mental equilibrium is transcendental with mental fusion and separa-

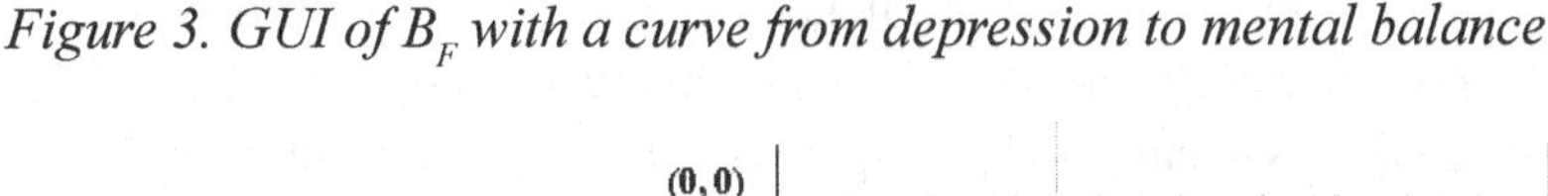

Figure 3. GUI of B_F with a curve from depression to mental balance

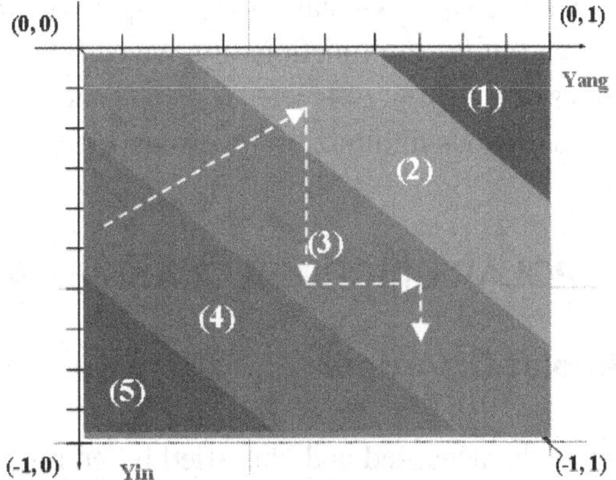

Figure 4. From mania to equilibrium – mental fusion

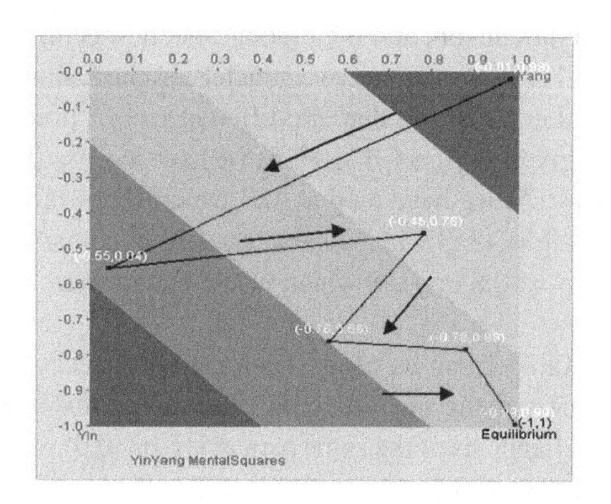

tion measures. While a typical mental (mood) fusion can be expressed with the operation (-1,0) ⊕ (0,1) = (-1,+1), a typical mental (mood) separation can be defined with the logical operation *Sep(x,y) = {(x,0),(0,y)}*. For instance, the mental fusion *(-1,0)⊕ (0,1) = (-1,+1)* characterizes a strong mental equilibrium, *Sep(-1,1) = {(-1,0),(0,1)}* characterizes the total mental separation.

Equilibrant, Balancer, Oscillator, and Life Extinguisher

Recall from Chapter 4 that, balance, energy, and stability are important properties of bipolar quantum lattices (BQLs) (Ch. 4). Since a BQL has balanced, almost balanced, and unbalanced elements, the concept of equilibrium has full, quasi- and/or non-equilibrium aspects self-contained. Let $W=\{\varphi_i\}$ be a set of bipolar equilibria (including quasi- or non-equilibria) or bipolar L-sets, a truth object A exists in W iff $\exists\ \varphi \in W$ such that the equilibrium energy of A in φ is greater than zero. Formally, we have: agent A exists *iff* $\{\exists\ \varphi \in W|\ (|\varphi^-(A)| + \varphi^+(A)) > 0\}$, where $|\varphi^-(A)| + \varphi^+(A)$ identifies the energy total of A. Let bipolar reflexivity be a bipolar equilibrium function $\varphi \in W$ and $\varphi = (\varphi^-, \varphi^+) =$ *(self-negation, self-assertion)*, an agent A has life in W iff $\varphi(A) \neq (0,0)$.

The definitions of energy, life, and existence follow the commonsense that anything has to exist in some type of equilibrium and any agent has to have certain self negation or self assertion ability. Based on the notions of life and energy, we consider $(N,0)$, $N<0$, as an **equilibrant** for its property $(N,0) \oplus ((N,0) \otimes (N,0)) \equiv (N,0) \oplus (0,|N|) \equiv (N,|N|)$. We call the blub operator ⊕ a life/energy **maximizer** or **balancer** because of its bipolar binding/fusion functionality; we call the cglb operator ⊗ a life/energy **oscillator** because of its semantics of – – = +; – + = –; + – = –; and ++ = +. We call & a life/energy **minimizer** or **extinguisher** for its obvious minimizing functionality.

EQUILIBRIUM-BASED BIPOLAR DISORDER CLASSIFICATION AND ANALYSIS

Three Types of Major Bipolar Disorders

Traditionally a bipolar disorder is characterized and classified based on the standard one-dimensional spectrum as depicted in Figure 1. With this spectrum, classical logic and sets (crisp or fuzzy) can be used for mood modeling. It can be observed, however, that classical logic and sets have a number of disadvantages: (1) they are unipolar in nature and can't support bipolar coexistence and equilibrium for bipolar visualization; (2) a unipolar logic does not support bipolar energy and stability analysis; (3) a unipolar logic does not support time-variant series of symptoms with bipolar interaction; and (4) BUMP has no equivalent in a unipolar logic.

For instance, given a bipolar disorder patient set X, a bipolar diagnostic function D for X can be defined as a bipolar mapping $D{:}X \Rightarrow B$, where B can be B_1, B_F, B_n, or any bipolar quantum lattice. Then BUMP can be used for bipolar inference. With a unipolar model bipolar mapping and inference would be impossible simply because a unipolar model doesn't support bipolarity in knowledge representation. In other words, classical logic and sets are truth-based with a true side and a false side that form a contradiction if both sides are true. But a contradiction is not an equilibrium. Evidently, the coexistence of a +side and a –side are inevitable for the representation of a bipolar dynamic equilibrium.

With bipolar logic and sets major depressive and bipolar disorders can be classified and mathematically characterized as three major categories:

Bipolar I Disorder (American Psychiatric Association, 2000). This type is characterized by the occurrence of one or more manic episodes or mixed episodes. Often individuals have also had one or more major depressive episodes. With a bipolar crisp set, a bipolar I disorder of patient p can be characterized as $(\psi^-,\psi^+)(p)$ = *(self-negation, self-assertion)(p)* = $(0,1)$ which stands for *"with excessive self-assertion ability but no self-negation ability."* Using a bipolar fuzzy set, this type can be more precisely characterized with a bipolar fuzzy value (n,p) where p is significantly larger than |n|.

Bipolar II Disorder (American Psychiatric Association, 2000). This type is characterized by the occurrence of one or more major depressive episodes accompanied by at least one hypomanic episode. Using a bipolar crisp set, a bipolar II disorder of a patient p can be characterized as $(\psi^-,\psi^+)(p)$ = *(self-negation, self-assertion) (p)* = $(-1,0)$ which means *"with excessive self-negation ability but no self-assertion ability."* Using a bipolar fuzzy set, this can be more precisely characterized with a bipolar fuzzy value (n,p) where p is significantly smaller than $|n|$.

Cyclothymic Disorder (American Psychiatric Association, 2000). This type is a chronic fluctuating mood disturbance involving numerous periods of hypomanic symptoms and numerous periods of depressive symptoms. This disorder can be characterized as the vacillating sequence $(\psi^-, \psi^+)(p)$ = *(self-negation, self-assertion) (p)* = $(-1,0)\otimes(-1,0)\otimes..\otimes(-1,0)=(-1,0)^N$ or $(n,p)\otimes(n,p)\otimes..\otimes(n,p) = (n,p)^N$, where (n,p) is significantly unbalanced in medical terms. When N is even $(-1,0)^N = (0,1)$, when N is odd, $(-1,0)^N = (-1,0)$. $(n,p)^N$ results in a vacillating sequence with different levels of granularities.

Scenarios in Bipolar Diagnostic Analysis (Adapted from Zhang, Pandurangi & Peace 2007)

1. Scenario for a mentally healthy person P:

Let P be a person and let
$(\varphi^-,\varphi^+)(P)$ = *(negative-trigger, positive-trigger)(P)* = $(-1,0)$;
(Note: $(-1,0)$ is a negative trigger to P such as economic hardship or other negative event)
$(\phi^-,\phi^+)(P)$ = (negative-feelings, positive-feelings)(P);
and, $\forall P,\ (\varphi^-,\varphi^+)(P) \Rightarrow (\varphi^-,\varphi^+)(P)$.

If *P* is a mentally healthy person with balanced *(self-negation, self-assertion)* ability, we have

$(\psi^-,\psi^+)(P)$ = (self-negation, self-assertion)(P) = (-1,1).

Based on BUMP, we have,

$[(\varphi^-,\varphi^+)(p)\otimes(\psi^-,\psi^+)(P) = (-1,0) \otimes (-1,1)=(-1,1)]$
$\Rightarrow [(\phi^-,\phi^+)(p)\otimes(\psi^-,\psi^+)(P) = (-1,0) \otimes (-1,1) = (-1,1)]$

This scenario can be interpreted as *"a negative trigger brings negative feelings to P. Since P is mentally healthy with balanced self-negation and self-assertion abilities, P can adjust his/her emotion to the equilibrium (-1,1)."*

2. ***Scenario for a bipolar I patient P.*** Let *P* be a bipolar I patient and let

$(\varphi^-,\varphi^+)(P)$ = *(negative-trigger, positive-trigger)(P) = (-1,0);*
(Note: *(-1,0)* is a negative trigger to P such as economic hardship or negative medication)
(φ^-,φ^+) *(P) = (negative-feelings, positive-feelings)(P);* and,
$\forall P, (\varphi^-,\varphi^+)(P)\Rightarrow (\phi^-,\phi^+)(P)$.

Since P is a bipolar I patient, we have

$(\psi^-,\psi^+)(P)$ = *(self-negation, self-assertion)(P) = (0,1).* (Note: *(0,1)* indicates the lack of self-negation)

Based on BUMP, we have,

$[(\varphi^-,\varphi^+)(P)\otimes(\psi^-,\psi^+)(P) = (-1,0) \otimes (0,1)=(-1,0)]$
$\Rightarrow [(\phi^-,\phi^+)(P)\otimes(\psi^-,\psi^+)(P) = (-1,0) \otimes (0,1) = (-1,0)]$

This scenario can be interpreted as *"a negative trigger brings negative feelings to patient P. Since P lacks self-negation ability, P can't adjust his/her feelings to the equilibrium (-1,1)."* This example indicates that bipolar I disorder patient may also have one or more major depressive episodes.

3. ***Scenario for a bipolar II patient P.*** Let *P* be a bipolar II patient and let

$(\varphi^-,\varphi^+)(P)$ = *(negative-trigger, positive-trigger)(P) = (0,1);*
(Note: *(0,1)* is a positive trigger such as a major victory or winning a lottery)
(φ^-,φ^+) *(P) = (negative-feelings, positive-feelings)(P);* and,
$\forall P,(\varphi^-,\varphi^+)(P)\Rightarrow (\phi^-,\phi^+)(P)$.

Since *P* is a bipolar II patient, we have

$(\psi^-,\psi^+)(P)$ = *(self-negation, self-assertion)(P) = (-1,0).* (Note: *(-1, 0)* shows the lack of self-assertion)

Based on BUMP, we have,

$[(\varphi^-,\varphi^+)(P)\otimes(\psi^-,\psi^+)(P) = (0,1) \otimes (-1,0)=(-1,0)]$
$\Rightarrow [((\phi^-,\phi^+)\otimes(\psi^-,\psi^+))(P) = (0,1) \otimes (-1,0) = (-1,0)]$.

This scenario can be interpreted as *"P is in deep depression and a positive trigger is unable to bring positive feelings to patient P. Moreover, since P has excessive self-negation without self-assertion ability, negative feelings can be triggered to positive feelings and then to negative feelings again."* That is why a bipolar II disorder patient may also have one or more hypomanic episodes. For instance,

$[(\varphi^-,\varphi^+)(P)\varnothing(\psi^-,\psi^+)(P) = (0,1) \varnothing (-1,0)=(0,1)]$
$\Rightarrow [((\phi^-,\phi^+)\varnothing (\psi^-,\psi^+))(P) = (0,1) \varnothing (-1,0) = (0,1)]$.

4. ***Scenario for cyclothymic disorder.*** In this type of disorder, the patient has the vacillating sequence.

$(-1,0)^N \in B_I$, $N\geq 1$. When N is odd, $(-1,0)^N = (-1,0)$. When N is even, $(-1,0)^N = (0,1)$.

Fuzzified Bipolar Diagnostic Analysis (Adapted from Zhang, Pandurangi & Peace 2007)

Note that $B_I = \{-1,0\}\times\{0,1\}$ doesn't support different gray levels on the negative and positive poles for representing quasi- or fuzzy-equilibria. This limitation is circumvented with $B_F = [-1,0]\times[0,1]$ to provide different granularities for seriousness measures of bipolar symptoms. For instance, if $(x,y) = (-0.9, 0.2)$, the vacillating sequence would be $(-0.9, 0.2)^N$. Assuming $\otimes_{II}$ neurobiological functionality, when N is even, $(-0.9,0.2)^N = (-0.2,0.9)$; when N is odd $(-0.9,0.2)^N = (-0.9,0.2)$.

Figure 5 shows a 2-D YinYang bipolar geometry of B_F with linguistic bipolar fuzzy sets. The $(0,0)$ corner region can be best described as *"Negative Small and Positive Small"* or *(NS,PS)*; the $(-1,1)$ corner region can be best described as *"Negative Large and Positive Large"* or *(NL,PL)*; the $(-1,0)$ corner region can be best described as *"Negative Large and Positive Small"* and *(NL,PS)*; the $(0,1)$ corner region can be best described as *"Negative Small and Positive Large"* or *(NS, PL)*; the center $(-0.5,+0.5)$ can be

Figure 5. Three types of bipolar disorders (Adapted from Zhang, Pandurangi & Peace, 2007)

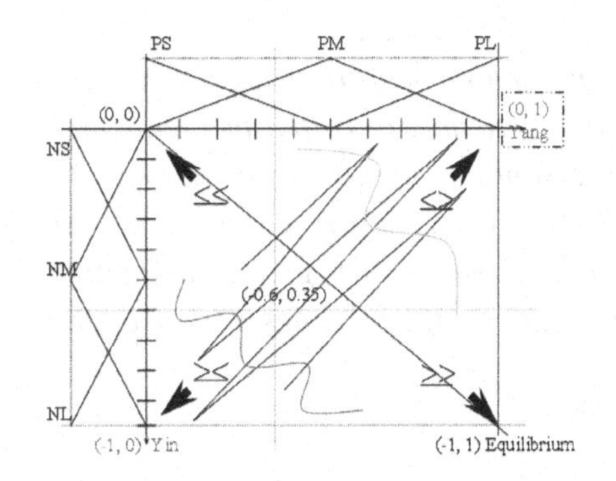

best described as *"Negative Medium and Positive Medium"* or *(NM,PM)*; where the Cartesian product *[NL,NM,NS,0]* × *[0,PS, PM,PL]* forms a bipolar linguistic fuzzy lattice. Any bipolar threshold α, for instance, *(-0.6, +0.35)*, divides the 2-D space into four regions. Each bipolar value in the figure can be fuzzified with a linguistic fuzzy set that generalizes Zadeh's extension principle (Zadeh, 1971) from *[0,1]* to B_F. For instance, *(-0.6, +0.35)* can be described as *(NL(0.2)/NM(0.8), PS(0.3)/PM(0.7))* as in Figure 5. It should be remarked that the C-type bipolar α-level sets (Zhang, 2006a) can be considered as 2nd order fuzzy sets with the bipolar linguistic terms *Severe Mania, Mild to Moderate Mania, Normal Balanced Mood, Mild to Moderate Depression,* and *Severe Depression* defined in the 2-dimensional lattice B_F = *[-1,0]* × *[0,1]*. Thus, bipolar fuzzy sets and BDFL enable bipolar computing with linguistic variables (or words) such as *"severe"*, *"mild"*, and *"normal"*. In Figure 5, the *(0,1)* corner typically characterizes severe symptoms of **bipolar I disorder**. The *(-1,0)* corner typically characterizes severe symptoms of **bipolar II disorder**. The vacillation amplitude and frequency with respect to the *(0,0)−(-1,1)* diagonal typically characterizes severe **cyclothymic disorder.***(-1,1)* is the perfect equilibrium state and *(0,0)* is the zero energy or external equilibrium state.

Bipolar Inference for Clinical Psychopharmacology (Adapted from Zhang, Pandurangi & Peace, 2007)

To this point, all the diagnostic scenarios may be considered normal, rational, and intuitive. Some big questions remain unanswered such as:

- How could a person in deep depression become suicidal or even kill his/her children?
- How could a bipolar disorder patient P show no feelings or even laugh at a tragic event?
- What would happen if a bipolar II (depression) patient P is given a negative trigger (treatment)?
- Why do some patients recover to bipolar equilibrium but some others do not?

Analytical answers to the above questions can be provided with BUMP. (Note: The universal operator * can be bound to infinite number of bipolar dynamic t-norms or t-conorms (Zhang, 2006b) (Ch. 4). In this section we focus on the three basic operators ⊕, *&*, and ⊗ only.)

If * is unknown, it can be determined with operator instantiation using trials (Note: Drugs in clinical trials in pharmaceutical research could produce side effects that are life threatening). The determination of * forms a bipolar diagnostic process, that is useful in bipolar neurobiological analysis and clinical psychopharmacology. Once * is determined for some patient with certain medication, a specific bipolar modus pones (linear or non-linear) is determined for dynamic inference. We illustrate the use of BUMP with the following public domain clinical cases:

Facts: *"Negative medication provides a negative trigger to un-excite the nervous system of a bipolar disorder patient; positive medication provides a positive trigger to un-depress the nervous system. Patient A has mania and gets into depression on negative medication. Patient B has serious depression and wants to kill himself for relief. Patient C has depression and recovered to mental balance after positive medication. Patient D had mania and died after negative medication. "*

Rules: *"For all patients, a negative trigger brings negative feelings and a positive trigger brings positive feelings."*

Questions: *"For each patient which neurobiological function is working or did work with a medication? Which one is not working or did not work?"*

Let p be any bipolar disorder patient and

let $(\varphi^-,\varphi^+)(p)$ = *(negative-trigger, positive-trigger)(p)*;
let $(\phi^-,\phi^+)(p)$ = *(negative-feelings, positive-feelings)(p)*.

We have $\forall p,(\varphi^-,\varphi^+)(p)\Rightarrow(\phi^-,\phi^+)(p)$ which is a bipolar knowledge fusion of the sentential rules. We then can represent the sentential facts as

$(\psi^-,\psi^+)(A)$ = *(self-negation, self-assertion)(A)* = *(0,1)* *(Note: (0,1)* characterizes mania – bipolar I disorder; and $(\varphi^-,\varphi^+)(A)$ = *(-1,0)* indicates a negative trigger or medication to *A*).

$(\psi^-,\psi^+)(B)$ = *(self-negation, self-assertion)(B)* = *(-1,0)* *(Note: (-1,0)* characterizes depression – bipolar II disorder; and suicide is a negative trigger).

$(\psi^-,\psi^+)(B)$ =*(self-negation, self-assertion)(C)* = *(-1,0)* (Note: *(-1,0)* is bipolar II disorder; and $(\varphi^-,\varphi^+)(B)$ = *(0,1)* indicates a positive trigger or medication to *A*).

First, using BUMP for patient A we have

$[(\psi^-,\psi^+)(A)\oplus(\varphi^-,\varphi^+)(A) = (0,1)\oplus(-1,0) = (-1,1)]$
$\Rightarrow [(\psi^-,\psi^+)(A)\oplus(\phi^-,\phi^+)\ (A) = (0,1)\oplus(-1,0) = (-1,1)]$;
$[(\psi^-,\psi^+)(A)\&(\varphi^-,\varphi^+)(A) = (0,1)\&(-1,0) = (0,0)]$
$\Rightarrow [(\psi^-,\psi^+)(A)\&(\phi^-,\phi^+)\ (A) = (0,1)\&(-1,0) = (0,0)]$;
$[(\psi^-,\psi^+)(A)\otimes(\varphi^-,\varphi^+)(A) = (0,1)\otimes(-1,0) = (-1,0)]$
$\Rightarrow [(\psi^-,\psi^+)(A)\otimes(\phi^-,\phi^+)\ (A) = (0,1)\otimes(-1,0) = (-1,0)]$.

A first impression from the above three implications could be that all three neurobiological functions (balancer $\oplus$, oscillator $\otimes$, and/or minimizer &) worked for A with the specific trigger or medication. A closer examination reveals that only the oscillator $\otimes$ worked for A because A gets into a depressive episode on negative medication that should be *(-1,0)* as a result.

Secondly, using BUMP for patient B we have:

$[(\psi^-,\psi^+)(B)\oplus(\varphi^-,\varphi^+)(B) = (-1,0)\oplus(-1,0) = (-1,0)]$
$\Rightarrow [(\psi^-,\psi^+)(B)\oplus(\phi^-,\phi^+)\ (B) = (-1,0)\oplus(-1,0) = (-1,0)]$;
$[(\psi^-,\psi^+)(B)\&(\varphi^-,\varphi^+)(B) = (-1,0)\&(-1,0) = (-1,0)]$
$\Rightarrow [(\psi^-,\psi^+)(B)\&(\phi^-,\phi^+)\ (B) = (-1,0)\&(-1,0) = (-1,0)]$;
$[(\psi^-,\psi^+)(B)\otimes(\varphi^-,\varphi^+)(B) = (-1,0)\otimes(-1,0) = (0,1)]$
$\Rightarrow [(\psi^-,\psi^+)(B)\otimes(\phi^-,\phi^+)\ (B) = (-1,0)\otimes(-1,0) = (0,1)]$.

The three implications for B are counterintuitive. However, if we follow the sentential facts precisely, we can determine that only the oscillator $\otimes$ worked for patient B because "relief" after suicide is for a positive feeling that should be (0,1).

Thirdly, using BUMP for patient C we have:

$[(\psi^-,\psi^+)(C)\oplus(\varphi^-,\varphi^+)(C) = (-1,0)\oplus(0,1) = (-1,1)]$
$\Rightarrow [(\psi^-,\psi^+)(C)\oplus(\phi^-,\phi^+)(C) = (-1,0)\oplus(0,1) = (-1,1)];$
$[(\psi^-,\psi^+)(C)\&(\varphi^-,\varphi^+)(C) = (-1,0)\&(0,1) = (0,0)]$
$\Rightarrow [(\psi^-,\psi^+)(C)\&(\phi^-,\phi^+)(C) = (-1,0)\&(0,1) = (0,0)];$
$[(\psi^-,\psi^+)(C)\otimes(\varphi^-,\varphi^+)(C) = (-1,0)\otimes(0,1) = (-1,0)]$
$\Rightarrow [(\psi^-,\psi^+)(C)\otimes(\phi^-,\phi^+)(C) = (-1,0)\otimes(0,1)(C) = (-1,0)].$

The three implications for C are clear and sound where only the balancer function $\oplus$ worked because C recovered to mental balance after medication that should be (-1,1).

Fourth, using BUMP for patient D we have:

$[(\psi^-,\psi^+)(D)\oplus(\varphi^-,\varphi^+)(D) = (0,1)\oplus(-1,0) = (-1,1)]$
$\Rightarrow [(\psi^-,\psi^+)(D)\oplus(\phi^-,\phi^+)(D) = (0,1)\oplus(-1,0) = (-1,1)];$
$(\psi^-,\psi^+)(D)\&(\varphi^-,\varphi^+)(D) = (0,1)\&(-1,0) = (0,0)]$
$\Rightarrow [(\psi^-,\psi^+)(D)\&(\phi^-,\phi^+)(D) = (0,1)\&(-1,0) = (0,0)];$
$[(\psi^-,\psi^+)(D)\otimes(\varphi^-,\varphi^+)(D) = (0,1)\otimes(-1,0) = (-1,0)]$
$\Rightarrow [(\psi^-,\psi^+)(D)\otimes(\phi^-,\phi^+)(D) = (0,1)\otimes(-1,0) = (-1,0)].$

The three implications indicate that the energy minimizer or life extinguisher & worked for D because D died after negative medication and death should be eternal equilibrium.

We summarize the above diagnostic analysis in the following:

- **Bipolar $\oplus$ Intrinsics** can be uncovered from symptoms whether a patient has any bipolar fusion capability of the medical effects and improves his/her mental balance.
- **Bipolar & Intrinsics** can be uncovered from symptoms whether a bipolar patient becomes significantly weaker with certain medication.
- **Bipolar $\otimes$ Intrinsics** can be uncovered from symptoms whether a bipolar I or II or III patient has oscillating episodes.

Bipolar $\otimes$ intrinsics is counterintuitive which deserves special attention. $(\psi^-,\psi^+)(P) = (-1,0)$ indicates P is in deep depression. $(\varphi^-,\varphi^+)(P)=(-1,0)$ indicates a strong negative trigger such as suicide. $(\varphi^-, \varphi^+)(P) \otimes (\psi^-,\psi^+)(P) =(0,1)$ indicates that P would have a positive feeling of relief with death. It is illogical and counterintuitive for any normal person to have positive feelings with a negative trigger. Nevertheless, this observation does lead to the diagnosis that the mental or neurobiological operator $\otimes$ of P is correctly functioning. This may well-explain the fact that (1) any closed dynamic system tends to become an equilibrium; and (2) a person in deep depression may become suicidal or even kill his/her loved ones to pursue positive feelings. Similarly, a serious bipolar disorder patient P may show no feelings of relief to a sad event or even get excited.

Unified Measures

The intrinsics revealed above provide leads to further clinical, therapeutic, and/or pharmacological research and application. Unified measures can be formulated as in the following for optimization, coherence, and coordination in integrated medical care of major depressive and bipolar disorders.

Balancer, extinguisher, and oscillator. The intrinsics revealed in the early analysis all involve the key concepts of bipolar balancer (⊕), extinguisher (&), and oscillator (⊗). A balancer is a neurobiological component for bipolar mental fusion; an extinguisher (&) can be used to minimize excessive mania and/or depression. It can also extinguish life and energy. An oscillator brings dynamics to an equilibrium or quasi-equilibrium. All three are essential neurobiological components for a bipolar patient to adapt to a bipolar (mental) equilibrium. Then, a medical challenge is to determine the neurobiological defects of a patient in these aspects and to produce different biomedicines for different treatments.

A unified diagnosis for all three disorder types. In the above, self-correction is defined as negative reflexivity characterized by $(-1,0)$ such that $(-1,0) \oplus (-1,0)^2 = (-1,0) \oplus (0,1) = (-1,1)$, which is a self-adaptation to YinYang bipolar equilibrium. The key here is that an adaptive agent should have the physical and mental fusion ability of two opposing feelings with balancer ⊕. Otherwise, the agent will be unstable with the vacillating sequence of symptoms $(-1,0)^N$ or $(n,y)^N$. *Therefore, all three types of disorders with certain vacillating symptoms can be diagnosed as the lack of physical or mental fusion ability.*

EXPLORATORY NEUROBIOLOGICAL DATA MINING

Neurobiological Engine of Bipolar Relativity

BUMP provides a basic neurobiological building block or engine for bipolar relativity in brain and behavior. A bipolar bidirectional 3-tier graph representation of the engine is depicted in Figure 6, which can be considered a unification of the genetic, neurobiological, and the mental (including mood and behavior) levels of a brain (Zhang, 2009b), here a solid bidirectional link ↔ can be interpreted as an

Figure 6. BUMP – 3-tier basic building block for brain and mind

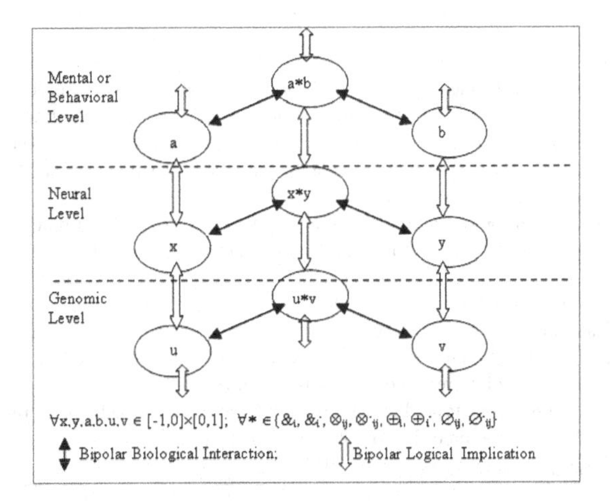

internal neurobiological connection; a bipolar logical implication $\Leftrightarrow$ can be interpreted as a quantum bipolar interaction through brain neuro-bio-electromagnetic field. A dotted link can be interpreted as a sensory connection to the outside world.

The neurobiological engine can be used in equilibrium-based hypothesis-driven exploratory knowledge discovery in computational psychiatry. A popular theory in psychiatric medicine is that serotonin deficiency causes mental depression and norepinephrine excess causes mania. With the bipolar inference engine, different equilibrium-based theories can be formulated and tested (Table 2).

Regardless of the validity of an equilibrium-based theory (Table 2), it can be tested, validated, or invalidated with BUMP from a computational point of view. If $p = A$ has minor depression, we may have $* = \oplus_1$ which indicates A still has neurobiological or mental balancing capability (see Figure 6). Let $x=\psi(A)= (-0.8, 0.5)$, $y = \varphi(m) = (0,0.8)$, $a = \chi(A) = (-0.8,0.5)$, and $b = \varphi(m) = (0,0.8)$, then we could have $[x \oplus y=(-0.8, 0.8)] \Rightarrow [a \oplus b = (-0.8, 0.8)]$ that characterizes a recovery to a bipolar balance at both the physical level and the mental level. If $p = B$ and B has deep depression we may have $*= \otimes$, $x = \psi(B) = (-0.9,0)$, $y=\varphi(m)=(0,0.9)$, $a=\chi(A)=(-0.9,0)$, and $b=\varphi(m) = (0,0.9)$. Then, we could have $[x \otimes y = (-0.9,0)] \Rightarrow [a \otimes b = (-0.9,0)]$ that characterizes a failed medical intervention. If $* = \oslash$ we could have $[x \oslash y = (0,0.9)] \Rightarrow [a \oslash b = (0,0.9)]$ that would show a mood switch from depression to mania.

Internal and External Unification

Since the bipolar links in Figure 6 are bidirectional, a trigger that causes bipolar disorder can be internal or external. An internal trigger can be genetic or biochemical. An external trigger can be environmental or social such as economic hardship may cause depression. On the one hand, genetic factors can be the major cause of bipolar disorders. On the other hand, there would be no healthy genetic mutation and "no brain" without mental equilibrium. Thus, different theories can be formulated and tested. If a hypothesis is valid in 100%, the universal operator * in Equation 1 and Figure 6 should always match at both the physical and mental levels. Any mismatch would be an invalidation of the hypothesis. Furthermore, a hypothesis can be partially validated or invalidated. The conditions of a partial validation could be very

Table 2. A Hypothesis in Neuroscience

Hypothesis:
1. The balance or imbalance of the genetic agents x and y denoted (g-x, g-y) at the genetic level lead to the neurobiological balance or imbalance of the bio agents b-x and b-y denoted (b-x, b-y) at the neuro- chemical or neurobiological level
2. The neurobiological balance or imbalance (b-x, b-y) leads to mental equilibrium or non-equilibrium at the mood or behavior level characterized as the fusion of self-negation and self-assertion abilities of a person denoted (self-negation, self-assertion).
3. Negative gene therapy un-excites the nervous system of a manic patient at the genetic level; positive gene therapy un-depresses the nervous system of a depressed patient at the genetic level.
4. Negative biomedicine un-excites the nervous system of a manic patient at the biological or biochemical level; positive medication un-depresses the nervous system of a depressed patient at the biological or biochemical level.
5. {p} is a set of bipolar patients;
6. {m} is a set of medicines or gene therapies for bipolar disorders;
7. $\psi = (\psi^-,\psi^+)$ is a bipolar predicate for the genetic level (g-x, g-y) or biological level (b-x, b-y) and $\psi(p) \in B_F$;
8. $\chi = (\chi^-,\chi^+)$ = (depression, mania) is a bipolar predicate at the mental/mood level and $\varphi(p) \in B_F$;
9. $\phi = (\phi^-,\phi^+)$ = (negative, positive) is a bipolar predicate at the physical level and $\phi(m) \in B_F$;
10. $\varphi = (\varphi^-,\varphi^+)$=(un-excite, un-depress) is a bipolar predicate for the effectiveness of a medicine or therapy and $\varphi(m) \in B_F$.
11. $\forall p,m,\ p \in \{p\}\ m \in \{m\}$, we have the instantiation of BUMP as shown in Eq. (10.1): $[\psi(p) \Rightarrow \chi(p)] \& [\varphi(m) \Rightarrow \varphi(m)] \Rightarrow [\psi(p) * \varphi(m) \Rightarrow \chi(p) * \varphi(m)]$. (10.1)

interesting. Thus, BDL/BDFL can be used for equilibrium-based hypothesis-driven exploratory scientific knowledge discovery in an open-world of equilibria.

Patient-Drug Neurobiological Reaction Classification and Data Mining

The bipolar lattices (B_1, ⇒, ¬, −, *) and (B_F, ⇒, ¬, −, *), where * can be instantiated to &, ⊕, &⁻, ⊕⁻, ⊗, ∅, ⊗⁻, ∅⁻, or any commutative and monotonic bipolar dynamic operator, provide a theoretical foundation for computational psychopharmacology. The key for such computation is the bipolar universal modus ponens (BUMP) (Table 3.4, Ch3). BUMP follows the pattern "IF premise1 and premise2; THEN consequent" which states that "If bipolar interaction * occurs (e.g. between a BPD patient and a bipolar drug at the neurobiological level) in premise1 the same interaction * also occurs (at the functional or behavioral level) in the consequent provided premise2 is bipolar true or a bipolar tautology."

With B_1, (-1,1) can be used for characterizing a balancer medicine as well as a bipolar mental equilibrium; (-1,0) can be used for characterizing a negative medicine as well as mental depression, (0,1) can be used for characterizing a positive medicine as well as mania, and (0,0) can be used for non-existence. Thus, a BPD patient set {p} and a psychiatric drug set {m} can be bipolar interactive sets.

Recall that we use ⊕ or ⊕⁻ a balancing operator, ⊗ or ∅ an intuitive oscillator, ⊗⁻ or ∅⁻ a counter-intuitive oscillator, and & or &⁻ an energy minimizer based on their functionalities. Given a set {p} of BPD patients and a set {m} of medical treatments, a bipolar diagnostic function f_1:{p_{t0}}⇒B_1 and a bipolar drug function f_2:{m}⇒B_1, we have the important pharmacological question: *What is the actual neurobiological reaction * for the treatment* $f_1(p_{t0})*f_2(m)$*?*

Once a theory has been proven, it can be applied for data mining. For instance, assume the hypothesis in Table 2 has been proven, given the text data in Table 3, a data mining process with pre- and post processing abilities would be able to apply BUMP to determine the neurobiological reactions of each patient to his/her medication as shown in Table 3. Such reactions can be used for personalized medicine and further knowledge discovery such as the possibility that patients with the same reactions might share the same gene.

Table 3. Personalized neurobiological text data mining (Case 1)

Case1–Text Data: *"Negative medication provides a negative trigger to un-excite the nervous system of a BPD patient; positive medication provides a positive trigger to un-depress the nervous system." "For all neurobiological patients a negative trigger brings negative feelings and a positive trigger brings positive feeling and vice versa." "A suicidal patient is deeply depressed."*
(1) *"Patient A had depression and became suicidal after taking pediatric antidepressant M."*
(2) *"Patient B had depression and became manic after taking pediatric antidepressant M."*
(3) *"Patient C had depression and recovered after taking pediatric antidepressant M."*
(4) *"Patient D had depression and dead after taking pediatric antidepressant M."*
(5) *"Patient E used to be mentally healthy but got depressed after job-related pressure."*
(6) *"Patient F used to be mentally healthy but became manic after extreme hardship."*
(7) *"Patient G used to be mentally healthy but became manic after overwhelming joy."*
(8) *"Patient H became manic after a suicidal attempt."*
Question: What bipolar operator can be used to characterize the neurobiological reactions of the patients to their medical treatments or external stimuli.
Mining Result:
(1) Oscillator ⊗ (or ∅⁻) worked or is working for A. (2) Oscillator ∅ (or ⊗⁻) worked or is working for B.
(3) Balancer ⊕ (or ⊕⁻) worked for patient C. (4) Minimizer & (or &⁻) worked for patient D.
(5) Minimizer & worked for patient E. (6) Minimizer &⁻ worked for patient F.
(7) Minimizer & worked for patient G. (8) Oscillator ⊗ (or ∅⁻) worked for patient H.

It is clear from Case 1 in Table 3 that the bipolar dynamic operators can be used as bipolar classifiers for classifying patient-drug reactions that can lead to important bipolar statistical data. It can also be observed that the four bipolar crisp values in B_1 are inadequate for granular computing with "severe", or "mild" symptoms as defined in Figures 3 and 4. Bipolar fuzzy norms and linguistic fuzzy sets should be used in these cases. Moreover, as a computational psychopharmacology model, the prototype described so far is rather primitive as psychopharmacology can be a very complex matter. For instance, the addition of an antipsychotic to mood stabilizing drugs or a low-dose antidepressant to an existing regimen of drugs may lead to dramatic compound effects (Zhang *et al.*, 2010). This falls into sequential bipolar data mining which is illustrated in Table 4.

It should be remarked that: (1) It would be difficult if not impossible to solve the above problem with Boolean logic or fuzzy logic defined in the unipolar lattice {0,1} or [0,1] due to the lack of bipolarity. (2) Although a resolution such as the sequential pattern $\oplus_3\otimes$ is not necessarily unique as for all inductive resolutions, it does provide important leads to a number of possibilities for further knowledge discovery in an uncertain environment.

While clinical trials and diagnostic analysis in medicine are not deterministic in many cases, bipolar data mining can support further clarification. This section is only intended to demonstrate the applicability of MSQs. The bipolar data mining example are used to illustrate the basic ideas. The new approach enables the structuring, storage, and retrieval of highly unstructured mental disorder data for knowledge discovery. While such bipolar database is not yet available, large scale knowledge discovery can be conducted when such bipolar database is available.

It should be remarked that the neurobiological data scenarios presented above are commonly encountered in clinical practice (Zhang *et al.*, 2010). Dr. Anand K. Pandurangi, Professor of Psychiatry at Virginia Commonwealth University (a collaborator in our articles on *MentalSquares*) with over 30-years of clinical experience, attests to how clinicians are challenged by such clinical occurrences. While the

Table 4. Sequential neurobiological data mining (Case 2)

Case 2 – Text Data: *Patient C (a child) had mild depression $\psi(C_{t0}) = (-0.6, 0.1)$ and became suicidal (deep depression $\psi(C_{t1})=(-1,0)$) after taking pediatric antidepressant M_1 and $\varphi(M_1) = (-0.4,0.9)$. After M_1 is stopped he was treated with another pediatric antidepressant M_2, $\varphi(M_2) = (0,1)$, and regained mental equilibrium ($\psi(C_{t2})=(-1,1)$).*

Question: *Which neurobiological functions of patient C worked with the medicines?*

Data Mining Steps and Results:

Step 1. Using M_1:
$(\psi^-,\psi^+)(P) \oplus_i [(\psi^-,\psi^+)(P) \otimes_{ij} (\varphi^-,\varphi^+)(M_1)]$
$= (-0.6, 0.1)?\oplus_i [(-0.6, 0.1) \otimes_{ij} (-0.4, 0.9)]$
//$\oplus_i$ can be determined as $\oplus_\Delta$ ($\oplus_3$) by applying BUMP
//$\otimes_{ij}$ can be determined as any $\otimes$ by applying BUMP. For instance
$= (-0.6,0.1) \oplus_3 [(-0.6, 0.1) \otimes_{13} (-0.5, 0.9)]$
$= (-0.6,0.1) \oplus_3 (-0.5,0) = (-1,0)$ // severe depression
Step 2. Using M_2:
$(\psi^-,\psi^+)(P) \oplus_i(\varphi^-,\varphi^+)(M_2) = (-1,0)\oplus(0,1) = (-1,1)$ //Mental equilibrium

Interpretation: If Step 1 and Step 2 are compounded we have the neurological reaction sequence $\oplus_3\otimes\oplus$. The sequence can be considered as a knowledge pattern or a bipolar association between patient C and psychiatric drugs M_1 and M_2. In contrast if we use B_1 instead of B_F, the granularity for mild depression and the compound sequential effect $\oplus_3\otimes$ would be hidden and undiscovered.

cases are presented in this chapter in logical terms, they are being further elaborated in specific clinical medical terms in our future works.

STABILITY ANALYSIS

From Figures 3, 4, and 5, it is evident that the square B_F provides a qualitative and quantitative model for bipolar neurobiological classification and computation. First, the new model supports more accurate classification. For instance, with Figure 5 a severe mania symptom can be associated with the quantitative bipolar value (0,1) or (0.2,0.9) or (-0.1, 0.8),.., etc. With the spectrum model, the bipolar quantitative characterization would be impossible. With the square, depressive and mania scales (such as the Goldberg scales) can be combined to yield a bipolar diagnosis. For instance, (-0.2, 0.8) in Figure 5 can well indicate that a patient has mania but still has certain mental balance because (-0.2,0.8) = (-0.2, +0.2) + (0,+0.6), where (-0.2,+0.2) is the balancing part, (0,0.6) is the mania part, and + is defined as bipolar addition such that (a,b) + (c,d) =(a+c,b+d) (Ch. 4). On the other hand (0,+0.6) alone does not show any mental balance.

The new model retains all features of the spectrum model. Therefore, all existing quantification and statistical analysis methods can still be used. For instance, (-0.2,0.8) can be easily converted to a single number -0.2+0.8 = 0.6 for a mania symptom, and (-0.8,0.2) can be easily converted to -0.8+0.2 = -0.6 for a depression symptom that fall in a traditional spectrum model scaled from -1.0 to +1.0. Thus, upward compatibility is fully achieved with the extension.

Most importantly, the new model makes stability analysis possible based on energy imbalance (Zhang, Pandurangi & Peace, 2007). Let $(\psi^-,\psi^+)(p)$ = *(self-negation, self-assertion)(p)* be a bipolar equilibrium function or mapping that maps a patient p's symptoms to a bipolar lattice; let P_{t0} and P_{t1} be a patient at time t_0 and t_1, respectively; let $(\psi^-,\psi^+)(P_{t0}) = (x_0,y_0)$, $(\psi^-,\psi^+)(P_{t1}) = (x_1,y_1)$, $\forall (x_0,y_0),(x_1,y_1) \in B_F$; let P_{t0} indicate that the treatment starts on P at t_0 and end at t_1. We say the bipolar disorder patient P is getting better if

1. $stability(x_1,y_1) :> stability(x_0,y_0)$ *if* $(1 - |x_1+y_1|/(|x_1|+y_1)) :> (1- |x_0+y_0|/(|x_0|+y_0))$; (10.1)
2. $total\_energy(P_{t1}) :> total\_energy(P_{t0})$ *if* $(|x_1|+y_1) :> (|x_0|+y_0)$; (10.2)

where :> stands for *"significantly greater."*

With Figures 3, 4 and 5, *"getting better"* can be characterized as a time series of symptoms moving significantly closer to the strong and healthy mental equilibrium state (-1,1). Stability alone can't determine the effectiveness of a treatment. A treatment should enhance both stability and energy levels over time. Otherwise, when a patient is getting more stable he/she may also be getting weaker. While condition (1) is easy to understand, the two conditions together guarantee enhanced mental equilibrium even for a mental switch. Thus stability and total-energy provide a unified measure for major depressive and bipolar disorders. With the unified measure, a switch from mania to depression or vice versa can also be described as "getting better" or "getting worse" in general terms. It can also be described in specific terms of stability and total energy. These are illustrated with the following cases:

Case1: A person with severe mania characterized with (0,1) became depressed characterized with (-0.7, 0.4) after a treatment. In this case, the total energy |-0.7|+0.4 = 1.1 which is greater than before the treatment, and stability got much better after the treatment (before: stability = 0.0; after: stability = (1.0 - |-0.7+0.4|)/1.0 = 0.7). Thus, we say the patient is "getting better" in general terms.

Case2: A person with severe mania characterized with (0,1) became depressed characterized with (-0.3, 0.0) after a treatment. Thus, after the treatment total energy = |-0.3|+0.0 = 0.3 and stability = (1.0 - |-0.3 + 0.0|)/1.0 = 0.3). In specific terms we say the stability is "getting better" but energy is getting "much worse" or near "death" (0,0) due to drug side effect. In this case, the side effect would be deemed unacceptable.

Case3: A person with severe bipolar depression characterized with (-1,0) became manic characterized with (-0.3,0.8) after a treatment. In this case, the total energy |-0.8|+0.3 = 1.1 which is greater than before the treatment. But stability got much better after the treatment (before: stability = 0.0; after: stability = (1.0 - |-0.8 + 0.3|)/1.0 = 0.5). We say the patient is "getting better" in general terms even though a mood switch is involved. It is interesting to notice that not only does the total energy increase guarantee the acceptable side effect, but also guarantees enhanced mood balance. On the other hand, had the mania after the treatment been measured as (-0.1, 0.8), the total energy would have been less than before the treatment. Even though the stability had been enhanced, but the patient had not got better in general term.

Case4: A person with severe bipolar depression characterized with (-1,0) became manic characterized with (-0.2,0.5) after a treatment. In this case, the total energy |-0.5|+0.2 = 0.7; but stability got much better after the treatment (before: stability = 0.0; after: stability = (1.0 - |-0.5 + 0.2|)/1.0 = 0.7). In specific terms we say the patient's stability is "getting better" but energy is "getting worse" due to a side effect of the drug. If the patient later regained lost energy with improved stability, we would say "the initial side effect is acceptable."

It can be observed that Figures 3, 4 and 5 can also be implemented as a computer graphical user interface (GUI) with a desktop, laptop, or palm computer with which a psychiatrist can click the screen and classify a patient's symptom. Therefore, the new features make bipolar diagnostic analysis operable with a computer (Zhang *et al.,* 2010). The clinical operability of MSQs will be further discussed in specific medical and bioinformatics terms in a forthcoming journal articles.

MIXED BPD CLASSIFICATION

So far mixed BPD has not been taken into consideration. A major limitation of the traditional spectrum model in Figure 1 is that it can't be used to visualize mixed states of manic-depressive or depressive-manic episodes. A mixed state is traditionally classified as a symptom of mania. It leaves a door open for the question *"Could there be mania mixed with depression (depression-major)?"* As affirmed by the Diagnostic and Statistical Manual of Mental Disorders (DSM-IV, 2000), a mixed state must meet the criteria for a major depressive episode and a manic episode nearly every day for at least one week. However, mixed episodes rarely conform to these qualifications; they can be described more practically as any combination of depressive and manic symptoms (Akiskal & Pinto, 1999; Perugi, Toni & Akiskal, 1999). The Merck Manual of Diagnosis and Therapy (Merck, 2006) splits the DSM-IV diagnosis into ***dysphoric mania*** and ***depressive mixed state***.

With MSQs not only can different mixed states be mathematically classified, they can also be numerically characterized through stability and equilibrium energy analysis. Figure 7 presents a MSQ classification of mixed states based on the concept of bipolar mental fusion and separation (Zhang & Peace, 2007). Figure 7a shows the YinYang coordinates for mental fusion and mental separation respectively; Figure 7b shows the superposition of the two coordinates. The two curves show mental fusion and the

Figure 7. (a) Mental fusion and separation geometries; (b) Superposition of the two geometries

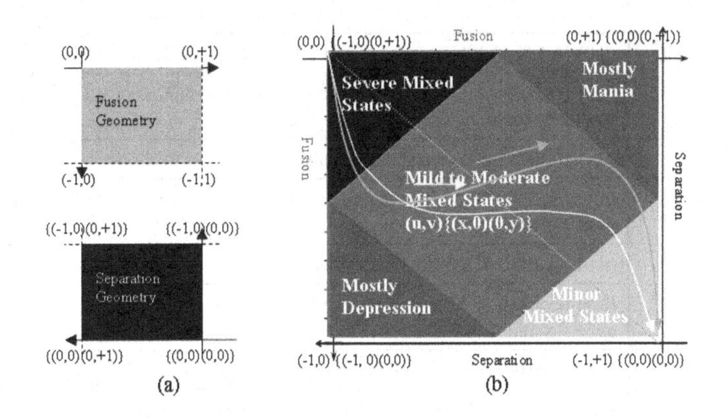

reduction of mental separation, respectively. Two curves are necessary because the absence of mental separation or mixed symptoms does not necessarily translate to strong mental fusion.

A typical mental fusion can be expressed with the logical operation $(-1,0)\oplus(0,1) = (-1,+1)$ and a typical mental separation can be expressed with the logical operation $Sep(-1,1) = \{(-1,0),(0,1)\}$. While (-1,+1) typically characterizes a strong mental equilibrium, {(-1,0),(0,+1)} typically characterizes the total mental separation with the most severe mixed mania and depression. Since the balancer operator $\oplus$ serves as a bipolar mental fusion operator for the two sides, a mixed state can be diagnosed as the result of neurobiological dysfunction of the bipolar mental fusion operator $\oplus$.

With Figure 7 mixed states classification is unified with major BPD classification. Figures 3, 4 and 5 represent the fusion of depression and mania into a mental equilibrium (including quasi- or non-equilibrium such as mania or depression); Figure 7 represents both mental fusion and separation of the two sides where $(u,v)\{(x,0)(0,y)\}$ stands for *"the bipolar mental fusion or equilibrium is measured as (u,v) and the mental separation is measured as {(x,0)(0,y)}."* Mental separation leads to mixed mania-depression states $\{(-x,0)(0,+y)\}$ with $\{(-1,0)(0,+1)\}$ as the most severe separation. The most severe separation can also be equivalently represented as no mental fusion $(0,0)$ or brain death. On the other hand the perfect mental fusion $(-1,+1)$ forbids any mental separation $\{(0,0)(0,0)\}$. Thus, medical treatment of major and mixed BPD should strive for the mental states $(-1,+1)$ and $\{(0,0)(0,0)\}$, respectively. The two are related but not the same. This is shown in Figure 7 and Table 5.

The severity of a mixed state can also be mathematically characterized based on equilibrium energy and stability analysis. The imbalance of a non-mixed state (x,y), $\forall(x,y)\in B_F$, can be characterized as

$$Imbalance(x,y) = |x+y|,$$

where x is negative and y is positive.

The severity of a mixed state $\{(x,y)(u,v)\}$ can be characterized by the total imbalance defined as the bipolar distance of the two as

$$Severity\{(x,y)(u,v)\} = |x + y| + |u + v|.$$

Table 5. Mental fusion and separation quantization $(\forall (x,y) \in B_F)$

Fusion	*Separation (Sep)*
(0,0) – mental death; (x,y), $\lvert x\rvert \approx y$ = small – weak equilibrium;	(0,0) {(-1,0),(0,1)} – worst mixed states with total mental separation without any mental fusion
	(u,v) {(x,0),(0,y)}, $\lvert x\rvert \approx y$=small – minor mixed symptoms with mental separation {(x,0),(0,y)} and mental fusion (u,v) where $u \le -1-x$ and $v \le 1-y$
(0,1) – worst mania; (x,y), $y > \lvert x\rvert$ – mania;	(0,y) {(0,0),(0,y)}, y>0 – mania only
	(u,v) {(x,0),(0,y)}, $y > \lvert x\rvert > 0$ – mixed mania with depression and with mental fusion (u,v) where $u \le -1-x$ and $v \le 1-y$
(-1,0) – worst depression; (x,y), $\lvert x\rvert > y$ – depression;	(x,0) {(x,0),(0,0)}, $\lvert x\rvert > 0$ – depression only
	(u,v) {(x,0),(0,y)}, $\lvert x\rvert > y > 0$ – mixed depression with mania and with mental fusion (u,v) where $u \le -1-x$ and $v \le 1-y$
(-1,1) – mental equilibrium; (x,y); $\lvert x\rvert \approx y$ = medium – moderate equilibrium; (x,y); $\lvert x\rvert \approx y$ = large – strong equilibrium	(-1,1) {(0,0)(0,0)} – no mixed symptoms or perfect mental fusion
	(u,v) {(x,0),(0,y)}, $\lvert x\rvert \approx y$=medium – moderate mixed states with mental separation {(x,0),(0,y)} and mental fusion (u,v) where $u \le -1-x$ and $v \le 1-y$.
	(u,v) {(x,0),(0,y)}, $\lvert x\rvert \approx y$=large – severe mixed states with mental separation {(x,0),(0,y)} and mental fusion (u,v) where $u \le -1-x$ and $v \le 1-y$.

For instance, we have *Severity{(-0.1,0)(0,0.1)}* = *0.2*; *Severity{(-1,0)(0,0)}* = *1*; *Severity{(0,0)(0,+1)}* = *1*; *Severity{(-1,0)(0,+1)}* = *2*; *Severity{(0,0)(0,0)}* = *0*. Evidently, 2 is the maximum severity for all mixed states that characterizes the mental separation of two strong poles of a bipolar equilibrium; 0 is the minimum severity for all mixed states that characterizes the mental fusion of two strong poles of a bipolar equilibrium. Interestingly, both severe mania and depression without mixed symptoms has a severity measure 1. Thus mixed and non-mixed BPDs are unified under MSQs (Zhang *et al.,* 2010). The unifying property is further discussed in specific medical and bioinformatics terms in a forthcoming reports.

BIPOLAR SCALABILITY TO SCHIZOPHRENIA CLASSIFICATION

Recent research results have shown that BPD and schizophrenia could be genetically related (Czobor *et al.,* 2007; Jaeger, Berns & Czobor, 2003; McDonald *et al.,* 2004). In the MSQs approach, the idea of mental fusion and separation provides a common logical ground for both mixed BPD and schizophrenia. While a severe mixed BPD can be characterized as an order-2 bipolar set {(-1,0), (0,1)} which shows the coexistence of both severe depression and mania, an order-N bipolar set as a N-dimensional separation could be a suitable mathematical representation for schizophrenia. After all, BPD and schizophrenia could be both caused by the dysfunction of the mental fusion operator ⊕. For instance, $(-1,0) \oplus (0,1) = (-1,1)$ which is a mental balance but ***Sep****(-1,1)* = *{(-1,0),(0,1)}* which is a severe mixed state. Could schizophrenia be caused by a N-dimensional bipolar separation $\mathbf{Sep}\{D_1,D_2,..D_n\} = \{[(-x,0),(0,x)],\ldots[(-y,0),(0,y)],\ldots[(-v,0),(0,v)]\}$ (e.g. the separation of imagination and observation in addition to the separation of mania and depression) (see Figure 8)? Here we do not attempt to answer this question but only present the possible logical linkage and a common mathematical foundation for both BPDs and schizophrenia. This observation well-illustrates the unifying generic dimensional nature of the equilibrium-based approach. In other words, bipolar equilibrium is a generic form of multidimensional equilibrium and MSQs may

Figure 8. Multidimensional equilibrium-based mental fusion and separation

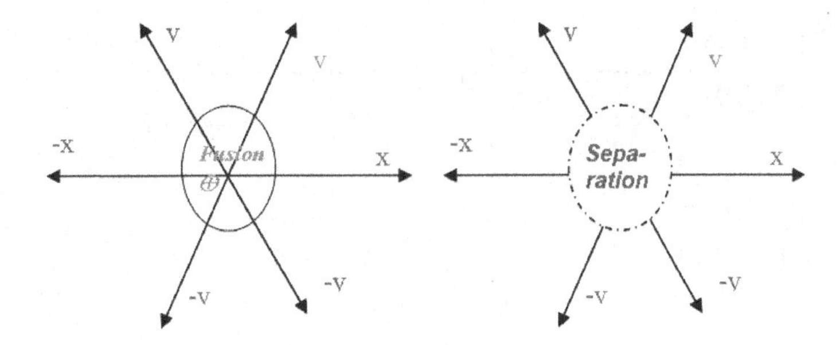

hold a key for unifying different multidimensional approaches to neuroscience, where each square has three dimensions: +pole, -pole, and equilibrium.

OTHER POTENTIAL APPLICATIONS

Neural and Genetic Implications

MSQs provide a unique logical basis for holistic equilibrium-based reasoning at the genetic, neural, mental, and behavioral levels. It links the internal biological activities, external stimuli, and behaviors of neurobiological systems with the bipolar linkage provided by BUMP. The basic neurobiological building block as shown in Figure 6 can be used as "a biological microscope" for deeper exploration into neurobiological systems with potential utilities in brain imaging, genetic regulation, neurobiological pattern analysis, emotion modeling, and psychopharmacological analysis.

For instance, it is common knowledge that genetically related patients show certain similar behaviors such as similar drug reactions (e.g. allergy). Thus, patients who show the same bipolar (e.g. suicidal) drug reaction and the same brain image pattern might also have the same genetic inheritance. In this case equilibrium-based behavioral patterns and drug effects in clinical psychiatry can be linked to other databases for computational psychiatry, computation psychopharmacology, computational neuroscience and genetic pattern classification and exploratory knowledge discovery as depicted in Figures 9, 10, 11 and 12. It is not unreasonable to predict that bipolar equilibrium and non-equilibrium patterns at the behavior and drug effect levels are somewhat related to those at the neurobiological and genomic levels. Such relations may hold the key for equilibrium-based epigenomic regulation as well as mental regulation.

Nano Implications

The concepts of bipolar coexistence, equilibrium, fusion, and separation supported by the bipolar dynamic logic provide an alternative approach to quantum mechanics and nanotechnology in medicine. It is too early to estimate the potential of the new approach in this important direction. However, it is easy to see that, theoretically, bipolar nanomedicine and nanotechnology can be used for mood regulation of

Figure 9. Integration of MSQ $B_F = [-1,0]\times[0,1]$

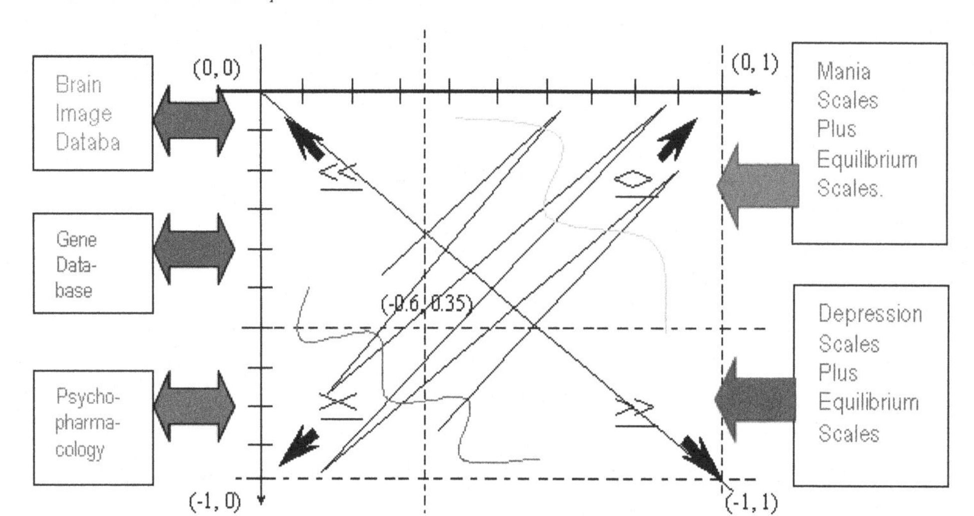

Figure 10. BPD classified as level sets

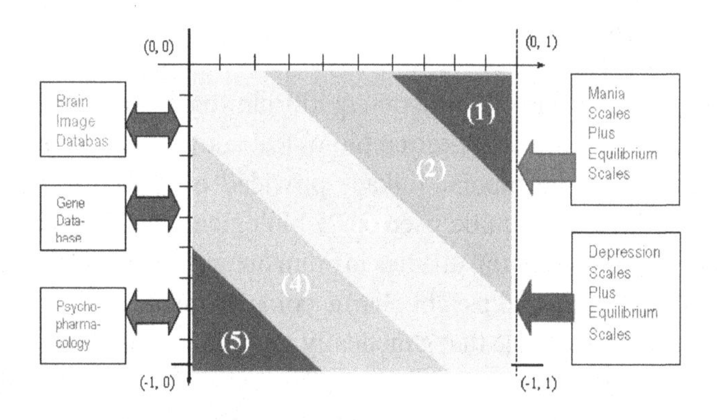

Figure 11. MSQ for Mixed States classification

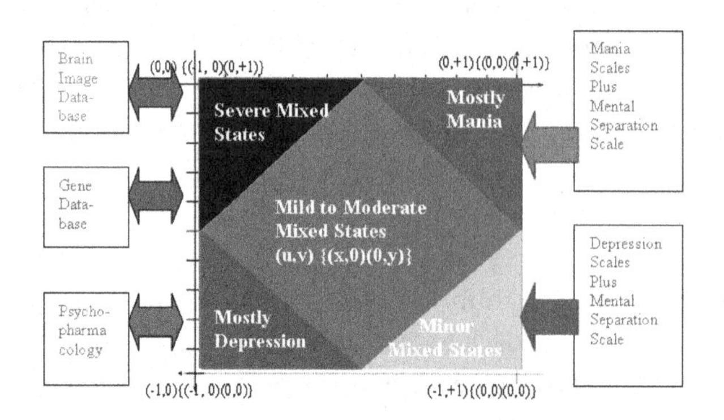

Figure 12. Integrated neurobiological data mining

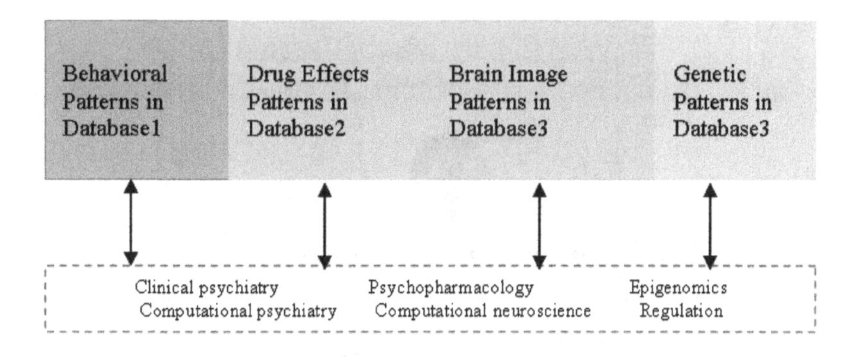

an individual or a cohort of psychiatric patients. The bipolar dynamic logic can be implemented with quantum mechanics for nanobiomedicine (Ch. 6) (Zhang, 2009a, 2009b, 2009c, 2009d, 2010).

COMPUTER OPERABILITY

An experimental web-based computer system has been designed and prototyped for testing the computer operability of the unifying dimensional model. Java Server Pages, Java Applets, and MySQL Database Server technologies are used to maximize the portability and extensibility of the system. The current prototype consists of a relational database for bipolar patient information, a set of Java Server Pages, and a set of Java Applets. The prototype has been implemented based on the well-known Goldberg and Hamilton depression and mania scales with the addition of an equilibrium scale. This extension allows the complementary assessment of a bipolar disorder patient's mental equilibrium in addition to his/her disorder. The computer system enforces the consistency of the equilibrium measure and the severity of mania or depression. Such consistency is expected to reduce misdiagnosis and overdose to a patient to prevent medical accident, especially to pediatric mental disorder patient (Note: Such medical accidents have been reported on national TV). For instance, if the mania scale results in (0, 0.7) and the equilibrium scale results in (-0.5, 0.5) we would have the invalid diagnosis (-0.5, 1.2). The computer system would alarm such inconsistency for further resolution before any final diagnostic decision is reached. A few snapshots are shown in Figures 13, 14, 15, 16, 17, 18, 19 and 20 to illustrate the operability and computability of the system. An equilibrium scale in Table 1 has been implemented as a graphical user interface as shown Figure 14. The equilibrium scale consists of a total of 20 questions each of which is given a score 1 to 5. The total score would naturally provide a percentage measure for mental equilibrium.

RESEARCH TOPICS

This focus of this chapter is bipolar neurobiological modeling at the mental and behavior levels using YinYang bipolar relativity. Other research topics include but are not limited to the following:

Figure 13. MSQs – Front Page

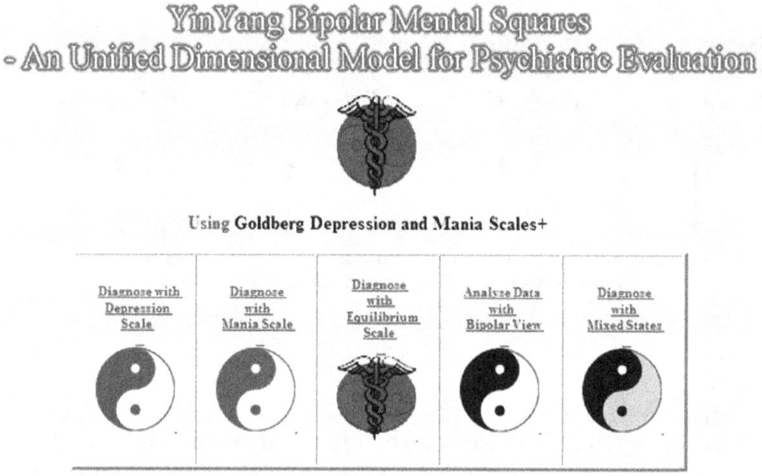

Figure 14. Mental equilibrium scale

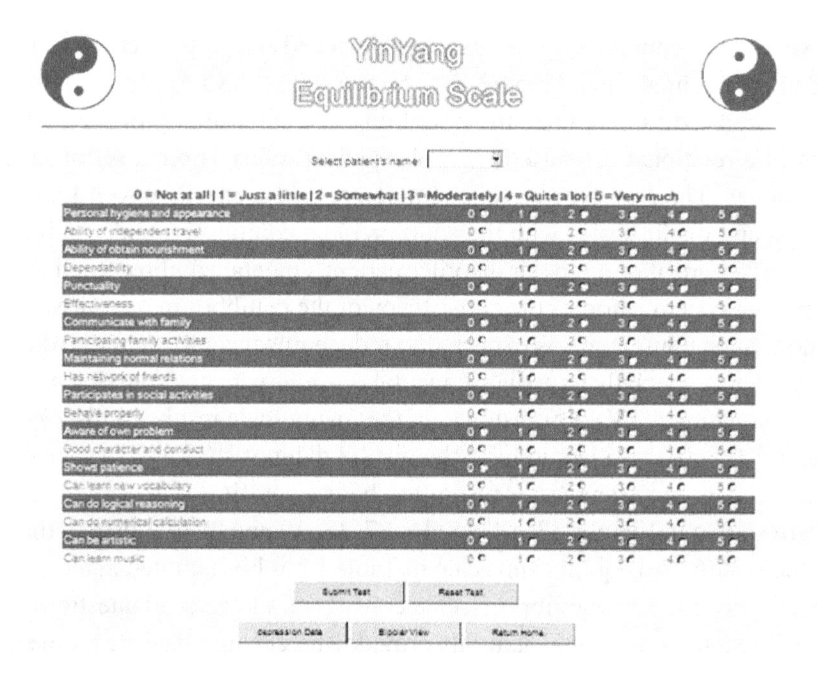

- Combine BDL and BDFL with BQLA and BQCA for application in equilibrium-based neurobiological simulation and regulation.
- Combine BDL and BDFL with BQLA and BQCA for application in equilibrium-based genomics simulation and regulation.
- Combine BDL and BDFL with BQLA and BQCA for application in equilibrium-based molecular interaction simulation and regulation.

Figure 15. Goldberg depression scale with equilibrium-based extension as in Table 4

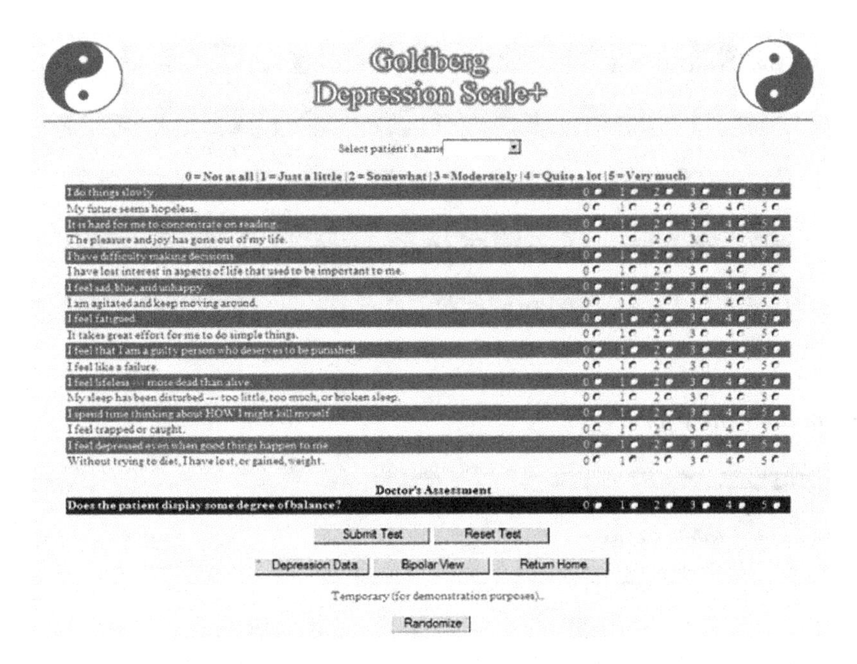

Figure 16. Segment of Goldberg mania scale with equilibrium-based extension as in Table 4

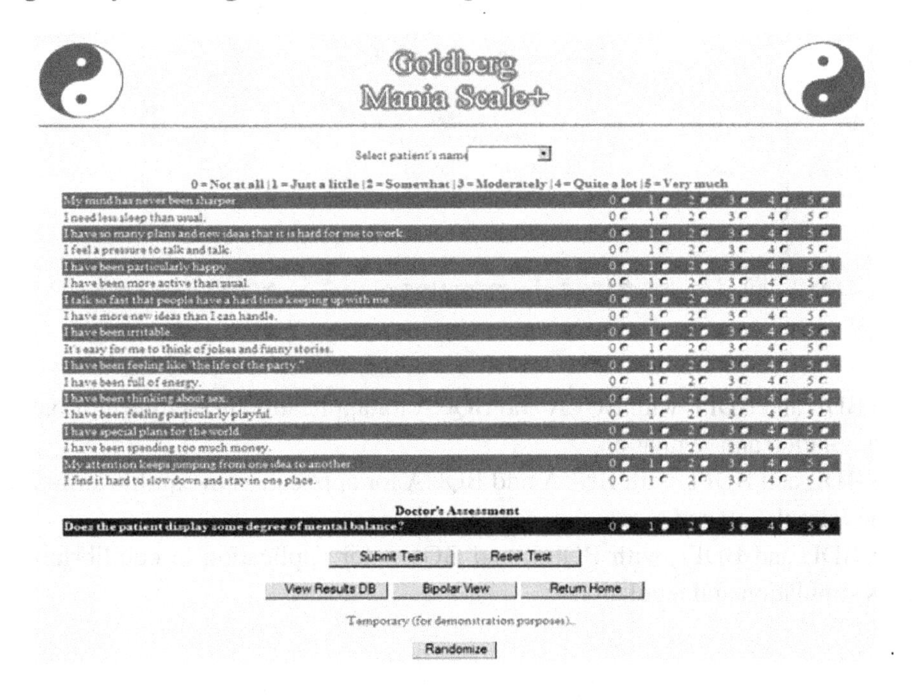

Figure 17. Segment of Mental separation scale

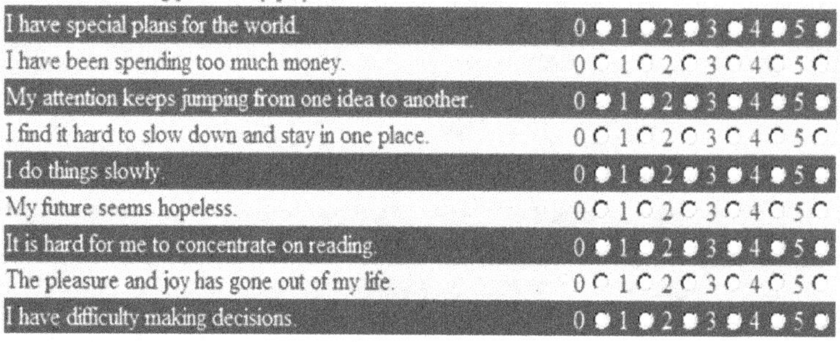

I have special plans for the world.
0 1 2 3 4 5
I have been spending too much money.
0 1 2 3 4 5
My attention keeps jumping from one idea to another.
0 1 2 3 4 5
I find it hard to slow down and stay in one place.
0 1 2 3 4 5
I do things slowly.
0 1 2 3 4 5
My future seems hopeless.
0 1 2 3 4 5
It is hard for me to concentrate on reading.
0 1 2 3 4 5
The pleasure and joy has gone out of my life.
0 1 2 3 4 5
I have difficulty making decisions.
0 1 2 3 4 5

Figure 18. Patient data entry form

New Patient Information
Please enter patient information below:
*First Name:
MI:
*Last Name:
Suffix:
*Sex: Female Male *Age:
*Medications:
Enter the name of the patient's medications here.

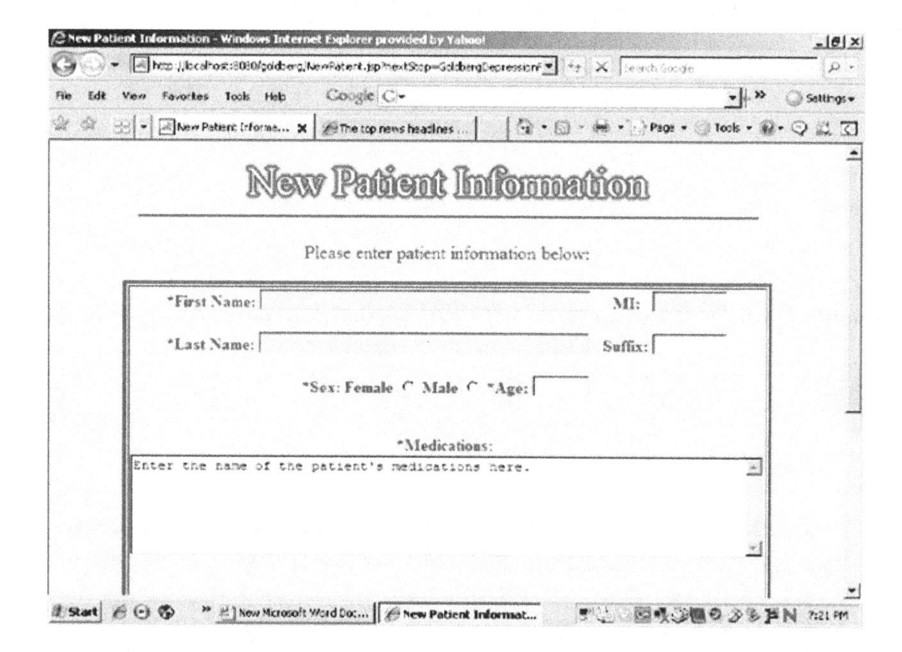

- Combine BDL and BDFL with BQLA and BQCA for application in equilibrium-based macroeconomics simulation and regulation.
- Combine BDL and BDFL with BQLA and BQCA for application in equilibrium-based cognitive informatics simulation and regulation.
- Combine BDL and BDFL with BQLA and BQCA for application in equilibrium-based social dynamics simulation and regulation.

SUMMARY

Based on the theory of YinYang bipolar relativity, *MentalSquares* (MSQs) – an equilibrium-based dimensional approach has been presented for the classification and diagnostic analysis of bipolar disorders.

Figure 19. Simulated patient data

	Entry Date	Patient Name	Score
	2007-06-18 23:55:50.0	John F. Doe	(-0.20, 0.71)
	2007-06-18 23:58:07.0	John F. Doe	(-0.40, 0.86)
	2007-06-19 00:12:59.0	John Boo	(-0.93, 0.40)
	2007-06-19 00:13:09.0	John Boy	(-0.63, 0.20)
	2007-06-19 00:13:22.0	John F. Doe	(-0.44, 0.00)
	2007-06-19 00:13:36.0	John Boo	(-0.74, 0.40)
	2007-06-19 00:14:11.0	John F. Foo	(-0.67, 0.00)
	2007-06-19 00:14:32.0	John Boy	(-0.41, 0.00)
	2007-06-19 00:14:48.0	John Boo	(-0.64, 0.00)
	2007-06-19 09:21:25.0	John F. Doe	(-0.97, 0.20)
	2007-06-19 10:25:22.0	John Boy	(0.00, 0.94)
	2007-07-05 18:57:59.0	John F. Foo	(-0.72, 0.20)
	2007-07-05 18:58:29.0	John F. Foo	(-0.40, 0.91)
	2007-07-05 18:59:21.0	John F. Foo	(-0.20, 0.59)
	Delete Selected	Unselect All	Return Home

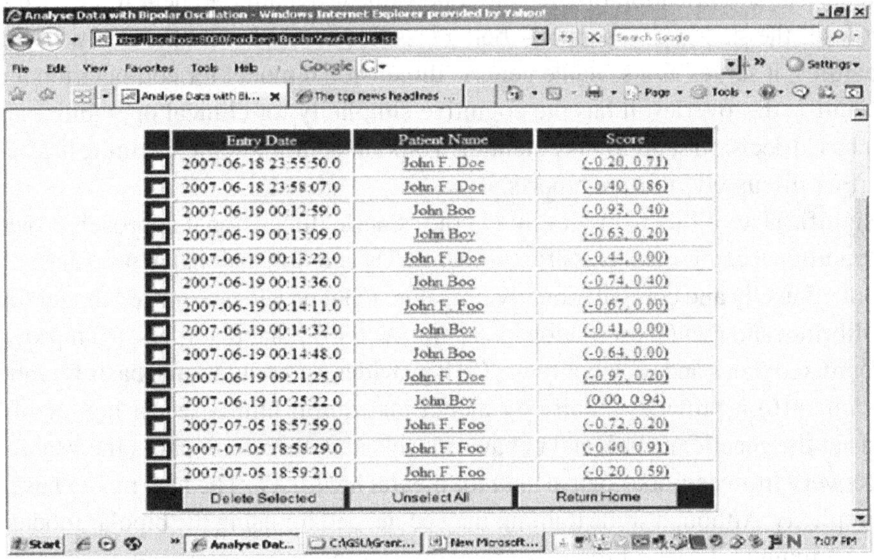

Figure 20. Simulated patient data with graph

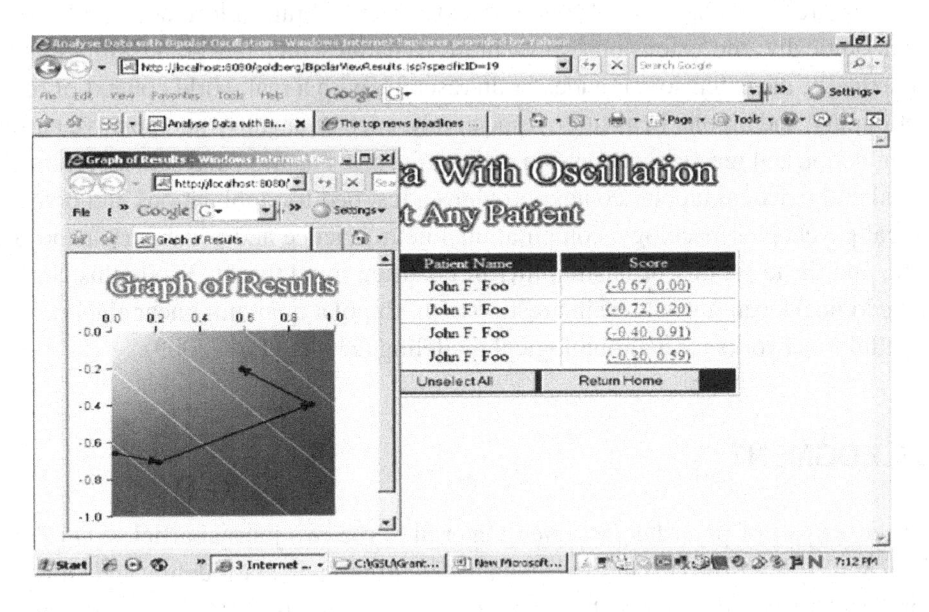

While a support vector machine is defined in Hilbert space, MSQs can be considered a generic approach to support vector machinery. MSQs are focused on mental balance or imbalance of two opposite but bipolar interactive poles. A MSQ is dimensional because two opposite poles are two dimensions from which a third dimension – equilibrium or non-equilibrium – is transcendental with essential mental fusion or mental separation measures. It is generic because any multidimensional mental equilibrium or non-equilibrium can be deconstructed into one or more bipolar equilibria which can then be represented as a MSQ. Different mental squares have been illustrated for BPD classification and diagnostic analysis based on the concept of mental fusion and separation. It has been shown that MSQ extends the traditional categorical classification of BPDs to a non-linear dynamic logical model while preserving all the properties of the standard; it supports both classification and visualization with qualitative and quantitative features; it serves as a scalable generic dimensional model for computational neuroscience for broader scientific discoveries; it has the cognitive simplicity for clinical operability and scalability with computer based decision support. Its potential in bipolar neurobiological mining has been illustrated and will be further discussed in future reports.

Thus, the significance of this work lies in: (1) the generic dimensional approach provides a unified model for the traditional categorical classification of BPDs and dimensional approaches; (2) the generic approach is both clinically and computationally operable; (3) it provides a unified model for both mental fusion for equilibrium and mental oscillation or separation for bipolar disorders; (4) it provides a unified logical basis for mixed states and schizophrenia; (5) it provides a computational basis for mood regulation in nanobiomedcine; (6) it provides a unifying model for equilibrium-based or non-equilibrium-based pattern analysis at the genetic, neural, and behavior levels and, therefore, paves the way for explorative knowledge discovery from neurobiological data for mental health; (7) The new model has the advantage of cognitive simplicity. Although some training may be needed for using the bipolar model, the required learning effort is minimal. From a broader perspective, the agent-oriented nature of MSQs provides a basis for multiagent data mining (Zhang & Zhang, 2004b) and cognitive informatics (Wang, 2004).

The theory of bipolar agents and bipolar relativity opened the door for bipolar brain dynamics and mental health research, which can be deemed an extension of brain informatics with the addition of bipolarity. Theoretically, the extension brings brain dynamics into the grand unification of bipolar relativity; practically, the extension provides brain research with a real world context of bipolar agents, bipolar functionality, and bipolar behaviors, respectively, such as negative and positive particles, matter and antimatter, action and reaction, depression and mania, self-negation and self-assertion abilities, etc. The theoretical and practical bipolar context provides a test bed for applications and potential applications in clinical psychopharmacology, computational neuroscience and psychiatry, nano-biomedicine, mental health, and the testability or falsifiability of quantum mind theory. While this chapter has been focused on the logical level, it would be interesting to use bipolar quantum linear algebra and YinYang-N-Element cellular networks for neurobiological modeling (Zhang *et al.,* 2009).

ACKNOWLEDGMENT

This chapter reuses part of published/accepted material in the two journal articles: (1) Zhang, W. -R., A. Pandurangi & K. Peace (2007). YinYang Dynamic Neurobiological Modeling and Diagnostic Analysis of Major Depressive and Bipolar Disorders. *IEEE Trans. on Biomedical Engineering,* Oct. 2007 54(10):1729-39; (2) Zhang, W. -R., Pandurangi, A., Peace, K., Zhang, Y & Zhao, Z. (2010), A Generic Bipolar Support Vector Machine for Mental Disorder Classification, Diagnostic Analysis, and Neurobiological Data Mining. Inderscience: *Int'l J. on Data Mining and Bioinformatics (IJDMB),* accepted, 2010. Permission to reuse is acknowledged.

REFERENCES

Akiskal, H. S., & Pinto, O. (1999). The evolving bipolar spectrum. Prototypes I, II, III, and IV. *The Psychiatric Clinics of North America, 22*(3), 517–534. doi:10.1016/S0193-953X(05)70093-9

American Psychiatric Association. (2000). *Diagnostic and statistical manual of mental disorders* (4th ed.).

Andrews, G., Brugha, T., Thase, M. E., Duffy, F. F., Rucci, P., & Slade, T. (2007). Dimensionality and the category of major depressive episode. *International Journal of Methods in Psychiatric Research, 16*(S1), S41–S51. doi:10.1002/mpr.216

Bearden, C. E., Hoffman, K. M., & Cannon, T. D. (2001). The neuropsychology and neuroanatomy of bipolar affective disorder: A critical review. [BPDs]. *Bipolar Disorders, 3*, 106–150. doi:10.1034/j.1399-5618.2001.030302.x

Czobor, P., Jaeger, J., Berns, S. M., Gonzalez, C., & Loftus, S. (2007). Neuropsychological symptom dimensions in BPD and schizophrenia. [BPDs]. *Bipolar Disorders, 9*(1-2), 71–92. doi:10.1111/j.1399-5618.2007.00428.x

Glenn, T., Whybrow, P. C., Rasgon, N., Grof, P., Alda, M., & Baethge, C. (2006). Approximate entropy of self-reported mood prior to episodes in bipolar disorder. [BPDs]. *Bipolar Disorders, 8*(51), 424–429. doi:10.1111/j.1399-5618.2006.00373.x

Hudziak, J. J., Achenbach, T. M., Althoff, R. R., & Pine, D. S. (2007). A dimensional approach to developmental psychopathology. *International Journal of Methods in Psychiatric Research, 16*(S1), S16–S23. doi:10.1002/mpr.217

Jaeger, J., Berns, S., & Czobor, P. (2003). The multidimensional scale for independent functioning: A new instrument for measuring functional disability in psychiatric populations. *Schizophrenia Bulletin, 29*, 153–168.

Kessler, R. C., Akiskal, H. S., Ames, M., Birnbaum, H., Greenberg, P., & Robert, M. A. (2006). Prevalence and effects of mood disorders on work performance in a nationally representative sample of U.S. workers. *The American Journal of Psychiatry, 163*, 1561–1568. doi:10.1176/appi.ajp.163.9.1561

Kraemer, H. C. (2007). DSM categories and dimensions in clinical and research contexts. *International Journal of Methods in Psychiatric Research, 16*(S1), S8–S15. doi:10.1002/mpr.211

Krueger, R. F., Andrew, E., Skodol, W., Livesley, J., Shrout, P. E., & Huang, Y. (2007). Synthesizing dimensional and categorical approaches to personality disorders: Refining the research agenda for DSM-V Axis II. *International Journal of Methods in Psychiatric Research, 16*(S1), S65–S73. doi:10.1002/mpr.212

Krug, K., Brunskill, E., Scarna, A., Goodwin, G. M., & Parker, A. J. (2008). Perceptual switch rates with ambiguous structure-from-motion figures in bipolar disorder. *Proceedings. Biological Sciences, 275*(1645), 1839–1848. doi:10.1098/rspb.2008.0043

Lopez, M. F., Compton, W. M., Grant, B. F., & Breiling, J. P. (2007). Dimensional approaches in diagnostic classification: A critical appraisal. *International Journal of Methods in Psychiatric Research, 16*(S1), S6–S7. doi:10.1002/mpr.213

McDonald, C., Bullmore, E. T., Sham, P. C., Chitnis, X., Wickham, H., & Bramon, E. (2004). Association of genetic risks for schizophrenia and bipolar disorder with specific and generic brain structural endophenotypes. *Archives of General Psychiatry*, *61*, 974–984. doi:10.1001/archpsyc.61.10.974

Merck. (2006). *The Merck manual of diagnosis and therapy*. Retrieved from http://www.merck.com/mmpe/index.html

NIMH. (2001). *BPD–a detailed booklet that describes symptoms, causes, and treatments, with information on getting help and coping*. Retrieved from http://www.nimh.nih.gov/publicat/anxiety.cfm

Perugi, G., Toni, C., & Akiskal, H. S. (1999). Anxious-bipolar comorbidity. Diagnostic and treatment challenges. *The Psychiatric Clinics of North America*, *22*(3), 565–583. doi:10.1016/S0193-953X(05)70096-4

Pettigrew, J. D., & Miller, S. M. (1998). A sticky interhemispheric switch in bipolar disorder? *Proceedings. Biological Sciences*, *265*(1411), 2141–2148. doi:10.1098/rspb.1998.0551

Pincus, S.M. (2006). Approximate entropy as a measure of irregularity for psychiatric serial metrics. *Bipolar Disorders,* 8(5p1), 430–440.

Rao, V.S.H., Rao, C.R. & Yeragani, V.K. (2006). A novel technique to evaluate fluctuations of mood: Implications for evaluating course and treatment effects in bipolar/affective disorders. *Bipolar Disorders (BPDs),* 8(5p1), 453–466.

Regier, D. A. (2007). Dimensional approaches to psychiatric classification: Refining the research agenda for DSM-V: An introduction. *International Journal of Methods in Psychiatric Research, 16*(S1), S1–S5. doi:10.1002/mpr.209

Shi, Y., Seto, E., Chang, L.-S., & Shenk, T. (1991). Transcriptional repression by YY1, a human GLI-Kruppel-related protein, and relief of repression by adenovirus E1A protein. *Cell, 67*(2), 377–388. doi:10.1016/0092-8674(91)90189-6

Singer, E. (2008). Mind reading with functional MRI -scientists use brain imaging to predict what someone is looking at. *Technology Review.* Retrieved from http://www.technologyreview.com/biomedicine/20380/

Vieta, E., Cieza, A., Stucki, G., Chatterji, S., Nieto, M., & Sánchez-Moreno, J. (2007). Developing core sets for persons with BPD based on the international classification of functioning, disability and health. [BPDs]. *Bipolar Disorders*, *9*(1-2), 16–24. doi:10.1111/j.1399-5618.2007.00322.x

Wang, Y. (2004). On cognitive informatics. *Brain and Mind*, *4*(2), 151–167. doi:10.1023/A:1025401527570

Zadeh, L. (1971). Similarity relations and fuzzy orderings. *Information Sciences*, *3*, 177–200. doi:10.1016/S0020-0255(71)80005-1

Zhang, W.-R. (2005a). YinYang bipolar lattices and L-sets for bipolar knowledge fusion, visualization, and decision. *International Journal of Information Technology and Decision Making*, *4*(4), 621–645. doi:10.1142/S0219622005001763

Zhang, W.-R. (2005b). YinYang bipolar cognition and bipolar cognitive mapping. *International Journal of Computational Cognition*, *3*(3), 53–65.

Zhang, W.-R. (2006a). YinYang bipolar fuzzy sets and fuzzy equilibrium relations for bipolar clustering, optimization, and global regulation. *International Journal of Information Technology and Decision Making*, *5*(1), 19–46. doi:10.1142/S0219622006001885

Zhang, W.-R. (2006b). YinYang bipolar T-norms and T-conorms as granular neurological operators. *Proceedings of IEEE International Conference on Granular Computing*, (pp. 91-96). Atlanta, GA.

Zhang, W.-R. (2007). YinYang bipolar universal modus ponens (bump) – a fundamental law of nonlinear brain dynamics for emotional intelligence and mental health. *Walter J. Freeman Workshop on Nonlinear Brain Dynamics, Proceedings of the 10th Joint Conference of Information Sciences*, (pp. 89-95). Salt Lake City, Utah.

Zhang, W.-R. (2009a). Six conjectures in quantum physics and computational neuroscience. *Proceedings of 3rd International Conference on Quantum, Nano and Micro Technologies (ICQNM 2009)*, (pp. 67-72). Cancun, Mexico.

Zhang, W.-R. (2009b). YinYang Bipolar Dynamic Logic (BDL) and equilibrium-based computational neuroscience. *Proceedings of International Joint Conference on Neural Networks (IJCNN 2009)*, (pp. 3534-3541). Atlanta, GA.

Zhang, W.-R. (2009c). YinYang bipolar relativity–a unifying theory of nature, agents, and life science. *Proceedings of International Joint Conference on Bioinformatics, Systems Biology and Intelligent Computing (IJCBS)*. (pp. 377-383). Shanghai, China.

Zhang, W.-R. (2009d). The logic of YinYang and the science of TCM–an Eastern road to the unification of nature, agents, and medicine. *International Journal of Functional Informatics and Personal Medicine*, *2*(3), 261–291. doi:10.1504/IJFIPM.2009.030827

Zhang, W.-R. (2010). YinYang bipolar quantum entanglement–toward a logically complete quantum theory. *Proceedings of the 4th Int'l Conf. on Quantum, Nano and Micro Technologies (ICQNM 2010)*, (pp. 77-82). St. Maarten, Netherlands Antilles.

Zhang, W.-R., Pandurangi, A., & Peace, K. (2007). YinYang dynamic neurobiological modeling and diagnostic analysis of major depressive and bipolar disorders. *IEEE Transactions on Bio-Medical Engineering*, *54*(10), 1729–1739. doi:10.1109/TBME.2007.894832

Zhang, W.-R., Pandurangi, K. A., Peace, K. E., Zhang, Y., & Zhao, Z. (in press). MentalSquares–a generic bipolar support vector machine for psychiatric disorder classification, diagnostic analysis and neurobiological data mining. *International Journal on Data Mining and Bioinformatics.*

Zhang, W.-R., & Peace, K. E. (2007). YinYang MentalSquares–an equilibrium-Based system for bipolar neurobiological pattern classification and analysis. *Proceedings of IEEE BIBE*, (pp. 1240-1244). Boston.

Zhang, W.-R., Zhang, H. J., Shi, Y., & Chen, S. S. (2009). Bipolar linear algebra and YinYang-N-Element cellular networks for equilibrium-based biosystem simulation and regulation. *Journal of Biological System*, *17*(4), 547–576. doi:10.1142/S0218339009002958

Zhang, W.-R., & Zhang, L. (2004a). YinYang bipolar logic and bipolar fuzzy logic. *Information Sciences*, *165*(3-4), 265–287. doi:10.1016/j.ins.2003.05.010

Zhang, W.-R., & Zhang, L. (2004b). A Multiagent Data Warehousing (MADWH) and Multiagent Data Mining (MADM) approach to brain modeling and NeuroFuzzy control. *Information Sciences*, *167*, 109–127. doi:10.1016/j.ins.2003.05.011

ADDITIONAL READING

Chen, X.-W., Han, B., Fang, J., & Haasl, R. J. (2008). Large-scale Protein-Protein Interaction prediction using novel kernel methods. *Int'l J. of Data Mining and Bioinformatic*, *2*(2), 145–156. doi:10.1504/IJDMB.2008.019095

Petoukhov, S., & He, M. (2009). *Symmetrical Analysis Techniques for Genetic Systems and Bioinformatics: Advanced Patterns and Applications*. Hershey, PA: IGI Global.

Wang, J. T. L., & Wu, X. (2006). Kernel design for RNA classification using Support Vector Machines. *Int'l J. of Data Mining and Bioinformatics*, *1*(1), 57–76. doi:10.1504/IJDMB.2006.009921

KEY TERMS AND DEFINITIONS

Bipolar Neurobiological Data Mining: Knowledge discovery from neurobiological data based on BDL and bipolar relativity. It is bipolar equilibrium-based instead of truth-based.

Bipolar Support Vector Machine (BSVM): SVM with two opposite poles.

BPD: Bipolar disorder, also known as manic depressive disorder, manic depressive psychosis, manic depression or bipolar affective disorder, is a psychiatric diagnosis that describes a category of mood disorders defined by the presence of one or more episodes of abnormally elevated mood. (American Psychiatric Association, 2000)

DSM IV: Diagnostic and Statistical Manual of Mental Disorders version IV. (American Psychiatric Association, 2000)

Mental Quantum Gravity: A theory on the interplay of quantum mechanics and brain dynamics which is part of the Q5 paradigm (Ch. 7).

Mixed Episode or Mixed-state: In the context of mental disorder, a mixed state (also known as dysphoric mania, agitated depression, or a mixed episode) is a condition during which symptoms of mania and depression occur simultaneously (e.g., agitation, anxiety, fatigue, guilt, impulsiveness, irritability, morbid or suicidal ideation, panic, paranoia, pressured speech and rage) (American Psychiatric Association, 2000).

Support Vector Machine (SVM): A set of related supervised learning methods used for classification and regression (Chen *et al.*, 2008; Wang & Wu, 2006).

Chapter 11
Bipolar Cognitive Mapping and Decision Analysis:
A Bridge from Bioeconomics to Socioeconomics

ABSTRACT

The focus of this chapter is on cognitive mapping and cognitive-map-based (CM-based) decision analysis. This chapter builds a bridge from mental quantum gravity to social quantum gravity. It is shown that bipolar relativity, as an equilibrium-based unification of nature, agent and causality, is naturally the unification of quantum bioeconomics, brain dynamics, and socioeconomics as well. Simulated examples are used to illustrate the unification with cognitive mapping and CM-based multiagent decision, coordination, and global regulation in international relations.

INTRODUCTION

The theory of YinYang bipolar relativity opened the door for bipolar quantum bioeconomics (BQBE) as discussed in Chapter 9, in turn, BQBE opened the door for bipolar quantum brain dynamics (Ch. 10) and bipolar quantum socioeconomics for decision, coordination, and global regulation. Indeed, as a grand unification of nature, agent and causality, YinYang bipolar relativity should naturally be the unification of quantum bioeconomics, brain dynamics, and socioeconomics. Under this unification, biological,

DOI: 10.4018/978-1-60960-525-4.ch011

mental, social, and economical accelerations/decelerations are directly or indirectly related to physical acceleration/deceleration which has been proven equivalent to gravitation in general relativity.

In this chapter we present the theory and application of *bipolar cognitive mapping* to illustrate the quantum nature of brain dynamics in socioeconomics. While decisions in the context of bioeconomics lead to material effects in the context of socioeconomics, brain dynamics can be deemed a symmetry which is defined in particle physics as an operation that doesn't change how something behaves relative to the outside world except in science fiction. With bipolar cognitive mapping, however, we show that brain dynamics and socioeconomics are unified with decision, coordination, and global regulation from a bioeconomics perspective.

According to YinYang theory everything has two poles: a biological agent society is the equilibrium or non-equilibrium of competition and cooperation; a political system is the equilibrium or non-equilibrium of the left and right wings; market economy is the equilibrium or non-equilibrium of the "bears" and "bulls"; the environment issue is the equilibrium or non-equilibrium of pollution and protection. Every relation between two agents or agencies is the equilibrium or non-equilibrium of conflict and common interests even for a married couple or for two allied countries. While objects and systems in the universe including the universe itself form a global equilibrium or non-equilibrium of action-reaction forces, international relations form a global equilibrium or non-equilibrium among human societies on the earth. If we consider the universe as an egg and equilibrium as a chicken, we have the question: *"which one came first and which one created the other in the very beginning?"*

Naturally, we need to ask the question: *If our brain is considered a neurobiological universe, how does it interact with its bipolar equilibrium and non-equilibrium context in decision, coordination, and global regulation?* Theoretically and practically, the bipolar context with logically definable causality and bipolar relativity provide a basis for cause-effect reasoning – a focal point in mind reading, cognitive mapping, and decision making as well as in quantum computing.

In this chapter we consider cognitive mapping as a process of mind reading from speeches, articles, behaviors, and activities of a person that is a different kind of mind reading from using fMRI technology. We use examples in cognitive map based (CM-based) decision, coordination, and global regulation in international relations to illustrate the bipolar equilibrium-based unification of brain dynamics, bioeconomics, and socioeconomics.

While BQCA can be typically used for bioeconomics simulation and regulation as discussed in Chapters 8 and 9, BQCA representation of cognitive maps (*CMs*) can be considered bipolar physical level CMs. Thus, brain dynamics, bioeconomics, and socioeconomics are all involved in CM-based decision, coordination, and global regulation. It is shown that (1) bipolar cognitive mapping can be used for cause-effect reasoning and bipolar clustering enables *conceptual cognitive maps* (*CCMs*) in bipolar relational forms be converted to *visual cognitive maps* (*VCMs*) more suitable for mind reading.

The remaining presentations and discussions of this chapter are organized in the following sections:

- **Mind reading, cognitive mapping, and quantum mind theory**. This section presents a brief background review and a classification of cognitive mapping.
- **Bipolar quantum brain dynamics.** This section presents a unifying theory and an architectural design for bipolar cognitive mapping.
- **Bipolar crisp cognitive map development**. This section presents an application of bipolar cognitive mapping and CM-based decision, coordination, and global regulation in international relations in crisp equilibrium or non-equilibrium relational forms including the topics: (1) An

Example Cognitive Map, (2) From Equilibrium Relations to Equivalence Relations, (3) Bipolar Partitioning and Visualization, (4) Local Equivalence and Bipolar Partitioning, (5) Equilibrium Classes, (6) Equilibrium Energy, (7) International Relations during the Cold War, and (8) The China "Card" and the "Release" of Energy.
- **Bipolar fuzzy CM development**. This section presents an application of bipolar fuzzy cognitive mapping and CM-based decision, coordination, and global regulation in international relations in fuzzy equilibrium or quasi-equilibrium relational forms.
- **Conceptual CM (*CCM*) vs. Visual CM (*VCM*)**. This section shows that CCMs can be converted to VCMs which are more suitable for mind reading.
- **Research Topics**. This section lists a few research topics.
- **Summary**. This section summarizes the key points of the chapter and draws a few conclusions.

This chapter is based on the ideas presented in (Zhang, Chen & Bezdek, 1989; Zhang *et al.*, 1992; Zhang *et al.*, 1994; Zhang, 1996, 2003a, 2003b, 2005a, 2005b, 2006a).

COGNITIVE MAPPING AND MIND READING

Cognitive Mapping

Cognitive map (CM) studies have been under way for more than a century. Urban development psychologists emphasize the importance of "landmarks" and "places" in CM studies (Downs & Stea, 1973). Neurophysiologists refer to CMs as place-coded neuron patterns in animal brains (O'Keefe & Nadal, 1979). Computer vision researchers refer to CMs as spatial closures and their high-level representations (Yeap, 1988). Researchers in the fields of international relations (e.g. Bonham, Shapiro & Trumble, 1979; Zhang, 2003a, 2003b), operational research (e.g. Klein & Cooper, 1982; Montibeller *et al.*, 2008; Montibeller & Belton, 2009), management science (e.g. Kwahk & Kim, 1999; Lee & Kwon, 2008), neural networks (e.g. Kosko, 1986), knowledge representation (e.g. Wellman, 1994; Noh *et al.*, 2000; Chaib-draa, 2002), decision and coordination (e.g. Zhang et al., 1989, 1992, 2006a) refer to CMs as conceptual graphs following Axelrod (1976).

Although the term *"cognitive map"* is used in many different ways, the target worlds of all CMs generally fall into two categories. One category is physical and visible; the other is conceptual and invisible. For instance, a landmark is physically visible, whereas, international relations are conceptual and imaginable but not physically visible. Based on their target worlds, CMs can be classified into *visual CMs* (VCMs) or *conceptual CMs* (CCMs) (Zhang *et al.,* 1992). A VCM is perceived from a physically visible target world that retains some spatial features of the target. A CCM is perceived from either a conceptual world or a physical world that does not retain any spatial features of the target. The fRMI mind reading approach evidently belongs to the VCM approach. It is shown later, however, that a CCM may be transformed to a VCM or vice versa with certain *CM transformation*.

Thus, a CM is defined as a representation of relations that are perceived to exist among the attributes and/or concepts of a given environment (Axelrod, 1976). Since relations in a CM can be neutral, negative and/or positive, CM is a more general representation than a classical binary relation. While BDL, bipolar relations, and equilibrium relations provide a set-theoretic logical basis for cognitive mapping, bipolar quantum linear algebra and bipolar quantum cellular automata provide an algebraic and physical basis for cognitive mapping. It is shown that, as mental representations of physical or conceptual worlds, CMs can be modeled in both ways for clustering, visualization, decision, coordination, and global regulation. We illustrate the ideas in online analytical processing (OLAP) and online analytical mining (OLAM) in international relations and strategic decision analysis.

Mind Reading – fMRI vs. EPR/EEG

The most stunning achievement in brain and mind research is perhaps mind reading. Now scientists can accurately predict which one of hundreds of pictures a person is looking at by analyzing brain activity using functional magnetic resonance imaging (fMRI) (Singer, 2008). The approach should shed light on how the brain processes visual information, and it might one day be used to reconstruct dreams and crime scenes for solving criminal cases. One of the most provocative potential applications for this type of "mind reading" technology has been in lie detection. Event-related potential/electroencephalography (EPR/EEG) is currently being intensively studied for event-related (versus image-related) brain research to further understand the cognitive processes of brain and mind.

As described in Singer:

fMRI detects blood flow in the brain, giving an indirect measure of brain activity. Most fMRI studies to date have used the technology to pinpoint the parts of the brain involved in different cognitive tasks, such as reading or remembering faces. The new study, however, adopts an emerging trend in fMRI: using the technology to analyze neural information processing. By employing computer models to analyze the kinds of information gathered from the neural activity, scientists can try to assess how neural signals are processed in different brain areas and ultimately fused to create a cohesive perception. Researchers have previously used this approach to show that some visual information can be gleaned from brain-imaging data, such as whether a person is looking at faces or houses. (Singer, 2008)

Based on the above description, *mind reading with fMRI technology can be considered a modern approach to visual cognitive mapping* (Downs & Stea, 1973). *On the other hand, mind reading with EPR/EEG technology can be deemed a modern approach to conceptual cognitive mapping of cause-effect reasoning.*

Quantum Mind Theory

Quantum mind theory has been based on the premise that (1) classical mechanics cannot fully explain consciousness and (2) quantum mechanics is necessary to fully understand the mind and brain, particularly concerning an explanation of consciousness. It is suggested that quantum mechanical phenomena such as quantum entanglement and superposition may play an important part in the brain's function and could form the basis of an explanation of consciousness. Since supporters of the theory are long overdue for submitting evidence to support its claims for peer review, the hypothesis has been shelved in the category of untested and also non-falsified theories. As a minority opinion in science, the theory has been unable to enter mainstream research in neuroscience.

BIPOLAR QUANTUM BRAIN DYNAMICS

Observations

Although mind reading is perhaps the most stunning achievement in brain and mind research, it is a top-down approach to the understanding of brain and mind. The intrinsic neurobiological interaction of human consciousness at the neurobiological level is so far not part of the picture. Moreover, the fMRI approach has so far focused on image-related mind reading about visual information. The event-related (versus image-related) potential/electroencephalography (EPR/EEG) approach for further understanding the cognitive processes of brain and mind in procedural cause-effect terms has so far made only limited progress.

While image-related mind reading can be considered visual cognitive map reading or VCM reading, event-related mind reading can be deemed conceptual cognitive map reading or CCM reading with cause-effect reasoning. Again, causality is unavoidable. Thus, CCM-based brain information processing may hold the key for furthering research in event-related mind reading.

From a different perspective, while quantum mind theory is struggling with the problem of testability and falsifiability, bipolar cognitive mapping brought a few important concepts into mental quantum gravity as enumerated in the following:

1. Bipolarity is inherently quantum in nature as it is a fundamental property of action-reaction force symmetry such as (-f,+f) and particle-antiparticle symmetry or quasi-symmetry (-q, +q).
2. While classical quantum entanglement is limited to a microscopic world that is not scalable to a macroscopic world, bipolar quantum entanglement as defined by bipolar relativity is scalable from a microscopic physical world to any macroscopic social, economical, biological, and mental world such as (self-negation, self-assertion), (competition, cooperation), and (repression, activation).
3. Based on the last observation, it is clear that while classical unipolar truth-based quantum entanglement and superposition do not relate the quantum level directly to high level mental activity, bipolar quantum entanglement has accomplished the mission impossible. For instance, (-q,+q) and (-f, +f) in a quantum world is logically equivalent to (competition, cooperation) in a socioeconomical world, (repression, activation) in a genomic world is logically equivalent to (self-negation, self-assertion) in a neurobiological world.
4. While the suggestion that unipolar truth-based quantum entanglement and superposition might play an important part in the brain's function and could form the basis of an explanation of consciousness has so far failed to deliver testable and falsifiable evidence, bipolar quantum cognitive mapping provide a test bed at both the physical and conceptual levels for falsifiability with an open-world and open-ended approach.
5. After all, BDL and bipolar relativity provide both a logical and physical basis for quantum cognitive mapping as a computational paradigm directly usable in CM-based decision, coordination, and global regulation (Zhang *et al.,* 1989, 1992, 1994; Zhang, 2003a, 2003b, 2005a, 2005b, 2006a, 2009a, 2009b, 2009c,, 2009d, 2010).

Theory and Architecture

The above observations lead to the following theory:

1. Bipolar equilibrium, quasi-equilibrium, and non-equilibrium at the microscopic neurobiological levels of the brain can record and process the conceptual information of bipolar equilibrium, quasi-equilibrium, and non-equilibrium in a macroscopic world, and bipolar quantum cellular automata (BQCA) networks serve as a major physical structure for such recording and processing.
2. More advanced conceptual information structures such as certain important or focal bipolar relations, equilibrium-relations, and other structures can be derived from BQCA networks at the physical level.
3. Based on (1) and (2), it can be postulated that BDL, BDFL, bipolar quantum linear algebra, and, especially, bipolar quantum entanglement, are among the major logical and mathematical tools for processing conceptual information.

While neurophysiologic and neurobiological testability is beyond the scope of this work, a 3-tier system architecture is designed as in Figure 1 for the simulation of bipolar cognitive mapping. The top tier is the environmental quantum world; the middle tier consists of conceptual bipolar cognitive maps; and the bottom tier consists of physical or mental cognitive maps. All three tiers are governed by bipolar relativity and all three tiers consist only of agents in equilibria, quasi-equilibria, and non-equilibria.

Figure 2 shows a three step procedural view of cognitive mapping. The first step is *CM composition* which may involve survey, analysis, and gathering opinions from other agents in a distributed environment. The second step is *CM derivation* and *focus generation* in which new and more advanced CMs can be developed. A typical example is the derivation of an equilibrium relation from which a number of foci can be generated (Zhang *et al.,* 1989, 1992; Zhang, 2003a, 2003b).

A first glance at the BQCA in Figure 2 gives us the impression that it is exactly the same as that in Chapter 8, Figure 5. Actually, it is essentially different. First, the number of cells of the BQCA in Figure 2 is not fixed reflecting the fact that human brain has many billions of neurons for constructing a large

Figure 1. A 3-tier architecture for bipolar cognitive mapping

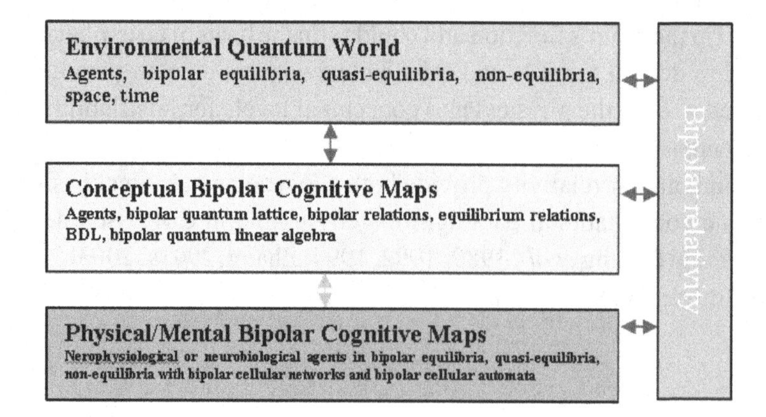

number of BQCA with a large number of cells and links for recording information about agents and agent relations in relational, graph, or matrix forms. Secondly, the link weights here can be randomly assigned, calculated, or perceived and recorded reflecting CM composition, CM derivation, and CM-based decision.

SIMULATION OF BIPOLAR CRISP CCM DEVELOPMENT (ADAPTED FROM ZHANG, 2003A)

An Example Cognitive Map

Following the fact that positive pole and negative pole reflexivity are not mutually exclusive, positive-pole reflexive and negative-pole reflexive equilibria are not mutually exclusive to each other. The word "equilibrium" is selected because a reflexive, symmetric, and transitive bipolar relation represents an equilibrium state of bipolar interrelations among a set of related concepts or agents in a physical or conceptual world. Similar to unipolar cases, a symmetric bipolar relation can be represented as a bi-directional graph (bigraph) and a non-symmetric bipolar relation can be represented as a directed graph (digraph). The edges in the graphs are marked with bipolar strengths. Similar to the unipolar case, the transitive closure of a reflexive and symmetric bipolar relation is still reflexive and symmetric.

It has been proven (Ch. 3) that: (1) the transitive closure of any negative pole reflexive and symmetric bipolar relation R or CM is a negative-pole reflexive or N-type equilibrium relation; (2) the transitive closure of any positive pole reflexive and symmetric bipolar relation R is a P-type equilibrium relation; and (3) the transitive closure of any bipolar reflexive and symmetric bipolar relation R is an NP-type equilibrium relation.

Figure 2. A three step iterative process for cognitive mapping

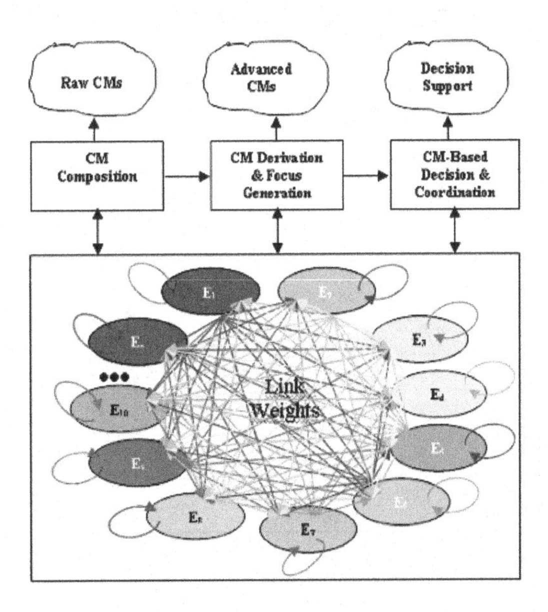

Equilibrium requires reflexivity, bipolar symmetry, and transitivity. Bipolar transitivity allows bipolar relationships to interact and to infer new relationships. Bipolar symmetry extends unipolar symmetry. The difference is that a bipolar symmetric edge represents a compound negative and positive bilateral relationship, while a unipolar symmetric edge is either related or unrelated only.

Bipolar reflexivity is quite different from unipolar reflexivity. In the unipolar case, reflexivity is used as a condition of equivalence. In the bipolar case, negative reflexivity can be used as an indicator of self-adjustability to external or internal changes, because a negative cycle in a transitive chain always adds a positive pole to a negative one and a negative pole to a positive one. Furthermore, negative reflexivity as a cycle in a transitive chain can also adjust itself by cycling one more time.

On the contrary, positive reflexivity does not show self-adjustability due to the fact that a positive cycle does not change the polarity of any negative or positive relationship. Positive pole reflexivity is essentially the same as unipolar reflexivity. It provides, however, an important tie back to classical equivalence relations and classical set theory.

Figure 3 shows a discrete bipolar cognitive map (CM) or bipolar relation in a set of eight concepts or agents {c1,c2,c3,c4,c5,c6,c7,c8} defined on $B_1 = \{-1,0\}\times\{0,1\}$. Note that a bipolar relation or CM can be the result of text data mining from the Web or the result of correlation analysis in a data mining process from a data warehouse. In this chapter we emphasize on CM derivation and focus generation for visualization, decision, and coordination in a post-mining process.

An equilibrium CM provides a data/knowledge fusion structure for clustering and visualization in OLAP and OLAM. Figure 4 shows the positive pole reflexive and symmetric bipolar relation of the CM. The bipolar graph in Figure 3 indicates the negative and positive ties between each other (for instance, the common interests and conflicts among eight agents or concepts (c1-c8). Its $\oplus$-$\otimes$ transitive closure (assuming P-type reflexivity) is computed as in Figure 5. The closure is obviously a P-type equilibrium relation.

Figure 3. A crisp bipolar cognitive map or conceptual graph (Zhang, 2003a) (© 2003 IEEE, used with permission)

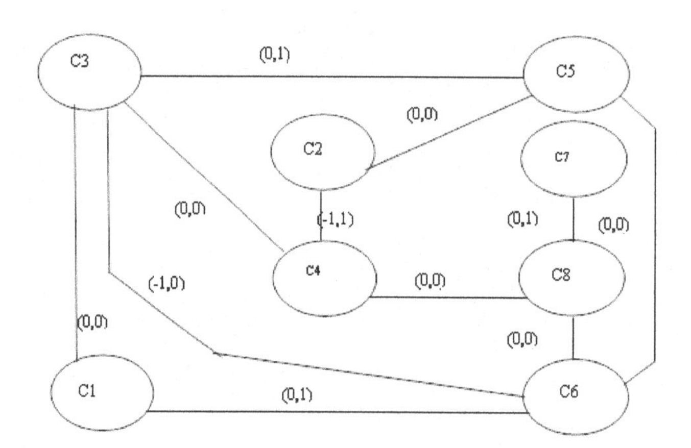

Strict bipolarity distinguishes *bipolar CMs* from those in (Axelrod, 1976). Without strict bipolarity, bipolar equilibrium would not have been brought to light. With bipolarity and equilibrium, the CM in Figure 5 can be considered a fusion of partial knowledge into a complete picture. The complete picture can be used for decision analysis or as a guide for further data mining.

It should be remarked that historically much attention has been given to coalition and conflict sets for competition, and not much research effort has been devoted to equilibrium and harmony. In reality, equilibrium and harmony hold the keys for many tough decision and coordination problems.

From Equilibrium Relations to Equivalence Relations

Bipolarity enriches the representational power, the reasoning power, and the flexibility of relational modeling. Polarized reflexive, symmetric, and transitive (r.s.t) properties (Ch. 3) make equilibrium relations more suitable for classification and multiagent coordination. However, a few theoretical aspects need to be illustrated regarding the relationship between equilibrium and equivalence: (1) How is equilibrium relation related to classical equivalence relations? (2) Is it just an ad hoc structure or a

Figure 4. A positive pole reflexive and symmetric bipolar relation (Zhang, 2003a) (© 2003 IEEE, used with permission)

	c1	c2	c3	c4	c5	c6	c7	c8
c1	(-0 +1)	(-0 +0)	(-0 +0)	(-0 +0)	(-0 +0)	(-0 +1)	(-0 +0)	(-0 +0)
c2	(-0 +0)	(-0 +1)	(-0 +0)	(-1 +1)	(-0 +0)	(-0 +0)	(-0 +0)	(-0 +0)
c3	(-0 +0)	(-0 +0)	(-0 +1)	(-0 +0)	(-0 +1)	(-1 +0)	(-0 +0)	(-0 +0)
c4	(-0 +0)	(-1 +1)	(-0 +0)	(-0 +1)	(-0 +0)	(-0 +0)	(-0 +0)	(-0 +0)
c5	(-0 +0)	(-0 +0)	(-0 +1)	(-0 +0)	(-0 +1)	(-0 +0)	(-0 +0)	(-0 +0)
c6	(-0 +1)	(-0 +0)	(-1 +0)	(-0 +0)	(-0 +0)	(-0 +1)	(-0 +0)	(-0 +0)
c7	(-0 +0)	(-0 +0)	(-0 +0)	(-0 +0)	(-0 +0)	(-0 +0)	(-0 +1)	(-0 +1)
c8	(-0 +0)	(-0 +0)	(-0 +0)	(-0 +0)	(-0 +0)	(-0 +0)	(-0 +1)	(-0 +1)

Figure 5. ⊕-⊗ transitive closure of Figure 4 (P-type crisp equilibrium relation) (Zhang, 2003a) (© 2003 IEEE, used with permission)

	c1	c2	c3	c4	c5	c6	c7	c8
c1	(-0 1)	(-0 0)	(-1 0)	(-0 0)	(-1 0)	(-0 1)	(-0 0)	(-0 0)
c2	(-0 0)	(-1 1)	(-0 0)	(-1 1)	(-0 0)	(-0 0)	(-0 0)	(-0 0)
c3	(-1 0)	(-0 0)	(-0 1)	(-0 0)	(-0 1)	(-1 0)	(-0 0)	(-0 0)
c4	(-0 0)	(-1 1)	(-0 0)	(-1 1)	(-0 0)	(-0 0)	(-0 0)	(-0 0)
c5	(-1 0)	(-0 0)	(-0 1)	(-0 0)	(-0 1)	(-1 0)	(-0 0)	(-0 0)
c6	(-0 1)	(-0 0)	(-1 0)	(-0 0)	(-1 0)	(-0 1)	(-0 0)	(-0 0)
c7	(-0 0)	(-0 0)	(-0 0)	(-0 0)	(-0 0)	(-0 0)	(-0 1)	(-0 1)
c8	(-0 0)	(-0 0)	(-0 0)	(-0 0)	(-0 0)	(-0 0)	(-0 1)	(-0 1)

Figure 6. (a) R^+ of Figure 5; (b). $|R^-| \vee R^+$ of Figure 5 (Zhang, 2003a) (© 2003 IEEE, used with permission)

	c1	c2	c3	c4	c5	c6	c7	c8
c1	1	0	0	0	0	1	0	0
c2	0	1	0	1	0	0	0	0
c3	0	0	1	0	1	0	0	0
c4	0	1	0	1	0	0	0	0
c5	0	0	1	0	1	0	0	0
c6	1	0	0	0	0	1	0	0
c7	0	0	0	0	0	0	1	1
c8	0	0	0	0	0	0	1	1

(a)

	c1	c2	c3	c4	c5	c6	c7	c8
c1	1	0	1	0	1	1	0	0
c2	0	1	0	1	0	0	0	0
c3	1	0	1	0	1	1	0	0
c4	0	1	0	1	0	0	0	0
c5	1	0	1	0	1	1	0	0
c6	1	0	1	0	1	1	0	0
c7	0	0	0	0	0	0	1	1
c8	0	0	0	0	0	0	1	1

(b)

generalization of an equivalence relation? Based on the ***equilibrium laws (ELaws)*** (Ch. 3), Figures 6a and 6b show, respectively, R^+ and $|R^-| \vee R^+$ of the P-type equilibrium relation R in Figure 5, which are both equivalence relations.

Bipolar Partitioning and Visualization

Evidently, the above example shows a bridge from an equilibrium relation back to classical equivalence relations. It is clear that an equilibrium relation is a natural generalization of an equivalence relation by adding polarity – a key property of nature and society for clustering and coordination. While the ***bipolar partitioning laws*** **(Ch. 3)** BPLaw1 and BPLaw2 provide two **necessary** conditions for a bipolar relation to be a crisp equilibrium relation, are the two conditions **sufficient** for a bipolar relation to be a crisp equilibrium relation? The answer is no. Although positive pole reflexivity of R follows the fact that R^+ is an equivalence relation and bipolar symmetry of R follows the fact that both R^+ and $|R^-| \vee R^+$ are equivalence relations, bipolar transitivity of R does not follow the two conditions in general. This can be illustrated by the following example, in which a bipolar relation R meets both the conditions but fails the bipolar transitivity requirement.

$$R = \begin{bmatrix} (-1 \;\; 1) & (0 \;\; 1) \\ (0 \;\; 1) & (-1 \;\; 1) \end{bmatrix}$$

Figure 7. Disjoint subset(s) in X (Zhang, 2003a) (© 2003 IEEE, used with permission)

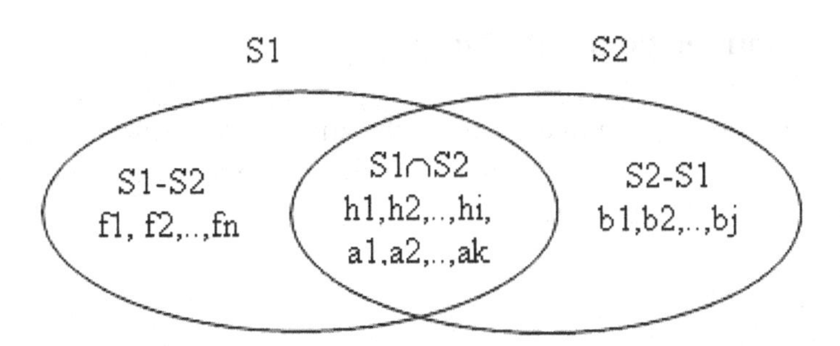

From the above, we can conclude that when ordering or clustering is solely based on R^+ or $|R^-|\vee R^+$ of any P-type equilibrium relation R, bipolar ordering degrades to unipolar ordering and bipolar classification can borrow all available techniques as with equivalence relations.

Following Chapter 3, let f and h be a conflict subset and a harmony subset, respectively; let a and b be a coalition subset not involved a conflict and a coalition subset involved in a conflict, respectively; BPLaw3 can be illustrated by Figure 7.

Based on the equivalence relation $|R^-|\vee R^+$ in Figure 7b, the set of eight concepts can be partitioned into a set S1 with three disjoint subsets {{c1, c3, c5, c6},{c2, c4},{c7, c8}}. This is illustrated in Figure 8a where {c2,c4} is a harmony subset, {c1, c3, c5, c6} is a conflict subset with two coalitions, and {c7, c8} is a coalition subset not involved in any conflict.

Based on the equivalence relation R^+ in Figure 6a, the set of concepts can be partitioned into a set S2 with four disjoint subsets {{c1,c6},{c2, c4},{c3, c5}, {c7,c8}}. This is illustrated in Figure 8b where {c1,c6}, {c3,c5}, and {c7,c8} are coalition subsets, and {c2,c4} is a harmony subset.

Evidently, we have

1. S3=S1∩S2={{c7,c8},{c2,c4}};
2. S4=S1∪S2={{c1,c3,c5,c6},{c1,c6},{c3,c5},{c2, c4},{c7, c8}};
3. S5= S1-S2={ {c1, c3, c5, c6}};

Figure 8. (a) Partitioning based on $|R^-|\vee R^+$ of Figure 6b; (b) Partitioning based on the R^+ of Figure 6a (Zhang, 2003a) (© 2003 IEEE, used with permission)

	{c1	c6	c3	c5}	{c2	c4}	{c7	c8}
c1	1	1	1	1	0	0	0	0
c6	1	1	1	1	0	0	0	0
c3	1	1	1	1	0	0	0	0
c5	1	1	1	1	0	0	0	0
c2	0	0	0	0	1	1	0	0
c4	0	0	0	0	1	1	0	0
c7	0	0	0	0	0	0	1	1
c8	0	0	0	0	0	0	1	1

(a)

	{c1	c6}	{c3	c5}	{c2	c4}	{c7	c8}
c1	1	1	0	0	0	0	0	0
c6	1	1	0	0	0	0	0	0
c3	0	0	1	1	0	0	0	0
c5	0	0	1	1	0	0	0	0
c2	0	0	0	0	1	1	0	0
c4	0	0	0	0	1	1	0	0
c7	0	0	0	0	0	0	1	1
c8	0	0	0	0	0	0	1	1

(b)

4. S6=S2-S1={{c1,c6},{c3,c5}}.

Local Equivalence and Bipolar Partitioning

Note that the above examples don't further distinguish harmony subsets from coalition subsets not involved in any conflict (see Figure 7). In Chapter 3 we proved the following **necessary and sufficient conditions** for R to be an equilibrium relation:

1. R^+ is an equivalence relation;
2. $|R^-|\vee R^+$ is an equivalence relation;
3. if $(R^+\wedge|R^-|)$ is not null, it must be a local equivalence;
4. if $(|R^-|\vee R^+)-(R^+\wedge|R^-|)$ is not null, it must be a local equivalence;
5. if $(|R^-|\vee R^+)-|R^-|$ is not null, it must be a local equivalence; and
6. $R^+-(R^+\wedge|R^-|)\equiv(|R^-|\vee R^+)-|R^-|$.

Figure 9. (a) $R^+\wedge|R^-|$ of R in Fig. 11.5; (b) $(|R^-|\vee R^+)-(R^+\wedge|R^-|)$ of R in Figure 5; (c). $(|R^-|\vee R^+)-|R^-| \equiv R^+-(R^+\wedge|R^-|)$ of R in Figure 5 (Zhang, 2003a) (© 2003 IEEE, used with permission)

	c1	c2	c3	c4	c5	c6	c7	c8
c1	0	0	0	0	0	0	0	0
c2	0	1	0	1	0	0	0	0
c3	0	0	0	0	0	0	0	0
c4	0	1	0	1	0	0	0	0
c5	0	0	0	0	0	0	0	0
c6	0	0	0	0	0	0	0	0
c7	0	0	0	0	0	0	0	0
c8	0	0	0	0	0	0	0	0

(a)

	c1	c2	c3	c4	c5	c6	c7	c8
c1	1	0	1	0	1	1	0	0
c2	0	0	0	0	0	0	0	0
c3	1	0	1	0	1	1	0	0
c4	0	0	0	0	0	0	0	0
c5	1	0	1	0	1	1	0	0
c6	1	0	1	0	1	1	0	0
c7	0	0	0	0	0	0	1	1
c8	0	0	0	0	0	0	1	1

(b)

	c1	c2	c3	c4	c5	c6	c7	c8
c1	1	0	0	0	0	1	0	0
c2	0	0	0	0	0	0	0	0
c3	0	0	1	0	1	0	0	0
c4	0	0	0	0	0	0	0	0
c5	0	0	1	0	1	0	0	0
c6	1	0	0	0	0	1	0	0
c7	0	0	0	0	0	0	1	1
c8	0	0	0	0	0	0	1	1

(c)

Figure 10. (a) A harmony set induced from $R^+\wedge|R^-|$; (b) A conflict subset {c1,c3,c5,c6} and a coalition subset {c7, c8} induced from $(|R^-|\vee R^+)-(R^+\wedge|R^-|)$; (c) Three coalition subsets induced from $(|R^-|\vee R^+)-|R^-| \equiv R^+-(R^+\wedge|R^-|)$ (Zhang, 2003a) (© 2003 IEEE, used with permission)

	c2	c4
c2	1	1
c4	1	1

(a)

	{c1	c3	c5	c6}	{c7	c8}
c1	1	1	1	1	0	0
c3	1	1	1	1	0	0
c5	1	1	1	1	0	0
c6	1	1	1	1	0	0
c7	0	0	0	0	1	1
c8	0	0	0	0	1	1

(b)

	{c1	c6}	{c3	c5}	{c7	c8}
c1	1	1	0	0	0	0
c6	1	1	0	0	0	0
c3	0	0	1	1	0	0
c5	0	0	1	1	0	0
c7	0	0	0	0	1	1
c8	0	0	0	0	1	1

(c)

These conditions are illustrated in Figure 9 – 10. Figsure 9(a-c) show, respectively, $(R^+ \wedge |R^-|)$, $(|R^-| \vee R^+)$-$(R^+ \wedge |R^-|)$, and R^+-$(R^+ \wedge |R^-|) \equiv (|R^-| \vee R^+)$-$|R^-|$ associated with the equilibrium relation R in Figure 5, which are all local equivalences. Figure 10(a-c) illustrate the partitions induced from Figure 9(a-c), which are the results of removing the zero rows and columns from the local equivalence relations from Figure 9(a-c) and rearranging the rows and columns, respectively.

From the above examples, it is clear that equilibrium relations provide a theoretical basis for bipolar partitioning. In practical applications, a scanning utility can be used effectively in a bipolar partitioning algorithm.

Equilibrium Classes

Equilibrium classes play similar roles with respect to an equilibrium relation as do equivalence classes in the case of an equivalence relation. While the meaning of an equivalence class is simple, *an equilibrium class carries the important information of bipolar relativity* of one to many. Specifically, let E be an equilibrium relation in X characterized by a bipolar membership function $\mu_E(x_i,x_j)$, where $X = \{x_i\}$, $1 \le i \le n$. With each x_i we associate *an equilibrium class* denoted by $E[x_i]$ or simply $[x_i]$. This class is a set in X characterized by the membership function

$$\mu_{E[xi]}(x_j) = \mu_E(x_i,x_j). \tag{11.1}$$

Equilibrium classes provide a unique theoretical basis for multiagent data classification, cooperation, competition, and coordination. To illustrate, the equilibrium classes in Figure 5 are listed in Figure 11. With an equilibrium class, a set of objects can be partitioned into four possible types of disjoint clusters: (1) objects related by (0,0) relationships; (2) objects related by (-1, 0) relationships; (3) objects related by (0,1) relationships; and (4) objects related by (-1,1) relationships.

For instance, with respect to c1, the set of concepts $\{c_i\}$ is partitioned into the following clusters based on E(c1):

1. (0,0){c2, c4, c7, c8}: unrelated or neutral concepts;
2. (0,1){c1, c6}: positively related concepts or coalition set;
3. (-1,0){c3, c5}: negatively related concepts or conflict set.

Figure 11. Equilibrium classes in the equilibrium relation of Figure 5 (Zhang, 2003a) (© 2003 IEEE, used with permission)

E(c1)	c1(-0,1),c2(-0,0),c3(-1,0),c4(-0,0),c5(-1,0),c6(-0,1),c7(-0,0),c8(-0,0)
E(c2)	c1(-0,0),c2(-1,1),c3(-0,0),c4(-1,1),c5(-0,0),c6(-0,0),c7(-0,0),c8(-0,0)
E(c3)	c1(-1,0),c2(-0,0),c3(-0,1),c4(-0,0),c5(-0,1),c6(-1,0),c7(-0,0),c8(-0,0)
E(c4)	c1(-0,0),c2(-1,1),c3(-0,0),c4(-1,1),c5(-0,0),c6(-0,0),c7(-0,0),c8(-0,0)
E(c5)	c1(-1,0),c2(-0,0),c3(-0,1),c4(-0,0),c5(-0,1),c6(-1,0),c7(-0,0),c8(-0,0)
E(c6)	c1(-0,1),c2(-0,0),c3(-1,0),c4(-0,0),c5(-1,0),c6(-0,1),c7(-0,0),c8(-0,0)
E(c7)	c1(-0,0),c2(-0,0),c3(-0,0),c4(-0,0),c5(-0,0),c6(-0,0),c7(-0,1),c8(-0,1)
E(c8)	c1(-0,0),c2(-0,0),c3(-0,0),c4(-0,0),c5(-0,0),c6(-0,0),c7(-0,1),c8(-0,1)

With respect to c2 (another agent), the set $\{c_i\}$ is partitioned into the following clusters based on E(c2):

1. (-1,1){c2, c4}: negatively and positively related concepts in a harmony set;
2. (0,0){c1, c3, c5, c6, c7, c8}: neutral or unrelated concepts.

With respect to c3 we have

1. (0,0){c2, c4, c7, c8}: unrelated concepts or neutral set;
2. (0,1){c3, c5}: positively related concepts or coalition set;
3. (-1,0){c1, c6}: negatively related concepts or conflict set.

It should be noted that the clusters of each equilibrium class are disjoint. It is, however, not clear how many disjoint clusters can be extracted from an equilibrium class. The answer is: each equilibrium class without a (-1,1) relationship induces at most three types of disjoint clusters which are characterized, respectively, by the compound relationships (0,0), (-1,0), (0,1). When a (-1,1) relationship exists, an equilibrium class induces at most two types of disjoint clusters characterized, respectively, by (-1,1) and (0,0) relationships.

Bipolar partitions of each equilibrium class with respect to a participating agent provide a basis for multiagent cooperation and competition. For instance, with respect to c1, c6 is a concept with a positive tie, which can indicate common interests or coalition; c3 and c5 are concepts with negative ties to c1, which means c1 can be a competitor to c3 and c5; c2, c4, c7, and c8 are neutral concepts to c1. Based on such a classification, different agents or concepts can find cooperators and competitors in OLAP/OLAM. To illustrate, two equilibrium classes (CMs that support bipolar views) E(c1) and E(c2) are depicted in Figure 12.

Equilibrium Energy

Apparently, given any bipolar cluster or any equilibrium class, the bipolar equilibrium relationships within the cluster or class form a subrelation of R. The bipolar relationships among such a subrelation can play

Figure 12. Graphical representations (CMs) of two equilibrium classes for C1 and C2 (Zhang, 2003) (© 2003 IEEE, used with permission)

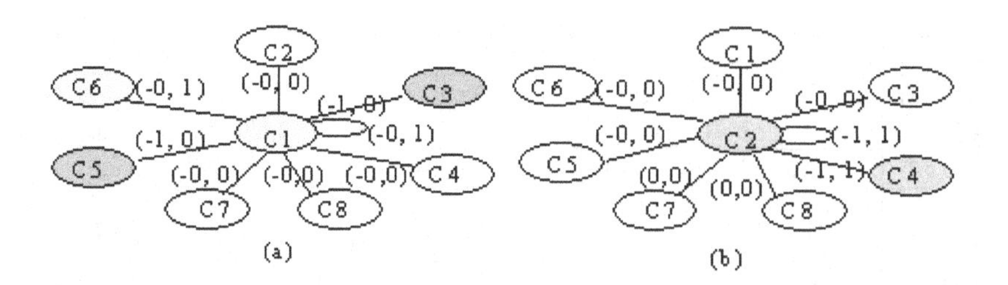

a key role in assessing the characteristics of the cluster or class. Note that a subrelation (e.g., equilibrium class) of an equilibrium relation may not be an equilibrium relation. We need the following relaxation:

Any principle subrelation of an equilibrium relation R is called a ***local equilibrium***. Any subrelation of an equilibrium relation R is called a ***partial equilibrium*** of R.

Based on the equilibrium energy definition in Chapter 6, it is not difficult to verify that in the crisp case (use the 4-valued bipolar lattice B_1), we have:

1. Given any n by n harmony relation R in a harmony set H with size n, we have $|\varepsilon|(R) \equiv 2n^2$, $|\varepsilon^-|(R) \equiv |\varepsilon^+|(R) \equiv n^2$, and $\varepsilon_{imb}(R) \equiv 0$.
2. Given any n by n coalition relation in a coalition set C with size n, we have $|\varepsilon|(R) \equiv \varepsilon_{imb}(R) \equiv n^2$.
3. Given any conflict relation R in a conflict set $F=C1 \cup C2$, where C1 and C2 are coalition sets with size m and n, respectively, we have $|\varepsilon|(R) = |\varepsilon^-|(R) + |\varepsilon^+|(R) \equiv \varepsilon_{imb}(R) \equiv (m+n)^2$ and $|\varepsilon^-|(R) \equiv (m+n)^2 - (m^2+n^2) = 2mn$.

From the above, it can be observed that, in the 4-valued case, energy imbalance is a measure of the lack of harmony. A harmony relation carries a large energy total ($2n^2$) but zero energy imbalance ($\varepsilon_{imb}(R) \equiv 0$). Therefore, it is a healthy state. A coalition relation carries an energy total $|\varepsilon|(R) \equiv n^2$, all of which is an imbalance. A conflict relation carries an energy total $|\varepsilon|(R) \equiv (m+n)^2$, all of which is an imbalance. Since the positive imbalance ε^+ of a conflict set is the total energy in two coalition sets that can be computed as m^2+n^2 assuming the two coalition sets have size m and n, the negative energy $|\varepsilon^-|$ of a conflict set is always $(m+n)^2 - (m^2+n^2) = 2mn$.

In the 4-valued case, the energy imbalance $\varepsilon_{imb}(R)$ can only be negative, zero, or positive. When it is zero, an equilibrium relation is defined on one or more harmony clusters; when it is positive, there are more coalition relations than conflicts; when it is negative, there are more conflict relations than coalition relations. Since any equilibrium relation in a conflict set must satisfy $|\varepsilon^-| \equiv (m+n)^2 - (m^2+n^2) = 2mn \leq (m^2+n^2)$, a negative equilibrium energy $\varepsilon_{imb}(R)$ can only occur in a partial equilibrium (e.g., an equilibrium class). This ensures that the energy cost of a conflict cannot exceed the energy total of both coalition sets.

Based on the above observations, we can conclude that the energy imbalance $\varepsilon_{imb}(R)$ built into coalition and conflict sets is a factor that contributes to the instability of an equilibrium relation. On the other hand, harmony is a factor that contributes to the stability of an equilibrium relation.

The equilibrium energy measures conform to the Yin-Yang theory in traditional Chinese medicine, where the health of a person is measured as the balance and harmonic levels of Yin (negative side) and Yang (positive side). Let the bipolar variable (x,y) = (level of Yin, level of Yang) and let *Health()* be a bipolar function that maps each patient to a bipolar variable, where a bipolar relation $\underline{\gg}$ defines a bipolar partial ordering among the patients.

For instance, given a set of patients {p1,p2,p3,p4} and the mapping function *Health()*; let *Health*(p1) = (0,0) = (no Yin, no Yang) ⇒ no life, *Health*(p2) = (-1,0) = (with Yin, no Yang) ⇒ life with Yang deficiency, *Health*(p3) = (0,1) = (no Yin, with Yang) ⇒ life with Yin deficiency, *Health*(p4) = (-1, 1) = (with Yin and Yang) ⇒ life with balanced YinYang in harmony; it is easy to see that *health*(p4) $\underline{\gg}$ *health*(p2) $\underline{\gg}$ *health*(p1) and *health*(p4) $\underline{\gg}$ *health*(p3) $\underline{\gg}$ *health*(p1). Since p2 cannot be compared with p3, $\underline{\gg}$ defines a bipolar partial ordering among the four patients. In the same light, the health of different objects or agents in cognitive mapping can be measured by their equilibrium energy.

Figure 13. A bigraph of international relationships of the East and West during the cold war (Zhang, 2003a) (© 2003 IEEE, used with permission)

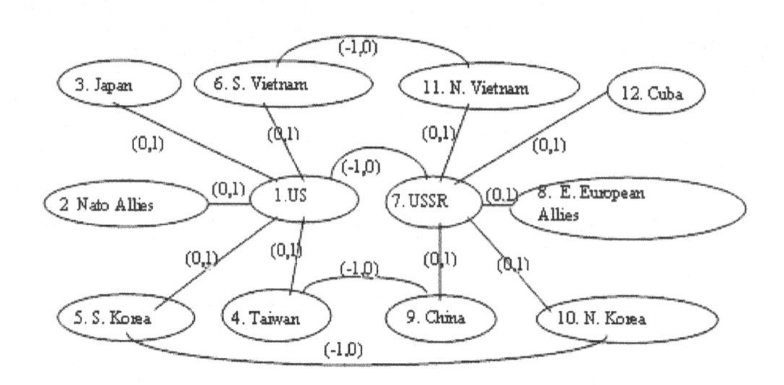

Based on the stability definition in Chapter 6, it is not difficult to verify that, given any harmony relation R in H with size n, coalition relation Q in C with size n, and conflict relation G in F=C1∪C2, C1 has size m and C2 has size n, we have *Stability*(R) $\equiv (2n^2-0)/(2n^2) \equiv 1$; *Stability*(Q) $\equiv (n^2- n^2)/n^2 \equiv 0$; *Stability*(G) $\equiv ((m+n)^2-(m+n)^2)/(m+n)^2 \equiv 0$.

International Relations during the Cold War

Figure 13 shows a bigraph (or *bipolar CM*) of international relations of the East and West in the cold war. A CM can be the result of text mining from the Web that is beyond the scope of this chapter. Here, we focus on bipolar visualization as a post-mining process in OLAP/OLAM. The bigraph can be represented with the reflexive and symmetric crisp bipolar relation R1, as in Figure 14a. The equilibrium relation E1 of R1 is computed by the bipolar relational query *Bclosure(R1),* as in Figure 14b. Partitions induced from $E1^+$ are shown in Figure 14c. The conflicts $E1^-$ are shown in Figure 14d.

This is a black and white example. It is apparent that {1,2,3,4,5,6,7,8,9,10,11,12} is a conflict set, in which two coalition sets are involved, {1,2,3,4,5,6} and {7,8,9,10,11,12}. Using the bipolar relational query *Partition*(E1) we have the result shown in Figure 14e.

The confrontation and armament race between the East and West accumulated a tremendous amount of energy during the cold war era. The energy might have led to a third world war (nuclear war). With the equilibrium relation E1 in Figure 14b, the dangerous energy can be measured by the energy imbalance. Notice that E1 is defined in a single conflict set, and no other sets are involved in E1. The energy measures are calculated as in Figure 14f.

The energy computation for this example is not that interesting; the interesting part lies in the release of the accumulated energy imbalance. Since a third world war was avoided, the energy imbalance did not get released through a nuclear war. Where did the energy go? This question is too important to neglect, and equilibrium relations hold the answers.

Figure 14. (a) The bipolar relation R1 of Figure 13; (b) The equilibrium relation E1 = Bclosure(R1); (c) Partitioning with E1⁺; (d) Conflicts in E1⁻; (e) Partition (E1); (f) Energy and Stability of E1 (Zhang, 2003a) (© 2003 IEEE, used with permission)

No.	1	2	3	4	5	6	7	8	9	10	11	12
1	(0,1)	(0,1)	(0,1)	(0,1)	(0,1)	(0,1)	(-1,0)	(0,0)	(0,0)	(0,0)	(0,0)	(0,0)
2	(0,1)	(0,1)	(0,1)	(0,1)	(0,1)	(0,1)	(0,0)	(0,0)	(0,0)	(0,0)	(0,0)	(0,0)
3	(0,1)	(0,1)	(0,1)	(0,1)	(0,1)	(0,1)	(0,0)	(0,0)	(0,0)	(0,0)	(0,0)	(0,0)
4	(0,1)	(0,1)	(0,1)	(0,1)	(0,1)	(0,1)	(0,0)	(0,0)	(-1,0)	(0,0)	(0,0)	(0,0)
5	(0,1)	(0,1)	(0,1)	(0,1)	(0,1)	(0,1)	(0,0)	(0,0)	(0,0)	(-1,0)	(0,0)	(0,0)
6	(0,1)	(0,1)	(0,1)	(0,1)	(0,1)	(0,1)	(0,0)	(0,0)	(0,0)	(0,0)	(-1,0)	(0,0)
7	(-1,0)	(0,0)	(0,0)	(0,0)	(0,0)	(0,0)	(0,1)	(0,1)	(0,1)	(0,1)	(0,1)	(0,1)
8	(0,0)	(0,0)	(0,0)	(0,0)	(0,0)	(0,0)	(0,1)	(0,1)	(0,1)	(0,1)	(0,1)	(0,1)
9	(0,0)	(0,0)	(0,0)	(-1,0)	(0,0)	(0,0)	(0,1)	(0,1)	(0,1)	(0,1)	(0,1)	(0,1)
10	(0,0)	(0,0)	(0,0)	(0,0)	(-1,0)	(0,0)	(0,1)	(0,1)	(0,1)	(0,1)	(0,1)	(0,1)
11	(0,0)	(0,0)	(0,0)	(0,0)	(0,0)	(-1,0)	(0,1)	(0,1)	(0,1)	(0,1)	(0,1)	(0,1)
12	(0,0)	(0,0)	(0,0)	(0,0)	(0,0)	(0,0)	(0,1)	(0,1)	(0,1)	(0,1)	(0,1)	(0,1)

a.

No.	1	2	3	4	5	6	7	8	9	10	11	12
1	(0,1)	(0,1)	(0,1)	(0,1)	(0,1)	(0,1)	(-1,0)	(-1,0)	(-1,0)	(-1,0)	(-1,0)	(-1,0)
2	(0,1)	(0,1)	(0,1)	(0,1)	(0,1)	(0,1)	(-1,0)	(-1,0)	(-1,0)	(-1,0)	(-1,0)	(-1,0)
3	(0,1)	(0,1)	(0,1)	(0,1)	(0,1)	(0,1)	(-1,0)	(-1,0)	(-1,0)	(-1,0)	(-1,0)	(-1,0)
4	(0,1)	(0,1)	(0,1)	(0,1)	(0,1)	(0,1)	(-1,0)	(-1,0)	(-1,0)	(-1,0)	(-1,0)	(-1,0)
5	(0,1)	(0,1)	(0,1)	(0,1)	(0,1)	(0,1)	(-1,0)	(-1,0)	(-1,0)	(-1,0)	(-1,0)	(-1,0)
6	(0,1)	(0,1)	(0,1)	(0,1)	(0,1)	(0,1)	(-1,0)	(-1,0)	(-1,0)	(-1,0)	(-1,0)	(-1,0)
7	(-1,0)	(-1,0)	(-1,0)	(-1,0)	(-1,0)	(-1,0)	(0,1)	(0,1)	(0,1)	(0,1)	(0,1)	(0,1)
8	(-1,0)	(-1,0)	(-1,0)	(-1,0)	(-1,0)	(-1,0)	(0,1)	(0,1)	(0,1)	(0,1)	(0,1)	(0,1)
9	(-1,0)	(-1,0)	(-1,0)	(-1,0)	(-1,0)	(-1,0)	(0,1)	(0,1)	(0,1)	(0,1)	(0,1)	(0,1)
10	(-1,0)	(-1,0)	(-1,0)	(-1,0)	(-1,0)	(-1,0)	(0,1)	(0,1)	(0,1)	(0,1)	(0,1)	(0,1)
11	(-1,0)	(-1,0)	(-1,0)	(-1,0)	(-1,0)	(-1,0)	(0,1)	(0,1)	(0,1)	(0,1)	(0,1)	(0,1)
12	(-1,0)	(-1,0)	(-1,0)	(-1,0)	(-1,0)	(-1,0)	(0,1)	(0,1)	(0,1)	(0,1)	(0,1)	(0,1)

b.

No.	{1	2	3	4	5	6}	{7	8	9	10	11	12}
1	1	1	1	1	1	1	0	0	0	0	0	0
2	1	1	1	1	1	1	0	0	0	0	0	0
3	1	1	1	1	1	1	0	0	0	0	0	0
4	1	1	1	1	1	1	0	0	0	0	0	0
5	1	1	1	1	1	1	0	0	0	0	0	0
6	1	1	1	1	1	1	0	0	0	0	0	0
7	0	0	0	0	0	0	1	1	1	1	1	1
8	0	0	0	0	0	0	1	1	1	1	1	1
9	0	0	0	0	0	0	1	1	1	1	1	1
10	0	0	0	0	0	0	1	1	1	1	1	1
11	0	0	0	0	0	0	1	1	1	1	1	1
12	0	0	0	0	0	0	1	1	1	1	1	1

c.

No.	{1	2	3	4	5	6}	{7	8	9	10	11	12}
1	0	0	0	0	0	0	-1	-1	-1	-1	-1	-1
2	0	0	0	0	0	0	-1	-1	-1	-1	-1	-1
3	0	0	0	0	0	0	-1	-1	-1	-1	-1	-1
4	0	0	0	0	0	0	-1	-1	-1	-1	-1	-1
5	0	0	0	0	0	0	-1	-1	-1	-1	-1	-1
6	0	0	0	0	0	0	-1	-1	-1	-1	-1	-1
7	-1	-1	-1	-1	-1	-1	0	0	0	0	0	0
8	-1	-1	-1	-1	-1	-1	0	0	0	0	0	0
9	-1	-1	-1	-1	-1	-1	0	0	0	0	0	0
10	-1	-1	-1	-1	-1	-1	0	0	0	0	0	0
11	-1	-1	-1	-1	-1	-1	0	0	0	0	0	0
12	-1	-1	-1	-1	-1	-1	0	0	0	0	0	0

d.

Partition(R) results in {H,F,C,A,B} where

H = {}; // no harmony sets
F = {{1,2,3,4,5,6,7,8,9,10,11,12}}; // one conflict sets
C = {{1,2,3,4,5,6} {7,8,9,10,11,12}}; // two coalition sets
A = {{1,2,3,4,5,6} {7,8,9,10,11,12}}; // two coalition sets in conflict
B = {}//no conflict-free coalition sets

e.

Total Energy	$\|\varepsilon\|(E1) \equiv \varepsilon_{max}(E1) \equiv (m+n)^2 = (6+6)^2 = 144$
Positive Energy	$\|\varepsilon^+\|(E1) \equiv m^2+n^2 = 6^2+6^2 = 72$
Negative Energy	$\|\varepsilon^-\|(E1) \equiv (m+n)^2 - (m^2+n^2) = 2mn = 72$
Energy Imbalance	$\|\varepsilon\|(E1) \equiv \varepsilon_{max}(E1) \equiv (m+n)^2 = (6+6)^2 = 144$
Stability	$Stability(E1) = 0$
Note: m is the size of the West coalition set and n is the size of the East coalition set, in this example m = n	

f.

Figure 15. A bigraph of international relationships in late 1960s (Zhang, 2003a) (© 2003 IEEE, used with permission)

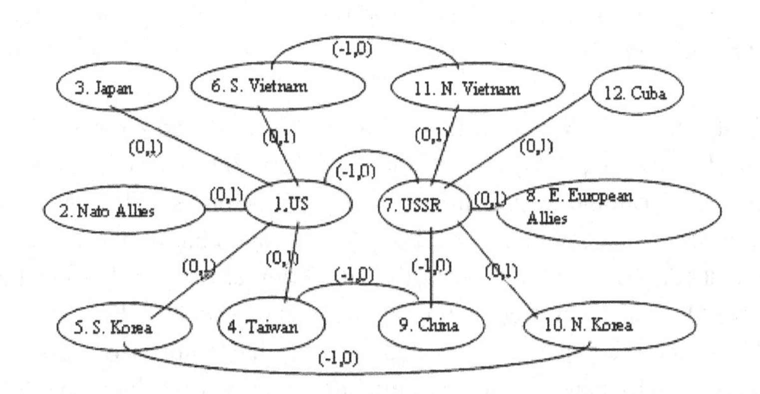

Figure 16. (a) The bipolar relation R2 of Figure 15; (b) The equilibrium relation E2 = Bclosure(R2); (c) Energy and Stability of E2 (Zhang, 2003a) (© 2003 IEEE, used with permission)

No.	1	2	3	4	5	6	7	8	9	10	11	12
1	(0,1)	(0,1)	(0,1)	(0,1)	(0,1)	(0,1)	(-1,0)	(0,0)	(0,0)	(0,0)	(0,0)	(0,0)
2	(0,1)	(0,1)	(0,1)	(0,1)	(0,1)	(0,1)	(0,0)	(0,0)	(0,0)	(0,0)	(0,0)	(0,0)
3	(0,1)	(0,1)	(0,1)	(0,1)	(0,1)	(0,1)	(0,0)	(0,0)	(0,0)	(0,0)	(0,0)	(0,0)
4	(0,1)	(0,1)	(0,1)	(0,1)	(0,1)	(0,1)	(0,0)	(0,0)	(-1,0)	(0,0)	(0,0)	(0,0)
5	(0,1)	(0,1)	(0,1)	(0,1)	(0,1)	(0,1)	(0,0)	(0,0)	(0,0)	(-1,0)	(0,0)	(0,0)
6	(0,1)	(0,1)	(0,1)	(0,1)	(0,1)	(0,1)	(0,0)	(0,0)	(0,0)	(0,0)	(-1,0)	(0,0)
7	(-1,0)	(0,0)	(0,0)	(0,0)	(0,0)	(0,0)	(0,1)	(0,1)	(-1,0)	(0,1)	(0,1)	(0,1)
8	(0,0)	(0,0)	(0,0)	(0,0)	(0,0)	(0,0)	(0,1)	(0,1)	(0,1)	(0,1)	(0,1)	(0,1)
9	(0,0)	(0,0)	(0,0)	(-1,0)	(0,0)	(0,0)	(-1,0)	(0,1)	(0,1)	(0,1)	(0,1)	(0,1)
10	(0,0)	(0,0)	(0,0)	(0,0)	(-1,0)	(0,0)	(0,1)	(0,1)	(0,1)	(0,1)	(0,1)	(0,1)
11	(0,0)	(0,0)	(0,0)	(0,0)	(0,0)	(-1,0)	(0,1)	(0,1)	(0,1)	(0,1)	(0,1)	(0,1)
12	(0,0)	(0,0)	(0,0)	(0,0)	(0,0)	(0,0)	(0,1)	(0,1)	(0,1)	(0,1)	(0,1)	(0,1)

(a)

No.	1	2	3	4	5	6	7	8	9	10	11	12
1	(-1,1)	(-1,1)	(-1,1)	(-1,1)	(-1,1)	(-1,1)	(-1,1)	(-1,1)	(-1,1)	(-1,1)	(-1,1)	(-1,1)
2	(-1,1)	(-1,1)	(-1,1)	(-1,1)	(-1,1)	(-1,1)	(-1,1)	(-1,1)	(-1,1)	(-1,1)	(-1,1)	(-1,1)
3	(-1,1)	(-1,1)	(-1,1)	(-1,1)	(-1,1)	(-1,1)	(-1,1)	(-1,1)	(-1,1)	(-1,1)	(-1,1)	(-1,1)
4	(-1,1)	(-1,1)	(-1,1)	(-1,1)	(-1,1)	(-1,1)	(-1,1)	(-1,1)	(-1,1)	(-1,1)	(-1,1)	(-1,1)
5	(-1,1)	(-1,1)	(-1,1)	(-1,1)	(-1,1)	(-1,1)	(-1,1)	(-1,1)	(-1,1)	(-1,1)	(-1,1)	(-1,1)
6	(-1,1)	(-1,1)	(-1,1)	(-1,1)	(-1,1)	(-1,1)	(-1,1)	(-1,1)	(-1,1)	(-1,1)	(-1,1)	(-1,1)
7	(-1,1)	(-1,1)	(-1,1)	(-1,1)	(-1,1)	(-1,1)	(-1,1)	(-1,1)	(-1,1)	(-1,1)	(-1,1)	(-1,1)
8	(-1,1)	(-1,1)	(-1,1)	(-1,1)	(-1,1)	(-1,1)	(-1,1)	(-1,1)	(-1,1)	(-1,1)	(-1,1)	(-1,1)
9	(-1,1)	(-1,1)	(-1,1)	(-1,1)	(-1,1)	(-1,1)	(-1,1)	(-1,1)	(-1,1)	(-1,1)	(-1,1)	(-1,1)
10	(-1,1)	(-1,1)	(-1,1)	(-1,1)	(-1,1)	(-1,1)	(-1,1)	(-1,1)	(-1,1)	(-1,1)	(-1,1)	(-1,1)
11	(-1,1)	(-1,1)	(-1,1)	(-1,1)	(-1,1)	(-1,1)	(-1,1)	(-1,1)	(-1,1)	(-1,1)	(-1,1)	(-1,1)
12	(-1,1)	(-1,1)	(-1,1)	(-1,1)	(-1,1)	(-1,1)	(-1,1)	(-1,1)	(-1,1)	(-1,1)	(-1,1)	(-1,1)

(b)

Total Energy	$\lvert\varepsilon\rvert(E2) = 2*12^2 = 288$
Positive Energy	$\lvert\varepsilon^+\rvert(E2) = 12^2 = 144$
Negative Energy	$\lvert\varepsilon^-\rvert(E2) = 12^2 = 144$
Energy Imbalance	$\varepsilon_{imb}(E1) = 0$
Stability	*Stability*(E2) = 1

(c)

The China "Card" and the "Release" of Energy

The China card played an important role in ending the cold war. The coalition between China and the USSR intensified the cold war in the 1950s. However, China and the USSR became enemies in the late 1960s. This change is captured in Figure 15. Note that the only difference between Figure 15 and Figure 13 is the relation polarity change between China and the USSR from positive to negative. The bigraph in Figure 15 can be represented with the reflexive and symmetric crisp bipolar relation R2 shown in Figure 16a.

The single polarity change between China and the USSR provided a turning point in the cold war. This can be visualized with the equilibrium relation E2 of R2 as in Figure 16b. Even today, it is a surprise to see that E2 is entirely harmonic, which induces a single harmony set. That is, the China factor made every pair of the East and West countries have both conflict and common interests. The US and China had a common interest because they were facing the same competitor – the USSR; the US and the USSR had a common interest because they were facing the same competitor – China; and similarly China and the USSR were facing the s competitor – the US. The turning point also created both common and conflict interests between any pair of other Western and Eastern countries through transitive or causal paths.

The 37th US president Nixon chose to play the China card with the USSR. He realized that besides ideology difference, the only major obstacle between the US and China was the Taiwan problem. He found a way to open the door to China and brought the US-China relationship into a normalization course, which was soon followed by Japan and the Western allies.

In the last three decades of the 20th century, many events took place; the most remarkable event was perhaps the collapse of the USSR that marked the end of the cold war era. No matter how dramatic it seemed the general trend can be described as a course of energy rebalancing between the bipolar ties among the East and the West countries. Now the harmonic relation E2 has become a reality among most of the countries once involved in the cold war, except that the USSR has been replaced with a new Russia, and Vietnam and Germany are united countries (Zhang, 2003a).

We know that the energy imbalance has been converted into both cooperation and competition in many areas among former cold-war enemies, notably, economic cooperation and competition. Such cooperation and competition led to the political stability and economic prosperity in the last decade of the 20th century. The healthy state of an expected harmonic world during Nixon's presidency could be measured by the energy analysis in Figure 16c. Comparing to the energy and stability in E1 of Figure 14b, the total energy, and the negative and positive energy are doubled, but there is no energy imbalance. Although it is questionable that Nixon expected to have a harmonic world similar to the one in Figure 16b when he planned his visit to China, his trip to China in the early 1970s undoubtedly contributed to the world harmony that was achieved 20 years later (Zhang, 2003a).

How was the energy imbalance "released" from E1? Where did it go? The answer is: *"It is not released; it did not go anywhere; it is rebalanced with bipolar ties."* The nuclear armament race is balanced with economic cooperation (e.g., among the US, China, Russia, Japan, and European countries). Coalitions are balanced with economic competition (e.g., between Boeing and AirBus). However, nuclear weapons still post a threat to mankind. If terrorists were able to gain control of nuclear weapons or if global harmony were jeopardized by local conflicts, a release of energy imbalance through a nuclear war would not be impossible (although unthinkable by any peaceful person) (Zhang, 2003a).

Figure 17. A digraph of international relationships among some countries in America, Asia, and Europe in the post cold war era (Zhang, 2003b)

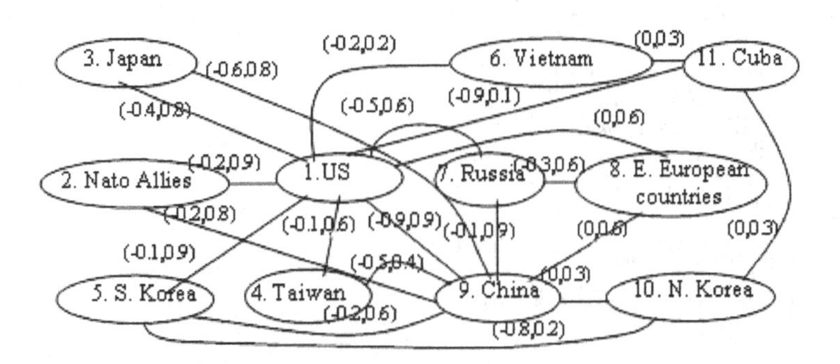

SIMULATION OF BIPOLAR FUZZY CM DEVELOPMENT (ADAPTED FROM ZHANG, 2003B)

A Fuzzified Bipolar CM of International Relations in the Post Cold War Era

Crisp cognitive maps suffer from the bivalent limitation as all crisp models do. That is, a *crisp CM* cannot capture the different degrees of gray levels in a CM. Based on this observation, Kosko (1986) extended unipolar crisp CMs to *fuzzy CMs* (*FCMs*).

Similarly in a bipolar CM a bipolar link cannot capture different degrees of gray levels on its two poles. For instance, a crisp harmonic pair (-1,1) carries no information about what pole is stronger. This limitation is circumvented with bipolar fuzzy CMs (Zhang, 2003b). Bipolar fuzzy relations or CMs provide different gray levels for assessing the strength of coalition, conflict, and harmony with energy and stability analysis. Bipolar fuzzy logic and equilibrium relations are detailed in future reports.

Figure 18. (a) A reflexive and symmetrical bipolar fuzzy relation (R3) of Figure 17 (Zhang, 2003b). (b) E1: ⊕-∧ Bipolar Transitive Closure of R3. (c) E2: ⊕-⊗ Bipolar Transitive Closure of R3 (Zhangm 2003b). (d) E3: ⊕-Δ Bipolar Transitive Closure of R3 (Zhangm 2003b). (e) Bipolar ⊕-⊗ transitive paths (Zhangm 2003b)

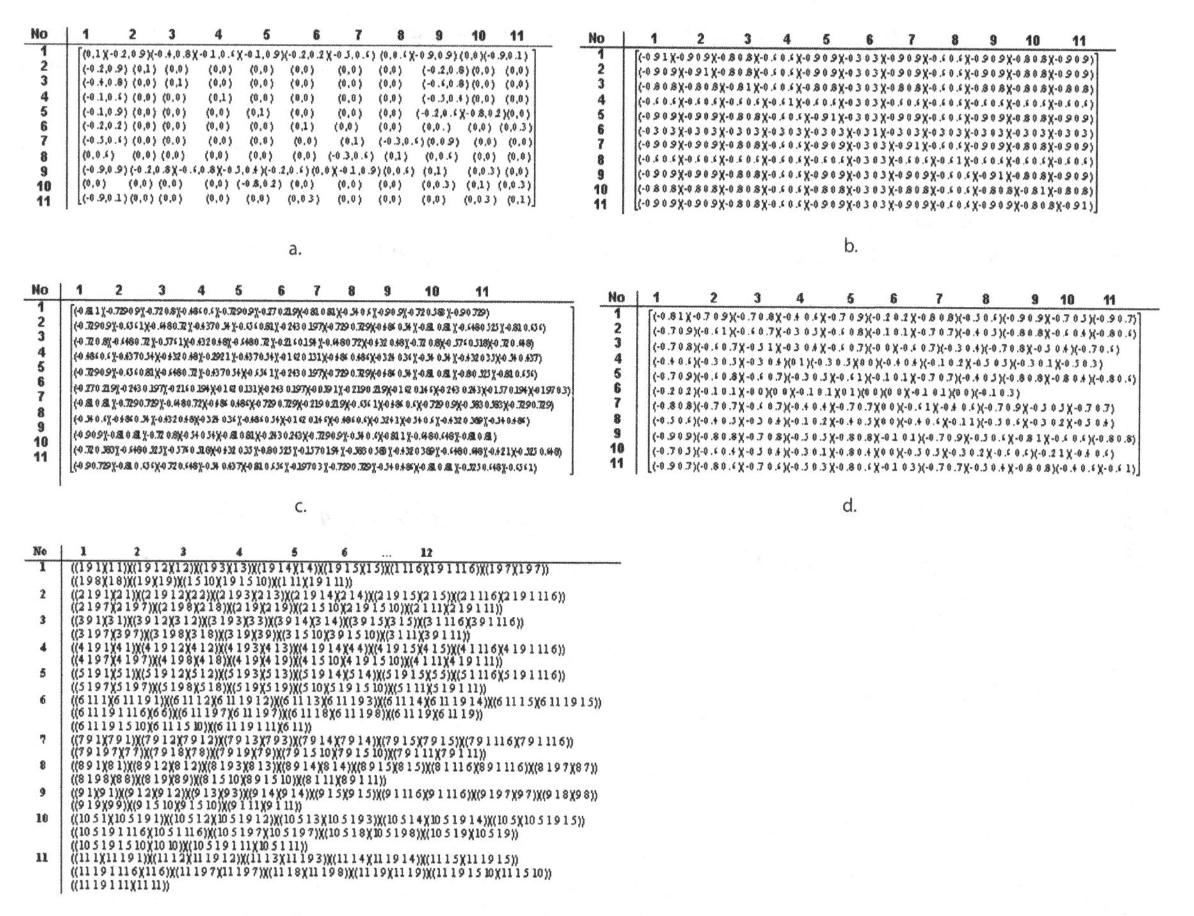

a.

b.

c.

d.

e.

Figure 17 shows a fuzzy bigraph of international relationships in the post cold war era. The bipolar relationship strengths can be asserted objectively based on a number of factors including economical, ideological, historical, and geographical ones. Here the strength is asserted subjectively to illustrate the use of fuzzy equilibrium relations in international relations. Figure 18(a) shows a positive pole reflexive bipolar relation R3 of the bigraph in Figure 17. The $\oplus$-$\wedge$, $\oplus$-$\otimes$, and $\oplus$-Δ transitive fuzzy equilibrium relations E1, E2, and E3 of R3 are computed, respectively, in Figure 18(b-d). The transitive or causal paths are shown in Figure 18e. The paths are also called heuristic paths (Zhang, Chen & Bezdek, 1989) for focus generation in heuristic decision, coordination, and global regulation.

Fuzzy Equilibrium Classes in International Relations

An equilibrium class E(x) provides decision support for agent x. For instance, Figure 19 shows three fuzzy equilibrium classes from the $\oplus$-Δ transitive equilibrium relation in Figure 18d which indicate the potential development of equilibrium among the US, Russia, and China, respectively. It is surprising to see that the US and North Korea are related with the bipolar value (-0.7,0.5) in the equilibrium relation; the US and Cuba are related with the bipolar value (-0.9,0.7). The positive poles seem to be nonsense at first glance. However, a meaningful explanation can be given by the corresponding transitive paths from Figure 18e.

The two corresponding paths for the above two relationships are ((1 5 10)(1 9 1 5 10)) and ((1 11)(1 9 1 11)). Except the direct US-Cuba negative relationship, all others are indirect paths. The path (1 5 10) indicates that the negative relationship -0.7 between the US (No. 1) and North Korea (No. 10) is due to the positive relationship between the US and South Korea (No. 5) and the negative relationship between South Korea and North Korea. The positive relationships are caused by the two paths (1 9 1 5 10) and (1 9 1 11) with North Korea and Cuba (No. 11), respectively. In both paths, there is a cycle (1 9 1). The cycle indicates that the US - China (No. 9) - US cycle leads to the bipolar reflexive relationship (-0.8,1) for the US.

It can be observed that the stronger the negative pole reflexivity of a symmetric bipolar relation R the more harmonic the equilibrium relation of R. Negative reflexivity suggests self-adjustability. Self-adjustability can find common interest from conflict interest. The cycle (1 9 1) (US - China - US) has demonstrated such US self-adjustability. The two equilibrium relationships suggest that, if the US really has strong conflict with Cuba and North Korea, the US will find strong common interests with the two countries sooner or later in some way that follows the US-China model. On the other hand, if the conflicts were trivial, there would be no basis for common interest. (Note: The CM was first constructed as a simulation in 1996. Pope John Paul II visited Cuba from January 21 to 25, 1998. Since then, the US-

Figure 19. Three fuzzy equilibrium classes form Figure 18d (Zhang, 2003b)

E(US)	{US(-0.8 1), NATO-Allies(-0.7 0.9), Japan(-0.7 0.8), Taiwan(-0.4 0.6), South-Korea(-0.7 0.9), Vietnam(-0.2 0.2), Russia(-0.8 0.8), East-European-Countries(-0.5 0.6), China(-0.9 0.9), North-Korea(-0.7 0.5), Cuba(-0.9 0.7)}
E(Russia)	{US(-0.8 0.8), NATO-Allies(-0.7 0.7), Japan(-0.6 0.7), Taiwan(-0.4 0.4), South-Korea(-0.7 0.7), Vietnam(0,0), Russia(-0.6 1), East-European-Countries(-0.4 0.6), China(-0.7 0.9), North-Korea(-0.5 0.5), Cuba(-0.7 0.7)}
E(China)	{US(-0.9 0.9), NATO-Allies(-0.8 0.8), Japan(-0.7 0.8), Taiwan(-0.5 0.5), South-Korea(-0.8 0.8), Vietnam(-0.1 0.1), Russia(-0.7 0.9), East-European-Countries(-0.5 0.6), China(-0.8 1), North-Korea(-0.6 0.6), Cuba(-0.8 0.8)}

Cuba relation has improved to a certain degree following the US-China-US model (1 9 1). On the other hand, the US and North Korea relation has seesawed with periodic improvements and deteriorations.)

Fuzzy Harmony in International Relations

With a fuzzy equilibrium relation we can distinguish different harmonic levels. For instance, with Figure 18b, each harmony condition $\geq\geq(-\alpha,\alpha)$ leads to a harmony clustering $H_{\geq\geq(-\alpha,\alpha)}$. Given $\alpha = 0.9, 0.8, 0.6$, and 0.3, respectively, Figure 18b induces the following harmony clusters:

1. $H_{\geq\geq(-0.9,0.9)}$ = {US, NATO Allies, South Korea, Russia, China, Cuba};
2. $H_{\geq\geq(-0.8,0.8)}$ = {US, NATO Allies, Japan, South Korea, Russia, China, North Korea, Cuba};
3. $H_{\geq\geq(-0.6,0.6)}$ = {US, NATO Allies, Japan, Taiwan, South Korea, Russia, East European Countries, China, North Korea, Cuba};
4. $H_{\geq\geq(-0.3,0.3)}$ = the whole set of 11 countries.

What-If Strategic Decision Analysis in International Relations

Since all countries can be modeled as virtual agents, WHAT-IF decision analysis can be easily conducted to exercise different thoughts. For instance, we can ask the question: "What if the bipolar relationship strength between the US and China in Figure 17 is changed to (0,0), (0,1) or (-1,0)?"

To answer the question, we can change the US-China relationship strength in Figure 17 to (0,0), (0,1), and (-1,0), respectively. We can then compute the equilibrium relations to see the impact of each change. Corresponding to the three changes, three $\oplus$-Δ transitive fuzzy equilibrium relations are computed, respectively, in Tables 5(a)-(c). Four equilibrium classes with respect to the US are compared in Figure 21.

From Figure 21 we see that, if the US-China relationship strength in Figure 17 is changed to (0,0), the maximal transitive bipolar relationship strength between the two countries would be (-0.4, 0.7). If the strength is changed to (0,1), the maximal transitive bipolar relationship strength between the two countries would be (-0.4, 1). If the strength is changed to (-1,0), the maximal transitive bipolar relationship strength between the two countries would be (-1, 0.7). Evidently, if the bipolar strength is set to (a,b), (c,d), and (e,f), where $(0,0) \leq\leq (a,b) \leq\leq (-0.4,0.7)$, $(0,1) \leq\leq (c,d) \leq\leq (-0.4,1)$, $(-1,0) \leq\leq (e,f) \leq\leq (-1, 0.7)$, the equilibrium relations would be the same as setting the strength to (0,0), (0,1), and (-1,0), respectively. Therefore, a choice can be made from the bipolar interval. Energy levels are compared in Figure 22.

From the equilibrium energy levels in Figure 22, we found that (1) the second and third choices lead to the lowest stability; (2) the first choice leads to the most absolute total energy and the highest stability; and (3) the last choice leads to the most negative energy. Based on stability analysis, the first choice is clearly the best choice assuming no other policy changes. The second and third choices are clearly the worst ones.

Similarly, energy analysis can be used for coalition set, conflict set, harmony set, or an entire equilibrium relation, respectively. The results of energy analysis can be used for achieving stronger coalition or cooperation, suitable levels of conflict for competition, suitable harmony levels for both purposes, or for achieving global stability in international relations.

Figure 20.(a) ⊕-Δ transitive fuzzy equilibrium relation if US-China has a (0,0) bipolar relationship strength (Zhang, 2003b). (b) ⊕-Δ transitive fuzzy equilibrium relation if US-China has a (0,1) bipolar relationship strength (Zhang, 2003b)., (c) ⊕-Δ transitive fuzzy equilibrium relation if US-China has a (-1,0) bipolar relationship strength (Zhang, 2003b)

No	1	2	3	4	5	6	7	8	9	10	11
1	[illegible]	[illegible]	[illegible]	[illegible]	[illegible]	[illegible]	[illegible]	[illegible]	[illegible]	[illegible]	[illegible]
2	[illegible]	[illegible]	[illegible]	[illegible]	[illegible]	[illegible]	[illegible]	[illegible]	[illegible]	[illegible]	[illegible]
3	[illegible]	[illegible]	[illegible]	[illegible]	[illegible]	[illegible]	[illegible]	[illegible]	[illegible]	[illegible]	[illegible]
4	[illegible]	[illegible]	[illegible]	[illegible]	[illegible]	[illegible]	[illegible]	[illegible]	[illegible]	[illegible]	[illegible]
5	[illegible]	[illegible]	[illegible]	[illegible]	[illegible]	[illegible]	[illegible]	[illegible]	[illegible]	[illegible]	[illegible]
6	[illegible]	[illegible]	[illegible]	[illegible]	[illegible]	[illegible]	[illegible]	[illegible]	[illegible]	[illegible]	[illegible]
7	[illegible]	[illegible]	[illegible]	[illegible]	[illegible]	[illegible]	[illegible]	[illegible]	[illegible]	[illegible]	[illegible]
8	[illegible]	[illegible]	[illegible]	[illegible]	[illegible]	[illegible]	[illegible]	[illegible]	[illegible]	[illegible]	[illegible]
9	[illegible]	[illegible]	[illegible]	[illegible]	[illegible]	[illegible]	[illegible]	[illegible]	[illegible]	[illegible]	[illegible]
10	[illegible]	[illegible]	[illegible]	[illegible]	[illegible]	[illegible]	[illegible]	[illegible]	[illegible]	[illegible]	[illegible]
11	[illegible]	[illegible]	[illegible]	[illegible]	[illegible]	[illegible]	[illegible]	[illegible]	[illegible]	[illegible]	[illegible]

a.

No	1	2	3	4	5	6	7	8	9	10	11
1	[illegible]	[illegible]	[illegible]	[illegible]	[illegible]	[illegible]	[illegible]	[illegible]	[illegible]	[illegible]	[illegible]
2	[illegible]	[illegible]	[illegible]	[illegible]	[illegible]	[illegible]	[illegible]	[illegible]	[illegible]	[illegible]	[illegible]
3	[illegible]	[illegible]	[illegible]	[illegible]	[illegible]	[illegible]	[illegible]	[illegible]	[illegible]	[illegible]	[illegible]
4	[illegible]	[illegible]	[illegible]	[illegible]	[illegible]	[illegible]	[illegible]	[illegible]	[illegible]	[illegible]	[illegible]
5	[illegible]	[illegible]	[illegible]	[illegible]	[illegible]	[illegible]	[illegible]	[illegible]	[illegible]	[illegible]	[illegible]
6	[illegible]	[illegible]	[illegible]	[illegible]	[illegible]	[illegible]	[illegible]	[illegible]	[illegible]	[illegible]	[illegible]
7	[illegible]	[illegible]	[illegible]	[illegible]	[illegible]	[illegible]	[illegible]	[illegible]	[illegible]	[illegible]	[illegible]
8	[illegible]	[illegible]	[illegible]	[illegible]	[illegible]	[illegible]	[illegible]	[illegible]	[illegible]	[illegible]	[illegible]
9	[illegible]	[illegible]	[illegible]	[illegible]	[illegible]	[illegible]	[illegible]	[illegible]	[illegible]	[illegible]	[illegible]
10	[illegible]	[illegible]	[illegible]	[illegible]	[illegible]	[illegible]	[illegible]	[illegible]	[illegible]	[illegible]	[illegible]
11	[illegible]	[illegible]	[illegible]	[illegible]	[illegible]	[illegible]	[illegible]	[illegible]	[illegible]	[illegible]	[illegible]

b.

No	1	2	3	4	5	6	7	8	9	10	11
1	[illegible]	[illegible]	[illegible]	[illegible]	[illegible]	[illegible]	[illegible]	[illegible]	[illegible]	[illegible]	[illegible]
2	[illegible]	[illegible]	[illegible]	[illegible]	[illegible]	[illegible]	[illegible]	[illegible]	[illegible]	[illegible]	[illegible]
3	[illegible]	[illegible]	[illegible]	[illegible]	[illegible]	[illegible]	[illegible]	[illegible]	[illegible]	[illegible]	[illegible]
4	[illegible]	[illegible]	[illegible]	[illegible]	[illegible]	[illegible]	[illegible]	[illegible]	[illegible]	[illegible]	[illegible]
5	[illegible]	[illegible]	[illegible]	[illegible]	[illegible]	[illegible]	[illegible]	[illegible]	[illegible]	[illegible]	[illegible]
6	[illegible]	[illegible]	[illegible]	[illegible]	[illegible]	[illegible]	[illegible]	[illegible]	[illegible]	[illegible]	[illegible]
7	[illegible]	[illegible]	[illegible]	[illegible]	[illegible]	[illegible]	[illegible]	[illegible]	[illegible]	[illegible]	[illegible]
8	[illegible]	[illegible]	[illegible]	[illegible]	[illegible]	[illegible]	[illegible]	[illegible]	[illegible]	[illegible]	[illegible]
9	[illegible]	[illegible]	[illegible]	[illegible]	[illegible]	[illegible]	[illegible]	[illegible]	[illegible]	[illegible]	[illegible]
10	[illegible]	[illegible]	[illegible]	[illegible]	[illegible]	[illegible]	[illegible]	[illegible]	[illegible]	[illegible]	[illegible]
11	[illegible]	[illegible]	[illegible]	[illegible]	[illegible]	[illegible]	[illegible]	[illegible]	[illegible]	[illegible]	[illegible]

c.

CONCEPTUAL CM (CCM) VS. VISUAL CM (VCM)

As discussed earlier, image-related mind reading has been made possible through fMRI technology but event-related cause-effect mind reading is much harder to achieve. With bipolar cognitive mapping, CCM can be converted to VCM. The conversion is made possible by YinYang bipolar relativity and its geometry.

Figure 21. Four fuzzy equilibrium classes with bipolar relativity (Zhang, 2003b)

E(US) from Fig. 11.20(b) if R(US,China) = (-0.9,0.9)	{US(-0.8 1), NATO-Allies(-0.7 0.9), Japan(-0.7 0.8), Taiwan(-0.4 0.6), South-Korea(-0.7 0.9), Vietnam(-0.2 0.2), Russia(-0.8 0.8), East-European-Countries(-0.5 0.6), China(-0.9 0.9), North-Korea(-0.7 0.5), Cuba(-0.9 0.7)}
E(US) from Fig. 11.20(a) if (0,0) ≤≤ R(US,China) ≤≤(-0.4,0.7)	{US(-0.2 1), NATO-Allies(-0.2 0.9), Japan(-0.4 0.8), Taiwan(-0.2 0.6), South-Korea(-0.1 0.9), Vietnam(-0.2 0.2), Russia(-0.5 0.6), East-European-Countries(-0.1 0.6), China(-0.4 0.7), North-Korea(-0.7 0.1), Cuba(-0.9 0.1)}
E(US) from Fig. 11.20(c) if (0,1) ≤≤ R(US,China) ≤≤ (-0.4,1)	{US(-0.4 1), NATO-Allies(-0.3 0.9), Japan(-0.6 0.8), Taiwan(-0.5 0.6), South-Korea(-0.3 0.9), Vietnam(0.2,0.2), Russia(-0.5 0.9), East-European-Countries(-0.2 0.6), China(-0.4 1), North-Korea(-0.7 0.3), Cuba(-0.9 0.3)}
E(US) from Fig. 11.20(c) if (-1,0) ≤≤ R(US,China) ≤≤ (-1,0.7)	{US(-0.7 1), NATO-Allies(-0.8 0.9), Japan(-0.8 0.8), Taiwan(-0.4 0.6), South-Korea(-0.6 0.9), Vietnam(-0.2 0.2), Russia(-0.9 0.6), East-European-Countries(-0.6 0.6), China(-1 0.7), North-Korea(-0.7 0.4), Cuba(-0.9 0.6)}

Figure 22. Equilibrium energy analysis of four What-If options (Zhang, 2003b)

	$(\varepsilon^-,\varepsilon^+)$ (E(US))	$\lvert\varepsilon\rvert$(E(US)) $=\lvert\varepsilon^-\rvert+\varepsilon^+$	ε_{Imp}(E(US))	Stability
E(US) from Fig. 11.20(b) if (US,China) = (-0.9,0.9)	(-7.3, 7.9)	15.2	1.4	(15.2-1.4)/15.2 = 0.908
E(US) from Fig. 11.20(a) if (0,0) ≤≤ R(US,China) ≤≤(-0.4,0.7)	(-3.9, 6.5)	10.4	4.5	(15.2-4.5)/15.2 = 0.704
E(US) from Fig. 11.20(c) if (0,1) ≤≤ R(US,China) ≤≤ (-0.4,1)	(-5.0, 7.3)	12.3	4.5	(15.2-4.5)/15.2 = 0.704
E(US) from Fig. 11.20(c) if (-1,0) ≤≤ R(US,China) ≤≤ (-1,0.7)	(-7.6, 7.3)	14.9	2.1	(15.2-2.1)/15.2 = 0.862

After the cold war in the last century and before the 9-11 terrorist attack in 2001, the world enjoyed a harmonic environment for a while and the global black-white partitioning as in the cold war era seems to have lost its basis. As a result, most countries belong to a harmony cluster based on certain harmony criteria. In such an environment, bipolar fuzzy clustering with fuzzy equilibrium classes respect to the corresponding countries is more meaningful.

For instance, the equilibrium class E(US) in Figure 20a can be roughly plotted in the 2-D geometry of bipolar relativity as in Figure 23a. From this plotting, it is clear that, except North Korea (No. 10) and Cuba (No. 11), all the other countries are either on the main diagonal or on the positive side of the 2-D space. Therefore, to the interest of the US (No. 1), the 11 countries can be roughly clustered with a bipolar α-cut as

F = {10,11}; // those with more conflicts than common interests
C = {1,2,3,4,5,8}; //those with more common interests than conflicts
H1 = {7,9}; //those with significant equal common interests and conflicts
H2 = {6}; //those with insignificant equal common interests and conflicts

Figure 23(a). A plot of the equilibrium class E(US) (Bipolar relativity of US to the world) (b). A plot of the equilibrium class E(Russia) (Bipolar relativity of Russia to the world) (c). A plot of the equilibrium class E(China) (Bipolar relativity of China to the world)

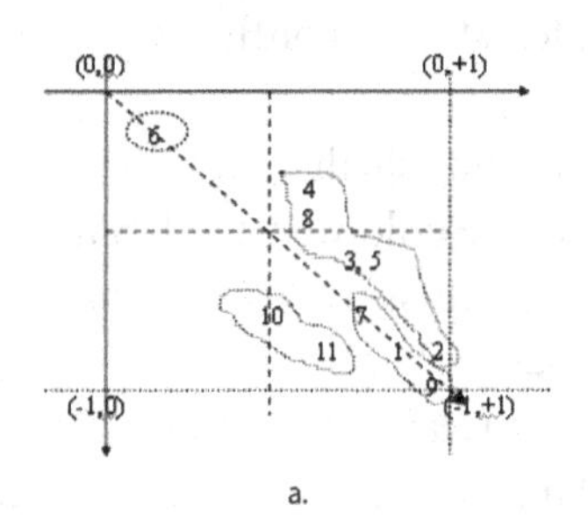

a.

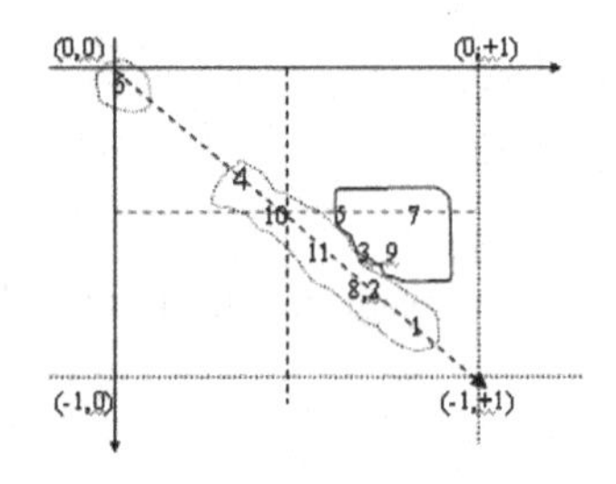

b.

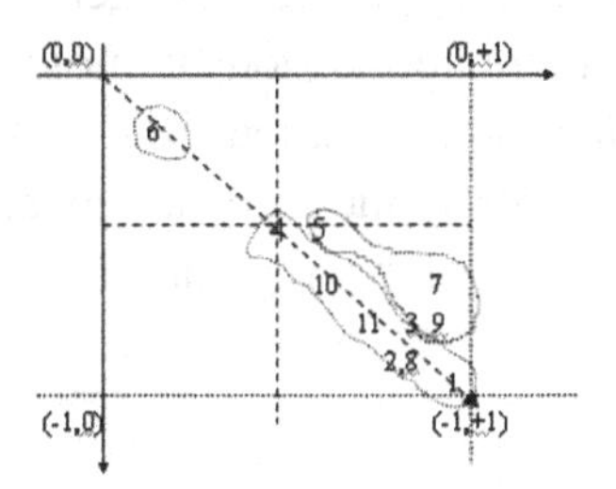

c.

The equilibrium class E(Russia) and E(China) in Figure 19 are roughly plotted in Figure 23b and Figure 23c, respectively. From E(Russia), we may conclude that, with respect to the interest of Russia (No. 7), the 11 countries can be roughly clustered with a bipolar α-cut as

F = {}; // those with more conflicts than common interests
C = {3,7,8,9}; // those with more common interests than conflicts
H1 = {1,4,5,8,10,11}; // those with significant equal common interests and conflicts
H2 = {6}; // those with no common interests and conflicts

Similarly, from E(China), we may conclude that, respect to the interest of China(No. 9), the 11 countries can be clustered with a bipolar α-cut as

F = {}; // those with more conflicts than common interests
C = {3,7,8,9}; // those with more common interests than conflicts
H1 = {1,4,5,8,10,11}; // those with significant equal common interests and conflicts
H2 = {6}; // those with no common interests and conflicts

Based on the above analysis we can conclude that more common interests among old enemies in the cold war can be developed. Although in the cold war, the US and China once faced the USSR together as their common enemy, after the cold war, Russia and China have more common interests than conflicts so do US and China and US and Russia.

From a technical perspective, bipolar relativity provides the YinYang bipolar geometry for CCM-VCM unification. It can be postulated that not only can such unification be used for image-related mind reading, but it can also be used for even-related mind reading with EPR/EEG technology.

RESEARCH TOPICS

This focus of the chapter is cognitive modeling in international relations. Other research topics include but are not limited to the following:

- Develop BQCA-based algorithms for equilibrium-based bipolar cognitive mapping.
- Apply bipolar cognitive mapping for applications in brain dynamics and cognitive informatics.
- Apply bipolar cognitive mapping in mind reading.
- Apply bipolar cognitive mapping for applications in genomics.
- Apply bipolar cognitive mapping for applications in molecular interaction.
- Apply bipolar cognitive mapping for applications in social dynamics for social harmony.
- Apply bipolar cognitive mapping for applications in operational research.
- Apply bipolar cognitive mapping for global regulation in environmental protection.

SUMMARY

Bipolar cognitive mapping has been discussed. It has been shown that bipolar relativity, as a grand unification of nature, agent and causality leads to the notion of equilibrium-based quantum gravity which as a physical theory can be the unification of quantum bioeconomics, brain dynamics, and socioeconomics as well. Simulated application examples in cognitive mapping and cognitive map based (CM-based) decision, coordination, and global regulation in international relations have been used to illustrate the unification of bipolar quantum brain dynamics, bipolar quantum bioeconomics, and bipolar socioeconomics. It has been shown that conceptual cognitive maps (CCMs) in bipolar relational forms can be converted to visual CMs (VCMs) in YinYang geometry of bipolar relativity.

Simulated examples of international relationships in the cold war era have been used for bipolar (crisp) partitioning; examples in the post cold war era have been used to illustrate fuzzy clustering and bipolar decision analysis. It is shown that WHAT-IF strategic decision analysis can be conducted. It has been illustrated that energy analysis can be used to support decision making regarding coalition, conflict, harmony, and global stability in international relations.

Bipolar cognitive mapping and visualization in OLAP and OLAM have a number of advantages over unipolar cognitive maps. First, it combines a mathematical model with a cognitive model into a unified representation that is ideal for clustering and visualization. Secondly, strict bipolarity leads to the notion of equilibrium relations that provides a theoretical basis for bipolar data/knowledge fusion and coordination. Thirdly, an equilibrium cognitive map induces bipolar partitioning or clustering for visualizing bipolar relativity of one to many. Fourthly, bipolar relativity distinguishes disjoint coalition subsets not involved in a conflict, disjoint coalition subsets involved in a conflict, disjoint conflict subsets, and disjoint harmony subsets. Fifthly, equilibrium energy analysis provides a basis for predictions in strategic decision analysis.

Crisp cognitive maps, however, suffer from the limitation that all crisp models do and, therefore, need to be complemented with fuzzy cognitive maps. That is, a crisp bipolar link in a CM cannot capture different degrees of gray levels on its two poles. This limitation is circumvented with bipolar fuzzy CMs. Bipolar fuzzy CMs provide different gray levels for assessing the strength of coalition, conflict, and harmony with energy and stability analysis. Bipolar fuzzy cognitive mapping, visualization, clustering, and coordination have been illustrated with examples in post cold-war international relations. Thus, this work bridges a theoretical gap between bioeconomics and socioeconomics with cognitive-map-based brain dynamics.

ACKNOWLEDGMENT

This chapter reuses part of published material in the two articles: (1) Zhang, W. -R. (2003a). Equilibrium Relations and Bipolar Cognitive Mapping for Online Analytical Processing with Applications in International Relations and Strategic Decision Support. *IEEE Trans. on SMC, Part B, Vol. 33. No. 2,* April 2003. pp.295-307. (2) Zhang, W. -R. (2003b). Equilibrium Energy and Stability Measures for Bipolar Decision and Global Regulation. *Taiwan Fuzzy Systems Association:International Journal of Fuzzy Systems. Vol. 5, No. 2,* June 2003, pp.114-122. Permission to reuse is acknowledged.

REFERENCES

Axelrod, R. (1976). *Structure of decision.* Princeton, NJ: Princeton University Press.

Bonham, G. M., Shapiro, M. J., & Trumble, T. L. (1979). The October war. *International Studies, 23,* 3–14. doi:10.2307/2600273

Chaib-draa, B. (2002). Causal maps: Theory, implementation, and practical applications in multiagent environments. *IEEE Transactions on KDE, 14*(6), 1201–1217.

Downs, R. M., & Stea, D. (1973). *Image and environment.* Chicago: Aldine.

Klein, J. H., & Cooper, D. F. (1982). Cognitive maps of decision-makers in a complex game. *The Journal of the Operational Research Society, 33*(1), 63–71.

Kosko, B. (1986). Fuzzy cognitive maps. *International Journal of Man-Machine Studies, 24,* 65–75. doi:10.1016/S0020-7373(86)80040-2

Kwahk, K.-Y., & Kim, Y.-G. (1999). Supporting business process redesign using cognitive maps. *Decision Support Systems, 25*(2), 155–178. doi:10.1016/S0167-9236(99)00003-2

Lee, K. C., & Kwon, S. (2008). A cognitive map-driven avatar design recommendation DSS and its empirical validity. *Decision Support Systems, 45*(3), 461–472. doi:10.1016/j.dss.2007.06.008

Montibeller, G., & Belton, V. (2009). Qualitative operators for reasoning maps: Evaluating multi-criteria options with networks of reasons. *European Journal of Operational Research, 195*, 829–840. doi:10.1016/j.ejor.2007.11.015

Montibeller, G., Belton, V., Ackermann, F., & Ensslin, L. (2008). Reasoning maps for decision aid: An integrated approach for problem-structuring and multi-criteria evaluation. *The Journal of the Operational Research Society, 59*(5), 575–589. doi:10.1057/palgrave.jors.2602347

O'Keefe, J., & Nadal, L. (1979). *The hippocampus as a cognitive map*. Oxford: Claredon Press.

Singer, E. (2008). Mind reading with functional MRI-scientists use brain imaging to predict what someone is looking at. *Technology Review*. Retrieved from http://www.technologyreview.com/biomedicine/20380/

Wellman, M. P. (1994). Inference in cognitive maps. *Mathematics and Computers in Simulation, 36*, 137–148. doi:10.1016/0378-4754(94)90028-0

Yeap, W. K. (1988). Towards a computational theory of cognitive maps. *Artificial Intelligence, 34*(3), 297–360. doi:10.1016/0004-3702(88)90064-1

Zhang, W.-R. (1996). NPN fuzzy sets and NPN qualitative algebra: A computational framework for bipolar cognitive modeling and multiagent decision analysis. *IEEE Transactions on SMC, 16*, 561–574.

Zhang, W.-R. (2003a). Equilibrium relations and bipolar cognitive mapping for online analytical processing with applications in international relations and strategic decision support. *IEEE Transactions on SMC. Part B, 33*(2), 295–307.

Zhang, W.-R. (2003b). Equilibrium energy and stability measures for bipolar decision and global regulation. *International Journal of Fuzzy Systems, 5*(2), 114–122.

Zhang, W.-R. (2005a). YinYang bipolar lattices and L-sets for bipolar knowledge fusion, visualization, and decision making. *International Journal of Information Technology and Decision Making, 4*(4), 621–645. doi:10.1142/S0219622005001763

Zhang, W.-R. (2005b). YinYang bipolar cognition and bipolar cognitive mapping. *International Journal of Computational Cognition, 3*(3), 53–65.

Zhang, W.-R. (2006a). YinYang bipolar fuzzy sets and fuzzy equilibrium relations for bipolar clustering, optimization, and global regulation. *International Journal of Information Technology and Decision Making, 5*(1), 19–46. doi:10.1142/S0219622006001885

Zhang, W.-R. (2009a). Six conjectures in quantum physics and computational neuroscience. *Proceedings of 3rd International Conference on Quantum, Nano and Micro Technologies (ICQNM 2009)*, 67-72. Cancun, Mexico.

Zhang, W.-R. (2009b). YinYang Bipolar Dynamic Logic (BDL) and equilibrium-based computational neuroscience. *Proceedings of International Joint Conference on Neural Networks (IJCNN 2009)*, (pp. 3534-3541). Atlanta, GA.

Zhang, W.-R. (2009c). YinYang bipolar relativity–a unifying theory of nature, agents, and life science. *Proceedings of International Joint Conference on Bioinformatics, Systems Biology and Intelligent Computing (IJCBS)*. (pp. 377-383). Shanghai, China.

Zhang, W.-R. (2009d). The logic of YinYang and the science of TCM–an Eastern road to the unification of nature, agents, and medicine. *International Journal Functional Informatics and Personal Medicine*, *2*(3), 261–291. doi:10.1504/IJFIPM.2009.030827

Zhang, W.-R. (2010). YinYang bipolar quantum entanglement–toward a logically complete quantum theory. *Proceedings of the 4th Int'l Conf. on Quantum, Nano and Micro Technologies (ICQNM 2010)*, (pp. 77-82). St. Maarten, Netherlands Antilles.

Zhang, W.-R., Chen, S., & Bezdek, J. C. (1989). POOL2: A generic system for cognitive map development and decision analysis. *IEEE Transactions on SMC*, *19*(1), 31–39.

Zhang, W.-R., Chen, S., Wang, W., & King, R. (1992). A cognitive map based approach to the coordination of distributed cooperative agents. *IEEE Transactions on SMC*, *22*(1), 103–114.

Zhang, W.-R., Wang, W., & King, R. (1994). An agent-oriented open system shell for distributed decision process modeling. *Journal of Organizational Computing*, *4*(2), 127–154. doi:10.1080/10919399409540220

ADDITIONAL READING

Kitchin, K., & Freundschuh, S. (Eds.). (2000). Cognitive Mapping*: Past, Present and Future*. Routledge, 2000.

Redish, A. D. (1999). *Beyond the Cognitive Map: From Place Cells to Episodic Memory*. MIT Press, 1999.

KEY TERMS AND DEFINITIONS

CM: Cognitive map – a conceptual representation of a perceived world. **CCM:** Conceptual CM in the form of conceptual graph (Zhang *et al.*, 1992).

VCM: Visual CM in the form of geometrical form (Zhang *et al.*, 1992).

Crisp CM (Crisp CCM): CCM with crisp link weights in {0,1} (Zhang *et al.*, 1992).

Fuzzy CM (Fuzzy CCM): CCM with fuzzy link weights in [0,1] (Kosko, 1986).

Bipolar CM: CCM with bipolar link weights (Zhang, 2003a).

CM Transformation: The transformation of a CCM to VCM or vice versa.

CM Composition: The process of constructing a bigger CM from more than one small ones (Zhang, 1992, 2003a).

CM Derivation: The process of extracting a different CM from an existing one (Zhang, 1992, 2003a).

Focus Generation: The process of generating foci from CMs (Zhang, 1992, 2003a).

Part 5
Discussions and Conclusions

Chapter 12
Causality is Logically Definable: An Eastern Road toward Quantum Gravity

ABSTRACT

This is the conclusion chapter. Bertrand Russell's view on logic and mathematics is briefly reviewed. An enjoyable debate on bipolarity and isomorphism is presented. Some historical facts related to YinYang are discussed. Distinctions are drawn between BDL from established logical paradigms including Boolean logic, fuzzy logic, multiple-valued logic, truth-based dynamic logic, intuitionist logic, paraconsistent logic, and other systems. Some major comments from critics on related works are answered. A list of major research topics is enumerated. The ubiquitous effects of YinYang bipolar quantum entanglement are summarized. Limitations of this work are identified. Some conclusions are drawn.

INTRODUCTION

Bertrand Russell pointed out:

Mathematics and logic, historically speaking, have been entirely distinct studies. Mathematics has been connected with science, logic with Greek. But both have developed in modern times: logic has become more mathematical and mathematics has become more logical. The consequence is that it has now become wholly impossible to draw a line between the two; in fact, the two are one. They differ as boy and man: logic is the youth of mathematics and mathematics is the manhood of logic. (Russell, 1919)

Russell's view on logic and mathematics has a few implications. First, the origin of mathematics is science and the origin of logic is Greek philosophy but not science. Therefore, certain historical logical or philosophical concepts could be socially constructed and not necessarily scientific in nature. Secondly,

DOI: 10.4018/978-1-60960-525-4.ch012

mathematics can become more logical and logic can become more mathematical or scientific. Then, we are entitled to ask the following questions:

1. As an Eastern tradition of equilibrium-based thinking, is YinYang bipolarity scientific?
2. Is isomorphism a scientific principle?
3. Can logic, mathematics, and physics be unified?
4. Can bipolarity as a mathematical physics concept be introduced into logical reasoning?
5. Can bipolar relativity transcend logic, mathematics, and physics?

This chapter presents discussions and conclusions to further clarify the above issues with the remaining sections:

- A Debate on Bipolarity and Isomorphism
- Pondering and Wondering
- Some Historical Facts
- Causality Is Logically Definable
- Bipolar Axiomatization for Physics
- About Ultimate Logic
- Logical Distinctions
- Answers to Critics
- On The Ubiquitous Effects of Quantum Entanglement
- Limitations
- Major Research Topics
- Summary
- References
- Additional Readings
- Key Terms and Definitions

A DEBATE ON BIPOLARITY AND ISOMORPHISM

It has been shown that YinYang bipolarity is indispensable in an equilibrium-based (or symmetry-based) axiomatization of physics because particle-antiparticle pairs form the basis of energy and equilibrium in the universe where electron-positron pair production denoted (e^-, e^+) as an example of the materialization of energy predicted by Einstein's special relativity has been accurately described by quantum electrodynamics (QED) (Dirac, 1927, 1928; Feynman, 1962, 1985). While the bipolar entangled physical reality can be accounted for by the two reciprocal energies of Yin and Yang, the coexistence and bipolar interactive nature, however, has so far been denied by truth-based mathematical abstraction.

A major argument of classical truth-based mathematical abstraction is that -1 and +1 are isomorphic and (-,+) bipolarity is unnecessary in mathematical abstraction. Ironically, no physicist would say electron and positron (e^-,e^+) are isomorphic. Therefore, the so-called -1 and +1 isomorphism has to be a historical blunder. The following is the digest of an enjoyable academic debate between a European colleague, journal reviewer (A), and the author of this monograph (B) that occurred a few years ago.

- *A (review comment):* All arguments stating that the bipolar lattice $[-1,0]\times[0,1]$ structurally differs from the unit square $[0,1]^2$ are dubious and/or wrong. It does not require a lot of mathematical skills to understand that the mapping $f:[-1,0]\times[0,1]\rightarrow[0,1]^2:(x,y)\rightarrow(-x,y)$ is an (isomorphic) bijection and, hence, the structure of $[0,1]^2 = [0,1]\times[0,1]$ is identical to the one of $[-1,0]\times[0,1]$ (i.e. there is a one to one correspondence).
- B (author question): *Can you develop the bipolar universal modus ponens (BUMP) without explicit YinYang (-,+) bipolar syntax and semantics?*
- A (firmly): *Yes. I can.*
- B (curiously): *How are you going to do it?*
- A (patiently): *I derive it from your axiomatization by stripping bipolarity off your logic.*
- B (astonished): *Can you originate yours without bipolarity?*
- A (honestly): *I don't think anyone in the world can do that.*
- *B (speechlessly):* Would you be willing to ask your children to learn math without the negative sign? Or would you like to see the two poles of your car battery being labeled with '+,+'?
- *A (honestly):* Only a fool would do that.
- *B (speechlessly):* How would you feel if you derive your unipolar result from my bipolar original and then you say the original is isomorphic to your derivation?
- *A (authoritatively):* Everyone in the fuzzy set community knows isomorphism. Go to find the definition from the web and read it.
- B (luckily): *Professor Zadeh recognized my bipolar fuzzy set theory in Scholarpedia.*
- A (impulsively): *Zadeh's words do not count. Dr. X has ...*

PONDERING AND WONDERING

Some historical events could be shocking, astonishing, or puzzling when they are recalled centuries later. According to the BBC the following is such a historical event: In 1759 the British mathematician Francis Maseres wrote that negative numbers *'darken the very whole doctrines of the equations and make dark of the things which are in their nature excessively obvious and simple.'*.. Why were negative numbers considered with such suspicion? Why were they such an abstract concept? And how did they finally get accepted? (BBC, 2006) Clearly, the current debate on bipolarity and isomorphism is a continuation of the debate on negative numbers extended to the logical arena. In Greek isos stands for "equal" and morphe stands for "shape." Clearly, isomorphism was conceptualized long before electron-positron pairs were discovered and long before knowledge representation (KR) became a major research area in cognitive and computer sciences. Since without semantics there would be no need for syntax and without syntactic representation we cannot reason on the semantics of any knowledge structure, without (-,+) bipolarity, we cannot reason on bipolar equilibrium or non-equilibrium. The negative sign "-" should not be forbidden as part of a bipolar equilibrium-based logic unless (1) mathematicians and logicians are willing to ask their children to learn math without it; (2) scientists would say particle and antiparticle are isomorphic; (3) car owners would like the two poles of their car batteries being labeled with +,+; (4) philosophers would say Yin is isomorphic to Yang. Otherwise, the above so-called isomorphism is not a scientific principle that enhances the strictness of science but a kind of socially constructed entrenched noble hypocrisy that is hindering a new mathematical abstraction. Bipolar sets provide a holistic equilibrium-based approach to mathematical abstraction based on the established scientific observations of nature's bipolarity. Since two different mathematical abstractions are involved here, saying two abstractions are isomorphic is like saying apple and orange are the same. After all, no one could possibly answer the questions: Equilibrium or universe, which one came first? Electron or positron, which one came first? Matter or antimatter, which one came first? Negative pole or positive pole, which one came first? If they coexist in equilibrium or non-equilibrium with opposite polarity, there is no way to represent bipolar fusion, interaction, oscillation, and entanglement without (-, +) bipolar syntax and semantics.

From a classical mathematical perspective, an isomorphism is a bijective map f such that both f and its inverse f^{-1} are homomorphisms. A homomorphism is a map from one algebraic structure to another of the same type that preserves all the relevant structure; i.e. properties like identity elements, inverse elements, and binary operations. In other words, isomorphism is a one-to-one correspondence between the elements of two sets such that the result of an operation on elements of one set corresponds to the result of the analogous operation on their images in the other set. It is easy for a critic to notice the word *bijective mapping* but not easy to notice that *both f and its inverse f^{-1} are homomorphisms* or maps from one algebraic structure to another of the ***same type*** that preserve all the relevant structure, i.e. ***properties like identity elements, inverse elements, and binary operations***. It can be observed that -1 and +1 as identity elements both present in the bipolar lattice $B = \{-1,0\} \times \{0,1\}$ that identify negative and positive elements, respectively. With the unipolar lattice $\{0,1\}^2 = \{0,1\} \times \{0,1\}$, -1 as an identity element is not preserved and bipolarity is lost. Therefore, we may say $B = \{-1,0\} \times \{0,1\}$ is not really isomorphic to $\{0,1\}^2$ unless we say -1 is identical to +1. But for a bipolar equilibrium, they are not. For instance, we can't label the two poles of our car battery with +,+; we can't say electron and positron are the same type. Furthermore, we examine whether the result of an operation on elements of one set corresponds to the result of the analogous operation on their images in the other set. Without bipolarity, the equilibrium-based non-linear bipolar dynamic cross-pole interaction or oscillation $\otimes$ can not be meaningfully defined on $\{0,1\}^2$ for the bipolar interaction/oscillation semantics $--=+$, $-+=-$, $+-=-$, and $++=+$ on each pole. Therefore, the bijective map f and its inverse f^{-1} are not really homomorphisms. Someone might say I can use 1 for -1 and make $(1,0)\otimes(1,0)=(0,1)$ and $(0,1)\otimes(0,1)=(0,1)$ for the same bipolar semantics. Unfortunately, the proposal is a derivation from $B = \{-1,0\} \times \{0,1\}$ not from $\{0,1\}$. Without bipolarity, it is at most a left-right logical structure with $(1,0)\otimes(1,0)=(0,1)$ for left×left = right and $(0,1)\otimes(0,1)=(1,0)$ for right×right = left, where the bipolar semantics -- = +, -+ = -, +- = -, and ++ = + can't be syntactically distinguished especially when the two lattices $\{-1,0\} \times \{0,1\}$ and $\{0,1\} \times \{-1,0\}$ are used together as equivalent or non-equivalent logical structures.

Structurally, the unit square lattice $\{0,1\}^2$ or $[0,1]^2$ has been for static truth-based reasoning where the keywords bipolar sets, dynamic equilibrium, and equilibrium-based mathematical abstraction have not been conceptualized and bipolar dynamic interaction and oscillation have not been observed. It is natural to use $\{0,1\}^2$ for characterizing the *truth* of x and y if they are not the two sides of one matter in opposite polarity. It would be meaningless, however, if we use (+1,+1) or (+,+) for bipolar reasoning.

Someone might still argue that I can use (0,1) to label the two poles of my car battery instead of (-,+). The answer is Yes, You can, but only after you ground the negative pole. It is well-known in digital logic design that a positive pole can also be neutralized to obtain a negative voltage. With NP or PN BDL, such grounding operations can be logically expressed as (-1,1) & (0,1) = (0,1) or (1,-1) & (1,0) = (1,0); (-1,1) & (-1,0)=(-1,0) or (1,-1)&(0,-1)=(0,-1). With a combination of NP and PN BDL we even can have (1,-1) & (-1,0) ≡ (-1,1) & (-1,0). Without bipolarity, even such linear bipolar operations can't be logically described.

Could BUMP be originated from any other logical system without (-,+) bipolar semantics and syntax?

The answer from the above debate is *"I don't think anyone in the world can do that."* Thus, anyone should be welcome to come up with an equivalent or better version of BUMP without borrowing implicit or explicit (-,+) bipolar syntax and semantics. If it were possible, the 2nd quadrant would be isomorphic to the 1st. Otherwise, we can conclude that

1. It is well-known that classical modus ponens (MP) has been the only generic inference rule in logical deduction for thousands of years.
2. Equilibrium-based YinYang bipolar universal modus ponens (BUMP) as a fundamentally different inference rule could not have been brought to light without (-,+) bipolar semantics and syntax.
3. Without BUMP bipolar axiomatization and bipolar relativity would be impossible.
4. The so-called *isomorphism* in the above debate is NOT a scientific principle and could be a socially constructed historical blunder in logic and mathematics.

SOME HISTORICAL FACTS

From the debate, we can see that due to social quantum gravity it is easy for a person to assume or presume things that are nevertheless untrue. Here are more such presumptions:

1. For some years the author thought that, since George Boole invented Boolean logic, presumably, he also invented the binary numeral system. But that was false.
2. For some years the author believed that negative numbers and zero were invented by the ancient Greeks. That was false.
3. Quite a few reviewers informed the author anonymously and some authors even informed the author publicly in their published indexed journal article that YinYang bipolar lattice is the same or isomorphic to *bilattice* or *FOUR*. But that was untrue.
4. After the author acknowledged that YinYang bipolar lattice is semantically, syntactically, philosophically, and axiomatically different from other 4-valued models, the author was told that even though it is fundamentally different, it resembles the form of bilattice or *FOUR*. The author found that was technically correct, however, all 4-valued models resemble each others' form and all seem to resemble the form of Lukasiewicz's 4-valued structure. Since Lukasiewicz was the founder of many-valued logic, presumably, the lattice form of Lukasiewicz's 4-values in his 4-valued logic had to be the first of its kind. Surprisingly, that was incorrect.
5. Once the author believed that fuzzy logic got to be the ultimate logic because it is an infinite-valued logic. No logic can go beyond infinity. Later the author realized that uncertainty or fuzziness is one dimension of truth. There could be other more fundamental dimensions such as YinYang bipolarity.
6. Since Hilbert posted Problem 6 – Axiomatize all of physics, presumably, he was the first ever who tried to axiomatize physics but that was false again.
7. It was believed by many that an axiomatization of all physics had to be a theory of everything such as string theory. That was untrue.

Who invented the first binary numeral system? Who invented zero and negative numbers? Who invented the diamond form of 4-values? When was the earliest effort of axiomatizing physics recorded? Is there an axiomatization for all physics? Is there an ultimate logic? Is causality definable? What caused the big bang? Is quantum entanglement logically comprehendible? Is the quantum mystery resolvable?

Binary numeral system. According to undisputed historical record, the legendary German mathematician (co-founder of calculus) Leibniz (Aiton, 1985, pp. 245–248) invented the first binary numeral system in the 17th century (Leibniz, 1703) and attributed his invention to YinYang hexagrams (Ch. 1, Fig. 2) as recorded in *YiJing* or *I Ching* – the ancient Chinese *Book of Change* (Moran & Yu, 2001). The modern binary numeral system was fully documented by Leibniz in the 17th century in his article *Explication de l'Arithmétique Binaire* (Leibniz, 1703). Leibniz's system used 0 and 1 like the modern binary numeral system. As a Sinophile, Leibniz was aware of the *I Ching* and noted with fascination how its hexagrams correspond to the binary numbers from 000000 to 111111, and concluded that this mapping was evidence of major Chinese accomplishments in the sort of philosophical mathematics he admired (Aiton, 1985). Leibniz's interpretation of YinYang is shown in Chapter 1, Figure 2. We call the Leibniz interpretation *binary YinYang* where Yin = 0 and Yang = 1.

A century later, British mathematician George Boole published a landmark paper (Boole, 1854) detailing an algebraic system of logic that would become known as Boolean algebra. His logical calculus has become instrumental in the design of digital electronic circuitry for the commercialization of modern digital computers. Clearly, binary YinYang is a basis for both binary numeral system and Boolean logic.

Now, presumably, the origin of binary numeral system is settled. Again, the word "presumably" is to leave the door open for future discoveries on the origin of binary numeral system.

Number Zero. According to Kaplan (2000) and Gupta (1995), the oldest known text to use a decimal place-value system, including a zero, is the Jain text from India entitled the Lokavibhâga, dated 458 CE. The first known use of special glyphs for the decimal digits that includes the indubitable appearance of a symbol for the digit zero, a small circle, appears on a stone inscription found at the Chaturbhuja Temple at Gwalior in India, dated 876 AD.

Negative numbers. According to Temple (1986, p. 141), negative numbers appeared for the first time in history in the Chinese *Nine Chapters on the Mathematical Art* (*Jiu Zhang Suan-Shu* or *Nine Chapters*

Figure 1. Hasse Diagrams: (a) Four values of FOUR; (b) Bilattice

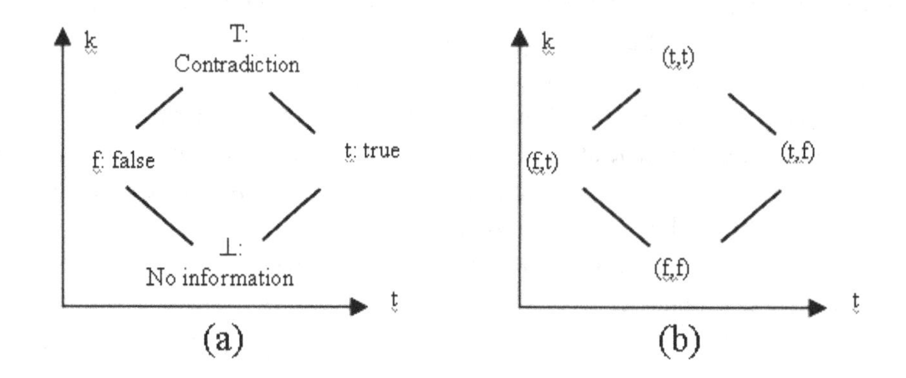

of Mathematics), which in its present form dates from the period of the Han Dynasty (202 BC – 220 AD), but may well contain much older material.

For a long time, negative solutions to problems were considered false in the West. Although Fibonacci allowed negative solutions in financial problems where they could be interpreted as debits and later as losses, European mathematicians, for the most part, resisted the concept of negative numbers until the 18th century (Martinez, 2006; BBC, 2006).

Four-valued diamond. According to the bilattice research community, Belnap proposed the diamond form of the 4-valued *paraconsistent logic FOUR* (Belnap, 1977) that is developed to bilattice (Ginsberg, 1990) with similar semantics and later extended by Fitting (1991) and Arieli and Avron (1998). According to the multiple-valued logic community, Lukasiewicz was the first one to propose the first 4-valued-logic (Rescher, 1969) which was later extended to fuzzy product lattices. The two logical systems are semantically and syntactically different. However, the author was told by anonymous reviewers that, in logical research, the diamond is considered essential for all 4-valued models. Belnap's *FOUR* and Ginsberg's bilattice are shown in Figure 1; Lukasiewicz 4-valued implication is shown in Figure 2.

Based on the YinYang philosophy, YinYang bipolar lattice $B_1=\{-1,0\}\times\{0,+1\}$ is proposed by the author (Zhang, 2005a) (See Ch. 3, Fig. 2). The bipolar lattice is based on the Daoist YinYang cosmology which claims that everything has two sides or two poles: a negative (-) pole and a positive (+) pole. Thus, in B_1 (-1,+1) stands for bipolar true or an equilibrium; (0,0) stands for bipolar false or an eternal equilibrium; (-1,0) stands for negative pole true but positive pole false or a non-equilibrium; (0,+1) stands for negative pole false but positive pole true or another non-equilibrium. While the semantics of many-valued logics have been problematic, the bipolar lattice holds concrete physical semantics. Not only does it satisfy the law of excluded middle but also four pairs of dual dynamic DeMorgan's laws. Above all, bipolar universal modus ponens (BUMP) can be defined on B_1, which presents for the first time a non-linear dynamic generalization of classical modus ponens (Zhang, 2005a, 2007) (Ch. 3).

After YinYang bipolar logic and bipolar lattice were published, a few researchers claimed that bipolar lattice is the same as or isomorphic to bilattice even though bilattice does not satisfy the law of excluded middle (LEM) but bipolar lattice does. When the author argued that bipolar lattice is fundamentally different, he was told that even though it was different, the form resembled bilattice. Since Lukasiewicz's 4-valued logic was much earlier, the author argued that bipolar lattice is more similar to Lukasiewicz's 4-valued logic; if bipolar lattice were the same as bilattice, bilattice would be the same as Lukasiewicz's 4-valued logic. Surprisingly, the author realized later that the argument was not the right answer to the question.

Actually, it is technically correct to say that all 4-valued logic models resemble each other's diamond form. However, a comparison of different 4-valued models with Leibniz interpretation makes it abundantly clear that the diamonds of all 4-valued logical models resemble YinYang-4-Images (Ch. 1, Fig. 2).

Figure 2. Lukasiewicz's 4-valued logic implication truth table

⇒	00 (0)	01 (1)	10 (2)	11 (3)
00 (0)	11	11	11	11
01 (1)	10	11	10	11
10 (2)	01	01	11	11
11 (3)	00	01	10	11

As discussed earlier, by his invention of binary numeral system and his attribution to YinYang hexagrams in *I Ching*, not only did Leibniz exhibit unprecedented creativity of a great mathematician and philosopher but also exemplary academic integrity of a great scientist. While Leibniz in the 17th century was able to attribute his invention to the oldest Chinese *Book of Changes*, modern time 21st century web surfers (including the author) are not doing any better in this regard.

Now, presumably, the origin of the 4-valued diamond is settled. Again, the word "presumably" is used here to leave the door open on the origin of the 4-valued diamond. But one thing is sure, based on Leibniz interpretation of YinYang, the origin of the 4-valued diamond can be traced thousands of years back to YinYang-4-Images. Therefore, it belongs to history and history belongs to mankind. No modern researcher should claim authorship of the diamond besides different logical interpretations and mathematical constructions. It is, therefore, childish and naive for anyone to use "isomorphism", "form", or "4-valued standard" as an academic bashing stick on philosophically different interpretations of YinYang-4-Images especially on the original YinYang bipolar cosmological interpretation.

Now it is legitimate to ask the question: *If modern derivations and extensions of YinYang are all considered scientific structures that have repeatedly appeared in prestigious journals, could the original concepts of YinYang be unscientific* (Ch. 1, Figs. 2 & 3)?

Of course, some interpretations of YinYang from ancient Chinese literature are indeed unscientific, especially the fortune telling interpretations, but that should not be a reason to ban YinYang from being an evolving scientific concept as Aristotelian science should not have been forbidden to evolve to Western science because Aristotle once claimed the Earth was the center of the universe and his cosmology was later borrowed for the brutal fire execution of Bruno and the jailing of Galileo. After all, it is common knowledge that YinYang is the oldest equilibrium-based philosophy and there is no dispute that equilibrium is a fundamental scientific concept central in thermodynamics.

CAUSALITY IS LOGICALLY DEFINABLE

Since without cause-effect relationship any theory would be incomplete, Aristotle's causality principle has been considered the doctrine of all sciences for more than two thousand years. Extending the critique on the causality principle by Scottish philosopher David Hume (1711 – 1776) (Blackburn, 1990), Professor Lotfi Zadeh continued the line of criticism and has become the strongest critic of the principle in modern times. A major argument of Professor Zadeh is that "*Causality is Undefinable*" (Zadeh 2001) and, therefore, uncertainty or fuzziness is unavoidable. Zadeh used the example "*Ann loves Paul and Paul loves Lisa does not imply that Ann loves Lisa*" (Zadeh, 2003). Zadeh's position on causality once caused an outcry from Judea Pearl's blog; it also cast doubt for a while on the author's mind about the validity of bipolar sets and BDL. Eventually, it helped the author to reaffirm and reinforce the YinYang bipolar approach.

From a classical unipolar static truth-based logical perspective, causality is indeed undefinable in a logical formula as asserted by Zadeh. For instance, following classical modus ponens (MP) we have IF A and A→B, THEN B where cause and effect have no position because no matter whether the premise A is true or false the consequent B can be true. Therefore, Zadeh's position on the definability of causality as well as his Ann-Paul-Lisa example has nothing wrong from a truth-based unipolar logical perspective.

From a holistic equilibrium or non-equilibrium-based dynamic point of view, however, causality is logically definable. Following equilibrium-based BUMP we have *IF A*C, A⇒B & C⇒D THEN B*D,* where A*C can cause B*D in a bipolar equilibrium-based quantum entanglement (Ch. 3). Although love and hatred are generally not transitive, friend-enemy relationship in war time international relations are clearly transitive where an enemy's enemy is a friend and an enemy's friend is an enemy (Zhang, 2003a). Furthermore, if we consider any of nature's basic forces as an action and reaction pair, could it be true that all forces in the universe are bipolar transitive to a certain extent?

Here the author must express unwavering respect for Professor Zadeh. After the author's work on YinYang bipolar fuzzy sets were formally published (Zhang, 2005a, 2006b), the author met him at IEEE GrC 2006 – Atlanta, GA, where he delivered a keynote talk and the author presented a paper on bipolar dynamic t-norms with some answers to critics (Zhang 2006b). Professor Zadeh attended the author's presentation. After the presentation, he asked the author a question: "*Is bipolar fuzzy set not the same as intuitionistic fuzzy set?*" The author answered: *Totally different. If both poles are true it would characterize a perfect bipolar equilibrium in YinYang bipolar logic which satisfies the law of excluded middle; if both sides are true it would characterize an indetermination in intuitionistic logic which doesn't satisfy the law of excluded middle.* Professor Zadeh commented: "*Very impressive.*"

Soon after the Atlanta encounter, Professor Zadeh became the first academic authority of the fuzzy set community to recognize (YinYang) bipolar fuzzy sets (Zhang, 1998) publicly in Scholarpedia (Zadeh, 2006). During 2007 and 2008, he mentioned bipolar fuzzy sets several times in his keynote talks (Zadeh, 2007, 2008). By this recognition, it is evident that Professor Zadeh already reconsidered his position on the definability of causality from an equilibrium-based bipolar dynamic logical perspective.

BIPOLAR AXIOMATIZATION OF PHYSICS

It is worthwhile to revisit the difference between the interpretation of YinYang by Leibniz and the bipolar equilibrium-based cosmology. First, both are consistent and both have a positive side and a negative side. In the former, the two sides are the 0 and the 1 sides that carry no physical semantics; in the latter, the two sides are the –pole and +pole sides that carry fundamental equilibrium-based physical semantics (Ch. 1, Figs. 1 – 3).

Among Hilbert's 23 mathematical problems (Hilbert, 1901), Problem 6 is *"Axiomatize all of physics."* Since Hilbert was the first one to post the question, the author once thought that Hilbert was the first one who tried to axiomatize physics. Actually, YinYang-1-2-4-8-64, with the theme Taiji (one) generates YinYang (two), YinYang generates four images, YinYang four images generate eight trigrams, and eight trigrams generates 64 changes (Wilhelm & Baynes, 1967, pp. 318-319), qualifies as one of the earliest recorded human efforts in axiomatizing physics. Some researchers including some Chinese probably would say that is unscientific. As a matter of fact, it used to be unscientific but is becoming more scientific in recent decades with new scientific discoveries and refined interpretations as evidenced by the appearance of the word YinYang in numerous scientific articles published in prestigious journals including *Science*, *Nature*, and *Cell*.

It should be remarked that YinYang bipolar relativity does not follow exactly the theme of YinYang-1-2-4-8-64. Instead, it follows the interpretation that "Every agent has two poles or two energies." It is contended that (1) the two poles or two energies define the agent or, alternatively, the agent is the realization of the two energies and (2) it is so far unclear which one came first in the very beginning.

Is there an answer to Hilbert's Problem 6? Although the problem is widely considered *unsolvable* now, Einstein believed it is solvable as evidenced by his quote "*Our experience hitherto justifies us in believing that nature is the realization of the simplest conceivable mathematical ideas. ... In a certain sense, therefore I hold it true that pure thought can grasp reality, as the ancients dreamed."* It is interesting to wonder whether YinYang-1-2-4-8-64 was considered by Einstein as such a dream by the ancients.

Nevertheless, YinYang bipolar mathematical abstraction has led to an equilibrium-based bipolar axiomatization as presented in the early chapters that can be considered a minimal but most general axiomatization of physics (Zhang, 2009a, 2009b, 2009c, 2009d)(Ch. 6). Evidently, on the one hand, the bipolar axiomatization can be deemed a continuation of the YinYang-1-2-4-8-64 effort initiated thousands of years ago and, on the other hand, since YinYang bipolarity forms a generic dimensional model in Hilbert space (Ch. 1, Fig. 1; Ch. 3, Fig. 1 – 2), it can also be considered a continuation of Hilbert's geometric approach to the axiomatization of physics initiated 100 years ago.

Why is there so far no truth-based axiomatization for physics? This question has been answered with the truth-based paradox *Logical Axiomatization for Illogical Physics* abbreviated as LAFIP or LAFIB (Zhang, 2009a) (Ch. 1). To remain a bipolar equilibrium-based logical space the bipolar lattice $B_1 = \{-1,0\} \times \{0,+1\}$ can no longer be further reduced; in contrast, to remain a unipolar truth-based logical space the bivalent lattice $\{0,1\}$ can no longer be further reduced. Therefore, Boolean logic can be regarded as the minimal but most general truth-based system; B_1 can be deemed the minimal but most general equilibrium-based logical space. Without bipolarity, however, the unipolar truth values 0 and 1 are incapable of carrying any direct physical semantics and syntax for an axiomatization of physics.

In other words, any hopeful approach to Hilbert's problem 6 has to be logical, mathematical, and physical in semantics and syntax, where certain physical properties such as bipolar disorder, particle-wave duality and equilibrium-based bipolar quantum entanglement are illogical in truth-based logical terms. Any truth-based logic is therefore too logical to be natural for reasoning on such illogical but nevertheless physical bipolar phenomena. The reification of (-,+) bipolarity makes it possible to bring the logical, mathematical, and illogical but nevertheless physical aspects all together into an equilibrium-based axiomatization.

Can bipolar mathematical abstraction and axiomatization be scaled up to bipolar mathematics? This question has been partially answered with bipolar quantum linear algebra (BQLA) (Ch. 8). Since the focus of this monograph is on the logical level, deeper mathematical extension is left for future research effort.

Is YinYang BDL the logic of everything? The answer is No. Even though it is the most general equilibrium-based logic for physics so far, it only provides a minimal and a forever open axiomatization without the assumption of any smallest fundamental element of the universe such as ether or strings.

NO ULTIMATE LOGIC IN AN OPEN WORLD

Nature and science seem to be open-ended and scientists have to be open-minded. But some journal editors in some areas tend to close their research areas. A few examples are listed as follows.

Example 1. For quite a few years the author was puzzled at why some anonymous reviewers in the AI community had been trying to bring YinYang bipolar lattice into the family of bilattice or *FOUR*. Latter the author realized that an article published in a top AI journal in 1990s claimed that by the generalization of *FOUR*, no one (in the future) could obtain *"something that is not already in FOUR."* Then, if the four values of the YinYang bipolar lattice are not part of *FOUR*, *FOUR* have to become *EIGHT* or even *TWELVE*. Here the problem is that once you imposed a closure on the value of four values you should no longer claim the value beyond the four values. Actually, bipolar YinYang and bilattice are like apple and orange. They should not be "enemy" to each other and they should not be said to be the same or isomorphic to each other either.

Example 2. For quite a few years the author was puzzled at why some anonymous reviewers in the fuzzy set community had been using isomorphism as a bashing stick trying to bring YinYang bipolar fuzzy sets into the family of so-called standard structure or the square diamond lattice $[0,1]^2 = [0,1] \times [0,1]$. Latter the author realized that an article published in a top fuzzy set journal a few years ago had a proposition that states *"Any lattice-ordered t-norm T on a product lattice $L = L_1 \times L_2$ is the direct product of two lattice ordered t-norms on L_1 and L_2, respectively."* Formally, $T_1 \times T_2((x_1,y_1),(x_2,y_2)) \equiv (T_1(x_1,x_2), T_2(y_1,y_2))$. Then, if the YinYang bipolar fuzzy lattice $B_F = [-1,0]\times[0,1]$ and the family of non-linear bipolar cross-pole t-norm ⊗ (Zhang, Chen & Bezdek, 1989; Zhang & Zhang, 2004, 2005a) were taken into consideration, the proposition would no longer hold. *"Are there other ways to define t-norms on product lattices which are not direct product?"* The answer is clearly "Yes" and ⊗ is such a t-norm on a YinYang bipolar (product) lattice. Now, our open question should be *"Are there other ways to define t-norms on a product lattice $L = L_1 \times L_2$ which are lattice ordered but not a direct product of two t-norms on L_1 and L_2, respectively?"*

Closed world. A major motivation driving an established journal to a "closed world" is to achieve a higher impact factor. If all authors work on established topics within the established boundary, a journal or a group of allied journals can cite each other's works more frequently than citing new ideas. To a certain extent, the impact factor favor is similar to the real-estate bubble. Both are influenced by social construction and social quantum gravity. Hopefully, modern science can avoid degeneration that is similar to the last economic recession caused by the burst of the real-estate bubble.

Is there an ultimate logic? In the author's opinion, the answer is No. The reason is that nature and science are open-ended, and anyone trying to close them will sooner or later be proven impossible. At the golden time of fuzzy set research, a well-respected authoritative colleague in the fuzzy set community attempted to regulate the terminologies of fuzzy sets, bipolar fuzzy sets, and intuitionistic fuzzy sets. The colleague seems to have believed that with his authority he could change the dictionary and redefine bipolarity as intuitionism. Lotfi Zadeh acted promptly by setting the record straight and recognized both bipolar fuzzy sets (Zhang, 1998) and intuitionistic fuzzy sets (Atanassov, 1986) together with his article *Fuzzy Logic* in *Scholarpedia* (Zadeh, 2006). The definition changing attempt has so far been more or less curbed.

Someone might have believed that the great value of fuzziness lies in its creation of an ultimate logic. The author used to believe that could be right because fuzzy logic is an infinite-valued logic and no other logic can go beyond infinity. Now the author believes that the great value of fuzzy sets lies in three major aspects: (1) it successfully challenged the principle of classical mathematical abstraction and opened a new research area, (2) it opened people's minds in academic and applied research, and (3) it is open for further theoretical and practical development. Truth-based logical systems, fuzzy or crisp, should take bipolarity into consideration.

LOGICAL DISTINCTIONS

YinYang BDL vs. DL. BDL is fundamentally different from *dynamic logic* (*DL*) (Pratt, 1976). DL is defined in the bivalent lattice $\{0,1\}$ for reasoning on computer programs that are dynamic but logical in truth-based terms. BDL is defined in the bipolar lattice $\{-1,0\}\times\{0,1\}$ for bipolar dynamic reasoning on equilibria/non-equilibria of physical systems with bipolar fusion, interaction, oscillation, and quantum entanglement which may be considered equilibrium-based logical but truth-based illogical.

YinYang BDL vs. Intuitionistic Logic. *Intuitionistic logic* (Brouwer, 1912) can be succinctly described as classical logic without the law of excluded middle (LEM) ($A \vee \neg A$). Therefore, intuitionistic logic is a static truth-based extension to Boolean logic whose logic values carry no physical semantics and syntax. Following equilibrium-based mathematical abstraction, BDL is for equilibrium-based dynamic reasoning whose truth values carry physical semantics and syntax. Bipolar LEM, BUMP, and bipolar quantum entanglement find no equivalent in intuitionism.

YinYang BDL vs. Lukasiewicz 4-Valued Logic. While Lukasiewicz 4-valued logic (Rescher, 1969) can be considered a linear extension to Boolean logic, BDL is a non-linear dynamic extension or polarization of Boolean logic. Lukasiewicz's 4-valued logic is truth-based, static, and unipolar in nature without (-, +) bipolar syntax and semantics. BDL, on the other hand, is equilibrium-based non-linear bipolar dynamic in nature. The bipolar syntax, physical semantics, non-linear bipolar dynamic DeMorgan's laws, BUMP, and quantum entanglement find no equivalent in Lukasiewicz 4-valued logic.

YinYang BDL vs. Fuzzy Logic. Fuzzy logic is a truth-based real-valued extension to Boolean logic; BDL and BDFL are non-linear dynamic polarization of Boolean logic and fuzzy logic, respectively. Fuzzy logic and sets (Zadeh, 1965; Goguen, 1967) do not satisfy LEM ($\neg 0.5 = (1\text{-}0.5) = 0.5$) and they rightfully belong to the same category of intuitionism while LEM is used as a criterion. This could be one of the reasons why the fuzzy set community is faced with the controversial terminological difficulties (Dubois *et al.*, 2005; Atanassov, 2005).

YinYang BDL vs. *FOUR* or Bilattice. Belnap's 4-valued logic *FOUR* (Belnap, 1977) provides a foundation for research in paraconsistency. *FOUR* does not satisfy the law of excluded middle (LEM) where contradiction is a permissible truth value. Thus, *FOUR* as the foundation of paraconsistency is a static truth-based extension to Boolean logic. The dynamic DeMorgan's laws, BUMP, and bipolar quantum entanglement feature of BDL find no equivalent in paraconsistency.

FOUR was later developed to Bilattice (Ginsberg, 1990; Fitting, 1991; Arieli & Avron, 1998). In Arieli and Avron (1998) the authors generalized Belnap's 4-valued logic, Ginsberg's bilattice, and Lukasiewicz 4-Valued Logic (Rescher, 1969) to the new model *FOUR*. The new *FOUR* did not change the static nature of the original *FOUR*. Despite its static truth-based properties, the new *FOUR* in general still does not satisfy the law of excluded middle (LEM) and therefore is not a tautological system but a collection of models for selection even though a classical 4-valued logical fragment is included as part of the collection that does satisfy LEM. On the contrary, despite its non-linear bipolar dynamic properties, the 4-valued BDL (Ch. 3) is a tautological system that satisfies LEM.

The truth values in *FOUR* (original or new) do not carry any (-,+) bipolar dynamic equilibrium-based physical semantics and syntax. Instead, with *FOUR* "*First, we have the standard partial order,* $\leq_t$*, which intuitively reflects differences in 'the measure of truth'... The other partial order,* $\leq_k$*, is understood (again, intuitively) as reflecting differences in the amount of 'knowledge or information' that each truth value exhibits.*" (Arieli & Avron, 1998). On the contrary, the bipolar lattice B_1 is a holistic bipolar poset (B_1,$\geq\geq$) with the key concepts equilibrium-based, bipolar sets, bipolar interaction and oscillation, non-linear dynamic, bipolar universal modus ponens (BUMP) and bipolar quantum entanglement that are not observed from the two partial orders in the new *FOUR* and any of its four consequence relations.

It is stated in (Arieli & Avron, 1998) that "*A natural question arises at this point is whether by this generalization one obtains something that is not already in FOUR.*" It is concluded that the answer is "*basically negative.*" The author of this book respectfully suggests that YinYang BDL is semantically, syntactically, physically, philosophically, and axiomatically different from *FOUR* and all other truth-based logical systems rooted from Greek philosophy. This position will ensure the "*basically negative*" conclusion remains valid for static truth-based cases. Otherwise, equilibrium-based YinYang bipolar sets, BDL, and BUMP as "*something that is not already in FOUR*" would be an overturn to the "*basically negative*" conclusion and *FOUR* would have to be changed to EIGHT or even TWELVE. The author believes that equilibrium and truth are like apple and orange that should not be an overturn to each other even if one does taste sweeter. On the other hand, the author does not think that apple and orange should be said the same even though both fruits are round in shape, sweet in taste, and similar in size. Historically, Greek philosopher Aristotle's truth-based logic school in 2300BC has been proven a major philosophical source for modern scientific computation; empirically the equilibrium-based Chinese YinYang, as a major theoretical basis for traditional Chinese medicine, has survived thousands of years of recorded human history. The two philosophies have to be complementary to each other and can no longer be an overturn to each other due to their self-evident survival history.

Someone insisted that, if YinYang bipolar lattice is not a bilattice "*it is at most an application of FOUR.*" This is untrue. YinYang is about equilibrium and symmetry. Even though the author of this book would not mind his work being an application of *FOUR*, it would be difficult to convince readers for the following reasons:

1. YinYang bipolar universal modus ponens (BUMP) (Zhang, 2005a, 2007) as a new modus ponens for reasoning on bipolar equilibria can be either original or derivative and valid or invalid but can't be an application.
2. Mental equilibrium of a human brain can't be application of the brain because there would be no healthy genetic mutation and no brain without mental equilibrium.
3. It is a commonly accepted view in modern physics that the universe is either an equilibrium or quasi- equilibrium. Then, equilibrium – the only physical concept that can form a chicken and egg philosophical paradox with the mighty universe in microcosmic or macrocosmic terms can't be application of the universe because no one knows exactly which one created the other.
4. Furthermore, without bipolar fusion, coupling, binding, interaction, oscillation, and bipolar entanglement there would be no electromagnetic field, no memory scanning, no human cognition, no logical reasoning, no brain, no equilibrium, no action and reaction, no quantum mechanics, no spacetime, and no universe.

"*Even if B_1 is fundamentally different from FOUR, it resembles the form of FOUR.*" This is technically a correct argument. Nevertheless, all 4-valued lattices resemble each other's form and all of them resemble the formation of the 2-bit binary numbers (00, 01, 10, and 11) that are part of the invention of the binary numeral system by Leibniz (Ch. 1, Fig. 2). Since Leibniz attributed his invention of the first binary numbering system to YinYang Hexagrams, it is clear that all 4-valued logical lattice structures resemble the form of YinYang-4-Images (Ch. 1, Fig. 2).

The 4-valued YinYang bipolar lattice B_1 (Ch. 3, Fig. 2) and the 4-valued bilattice (Fig. 1b) are actually like two individuals who look alike but share no common genes. We summarize and enumerate their fundamental differences as follows.

1. **Semantically**, the word bipolar is related to two poles while the two sides involved in a bilattice have nothing to do with bipolarity. Logically, bilattice is static truth-based in nature for reasoning

on truth, falsity, and contradiction w.r.t. truth object, where the four values assume, respectively, the semantics true (1,0), false (0,1), no information (0,0) and contradiction (1,1) (Ginsberg, 1990). The bipolar lattice is equilibrium-based in nature for holistic reasoning with dynamic fusion, interaction, oscillation, and quantum entanglement of the negative and positive poles of bipolar equilibria and non-equilibria where the two poles are reciprocal and interdependent truth objects with opposite polarities that find no equivalent semantics in bilattice.

2. **Syntactically**, (-,+) bipolarity distinguish the two types of lattices. Following paraconsistency (Belnap 1977), bilattice is a pair of bilateral partial orderings where the partial order $\leq t$ is a measure of truth and the other partial order $\leq k$ is a measure of the amount of knowledge or information that each truth value exhibits (Ginsberg, 1990). YinYang bipolar lattice B_1 is a bipolar poset with the holistic partial ordering (B_1, $\geq\geq$) where the two sides form an inseparable fusion, coupling, or binding. One side is a measure of +truth *w.r.t.* the positive pole and another is a measure of –truth w.r.t. the negative pole of a bipolar equilibrium or non-equilibrium. $\forall(n,p)\in B_1$, let (n,p) = (-pole, +pole) be a bipolar element, we have –(n,p) = (-p,-n), where – stands for arithmetic negation. For instance, –(–1, 0) = (0, 1) or –(electron, 0) = (0, positron). It was argued that the negation sign "–" was also defined and widely used on bilattices. That is, in fact, a notational misunderstanding. The operator "–" was defined on bilattices as a conflation operator (Fitting, 1991) as a weak complement fundamentally different from arithmetic negation. For instance, the conflation in bilattice –(t,t) = (¬t, ¬t) = (f,f) and –(f,f) = (¬f, ¬f) = (t,t); but the bipolar arithmetic negation in bipolar lattice –(-1,1) = –(-1,-(-1)) = (-1,1) and –(0,0) = (0,0). Therefore, even though the two dimensions k and t in a bilattice were mapped to the second quadrant, arithmetic negation would still be missing. Thus, in the YinYang bipolar lattice both logical negation (¬) and arithmetic negation (–) are indispensable strong negations with – – (n,p) = (n,p) and ¬¬(n,p) = (n,p); in bilattice, however, the arithmetic negation is not observed. This is just one of the basic syntactic differences.
3. **Philosophically**, bipolar lattice is rooted in the ancient Chinese YinYang; bilattice belongs to truth-based systems with paraconsistency. Cosmologically, equilibrium-based YinYang views everything in the universe including the universe itself as an equilibrium or non-equilibrium with bipolar equilibrium or non-equilibrium as its generic form. On the other hand, Aristotle's bivalent logic claims that the universe consists of truth objects that can be mapped to the bivalent lattice {0,1} or {false, true} which forms a basis for Boolean logic and its extensions including bilattice.
4. **Physically**, bipolar lattice is non-linear dynamic in nature carrying equilibrium-based quantum physics semantics and exhibits ubiquitous effects of quantum entanglement (Zhang, 2009c, 2009d, 2010) (Ch. 7). Bilattice, on the other hand, is "*as static as Boolean logic*" (direct quote from a reviewer) that carries no direct physical semantics. For instance, (i) $(-1,0) \oplus (0,1) = (-1,1)$ can be used to capture the dynamic fusion A patient previously had depression (-1,0) but regained mental equilibrium (-1,1) after an antidepressant treatment (0,1); (ii) $(-1,0) \otimes (-1,0)..\otimes(-1,0) = (-1,0)^N$ can capture an oscillation sequence between the two poles of an equilibrium or non-equilibrium such as cyclothymic mental disorder or a bipolar subatomic particle (Fermilab, 2006), when N is even the sequence results in (0,+1) and when N is odd it results in (-1,0). Such physical semantics is not observed in bilattice. Furthermore, bipolar quantum entanglement is undefined in bilattice.
5. **Axiomatically**, YinYang bipolar lattice has led to an equilibrium-based axiomatization for physics (Zhang, 2009c, 2009d) (Ch. 6); bilattice on the other hand, is not a tautological system because it doesn't satisfy the law of excluded middle (LEM) and has contradiction (t,t) as a permissible truth value. Despite the non-linear dynamic properties, BDL is a tautological system that satisfies LEM

where both side true (-1,1) characterize a healthy bipolar equilibrium such as mental equilibrium, the electromagnetic field of the earth or the heart-brain bioelectromagnetic field. LEM, BUMP, and equilibrium-based axiomatization of physics find no equivalent in bilattice.

YinYang BDL vs. Unipolar Quantum Logic. In mathematical physics and quantum mechanics, quantum logic can be formulated as a modified version of propositional logic with a set of rules for reasoning about propositions (Birkhoff & von Neumann, 1936). While quantum logic takes the principles of quantum theory into account, it is fundamentally truth-based and unipolar in nature because it does not consider the bipolar coexistence, interaction, and bipolar quantum entanglement among particles that form global or local dynamic equilibria or non-equilibria. Therefore, quantum logic does not go beyond truth-based mathematical abstraction and is still subjected to the LAFIP paradox.

YinYang Bipolar Quantum Entanglement vs. Unipolar Quantum Entanglement. Logically speaking, quantum entanglement can be generally categorized into two types: unipolar truth-based or bipolar equilibrium-based. A distinguishing factor is that bipolar quantum entanglement exhibits ubiquitous effects with logically definable causality in both physical and social sciences (Chs. 6-11) but unipolar quantum entanglement so far is limited to quantum mechanics and quantum computing where entanglement is considered a physical resource without ubiquitous effects.

ANSWERS TO CRITICS

Many anonymous reviewers and friends made critical and constructive comments on accepted or rejected submissions related to this monograph in more than 20 years; one authoritative reviewer made antagonistic comments but was unwilling to reveal his/her identity for a public debate. While the constructive comments helped the author to enhance the line of research, the antagonistic comments provided a surviving environment for this work to grow stronger and hopefully healthier. Thus, all critical comments, constructive or destructive, have helped in the development of this work one way or another. The author acknowledges the contributions of all reviewers who made the comments. The author deems it an obligation to share and discuss the following major comments with readers.

1. **Comments:** *Drop the word "YinYang" from titles of future submissions to avoid hurting the others' feelings.*

Discussion: This comment is from a well-established Chinese American friend. As stated in the preface, believe it or not, in the world-wide scientific community there could be more Chinese who emotionally resent the word "YinYang" than Westerners who scientifically oppose the YinYang cosmology. This may sound ironic but is actually a historical phenomenon with socioeconomic reasons. First, most modern Chinese want China to be part of the modern world and don't care much about YinYang which has been often deemed an unscientific concept of the old school. Secondly, some overseas Chinese are concerned that the word YinYang might offend Western colleagues.

Western scholars, on the other hand, are free from the above historical or socioeconomical baggage and curious about YinYang. While many Westerners consider YinYang as a philosophical word related to nature, society, and Chinese medicine, some Westerners expect YinYang to play a critical or even unifying role in modern science. For instances,

a. It is not Beijing University but Harvard Medical School where a ubiquitous genetic agent is discovered and named YinYang 1 which has been widely referenced by articles in top journals including but not limited to *Nature*, *Science*, and *Cell.*
b. It is not Tsinghua University but MIT campus where a YinYang Pavilion created by American Artist Dan Graham is housed.
c. It was not a Chinese politician but legendary German mathematician Leibniz who invented the modern binary numeral system and attributed his invention to YinYang hexagrams.
d. It was not a founding father of China but a founding father of quantum mechanics – legendary Danish Physicist Niels Bohr – who first brought YinYang into quantum theory for his complementarity principle regarding particle-wave duality.

2. **Comments:** *YinYang is not truth and is not related to the notion of truth. YinYang bipolar space is a lattice. YinYang bipolar sets are L-fuzzy sets.*

Discussion: It is evident that the three comments are in contradiction among themselves where the first contradicts the 2nd and 3rd. The answer to the first comment is that YinYang is not unipolar truth but bipolar holistic truth because YinYang bipolar equilibrium is natural or physical reality. Since any closed dynamic system tends to reach an equilibrium state, equilibrium is central in thermodynamics – the ultimate physical source of existence, energy, life, and information. Since YinYang bipolarity is a fundamental reality (Gore & van Oudenaarden, 2009), bipolar sets and bipolar lattice are valid physical, and mathematical structures. Furthermore, bipolarity is a key for avoiding the LAFIP paradox and leading to logically definable bipolar causality, bipolar relativity, and bipolar quantum entanglement (Chs. 6 &7).

The 2nd comment is correct. Indeed, YinYang bipolar logical space is a bipolar quantum lattice just as Boolean logic space is a bivalent lattice. The 3rd comment, however, is generally correct but not accurate. YinYang bipolar fuzzy set is a bipolar fusion of L-fuzzy sets (Goguen, 1977), however, bipolar lattice is not necessarily a fuzzy lattice and bipolar L-sets are not necessarily fuzzy. As discussed earlier, the crisp bipolar quantum lattice B_1 satisfies the law of excluded middle and leads to a bipolar zeroth- and 1st-order axiomatization that finds no equivalence in L-fuzzy sets.

After all, logic is the formal systematic study of the principles of valid inference and correct reasoning. Such reasoning does not have to be on bivalent truth. It could be on composite or holistic truth. It would be unreasonable if logical reasoning on bipolar equilibrium or non-equilibrium were forbidden as they are clearly fundamental physical concepts. Without bipolarity, bipolar relativity and bipolar quantum entanglement would have not been possible.

3. **Comments:** *The operator ⊗ is not a free operator.*

Discussion: First, even the free operators $\wedge$ and $\vee$ can not be used freely as we cannot use $\wedge$ when it should be $\vee$. Secondly, bivalent logic leads to the LAFIP paradox. Thirdly, Einstein envisioned a new logical foundation for physics. The non-linear conjunctive is necessary for the equilibrium-based axiomatization of physics (Ch. 6) with bipolar holistic truth. Since other well-established models allow undetermined or even contradiction to exist as a type of truth value to remedy the limitation of classical logic when/where the law of excluded middle is not a natural property to be imposed, it would be unreasonable if the operator ⊗ is banned for non-linear open-world dynamic reasoning with bipolar equilibria. Furthermore, the non-linear conjunctive ⊗ can be used for bipolar quantum entanglement and bipolar quantum computing. As an oscillator it is unavoidable. After all, regardless of its non-linear bipolar dynamic property, it does not compromise the basic law of excluded middle and provides the basic operation for revealing the ubiquitous effects of bipolar quantum entanglement in logical, physical, social, mental, and biological terms (Chs. 6-11).

4. **Comments:** *YinYang bipolar logic is all about balance and symmetry.*

Discussion: This comment is inaccurate. Firstly, equilibrium is a bipolar dynamic balance, not a static balance. In addition to the linear properties, YinYang bipolar dynamic logic is defined on dynamic equilibria as bipolar sets with four pairs of dual dynamic DeMorgan's laws and BUMP. Secondly, although both equilibrium and balance require symmetry, the words negative and positive can be used for false and true as well as for - and +, respectively. Without the concepts of negative pole (Yin) and positive pole (Yang) balanced fuzzy sets are concerned with the balance of the true side and the false side, but not YinYang bipolar dynamic equilibrium. Therefore, balanced fuzzy sets rightfully belong to Boolean YinYang, where the counterparts of bipolar quantum lattices and BUMP are not observed.

5. **Comments:** *Someone suggested to use the bipolar interval [-1,1], but [0,1] is already bipolar. 0 has a negative flavor; 1 has a positive flavor; and 0.5 has a neutral flavor.*

Discussion: It is believed that this suggestion is somewhat related to the so-called terminological difficulties and confusions on fuzzy sets (Zadeh, 1965), bipolar possibility (Benferhat et al., 2006), intuitionistic fuzzy sets (Atanassov, 1986), and bipolar YinYang (Zhang, 1998, 2005a, 2006a; Zhang & Zhang, 2004a). Although a bipolar equilibrium can be linguistically described as an antonym pair or linguistic fuzzy set, an antonym pair does not necessarily form a bipolar equilibrium. For instance, the pairs (small, large) and (low, high) are unipolar granules that do not form a bipolar equilibrium. Similarly, (true, false), (impossibility, possibility), (unlikelihood, likelihood), (unsafe, safe), and (negative preference or against, positive preference or in favor) are antonym pairs but not necessarily bipolar equilibrium variables. They are inherently bivalent in nature because if it is possible it is not impossible, if it is in favor of something it is not against it. If both are true, we have an indetermination, abstention, or contradiction. Therefore, bipolar possibility rightfully belongs to intuitionism. The author respectfully suggests that bipolar possibility with negative preference (against) and positive preference (in favor) is actually a misnomer. The proper name for bipolar possibility is intuitionistic possibility or paraconsistent possibility that can be classified as a Boolean YinYang model.

It is puzzling to notice that the authors of an article (Dubois et al., 2005) suggest that intuitionistic fuzzy sets should be called bipolar fuzzy sets. This suggestion evidently missed the key point that bipolar fuzzy sets are defined on the two poles of an equilibrium that are fundamentally different from intuitionistic fuzzy sets where the negative and positive sides are the false and true sides, respectively, but not two opposite poles or energies.

While the author of this book believes that bipolar possibility is intuitionistic or paraconsistent in nature, bipolar YinYang is equilibrium-based and hence is philosophically different. Without YinYang bipolarity, the two opposing poles of a bipolar equilibrium cannot be fully represented and visualized. For instance, (reaction, action), (competition, cooperation), and (self-negation, self-assertion) are YinYang bipolar equilibrium variables which can be mapped over to a bipolar quantum lattice such as B_1 or B_F. If both sides are true, we have a perfect equilibrium or bipolar fusion instead of indetermination or contradiction. Knowledge flavoring with the unipolar lattice [0,1] cannot replace YinYang bipolarity for knowledge representation and visualization because each pole in a YinYang bipolar equilibrium has its own likelihood and unlikelihood with non-linear bipolar interaction, fusion, oscillation, equilibrium and/or harmony.

Therefore, it is misleading and irresponsible to suggest that a YinYang bipolarity is unnecessary regardless of its vulnerability. It should not be banned. On the contrary, it is really doubtful that knowledge flavoring or, equivalently, knowledge tasting or smelling can become a new research area beyond knowledge representation and visualization, let alone providing a new relativity theory beyond spacetime.

It should be remarked that the misnomer "bipolar possibility" is not the first one of its kind. In classical logic and machine learning negative and positive examples refer to false and true examples, respectively, where the word negation is used for complement (¬) that is totally different from arithmetic or rational number negation (-). In digital logic, the original full name of bipolar logic is "bipolar transistor logic" where a physical transistor is bipolar with PN or NP junctions but the logic implemented is bivalent or digital logic with 0 for false and 1 for true which have nothing bipolar. Therefore, "bipolar (transistor) logic" really meant to be unipolar digital logic implemented with bipolar transistors.

The author only wishes to point out such historical misnomers and has no plea for other researchers to change history for resolving terminological difficulties. It is perfectly alright to the author if different valid logical models and set theories all are allowed to coexist in equilibrium or harmony. In this spirit, YinYang provides a framework and a catalyst for resolving terminological difficulties regarding fuzzy sets, intuitionism, and bipolarity.

6. **Comments:** *YinYang bipolar Lattice is the same as bilattice in a pure mathematical view point. YinYang bipolar fuzzy sets are the same as the wrongly-called "intuitionistic fuzzy sets".*

Discussion: At this point, it has been made clear that these comments are unreasonable bashing. To repeat, intuitionism and paraconsistency (including bilattice) are truth-based systems that don't satisfy the law of excluded middle (LEM). Not only does (crisp) BDL satisfy the law of excluded middle (Ch. 3), its crisp and fuzzy versions both enjoy a bipolar universal modus ponens (BUMP). Intuitionism and paraconsistency are concerned with truth, falsity, and/or contradiction (Moschovakis 2010); YinYang BDL is for dynamic reasoning on bipolar interaction and quantum entanglement. The two are different in both axiomatic forms and philosophical bases. Evidently, the critic failed to understand the basics before his bashing. The author must admit though that YinYang BDL can be coerced into an intuitionism or paraconsistency model if its bipolarity is removed. However, an intuitionism and paraconsistency model cannot be coerced to a YinYang bipolar model due to their lack of equilibrium-based bipolarity – the most fundamental property of Mother Nature (Ch. 1).

7. **Comments:** *Equilibrium can only be true or false. I have never heard of quasi- or fuzzy- equilibrium.*

Discussion: Although it is somewhat shocking to hear such comment from an authoritative reviewer in the fuzzy set community, the answer is simple. Quasi-equilibrium is a generally accepted concept in modern physics. Secondly, according to fuzzy set theory, everything can be fuzzy, why not equilibrium? Is it untrue that everything can be fuzzy?

8. **Comments:** *⊗ is severely flawed because not all relations are transitive. For instance: Ann loves Paul and Paul loves Lisa does not imply that Ann loves Lisa.*

Discussion: This comment is unreasonable because one cannot require all bipolar relations to be transitive while some unipolar relations are evidently not as stated in the example. Like in unipolar case, some bipolar relations are transitive and some are not. Although love and hatred are generally not transitive, friend and enemy relationship in wartime international relations are undoubtedly transitive. Could action and reaction forces in the universe be transitive? It seems to be an unanswered question.

9. **Comments:** *A subscript i is used to denote all (bipolar) t-norms. However, the number of t-norms is denumerable. Hence, such a notation cannot be used.*

Discussion: Just like digital signals are discrete representations of analog signals, indexed t-norms are discrete t-norms. Discrete unipolar t-norms have been widely used and discrete bipolar t-norms have to be used by humans following the magical number 7±2 and by machines following the finite rule. After all, the non-linear bipolar oscillator t-norm ⊗ is not continuous, it has to be discrete.

10. **Comments:** *YinYang is old baby.*

Discussion: One journal reviewer rejected one of the author's submissions with this comment that perhaps meant to be *"YinYang can never grow up."* This is a paradoxical comment because not only does the old baby refuse to get old but also keeps growing after more than five thousands of years (Moran & Yu 2001). That suggests some deep fundamental to be rediscovered. The author believes that the deep concept is equilibrium-based bipolarity that has led to logically definable causality and bipolar relativity. Since the word "YinYang" has appeared in many articles in *Science, Nature,* and *Cell,* the comment has to be from an uninformed prejudiced reviewer.

11. **Comments:** *YinYang bipolar logic is not useful. Boolean logic is superior.*

Discussion: The author considers these comments irresponsible and highly prejudiced views. YinYang bipolar dynamic logic has been shown a unification of Boolean logic and quantum logic that reveals the ubiquitous effects of quantum entanglement. On the other hand, it is a well-known fact that Boolean logic is a static logic and not quantum in nature.

12. **Comments:** *Einstein's special and general relativity should be reviewed in a separate chapter (not just in the introduction) to provide a basis for the intended unification.*

Discussion: As clarified in Chapter 1, after inventing his most celebrated general theory of relativity, Einstein pointed out that *"For the time being we have to admit that we do not possess any general theoretical basis for physics which can be regarded as its logical foundation"* and *"the axiomatic basis of theoretical physics cannot be extracted from experience but must be freely invented"*. Clearly, Einstein did not believe that his relativity theory could serve as a logical foundation for the grand unification.

Following the above guideline, this monograph has presented the theory of bipolar relativity as an equilibrium-based unifying logical theory based on bipolar causality instead of truth and singularity.

Due to the theoretical magnitude of this topic, the unifying properties have to focus primarily on the logical level with limited mathematical development and illustration. Further investigations into the possibility of a bipolar mathematical unification of Einstein's classical equations of general relativity and that of classical quantum mechanics have to be left for future research efforts. Hopefully, YinYang bipolar relativity has brought such possibility closer to reality due to its logically definable causality.

13. **Comments:** *BDL and Bipolar relativity seem to be too simplistic to be a unifying logical foundation for physics. BDL and BUMP are logically too complex.*

Discussion: According to Einstein, *"Pure thought can grasp reality, as the ancients dreamed"; "Nature is the realization of the simplest conceivable mathematical ideas"; "Evolution is proceeding in the direction of increasing simplicity of the logical basis (principles)."* Simplicity, therefore, could be its strength for a unifying theory. BDL and BUMP are indeed significantly more complex than Boolean algebra (Boole 1854) and classical MP, respectively. Bipolar relativity (Eq. 6.7), however, is evidently much simpler than string theory. Of course, it is questionable whether bipolar symmetrical and asymmetrical quantum entanglement $(a^-,a^+)\Leftrightarrow(b^-,b^+)$ (Eq. 7.2a) and $(a^-,a^+)\Rightarrow(b^-,b^+)$ (Eq. 7.2b) are as elegant and significant as $E = mc^2$. Perhaps only history can tell. Before then, they are expected to be regarded as either falsifiable predictions or as absurdity or even madness for years to come.

14. **Comments:** *Bipolar quantum theory is quite less than quantum mechanics.*

Discussion: While a classical single qubit register entails one Bloch sphere, a bipolar qubit register entails two Bloch spheres (See Ch. 7, Fig. 4). The information entropy is therefore expanded exponentially due to bipolar interaction. For instance a classical qubit has two basis states, a bipolar qubit has $2^2 = 4$ basis states (Ch. 7). The two basis states of a classical qubit alone are inadequate for quantum entanglement; the four basis states of a bipolar qubit alone are adequate for bipolar quantum entanglement. Chapter 7 is focused on the four basis bipolar states that can be extended to non-orthogonal states. Thus, bipolar quantum theory is the first deterministic coherent quantum theory with logically definable causality for quantum entanglement that shows the potential of bipolar quantum teleportation without conventional communication and the potential of quantum-digital cryptography with bitwise encryption.

If bipolar quantum entanglement can be physically realized it will bring quantum computer and communication much closer to reality. It has the potential to make both prime-number-based encryption and quantum factorization algorithms obsolete due to its illogical bitwise encryption. Even if the theory were physically impossible in the near future, bipolar quantum computing and quantum-digital cryptography are useful in pure software implementation where bipolar illogical computational aspect can be invaluable (Zhang, 2010) (Ch. 7).

15. **Comments:** *BDL is not general purpose logic but a semantic logic.*

Discussion: BDL is not truth-based logic but equilibrium-based logic. Whether truth-based logical semantics is more general than equilibrium-based bipolar semantics is debatable as we say that our universe is either an equilibrium or non-equilibrium but we don't say that equilibrium is a universe. We don't know which one, equilibrium or universe, came first in the very beginning. It can be observed though that classical modus ponens (MP) is derivable from BUMP but not vice versa; Boolean logic is derivable from BDL but not vice versa; BDL has quantum property but Boolean does not.

16. **Comments:** *BDL is philosophical logic.*

Discussion: Computability of BDL has been proven in Chapter 3 (Theorem 3.8). Recovery of bipolar quantum lattice to bivalent lattice has been proven in Chapter 4 (Theorem 4.20). Bipolar quantum computing has been discussed in Chapter 7. The illogical computing aspect of BDL could be invaluable to quantum-digital cryptography. After all, matter-antimatter and action-reaction bipolarity is the most fundamental physical reality but not philosophy. On the other hand, truth and falsity sound rather philosophical without any direct physical semantics. Due to its equilibrium-based bipolar nature, BDL can be regarded as logical, physical, mental, social, and biological. Thus, it provides a logical foundation for mathematical physics and biophysics that are not philosophical.

17. **Comments:** *YinYang is philosophy; philosophy should not be part of science or physics.*

Discussion: According to Einstein, history and philosophy provides the context for science and should be a significant part of science and physics education (Smolin, 2006, p. 310-311). For instance, difference in culture or language can be the result of different cosmologies and can make a major difference in the interpretation and understanding of nature (Alford, 1993).

18. **Comments:** *YinYang or not does not really matter, bipolar matters.*

Discussion: Linguistically, there is no other single English word that can match the word YinYang. YinYang symbolizes dynamic equilibrium, harmony, and unity; bipolarity without YinYang is often used in the West to indicate disorder, chaos, and dichotomy. "Bipolar" can't even qualify to be a noun. As stated by Alford (Alford 1993), YinYang *"represents a higher level of formal operations,.., which lies beyond normal Western Indo-European development."* For instance, the term "bipolar logic" has been widely used in digital circuit design that has really meant to be "binary logic implemented by bipolar transistors." Without YinYang the two usages cannot be distinguished. Evidently, YinYang does matter.

19. **Comments:** *The author's term "application" means "to apply the bipolar theory to explain the world." To the reviewer, "application" means "to apply the model to solve a real problem." People do not really mind whether YinYang is in their body, but they do care whether the theory helps medical doctors cure the patients. You need to prove your model is better.*

Discussion: The major application of any relativity theory has to be making significant predictions and interpretations of the world. It would be a trivial theory if such predictions and interpretations could be immediately validated or falsified. Requiring YinYang bipolar relativity to help a doctor to cure patients is, therefore, like requiring spacetime relativity to help a mechanic to fix a car or requiring macroeconomics to help a stock trader to make more money. Can they help? Sure, but not necessarily in a direct way. For instance, while acceleration is equivalent to gravity based on general relativity that must be taken into consideration in car design, YinYang bipolar thinking can help a person to maintain mental equilibrium by exercising positive thinking when depressed and negative thinking when excited.

Otherwise, more people would suffer from bipolar disorder. People do care what is in their body and how their body works because that belongs to system biology and without system biology there would be no biomedicine.

"You need to prove your model is better" is a typical example of truth-based positivist thinking that needs to be balanced with YinYang. As stated in Chapter 1, truth is subjected to observability and limited to certain spacetime. YinYang bipolar relativity promotes balanced thinking. For instances:

a. "My car has more horsepower and is better", but later that can become false due to higher gas price and environmental issues.
b. "A larger bank is better because it can earn more money by investing in real estate", but later the bank may get into bankruptcy due to "bursting of the real estate bubble".
c. "The scientific and technological establishment has been proven better", but the proof mat have to be reexamined if global warming is proven a deadly side effect.
d. "In 1759 the British mathematician Francis Maseres wrote that negative numbers *'darken the very whole doctrines of the equations and make dark of the things which are in their nature excessively obvious and simple'"* (BBC, 2006); but nowadays no parents would be willing to ask their children to learn math without negative numbers because the universe is fundamentally bipolar.

20. **Comments:** *YinYang is no parallel with Einstein; "relativity" for your book is unnecessary; Einstein's words do not really help YinYang much.*

Discussion: First, the author is not trying to find a parallel with Einstein but trying to explore the possibility of unifying spacetime relativity with quantum theory as envisioned by Einstein. Secondly, even without a formal logical foundation, YinYang has been a well-known relativistic cosmology for thousands of years. Thirdly, Niels Bohr, father figure of quantum mechanics besides Einstein, was the first one to bring YinYang into quantum theory for his quantum complementarity principle. Fourthly, the word "YinYang" has appeared in *Nature*, *Science*, *Cell* and other top scientific journals many times in recent decades. Einstein perhaps would have been happier had his colleagues not tried to avoid him in his late years (Smolin, 2006, pp. 49-50). Now, denying *"The yin and yang of nature"* (Gore & van Oudenaarden, 2009) and banning others to use the word "relativity" would make Einstein an isolated "God of Science" he never wanted to be, and that is undoubtedly unscientific.

According to Einstein: *"Put your hand on a hot stove for a minute, and it seems like an hour. Sit with a pretty girl for an hour, and it seems like a minute. That's relativity." "For the time being we have to admit that we do not possess any general theoretical basis for physics which can be regarded as its logical foundation."*

In the last quote, Einstein used sorrow and joy to hint the two sides of YinYang in general. Symbolically, the two sides can be paired up as bipolar variables and generalized to action-reaction forces denoted $(-f, +f)$, negative-positive electromagnetic charges denoted $(-q, +q)$, matter-antimatter particles $(-p, +p)$, the bipolar logical variable (e^-, e^+), or the Yin and Yang of nature for YinYang bipolar dynamic logic (BDL) and the theory of YinYang bipolar relativity. With Einstein's words, BDL and YinYang bipolar relativity are practically placed on the shoulders of Giants.

21. **Comments:** *Selling unifying theory is very difficult. As commented before, your models are subject to many unrealistic assumptions. Those assumptions might make your models worthless.*

Discussion: A closed system tends to reach equilibrium is not an assumption but the right foundation of modern science as manifested by the 2nd law of thermodynamics. YinYang bipolar relativity is not subjective unless action-reaction forces (-f,+f), electromagnetic particles (-q,+q), particle-antiparticles (-p, +p), competition-cooperation (comp, coop), input-output (in, out), genetic repression and activation (r, a), self-negation and self-assertion abilities (-self, +self), or the Yin and Yang of nature in general are deniable reality.

22. **Comments:** *The logic is sound. Not madness, but neither computer science nor mathematics. Is there another non-linear logic that satisfies the law of excluded middle and the law of non-contradiction with a bipolar universal modus ponens (BUMP)? The author brilliantly proposed bipolar t-norms... The non-linearity of BUMP could shed some new light on fuzzy logica (seems to be in Spanish) research. I cannot rule out the possibility that this work is substantive and valuable... Keep an eye on his work.*

Discussion: These are some inspiring comments the author received in a few years ago. From the above, it is clear that the controversy about YinYang bipolarity is a clash between new ideas and the scientific establishment, physics and mathematics, as well as between the East and the West philosophies. Despite the negative side, there is the positive side. Reality itself is a vindication of YinYang bipolar relativity.

ON THE UBIQUITOUS EFFECTS OF QUANTUM ENTANGLEMENT

Now YinYang equilibrium has entered modern science; it has been quite puzzling that mental equilibrium has not been considered observable in Western medicine. Without mental equilibrium mental disorder would be the disorder of 0 or big bang from nowhere. This is apparently paradoxical because there would be no universe and no brain without a certain form of equilibrium (Zhang, 2009a, 2009b).

It is a popular theory that the universe was created by the big bang. Someone has to ask, if the universe was created by the big bang, what was there before the big bang? What caused the big bang? Could it be true that dynamic equilibrium or non-equilibrium caused the big bang? Could it be true that the big bang is the explosion of a black hole or a collision of two black holes? Could it be true that dynamic equilibrium or non-equilibrium created the universe?

The debate once went astray from science to religion. Here are some interesting arguments and counter arguments from a religious perspective: *"God created the universe. It is wrong to say otherwise." "But would it be right to imply that equilibrium and non-equilibrium did not belong to God?" "Is equilibrium separable from the universe?" "Did God create equilibrium and non-equilibrium?" "If there was God and only God in the very beginning, did God use his magic power equilibrium or non-equilibrium to create and regulate the universe?"*

Coming back from the religious detour, we enumerate the fundamental differences of YinYang bipolar relativity from other approaches to quantum gravity (Smolin, 2000, 2006) in the following:

1. It logically defines causality for the first time based on the postulate that the most fundamental property of nature is YinYang bipolarity. Not only is this postulate supported by electron-positron pair (e^-,e^+) production as the most fundamental example of the materialization of energy predicated in special relativity and accurately verified in quantum electrodynamics (QED) (Dirac, 1927,,1928)

(Feynman, 1962,,1985) but it is also supported by Hawking Radiation which predicts that even when the universe ends in a black hole, bipolarity will miraculously survive due to particle and/or antiparticle emission from the black hole (Hawking, 1974).

2. It is an equilibrium-based open-world open-ended logical foundation for physics without assumption on the most fundamental physical element of the universe such as ether, strings, or loops (Smolin, 2000, 2006). Therefore it is meaningful even without spacetime – a condition for a pure quantum theory with complete background independence (Smolin, 2005, 2006).
3. It provides a logically coherent explanation to the cause-effect relation of big bang and black holes.
4. Its quantum entanglement feature leads to a logically complete quantum theory and a deterministic logical interpretation to the EPR paradox for the first time.
5. It leads to the properties of quantum-digital compatibility, bitwise encryption, and quantum controllability that are expected to bring quantum computer closer to reality.
6. It is rooted in YinYang but also rooted in Boolean logic because BDL is a nonlinear bipolar dynamic generalization of Boolean logic that doesn't compromise the basic law of excluded middle.
7. Bipolar universal modus ponens (BUMP) provides logically definable causality and bridges the gap between truth-based reasoning and equilibrium-based quantum entanglement.
8. The above salient features enable YinYang bipolar relativity to bring quantum entanglement from physics to logical, physical, social, mental, and biological worlds to form a Q5 quantum gravity paradigm for quantum computing in both physical and social sciences.

Now we are ready to revisit the two mysteries of quantum entanglement: The first mystery is the phenomenon itself; the second one, according to Penrose, is

How are we to come to terms with quantum entanglement and to make sense of it in terms of ideas that we can comprehend, so that we can manage to accept it as something that forms an important part of the workings of our actual universe?.. The second mystery is somewhat complementary to the first. Since according to quantum mechanics, entanglement is such a ubiquitous phenomenon – and we recall that the stupendous majority of quantum states are actually entangled ones – why is it something that we barely notice in our direct experience of the world? Why do these ubiquitous effects of entanglement not confront us at every turn? I do not believe that this second mystery has received nearly the attention that it deserves, people's puzzlement having been almost entirely concentrated on the first. (Penrose, 2005, p. 591)

Hopefully, the two quantum mysteries have been partially resolved logically (Ch. 3, 4, 7), physically (Ch. 6,8), mentally (Ch. 10), socially (Ch. 5, 11), and biologically (Ch. 9) with YinYang bipolar relativity and bipolar quantum entanglement.

LIMITATIONS

Mathematically, the theory of YinYang bipolar relativity as a pure invention is not derived from general theory relativity or quantum theory. Instead, it presents a fundamentally different approach to quantum gravity. As a first step, the monograph has been focused on the logical level of the theory and its applications in physical, social, brain, biological, and computing sciences with limited mathematical or algebraic extensions. Thus, equilibrium-based bipolar logical unification of gravity and quantum mechanics has been a topic within the scope of the book; the quantization of YinYang bipolar relativity and the mathematical unification of Einstein's equations of general relativity and that of quantum mechanics have been left for future research efforts because *"For the time being we have to admit that we do not possess any general theoretical basis for physics which can be regarded as its logical foundation"* (Einstein 1940).

Theoretically, YinYang bipolar relativity presents an open-world and open-ended approach to science that is not *"a theory of everything."* In this approach, the author doesn't attempt to define the smallest fundamental element such as ether or string. Instead, it has been postulated that YinYang bipolarity is the most fundamental property of the universe based on well-established observations in physical and social sciences. With the basic hypothesis, equilibrium-based logical constructions have been developed with a number of predictions for experimental verification or falsification. This approach actually has followed the principle of exploratory scientific knowledge discovery.

Practically, YinYang bipolar relativity is expected applicable where bipolar equilibrium or non-equilibrium is central. Since it is not *"a theory of everything"*, it does not claim universal applicability. As a quantum logic theory it is recoverable to Boolean logic and, therefore, is quantum computational. As a relativity theory, its major role is to provide predictions and interpretations about nature, agents, and causality. Simulated application examples have been presented in quantum computing, cognitive informatics, and life sciences to illustrate the utility of the theory. The examples, however, are not intended to be systematic and comprehensive applications but only sufficient illustrations. While the theory is logically proven sound, predictions or interpretations made in the book can be either validated or falsified in the future as usual.

MAJOR RESEARCH TOPICS

While modern science has been built upon truth-based unipolar cognition, this book has, hopefully, opened a new approach to scientific unification built upon equilibrium-based bipolar cognition. The new approach, however, is so far primarily focused on the logical level for qualitative reasoning with limited mathematical development. Many challenges lay ahead for building further on the logical foundation. A list of major research topics are enumerated in the following:

1. Quantize YinYang bipolar relativity and transform it from a qualitative model to a quantitative model. A basic unit in the quantization could be an electron-positron pair denoted $e = (e^-,e^+)$.
2. Experimentally test YinYang bipolar quantum entanglement with particle-antiparticle pairs to verify or falsify its validity. (Zhang, 2010)
3. Develop bipolar quantum computer technology that is compatible for both digital and quantum computing (Zhang, 2010).
4. Develop software algorithms for bitwise bipolar quantum encryption that is unbreakable even by quantum computing and make prime-number-based cryptograph and quantum factorization algorithms obsolete.
5. Develop YinYang bipolar calculus based on YinYang bipolar mathematical abstraction.
6. Develop YinYang bipolar probabilistic and statistical models based on YinYang bipolar mathematical abstraction.

7. Extend BQLA to YinYang *bipolar quantum tensor calculus* (*BQTC*) for equilibrium-based unification of general relativity and quantum mechanics at the mathematical level.
8. Extend BQLA to a *complex number-based BQLA* (*CBQLA*) for the unification of unipolar and bipolar quantum mechanics at the mathematical level (Zhang, 2010)
9. Apply YinYang bipolar relativity in epigenomics for equilibrium-based biosystem simulation and regulation. (Zhang *et al.*, 2009).
10. Apply YinYang bipolar relativity in computational neuroscience and psychiatry for equilibrium-based mental simulation and regulation.
11. Apply YinYang bipolar relativity in TCM for its unification with system biology.
12. Apply YinYang bipolar relativity in socio- and macro-economics for equilibrium-based global regulation and social harmony.
13. Apply YinYang bipolar relativity in the scientific exploration of outer space.
14. Apply YinYang bipolar relativity in global warming research.
15. Investigate into the logical and physical connections as well as their boundaries and limitations of bipolar relativity and the 2nd law of thermodynamic.

SUMMARY

From the above discussion, it is evident that the development of physical science is interrelated with that of social science. Tradition as a kind of social quantum gravity may help in advancing scientific research or hinder scientific advances. Occasionally, it had the tendency to kill new ideas with brutality. Isomorphism can be a scientific term when being used correctly. It can also become an academic bashing stick. The debate on isomorphism is actually a clash between new ideas and the established ones with deep philosophical differences regarding the universe.

From a different perspective, the debate can be deemed a clash between physics and mathematics or mathematics and logic. It is interesting to observe that, although mathematics is historically more related to science than logic as pointed out by Russell (Russell, 1919), the truth-based establishment of both logic and mathematics sometimes tends to ignore the bipolar physical reality. Denying the existence of the Solar system is one example and denying the necessity of negative numbers is another. Above all, denying the necessity of bipolar equilibrium-based mathematical abstraction is perhaps the longest denial in the history of science.

The last denial did not significantly affect the part of science that is directly related to actual human lives such as the science of air and water until recent decades. With the threat of global warming, worldwide economic crises, deep questions in quantum information and life sciences, the last denial has to come to an end. It is suggested that, regardless of the glorious scientific achievements of the West over many centuries, a resolution of the Eastern and the Western philosophical differences on bipolar equilibrium could be the key for furthering scientific explorations.

The YinYang bipolar resolution exhibits the following unifying features:

1. Unipolar Boolean logic is derivable from BDL but not vice versa – a property that qualifies BDL as a generalization of Boolean logic and, in the meantime, enables BDL to recover to Boolean logic for computability (Ch. 1);

2. Classical modus ponens (MP) is derivable from BUMP but not vice versa (Ch. 3) – a property that qualifies BUMP as a generalization of MP – the only generic inference rule in logic deduction for thousands of years;
3. BUMP leads to logically definable causality and quantum entanglement for the first time in simple logical terms;
4. Space and time emergence has been a difficult goal in the quest for quantum gravity (Smolin, 2006), which can be achieved with YinYang bipolar relativity (Ch. 6);
5. Bipolar equilibrium relation can be a non-linear bipolar fusion of many equivalence relations but not vice versa (Ch. 3,4,5);
6. While spacetime relativity leads to singularity, YinYang bipolar relativity supports a coherent bipolar process model of the universe (Ch. 6,8);
7. While space and time are not symmetrical, not quantum entangled, and not bipolar interactive, the Yin and Yang energies of nature are symmetrical and interactive that can be quantum entangled from a quantum mechanics perspective $(a^-,a^+)\Leftrightarrow(b^-,b^+)$ (Eq. 7.2a) and bipolar related from a relativity perspective $(a^-,a^+)\Rightarrow(b^-,b^+)$ (Eq. 7.2b). Thus, quantum entanglement and spacetime relativity are logically unified under YinYang bipolar relativity (Eq. 6.7).
8. While spacetime relativity is a unification of space, time, and gravity, bipolar relativity is an equilibrium-based logical unification of physics, socioeconomics, and life sciences;
9. With the postulate that bipolarity is the most fundamental property of nature it can be concluded that bipolar relativity is a deeper theory than spacetime relativity and unavoidable for the unification of logical, physical, social, mental, and biological quantum gravities (Q5);
10. YinYang bipolar relativity as a logical foundation for scientific unification does not exclude truth-based logical thinking; instead, it provides a symmetrical complementary or regulatory paradigm where agents, truth, singularity, and spacetime can be hosted as emerging and evolving aspects in equilibrium or non-equilibrium. The new paradigm is expected to foster holistic balanced thinking as well as mental, social, economical, biological, and environmental harmony in future scientific explorations.

From the above it is clear that YinYang missed the bus of Aristotelian science but still can catch the train of Einstein's unfinished grand scientific unification. It is contended that YinYang bipolar relativity has opened an Eastern road toward quantum gravity (Zhang, 2009d) in addition to the three Western roads as discussed by Professor Smolin (Smolin, 2000).

Before we conclude this monograph, let's recite Einstein:

Physics constitutes a logical system of thought which is in a state of evolution, whose basis (principles) cannot be distilled, as it were, from experience by an inductive method, but can only be arrived at by free invention. The justification (truth content) of the system rests in the verification of the derived propositions (a priori/logical truths) by sense experiences (a posteriori/empirical truths).... Evolution is proceeding in the direction of increasing simplicity of the logical basis (principles)... We must always be ready to change these notions – that is to say, the axiomatic basis of physics – in order to do justice to perceived facts in the most perfect way logically. (Einstein 1916)

To conclude the monograph, let's reiterate its theme: While scientists have been seeking truths from the universe for many centuries, mounting scientific evidence has shown that the universe is not truthful.

The most fundamental property of our universe is, therefore, not truth, not fuzziness, not space, not time, not spacetime relativity, not ether, not matter, not strings, and not singularity. The most fundamental property of Mother Nature should, instead, be equilibrium-based YinYang bipolarity which transcends spacetime. The fundamental postulate has led to the theory YinYang bipolar relativity with logically definable causality. Hopefully, the logical foundation, no matter how "simple", "complex", "complete", "incomplete", "sound", "flawed", or even "absurd", can serve as the first step forward in the right direction of a long journey for modern science to advance beyond spacetime and for quantum gravity to move closer to logical, physical, mental, social, and biological reality. If not, hopefully, it can serve as a piece of firework for the upcoming centennial celebration of Einstein's great theory of general relativity.

REFERENCES

Aiton, E. J. (1985). *Leibniz: A biography*. Hilger (UK).

Alford, D. M. (1993). *A report on the Fetzer Institute-sponsored dialogues between Western and Indigenous scientists*. A presentation for the Annual Spring Meeting of the Society for the Anthropology of Consciousness, April 11, 1993. Retrieved from http://www.enformy.com/dma-b.htm

Arieli, O., & Avron, A. (1998). The value of four values. *Artificial Intelligence*, *102*, 97–141. doi:10.1016/S0004-3702(98)00032-0

Atanassov, K. T. (1986). Intuitionistic fuzzy sets. *Fuzzy Sets and Systems*, *20*, 87–96. doi:10.1016/S0165-0114(86)80034-3

BBC. (2006). Negative numbers. Retrieved from http://www.bbc.co.uk/programmes/p003hyd9

Belnap, N. (1977). A useful 4-valued logic. In Epstein, G., & Dunn, J. M. (Eds.), *Modern uses of multiple-valued logic* (pp. 8–37). Reidel.

Benferhat, S., Dubois, D., Kaci, S., & Prade, H. (2006). Bipolar possibility theory in preference modeling: Representation, fusion and optimal solutions. *Information Fusion*, *7*(1), 135–150.

Birkhoff, G., & von Neumann, J. (1936). The logic of quantum mechanics. *The Annals of Mathematics*, *37*(4), 823–843. doi:10.2307/1968621

Blackburn, S. (1990). Hume and thick Connexions. *Philosophy and Phenomenological Research*, *50*, 237–250. doi:10.2307/2108041

Boole, G. (1854). *An investigation of the laws of thoughts*. London: MacMillan.

Brouwer, L. E. (1912). Intuitionism and formalism. *Bulletin of the American Mathematical Society*, *20*, 81–96. doi:10.1090/S0002-9904-1913-02440-6

Brouwer, L. E. (1912). Intuitionism and Formalism. English translation by A. Dresden in *Bull. Amer. Math. Soc.* 20 (1913): 81-96, reprinted in Benacerraf and Putnam, eds., 1983: 77-89; also reprinted in Heyting, ed., 1975: 123-138.

Dirac, P. A. M. (1927). The quantum theory of the emission and absorption of radiation. *Proceedings of the Royal Society of London. Series A, 114*, 243–265. doi:10.1098/rspa.1927.0039

Dirac, P. A. M. (1928). The quantum theory of the electron. *Proceedings of the Royal Society of London. Series A, Containing Papers of a Mathematical and Physical Character, 117*(778), 610–624. doi:10.1098/rspa.1928.0023

Dubois, D., Gottwald, S., Hájek, P., Kacprzyk, J., & Prade, H. (2005). Terminological difficulties in fuzzy set theory–the case of intuitionistic fuzzy Sets. *Fuzzy Sets and Systems, 156*(3), 485–491. doi:10.1016/j.fss.2005.06.001

Einstein, A. (1916). The Foundation of the General Theory of Relativity. Originally published in Annalen der Physik (1916), Collected Papers of Albert Einstein, English Translation of Selected Texts, Translated by A. *Engel, 6*, 146–200.

Einstein, A. (1934). *On The Method Of Theoretical Physics. The Herbert Spencer lecture, delivered at Oxford, June 10, 1933. Published in Mein Weltbild*. Amsterdam: Querido Verlag.

Einstein, A. (1940, May 24). Considerations Concerning The Fundaments of Theoretical Physics. *Science, 91*(2369), 487–491. doi:10.1126/science.91.2369.487

Einstein, A., Podolsky, B., & Rosen, N. (1935). Einstein, A., Podolsky, B. & Rosen N. (1935). Can quantum-mechanical description of physical reality be considered complete? *Physical Review, 47*(10), 777–780. doi:10.1103/PhysRev.47.777

Fermi National Accelerator Laboratory. (2006). *Press Release 06-19*, September 25, 2006. Retrieved from http://www.fnal.gov/pub/presspass/press_releases/CDF_meson.html

Feynman, R. P. (1962). *Quantum electrodynamics*. Addison Wesley.

Feynman, R. P. (1985). *QED: The strange theory of light and matter*. Princeton University Press.

Fitting, M. C. (1991). Bilattices and the semantics of logic programming. *The Journal of Logic Programming, 11*(2), 91–116. doi:10.1016/0743-1066(91)90014-G

Ginsberg, M. (1990). Bilattices and modal operators. [Oxford University Press.]. *Journal of Logic and Computation, 1*(1), 41–69. doi:10.1093/logcom/1.1.41

Goguen, J. (1967). L-fuzzy sets. *Journal of Mathematical Analysis and Applications, 18*, 145–174. doi:10.1016/0022-247X(67)90189-8

Gore, J., & van Oudenaarden, A. (2009). Synthetic biology: The yin and yang of nature. *Nature, 457*(7227), 271–272. doi:10.1038/457271a

Gupta, R. C. (1995). Who invented the zero? *Ganita-Bharati, 17*(1-4), 45–61.

Hawking, S. (1974). Black-hole evaporation. *Nature, 248*, 30–31. doi:10.1038/248030a0

Hilbert, D. (1901). Mathematical problems. *Bulletin of the American Mathematical Society, 8*, 437–479. doi:10.1090/S0002-9904-1902-00923-3

Kaplan, R. (2000). *The nothing that is: A natural history of zero*. Oxford: Oxford University Press.

Leibniz, G. (1703). *Explication de l'Arithmétique Binaire*; Gerhardt. *Mathematical Writings, VII*, 223.

Martinez, A. A. (2006). *Negative math: How mathematical rules can be positively bent*. Princeton University Press.

Moran, E., & Yu, J. (2001). *The complete idiot's guide to the I Ching*. New York: Alpha Books.

Moschovakis, J. (2010). Intuitionistic logic. *Stanford Encyclopedia of Philosophy*. Retrieved from http://plato.stanford.edu/entries/logic-intuitionistic/

Penrose, R. (2005). *The road to reality: A complete guide to the laws of the universe*. New York: Alfred A. Knopf.

Pratt, V. R. (1976). Semantical considerations on Floyd-Hoare logic. *Proceedings of the 17th Ann. IEEE Symposium on Foundations of Computer Science*, 109-121.

Rescher, N. (1969). *Many-valued logic*. New York: McGraw-Hill.

Russell, B. (1919). *Introduction to mathematical philosophy* (pp. 194–195). London: Routledge Inc.

Smolin, L. (2000). *Three roads to quantum gravity*. Basic Books.

Smolin, L. (2005). *The case for background independence*.

Smolin, L. (2006). *The trouble with physics: The rise of string theory, the fall of a science, and what comes next?*New York: Houghton Mifflin Harcourt.

Spain, B. (1965). *Tensor Calculus*. Oliver and Boyd, 1965.

Synge, J. L., & Schild, A. (1969). *Tensor Calculus*. Dover Publications, 1969.

Temple, R. (1986). *The genius of China: 3,000 years of science, discovery, and invention*. New York: Simon and Schuster.

Wilhelm, R., & Baynes, C. F. (1967). *The I Ching or book of changes. Bollingen Series XIX*. Princeton University Press.

Zadeh, L. A. (1965). Fuzzy sets. *Information and Control, 8*, 338–353. doi:10.1016/S0019-9958(65)90241-X

Zadeh, L. A. (2001). Causality is undefinable–toward a theory of hierarchical definability. *Proceedings of FUZZ-IEEE,* (pp. 67-68).

Zadeh, L. A. (2003). *Protoform theory and its basic role in human intelligence, deduction, definition and search*. 7th Joint Conference on Information Sciences, Sept. 26–30, 2003, Research Triangle Park, North Carolina.

Zadeh, L. A. (2006). Fuzzy logic. *Scholarpedia, 3*(3), 1766. doi:10.4249/scholarpedia.1766

Zadeh, L. A. (2007b). *From fuzzy logic to extended fuzzy logic–the concept of f-validity and the impossibility principle*. FUZZ-IEEE 2007, July 23, 2007, Imperial College, London, UK. Retrieved from www.fuzzieee2007.org/ZadehFUZZ-IEEE2007London.pdf

Zadeh, L. A. (2008). *Toward human-level machine intelligence–is it achievable*? WSEAS AIKED'08, WSEAS SEPADS'08, WSEAS EHAC'08, WSEAS ISPRA'08, February 21, 2008, University of Cambridge, UK. Retrieved from www.wseas.org/wseas-zadeh-2008.pdf

Zhang, W.-R. (1998). YinYang bipolar fuzzy sets. *Proceedings of IEEE World Congress on Computational Intelligence – Fuzz-IEEE*, (pp. 835-840). Anchorage, AK.

Zhang, W.-R. (2003a). Equilibrium relations and bipolar cognitive mapping for online analytical processing with applications in international relations and strategic decision support. *IEEE Transactions on SMC. Part B*, *33*(2), 295–307.

Zhang, W.-R. (2003b). Equilibrium energy and stability measures for bipolar decision and global regulation. *International Journal of Fuzzy Systems*, *5*(2), 114–122.

Zhang, W.-R. (2005a). YinYang bipolar lattices and L-sets for bipolar knowledge fusion, visualization, and decision making. *International Journal of Information Technology and Decision Making*, *4*(4), 621–645. doi:10.1142/S0219622005001763

Zhang, W.-R. (2005b). YinYang bipolar cognition and bipolar cognitive mapping. *International Journal of Computational Cognition*, *3*(3), 53–65.

Zhang, W.-R. (2006a). YinYang bipolar fuzzy sets and fuzzy equilibrium relations for bipolar clustering, optimization, and global regulation. *International Journal of Information Technology and Decision Making*, *5*(1), 19–46. doi:10.1142/S0219622006001885

Zhang, W.-R. (2006b). YinYang bipolar T-norms and T-conorms as granular neurological operators. *Proceedings of IEEE International Conference on granular computing*, (pp. 91-96). Atlanta, GA.

Zhang, W.-R. (2007). YinYang bipolar universal modus ponens (bump) – a fundamental law of non-linear brain dynamics for emotional intelligence and mental health. *Walter J. Freeman Workshop on Nonlinear Brain Dynamics, Proceedings of the 10th Joint Conference of Information Sciences*, (pp. 89-95). Salt Lake City, Utah.

Zhang, W.-R. (2009a). Six conjectures in quantum physics and computational neuroscience. *Proceedings of 3rd International Conference on Quantum, Nano and Micro Technologies (ICQNM 2009)*, (pp. 67-72). Cancun, Mexico.

Zhang, W.-R. (2009b). YinYang Bipolar Dynamic Logic (BDL) and equilibrium-based computational neuroscience. *Proceedings of International Joint Conference on Neural Networks (IJCNN 2009)*, (pp. 3534-3541). Atlanta, GA.

Zhang, W.-R. (2009c). YinYang bipolar relativity–a unifying theory of nature, agents, and life science. *Proceedings of International Joint Conference on Bioinformatics, Systems Biology and Intelligent Computing (IJCBS)*. (pp. 377-383). Shanghai, China.

Zhang, W.-R. (2009d). The logic of YinYang and the science of TCM–an Eastern road to the unification of nature, agents, and medicine. *International Journal of Functional Informatics and Personal Medicine*, *2*(3), 261–291. doi:10.1504/IJFIPM.2009.030827

Zhang, W.-R. (2010). YinYang bipolar quantum entanglement–toward a logically complete quantum theory. *Proceedings of the 4th International Conference on Quantum, Nano and Micro Technologies (ICQNM 2010)*, (pp. 77-82). St. Maarten, Netherlands Antilles. Additional Readings Belnap, N. (1977). A useful 4-valued logic. In *Modern Uses of Multiple-Valued Logic*, G. Epstein and J. M. Dunn (EDs), Reidel, 8-37.

Zhang, W.-R., Wang, P., Peace, K., Zhan, J., & Zhang, Y. (2008). On truth, uncertainty, equilibrium, and harmony–a taxonomy for YinYang scientific computing. *International Journal of New Mathematics and Natural Computing*, *4*(2), 207–229. doi:10.1142/S1793005708001033

Zhang, W.-R., Zhang, H. J., Shi, Y., & Chen, S. S. (2009). Bipolar linear algebra and YinYang-N-Element cellular networks for equilibrium-based biosystem simulation and regulation. *Journal of Biological System*, *17*(4), 547–576. doi:10.1142/S0218339009002958

Zhang, W.-R., & Zhang, L. (2004). YinYang bipolar logic and bipolar fuzzy logic. *Information Sciences*, *165*(3-4), 265–287. doi:10.1016/j.ins.2003.05.010

KEY TERMS AND DEFINITIONS

Truth: Under bipolar relativity, truth is considered a temporary static concept limited to observability and relative to spacetime.

Isomorphism: From a classical mathematical perspective, an isomorphism is a bijective map f such that both f and its inverse f^{-1} are homomorphisms. A *homomorphism* is a map from one algebraic structure to another of the same type that preserves all the relevant structure; i.e. properties like identity elements, inverse elements, and binary operations. In other word, isomorphism is a one-to-one correspondence between the elements of two sets such that the result of an operation on elements of one set corresponds to the result of the analogous operation on their images in the other set.

DL (Dynamic Logic): A truth-based dynamic extension of Boolean logic

Intuitionistic Logic: Intuitionistic logic was first used by L. E. J. Brouwer in developing his intuitionistic mathematics. Intuitionistic logic may be considered the logical basis of constructive mathematics. (Brouwer 1912)

Paraconsistent Logic: A logical system that attempts to deal with contradictions in a discriminating way. Alternatively, paraconsistent logic is the subfield of logic that is concerned with studying and developing paraconsistent (or inconsistency-tolerant) systems of logic. (Belnap 1977)

Bilattice: A lattice with four truth values for paraconsistent logical reasoning (Ginsberg 1990).

Tensors: Tensors are geometrical entities introduced into mathematics and physics to extend the notion of scalars, (geometric) vectors, and matrices. (Spain 1965) (Synge & Schild 1969)

Complex Numbers: A complex number, in mathematics, is a number comprising a real number and an imaginary number. It can be written in the form $a + bi$, where a and b are real numbers, and i is the standard imaginary unit with the property $i^2 = -1$. Complex numbers are used in quantum computing.

BQTC: Bipolar quantum tensor calculus to be developed based on bipolar relativity and tensor calculus.

CBQLA: Complex bipolar quantum linear algebra to be developed based on complex numbers for bipolar quantum computing.

About the Author

Professor Wen-Ran Zhang obtained his Ph.D. degree in Computer Engineering from the University of South Carolina - Columbia in 1986. Since 2001, he has been a professor of computer science, Georgia Southern University in Statesboro, GA. Before then, he had been on the faculty of computer science at the University of North Carolina-Charlotte, NC (1986-1988), Victoria University of Wellington, NZ (1988-1990), and Lamar University, Texas (1990-2001).

Professor Zhang's major research areas include agent interaction and coordination, multiagent data mining, cognitive mapping, bipolar neurobiological modeling and computational psychiatry, bipolar sets and bipolar fuzzy sets, YinYang bipolar dynamic logic, YinYang-N-Element Cellular Automata, bipolar relativity and quantum gravity, bipolar quantum computing, and communication.

In agent interaction, he proposed the use of multiagent data warehousing (MADWH) and multiagent data mining (MADM) for brain modeling and neurofuzzy control. It is shown that the new approach leads to the concepts of semiautonomous agents and coordinated computational intelligence (CCI).

In the set-theoretic front, he pioneered research in equilibrium-based YinYang bipolar sets, bipolar fuzzy sets, bipolar dynamic logic (BDL), and bipolar dynamic fuzzy logic (BDFL), showing that BDL and BDFL are non-linear bipolar dynamic fusions of Boolean logic, fuzzy logic, and bipolar quantum entanglement. Equilibrium relations as bipolar sets can be used for bipolar cognitive mapping, multiagent coordination, and decision support.

In bipolar neurobiological modeling, he and his co-authors pioneered research in computational psychiatry that has triggered a clinical test. The new approach is expected to extend the current bipolar disorder classification to a bipolar computational model for the integration of clinical psychiatry with bioinformatics and pharmacological databases.

Based on bipolar dynamic logic, he pioneered works in bipolar linear algebra and bipolar cellular networks for biosystem simulation and regulation. These works provide a unique mathematical basis for the modeling of YinYang-N-Element networks. These cellular structures are also expected to find applications in quantum computing.

In recent years he developed the theory of YinYang bipolar relativity and bipolar quantum entanglement that resulted in this book. The new theory has led to a number of predictions. If experimentally verified, the predictions may lead to major scientific unifications.

Central in Professor Zhang's research result is *YinYang Bipolar Universal Modus Ponens (BUMP)*. For the first time, BUMP provides logically definable quantum causality and quantum entanglement from a mathematical physics or biophysics perspective. For the first time, it resulted in a minimal but most general equilibrium or non-equilibrium-based axiomatization of physics. For the first time, it brings the ubiquitous effects of bipolar quantum entanglement into the real world of macroscopic and microscopic agent interactions in quantum computing, cognitive informatics, and life sciences.

His pioneering works are documented in more than 80 of authored and co-authored academic publications in refereed journals and conference proceedings. It is believed that his unique works have bridged a major gap between the East and the West traditions in philosophical and scientific thinking. It is expected that his works will foster further research and development for solutions of unsolved problems in modern science and technology especially in quantum computing, cognitive informatics and life sciences.

Professor Zhang enjoys teaching, research, and quiet life. He serves on the editorial board of *IJFIPM.* He serves on a number of program committees of international conferences and workshops. He served as Panel Chair of *IEEE ICDM-2005.* He served as an invited referee for the *British Engineering and Physical Sciences Research Council.* He has served as a reviewer for many journals or transactions. He received the *Award for Teaching Excellence* twice from Lamar University, Beaumont, TX, in 1995 and 1997, respectively. In 2008, he received *The Outstanding Research Award* from The College of Information Technology at Georgia Southern University, Statesboro, GA.

Index

A

B

C

R

S

T

U

V

W

Y